Lecture Notes of the Institute for Computer Sciences, Social Informatics and Telecommunications Engineering 356

More information about this series at http://www.springer.com/series/8197

Shuo Shi · Liang Ye · Yu Zhang (Eds.)

Artificial Intelligence for Communications and Networks

Second EAI International Conference, AICON 2020
Virtual Event, December 19–20, 2020
Proceedings

Editors
Shuo Shi
Harbin Institute of Technology
Harbin, China

Liang Ye
Harbin Institute of Technology
Harbin, China

Yu Zhang
Harbin Institute of Technology
Harbin, China

ISSN 1867-8211 ISSN 1867-822X (electronic)
Lecture Notes of the Institute for Computer Sciences, Social Informatics and Telecommunications Engineering
ISBN 978-3-030-69065-6 ISBN 978-3-030-69066-3 (eBook)
https://doi.org/10.1007/978-3-030-69066-3

This Springer imprint is published by the registered company Springer Nature Switzerland AG
The registered company address is: Gewerbestrasse 11, 6330 Cham, Switzerland

Preface

We are delighted to introduce the proceedings of the second edition of the European Alliance for Innovation (EAI) International Conference on Artificial Intelligence for Communications and Networks (AICON 2020). This conference is meant to stimulate debate and provide a forum for researchers working in related problems to exchange ideas and recent results (both positive and negative ones) in applying artificial intelligence to communications and networks. Both supervised learning and unsupervised learning, reinforcement learning, and recent developments in generative adversarial networks and game-theoretic setups are also of great interest.

The technical program of AICON 2020 consisted of 56 full papers, including 2 invited papers, in oral presentation sessions at the main conference tracks. The conference tracks were: Track 1 - Deep Learning/Machine Learning on Information and Signal Processing; Track 2 - AI in Ubiquitous Mobile Wireless Communications; Track 3 - AI in SAR/ISAR Target Detection; Track 4 - AI in UAV-Assisted Wireless Communications; Track 5 - Smart Education: Educational Change in the Age of Artificial Intelligence; Track 6 - Recent Advances in AI and their Applications in Future Electronic and Information Field. Aside from the high-quality technical paper presentations, the technical program also featured two keynote speeches. The two keynote speakers were Prof. Rui Zhang from National University of Singapore and Prof. Wei Xiang from La Trobe University, Australia.

Coordination with the steering chairs, Imrich Chlamtac, Xuemai Gu, and Cheng Li, was essential for the success of the conference. We sincerely appreciate their constant support and guidance. It was also a great pleasure to work with such an excellent organizing committee team for their hard work in organizing and supporting the conference. In particular, the Technical Program Committee, led by our TPC Co-Chairs, Dr. Wei Qu, Dr. Liang Ye, and Dr. Yu Zhang, completed the peer-review process of technical papers and made a high-quality technical program. We are also grateful to the Conference Manager Aleksandra Sledziejowska, for her support and to all the authors who submitted their papers to the AICON 2020 conference.

We strongly believe that the AICON conference provides a good forum for all researchers developers, and practitioners to discuss all scientific and technological aspects that are relevant to applying artificial intelligence to communications and networks. We also expect that future AICON conferences will be as successful and stimulating, as indicated by the contributions presented in this volume.

Shuo Shi
Liang Ye
Yu Zhang

Conference Organization

Steering Committee

Imrich Chlamtac	University of Trento, Italy
Xuemai Gu	Harbin Institute of Technology, China
Cheng Li	Memorial University of Newfoundland, Canada

Organizing Committee

General Chair

Qinyu Zhang	Peng Cheng Laboratory, China

General Co-chairs

Wei Xiang	La Trobe University, Australia
Lian Zhao	Ryerson University, Canada
Shuo Shi	Harbin Institute of Technology, China

TPC Chair and Co-chairs

Wei Qu	Space Engineering University, China
Liang Ye	Harbin Institute of Technology, China
Yu Zhang	Harbin Institute of Technology, China

Sponsorship and Exhibit Chair

Jian Jiao	Harbin Institute of Technology (Shenzhen), China

Local Chair

Yu Zhang	Harbin Institute of Technology, China

Workshops Chair

Shaohua Wu	Harbin Institute of Technology (Shenzhen), China

Publicity and Social Media Chair

Wei Li	Harbin Institute of Technology (Shenzhen), China

Publications Chair

Lu Jia	China Agricultural University, China

Web Chair

Ning Zhang	University of Windsor, Canada

Tutorials Chair

Zhenyu Xu	Huizhou Engineering Vocational College, China

Technical Program Committee

Xiaotian Zhou	Shandong University
Yongliang Sun	Nanjing Tech University
Nian Xia	National Cheng Kung University
Jian Sun	Shandong University
Shuyi Chen	Harbin Institute of Technology
Wanlong Zhao	Harbin Institute of Technology
Laiwei Jiang	Civil Aviation University of China
Jian Yang	Harbin Institute of Technology
Mo Han	Northeastern University, USA
Zheng Dong	Shandong University
Zhangyin Feng	Harbin Institute of Technology

Contents

Deep Learning/Machine Learning on Information and Signal Processing

AI in Ubiquitous Mobile Wireless Communications

AI in UAV-Assisted Wireless Communications

Recent Advances in AI and Their Applications in Future Electronic and Information Field

Deep Learning/Machine Learning on Information and Signal Processing

Cell Detection and Counting Method Based on Connected Domain of Binary Image

Junwen Si, Chuanchuan Zhu, and Xufen Xie(✉)

School of Information Science and Engineering, Dalian Polytechnic University, No. 1 Qinggongyuan, Ganjingzi District, Dalian 116034, China
xiexf@dlpu.edu.cn

Abstract. Cell counting plays an important role in biomedical research. There are always some phenomena such as indistinct intervals and target adhesion in cell images, which leads to poor segmentation effect and therefore inaccurate counting. In view of this situation, based on image binarization technology, this paper proposed a rapid cell count method combining mathematical morphology and connected domain labeling in which the cell images can be grayed, USM sharpened, binarized, morphologically processed, and connected domain labeled, and ultimately the number of cells could be calculated. The experimental results show that this method can effectively complete the segmentation of sparse cell images and intensive cell images, and the counting error is less than 5%.

Keywords: Binarization · Morphology · Connected domain · Cell counting

1 Introduction

Cell image counting is an important research direction in medical image processing. In recent years, with the development of image processing technology and machine learning methods, a large number of new theories and methods have emerged in the detection and counting of cells.

Literature [1, 2] proposed a method for automatic cell counting using Hough transform. Adherent cells were separated by morphological treatment, and cell count was realized by using circulating Hough transform. Literature [3] proposed a watershed segmentation method based on basin expansion, which can effectively segment adhesion objects and non-adhesion objects. Literature [4], the image noise is removed by the reconstruction operation of grayscale morphology, and the contrast of gradient image is enhanced by combining the top-hat transformation and the bottom-hat transformation. The cell image is segmented by the watershed algorithm, with significant segmentation effect. Literature [5], a new method of cell image segmentation based on wavelet transform and morphological watershed is proposed to solve the problems of over-segmentation when watershed segmentation algorithm is used. Literature [6] proposed a new blood cell image segmentation and counting algorithm by studying the PCNN and Autowave characteristics, which effectively eliminates the influence of small interference on cell image segmentation. Literature [7] proposed an image segmentation method that combines edge detection and mathematical

S. Shi et al. (Eds.): AICON 2020, LNICST 356, pp. 3–12, 2021.
https://doi.org/10.1007/978-3-030-69066-3_1

morphology, which effectively makes up for the shortcomings of a single segmentation algorithm, such as the lack of fine edge segmentation and the large number of holes in cells. In literature [8], K-means clustering was used to segment cell images under HIS color model, and the results showed that this method could remove background noise and retain target images well, thus achieving a good segmentation effect. Literature [9] discussed and evaluated four counting methods of K-means clustering, watershed transformation, histogram equalization and detection based on shape features. The experimental results showed the advantages and disadvantages of each method. Literature [10] proposed an automatic counting system of red blood cells based on fuzzy C-means clustering. Image filtering was carried out by fuzzy clustering method to separate red blood cells from background cells. After morphological processing, the counting accuracy was relatively high.

The above methods usually require a large amount of computation and take a long time to count. In this paper, a fast, intuitive and simple cell counting method is realized by programming graphical user interface (GUI) based on Matlab language, as shown in Fig. 1. This method can effectively conduct cell detection and more accurate cell count. The flow chart is shown in Fig. 2.

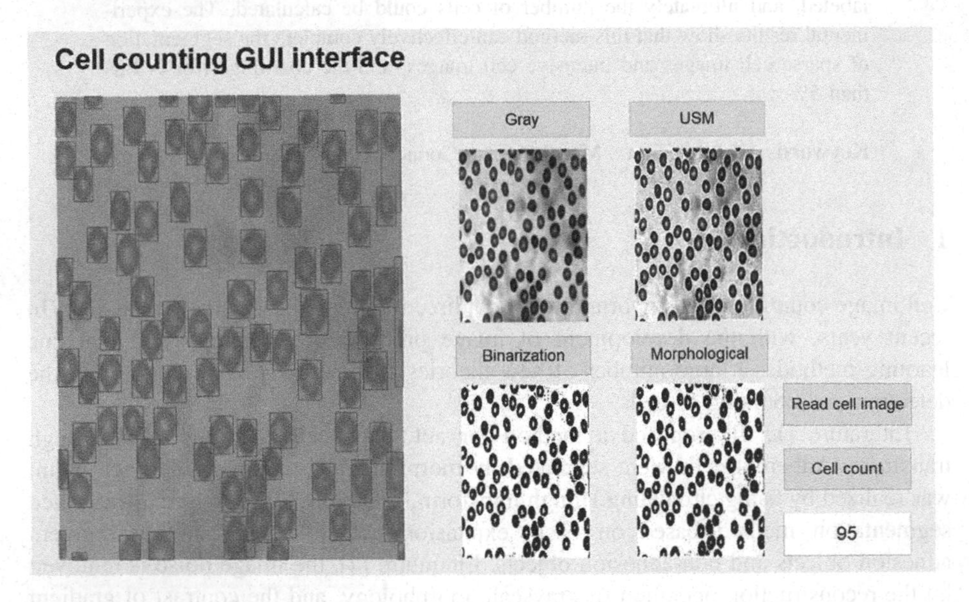

Fig. 1. GUI of cell counting

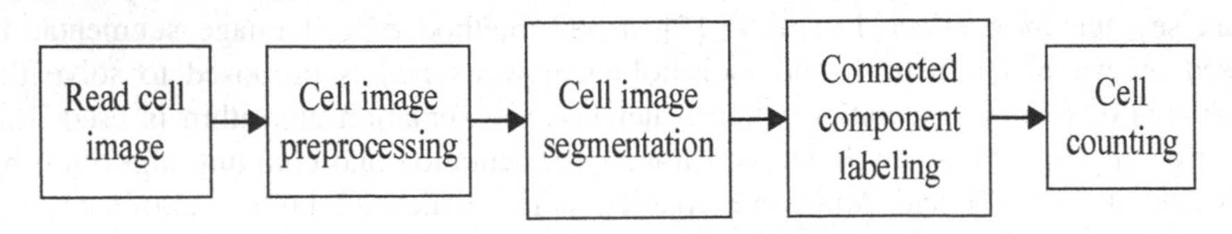

Fig. 2. Flow chart of cell counting

2 Cell Image Preprocessing

2.1 Gray

The cell microimage is an unprocessed color RGB image. In order to reduce the computational burden of image processing, this paper adopts the weighted average method for grayscale [11] processing, as shown in formula (1):

$$Gray = [R \quad G \quad B]\begin{bmatrix} 0.299 \\ 0.287 \\ 0.114 \end{bmatrix} \tag{1}$$

The image after graying processing is shown in Fig. 3.

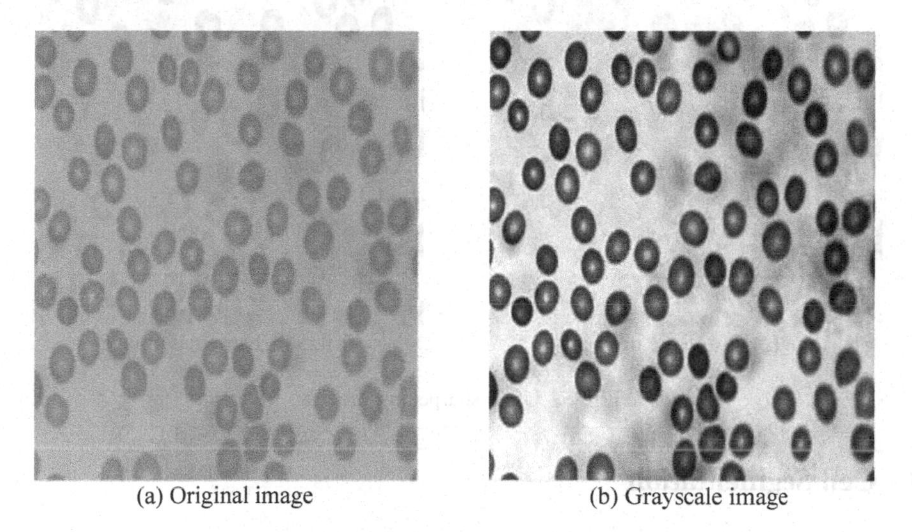

(a) Original image (b) Grayscale image

Fig. 3. Image graying processing

2.2 USM Sharpening Processing

Sharpness [12] is an attribute of image detail sharpness. In this paper, cell edge contrast was enhanced by sharpening the cell image. There is a big difference between the gray level of the target edge and that of the background, so the global binarization can be carried out better. The USM sharpening process is shown in Fig. 4:

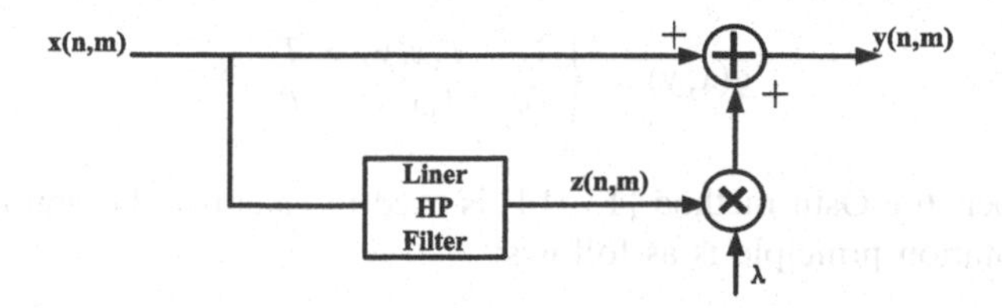

Fig. 4. USM model

The specific formula is expressed as Eq. (2):

$$y(n,m) = x(n,m) + \lambda z(n,m) \tag{2}$$

In the formula, $x(n,m)$ is the input image, $y(n,m)$ is the output image, $z(n,m)$ is the correction signal, which is obtained by high-pass filtering the original image, and λ is the scaling factor used to control the enhancement effect. The USM sharpening result of the cell image is shown in Fig. 5.

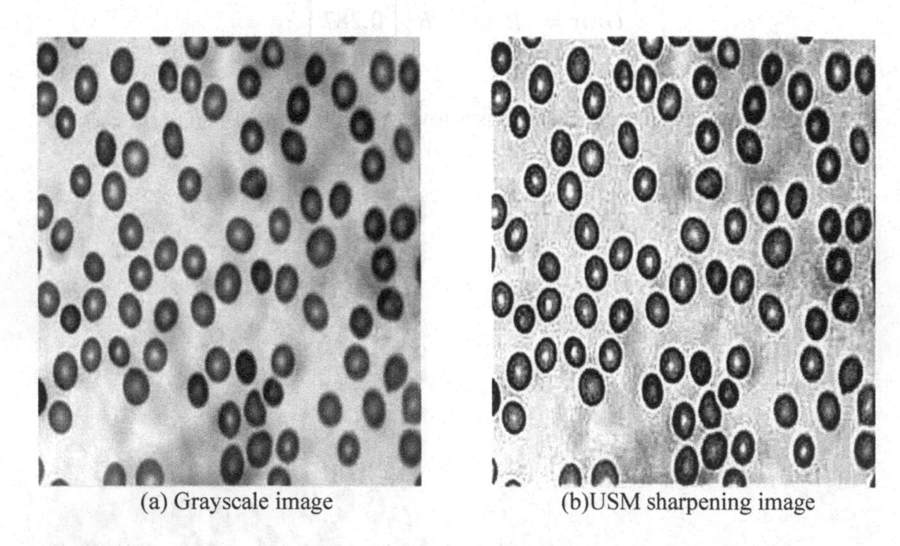

(a) Grayscale image (b)USM sharpening image

Fig. 5. USM sharpening process

3 Cell Segmentation

3.1 Binarization

In order to get the cell image from the background image better, the grayscale image needs to be binarized. Binarization can be divided into global binarization and adaptive binarization. The global binarization method has a very good effect on simple image processing of picture information. An image $f(x,y)$ consists of background pixels and target pixels. If you want to extract the target image from the background, you need to find the optimal threshold T. Make all the points (x,y) of $f(x,y) \leq T$ the target points, and vice versa as background points. The segmented image is shown in formula (3).

$$g(x,y) = \begin{cases} 1, & f(x,y) > T \\ 0, & f(x,y) \leq T \end{cases} \tag{3}$$

In this paper, the Ostu method [13, 14] is used to calculate the optimal threshold. The implementation principle is as follows:

1) Normalized histogram (Gray level L, total pixels N, probability of occurrence of each grayscale pixel is p_i)

$$N = \sum_{i=0}^{L-1} n_i, p_i = \frac{n_i}{N} \tag{4}$$

2) Calculate the occurrence probability of background pixel (p_A) and target pixel (p_B)

$$P_A = \sum_{i=0}^{t} p_i, P_B = \sum_{i=t+1}^{L-1} p_i = 1 - p_A \tag{5}$$

3) Calculate the inter-class variances of the background and target regions

$$\mu_A = \frac{\sum_{i=0}^{t} ip_i}{p_A}, \mu_B = \frac{\sum_{i=t+1}^{L-1} ip_i}{p_B} \tag{6}$$

$$\mu_0 = p_A\mu_A + p_B\mu_B = \sum_{i=1}^{L} ip_i \tag{7}$$

$$\sigma^2(t) = p_A(\mu_A - \mu_0)^2 + p_B(\mu_B - \mu_0)^2 \tag{8}$$

In the formula, μ_A and μ_B represent the average gray value of the target image and background respectively, and μ_0 represents the average gray value of the whole image. The image after binarization is shown in Fig. 6.

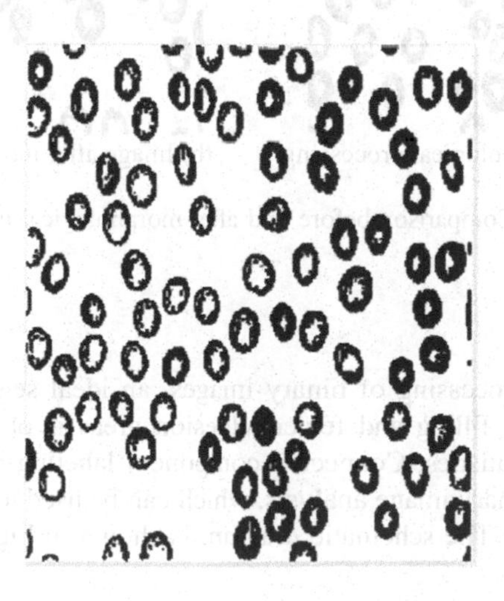

Fig. 6. Binarization image

3.2 Morphological Treatment

After image binarization, there will still be problems such as more corner burrs, cell. adhesion, extremely small noise points, and small holes. The open operation of morphological image processing [15–17] can solve these problems well. The open operation first performs the corrosion operation on the image, and then uses the same size structure element to perform the expansion operation on the corroded image.

The open operation of structure element B on set A, expressed as $A \circ B$, and its definition is shown in formula (9):

$$A \circ B = (A \Theta B) \oplus B \tag{9}$$

The cell adhesion is less after open operation, and the small boundary cavity is filled well. To further optimize the segmentation effect, the median filter is used to remove the minimal noise points, and finally a relatively ideal cell microsegmentation image is obtained, as shown in Fig. 7.

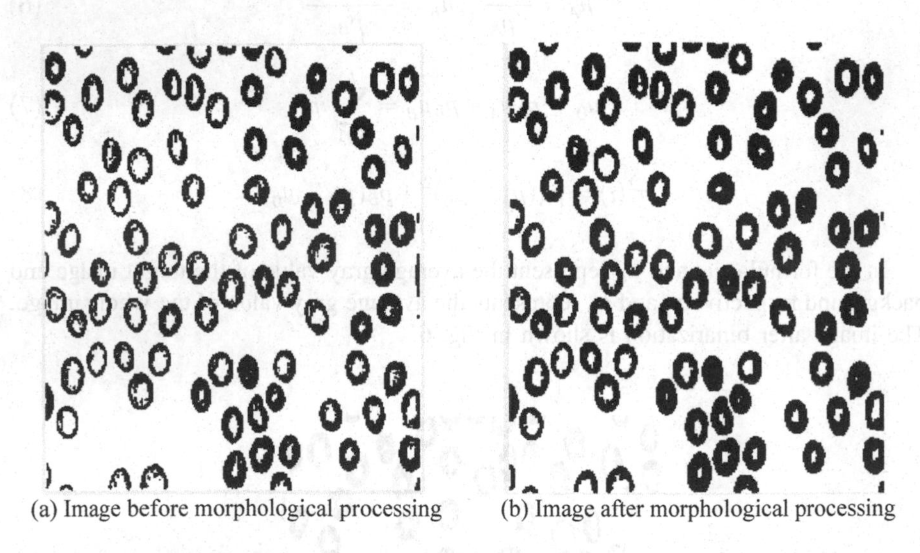

(a) Image before morphological processing (b) Image after morphological processing

Fig. 7. Comparison before and after morphological treatment

4 Cell Counting

After morphological processing of binary images, an ideal segmentation image with clear cell edges, cavity filling and fewer adhesion areas is obtained to facilitate the following cell count statistics. Connected component labeling algorithm [18–20] is a common method for binary image analysis, which can be used to extract the features of the target in the image, The schematic diagram is shown in Fig. 8.

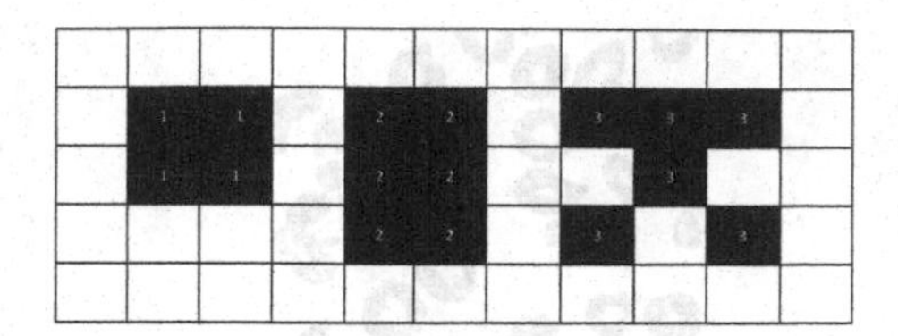

Fig. 8. Connected domain marking

In this paper, cell counting is achieved by counting the number of connected domains.

in a binary image. Taking 8-connected labeling as an example, the principle is:

1) Traverse the whole image, when the target pixel $f(x, y)$ is scanned, judge whether the pixel is marked, if the pixel point is not marked, save its coordinate value $M[x][y]$ to the queue, And mark the pixel point at the corresponding coordinate position of the marking matrix.
2) Scan 8 neighborhoods of pixel point $f(x, y)$. When the new unmarked target pixel point is scanned, the coordinate value $M[x+1][y]$ of the point $f(x+1, y)$ is saved in the queue and marked in the marking matrix.
3) When the 8 neighborhood scan marks are completed, $f(x, y)$ is listed, and the column head is $f(x+1, y)$. Then the 8-neighborhood scan and marks in Step 2 are performed again.
4) When a connected area is marked, the mark count is incremented by 1, the queue is cleared, and steps 1 to 3 are performed again to mark the new connected area.
5) After traversing the complete picture, the number of connected domain markers is the number of cells.

5 Experimental Results and Analysis

In this paper, the stained red blood cells is taken as the research object, and part of the original images is intercepted for image processing and counting,as shown in Fig. 9.

In order to verify the feasibility of the method in this paper, the statistical results are compared with the manual counting results. The comparison results are shown in Table 1. It can be seen from Table 1 that the method in this paper can count the number of cells well, whether it is a sparse cell image or a dense cell image. The results show that this method has low counting error and strong versatility.

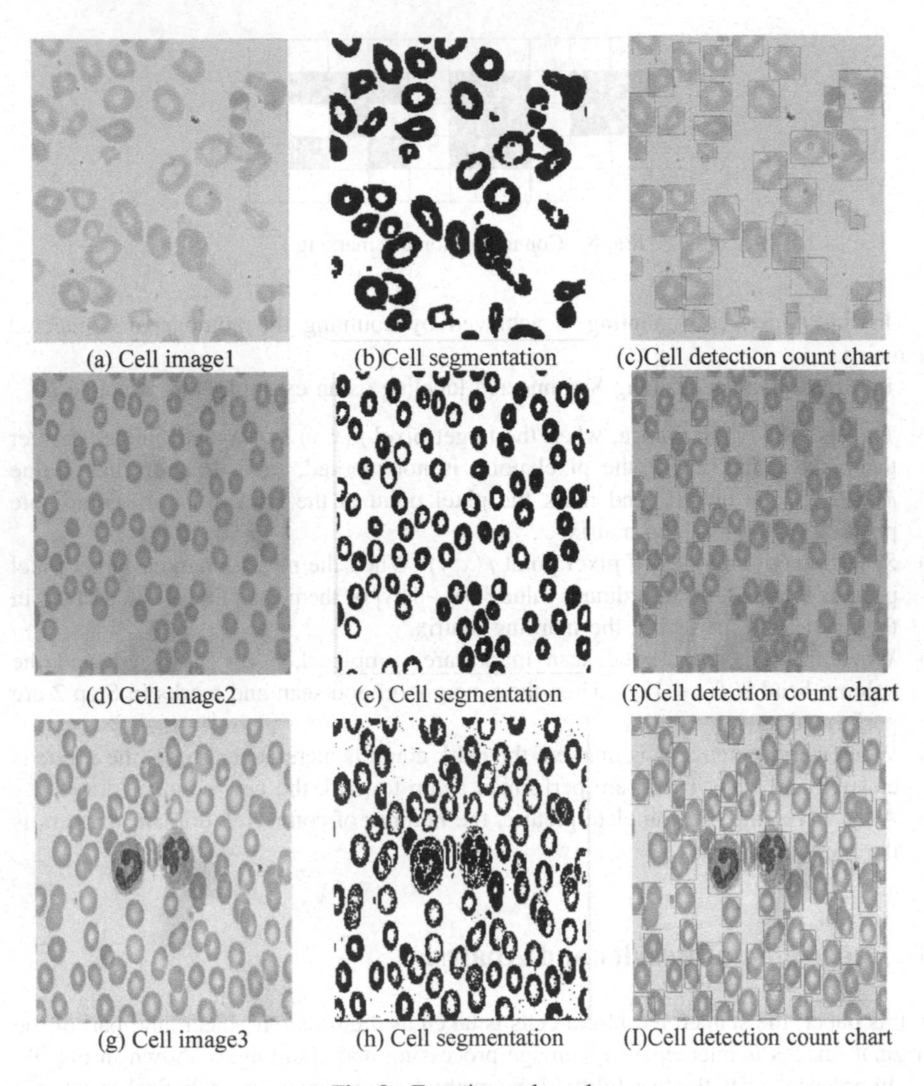

(a) Cell image1 (b)Cell segmentation (c)Cell detection count chart

(d) Cell image2 (e) Cell segmentation (f)Cell detection count chart

(g) Cell image3 (h) Cell segmentation (I)Cell detection count chart

Fig. 9. Experimental results

Table 1. Cell counting error table

Classifly	Cell image1	Cell image2	Cell image3
Manual counting results	43	93	100
System counting results	45	95	95
Absolute error	4%	2%	5%

6 Conclusions

This paper presents a method for cell counting based on Matlab language.This method firstly increases the contrast between the object and the background by USM sharpening the grayscale image, and then performs global threshold segmentation and mathematical morphology processing to obtain the microcell segmentation image with clear edge contour and less target adhesion.Finally, the cells are quickly labeled and counted. The results show that the method has strong adaptability and low counting error.The next step is to study the isolation of overlapping cells to improve the reliability of counting.

References

1. Reddy, V.H.: Automatic red blood cell and white blood cell counting for telemedicine system. Int. J. Res. Advent Technol. (2014)
2. Mazalan, S.M., Mahmood, N.H., Razak, M.A.A.: Automated red blood cells counting in peripheral blood smear image using circular hough transform. In: International Conference on Artificial Intelligence. IEEE Computer Society (2013)
3. Lin, X.Z., Wang, Y.M., Du, T.C.: A segment and count method based on watershed transformation. Comput. Eng. **32**(15), 181–183 (2006)
4. Zhang, S., Peng, D.L.: Cell image segmentation based on mathematical morphology. J. Hangzhou Dianzi Univ. **28**(6), 52–55 (2008)
5. Huang, Z.B., Liu, R.R., Liang, G.M.: Blood cell image segmentation based on wavelet transform and morphological watershed. Comput. Technol. Autom. **36**(3), 100–104 (2017)
6. Su, M.J., Wang, Z.B., Zhang, H.J.: A new method for blood cell image segmentation and counting based on PCNN and Autowave. In: 3rd International Symposium on Communications, Control, and Signal Processing (ISCCSP 2008), vol. 1, pp. 6–9 (2008)
7. Zhang, Y.H., Li, Y., Zhang, Y.N.:Research and implementation of cell image segmentation method based on mathematical morphology. Comput. Modernization (7), 135–137, 163 (2013)
8. Fatma, M., Sharma, J.: Leukemia image segmentation using K-means clustering and HSI color image segmentation. Int. J. Comput. Appl. **94**(12), 6–9 (2014)
9. Vaghela, H.P., Modi, H., Pandya, M.: Leukemia detection using digital image processing techniques. IJAIS **10**(1), 43–51 (2015)
10. Chourasiya, S., Rani, U.G., Purushotham, A.: A novel automatic red blood cell counting system using fuzzy C-means clustering. Int. J. Digital Appl. Contemp. Res. **2**(12), 1–5 (2014)
11. Chen, G.Q., Wang, B.X., Liu, M.: Linear projection grayscale algorithm based on structural information similarity. J. Jilin Univ. (Sci. Ed.) **58**(4), 877–884 (2020)
12. Luo, J., Teng, Q.Z., He, H.B.: Fuzzy image enhancement of rock thin slice based on USM. Mod. Comput. (2019)
13. Yue, Z.J., Qiu, W.C., Liu, C.L.: A self-adaptive approach of multi-object image segmentation. J. Image Graph. Ser. A **9**(6), 674–678 (2004)
14. Gao, H.B., Wang, W.X.: New connected component labeling algorithm for binary image. Comput. Appl. **27**(11), 2776–2777, 2785 (2007)

15. Zhang, X., Liu, M.J., Chen, L.W.: Method of picking up edge on the basis of the mathematics morphologic subject. J. Univ. Electron. Sci. Technol. China **31**(5), 490–493 (2002)
16. Liu, Q., Lin, T.S.: Image edge detection algorithm based on mathematical morphology. J. South China Univ. Technol. (Nat. Scie. Ed.) **36**(9), 113–116, 121 (2008)
17. Zhang, J.L., Ye, P.K., Sun, S.S.: The detection of ore particle size under morphological image processing. Mach. Des. Manuf. **3**, 68–71 (2008)
18. Liu, G.S., Lu, J.W., Xu, J.G.: A new algorithm for fast pixel labeling in binary images. Comput. Eng. Appl. **38**(004), 57–59 (2002)
19. Miao, L.Y., Yu, Z.L., Wang, Z.: Regional filling algorithm based on connected region labeling. J. Changchun Univ. Sci. Technol. (Nat. Sci. Ed.) **041**(004), 114–117 (2018)
20. Zhang, X.J., Guo, X., Jin, X.Y.: The pixel labeled algorithm with label rectified of connecting area in binary pictures. J. Image Arid Graph. Ser. A **8**(2), 198–202 (2003)

2D DOA Estimation Based on Modified Compressed Sensing Algorithm

Chang Fu[1](✉) and Jun Ma[2]

[1] AVIC Harbin Aircraft Industry Group Co. Ltd., Harbin 150066, Heilongjiang, China
508045450@qq.com
[2] Harbin Engineering University, Harbin 150001, Heilongjiang, China

Abstract. In order to realize high-precision DOA tracking in space, researches on two-dimensional DOA estimation have been conducted in recent years. The existing algorithms often need large snapshots for estimation accuracy, going against the fast solution. Considering the low sensitivity of DOA estimation algorithm based on compressed sensing theory to the number of snapshots and the correct estimation with less sampling data, a modified two-dimensional multitask compressed sensing algorithm based on SVD decomposition is proposed in this paper. This algorithm makes up for the drawbacks of existing compressed sensing algorithms in dealing with multi snapshot problem and reduces the unnecessary calculation. Simulation results show that the proposed algorithm can solve the off-grid problem in compressed sensing, and has better estimation performance than other algorithms under the condition of low SNR and few snapshots.

Keywords: 2D DOA estimation · Compressed sensing · SVD

1 Introduction

First proposed by Malioutov et al. in 2005, the concept of CS-DOA is called l1-SVD [1], with the core idea of transforming the DOA estimation problem into a sparse reconstruction problem for solving an underdetermined system of equations. It has become a classic algorithm in the field of CS-DOA. The array signal models of the following algorithms are roughly the same, and the differences mainly focus on the selection of optimal reconstruction algorithms. To reduce the calculation amount in OMP algorithm used in sparse signal reconstruction, Wang Shuhao et al. proposed DOA Estimation of LFM Signals Based on Compressed Sensing [2], which makes use of the bat algorithm's population search mode and excellent echo localization ability in flight to achieve fast optimization. To solve the problem of poor accuracy of DOA estimation based on compressed sensing when the array antenna has amplitude and phase errors, Zuo Luo et al. proposed a super-resolution DOA estimation method based on TLS-CS [3], combined with singular value decomposition (SVD) and greedy iterative pursuit algorithm for CS sparse reconstruction to obtain the azimuth information of the target. In order to solve the problem of low efficiency when orthogonal matching pursuit algorithm is used for sparse recovery of high dimensional signals, Zhao

S. Shi et al. (Eds.): AICON 2020, LNICST 356, pp. 13–25, 2021.
https://doi.org/10.1007/978-3-030-69066-3_2

Hongwei et al. proposed a DOA estimation algorithm combined with particle swarm optimization [4]. The algorithm uses PSO algorithm to solve the optimization problem, and improves the particle renewal mechanism and inertia weight.

At present, among the problems in DOA estimation algorithm based on compressed sensing theory, the main one is off-grid. No matter how fine the grid is divided, the target signal may be located between the grids, resulting in mismatch between the sparse base of the artificially constructed redundant dictionary and the real base of the target, so the off-grid problem is also known as the base mismatch problem. In the field of DOA, many researches on off-grid problem have been carried out, such as the alternating iteration method [5], and the sparse Bayesian method [6] used to solve the off-grid problem in DOA. Some variant algorithms on this basis also appear, such as the introduction of second-order Taylor polynomial, and the idea of noise subspace [7]. However, the limitation of these algorithms is that most of the signal models are for one-dimensional DOA estimation, which cannot be effectively extended to two-dimensional cases. And the two-dimensional off-grid problem is more universal. In order to improve the accuracy of DOA estimation, this paper will focus on the realization of two-dimensional DOA estimation based on compressed sensing and the off-grid problem.

2 Problem Formulation

In the case of one-dimensional space, there are Q incident signals with azimuth angles $\{\theta_1, \cdots \theta_Q\}$. Meanwhile, we divide the space $[0^{\circ}, 90^{\circ}]$ into N grids as $[\alpha_1, \alpha_2, \cdots, \alpha_N]$ with $Q << N$. When there is an incident signal at a certain grid, the value at the corresponding position of sparse representation signal is not zero; otherwise, it is zero.

The relationship between the division of spatial discrete grids and the angles of incident signals is shown in Fig. 1, where "•"is the real spatial incident signals, and "o"the potential ones.

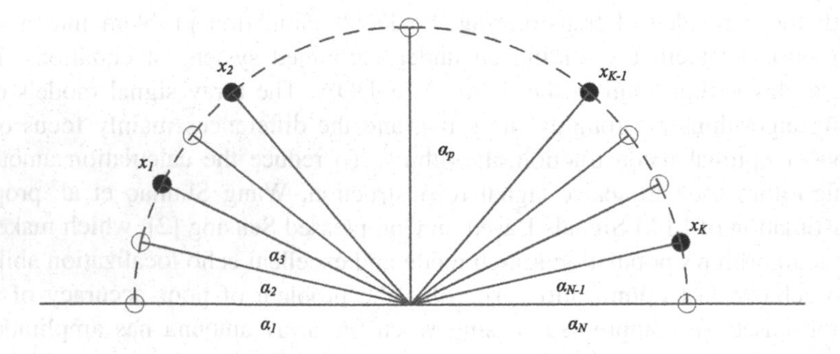

Fig. 1. Spatial discrete grids and the angles of incident signals

In practice, the direction of arrival of the signal is continuous in the spatial domain, but we discretized space angles in the process of DOA estimation. The incoming direction of the actual source may fall between the adjacent grid points in spite of dense grid sampling, as shown in Fig. 2. The off-grid problem hence appears, leading to errors in DOA estimation. Although the dense division of spatial grids can alleviate this problem, it will increase the dimension of redundant dictionary (array manifold) and the amount of calculation and slow down the solution, which restricts the application of DOA estimation algorithm based on compressed sensing theory in practice.

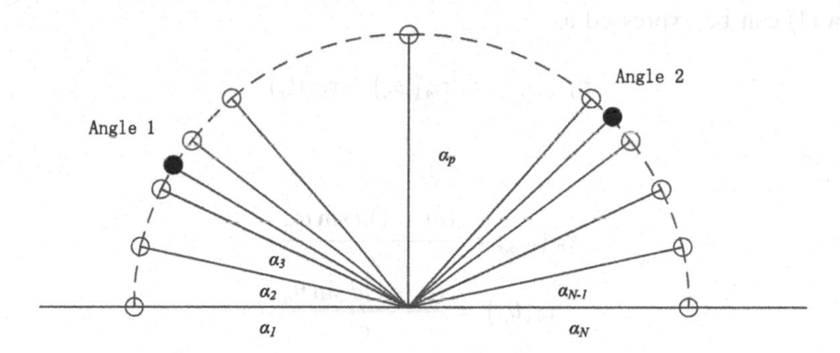

Fig. 2. Off-grid problem

Due to the coupling of pitch angles and azimuth angles in conventional array manifold, the steering matrix $\boldsymbol{A}$ cannot be decomposed. Hence, new array model is introduced, as shown in Fig. 3.

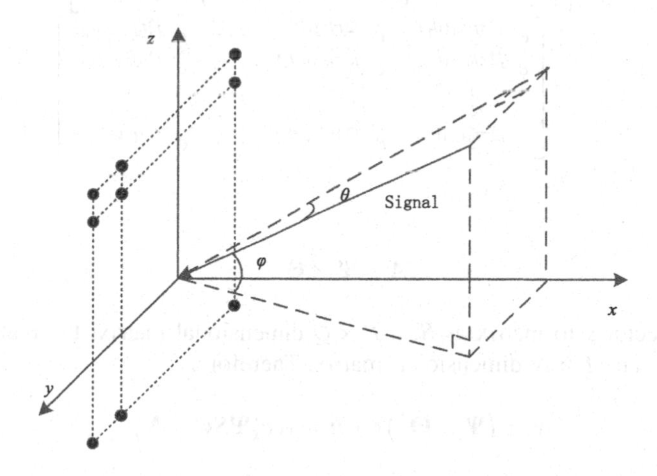

Fig. 3. Array model

The uniform square array is set as the object, with dimension as $M \times N$ in zoy plane. The number of spatial domain grids is shown as $P \times Q$, and the signal time delay $\tau_{(m,n),(p,q)}$ of the (p,q)-th grid received by the (m,n)-th array element as

$$\tau_{(m,n),(p,q)} = \frac{(n-1)d\sin\theta_{p,q} + (m-1)d\sin\varphi_{p,q}}{c} \tag{1}$$

Where p is the pitch dimension of the grid, and q is the azimuth dimension. For a clearer derivation, the azimuth angle $\theta_{p,q}$ is θ_q, and the pitch angle $\varphi_{p,q}$ is φ_p. The formula (1) can be expressed as

$$\tau_{(m,n),(p,q)} = \tau_m(\varphi_p) + \tau_n(\theta_q) \tag{2}$$

Where

$$\begin{aligned} \tau_m(\varphi_p) &= \frac{(m-1)d\sin\varphi_p}{c} \\ \tau_n(\theta_q) &= \frac{(n-1)d\sin\theta_q}{c} \end{aligned} \tag{3}$$

And the steering matrix A can be expressed as

$$\begin{aligned} A = &\begin{bmatrix} e^{-j2\pi f_0 \tau_1(\varphi_1)} & e^{-j2\pi f_0 \tau_1(\varphi_2)} & \cdots & e^{-j2\pi f_0 \tau_1(\varphi_P)} \\ e^{-j2\pi f_0 \tau_2(\varphi_1)} & e^{-j2\pi f_0 \tau_2(\varphi_2)} & \cdots & e^{-j2\pi f_0 \tau_2(\varphi_P)} \\ \vdots & \vdots & \ddots & \vdots \\ e^{-j2\pi f_0 \tau_M(\varphi_1)} & e^{-j2\pi f_0 \tau_M(\varphi_2)} & \cdots & e^{-j2\pi f_0 \tau_M(\varphi_P)} \end{bmatrix} \\ &\otimes \begin{bmatrix} e^{-j2\pi f_0 \tau_1(\theta_1)} & e^{-j2\pi f_0 \tau_1(\theta_2)} & \cdots & e^{-j2\pi f_0 \tau_1(\theta_Q)} \\ e^{-j2\pi f_0 \tau_2(\theta_1)} & e^{-j2\pi f_0 \tau_2(\theta_2)} & \cdots & e^{-j2\pi f_0 \tau_2(\theta_Q)} \\ \vdots & \vdots & \ddots & \vdots \\ e^{-j2\pi f_0 \tau_N(\theta_1)} & e^{-j2\pi f_0 \tau_N(\theta_2)} & \cdots & e^{-j2\pi f_0 \tau_N(\theta_Q)} \end{bmatrix} \end{aligned} \tag{4}$$

i.e.

$$\boldsymbol{A} = \boldsymbol{\Psi} \otimes \boldsymbol{\Theta}^T \tag{5}$$

Restore vector $\boldsymbol{s}$ to matrix as $\boldsymbol{S}$, a $P \times Q$ dimensional matrix; the matrix form of vector $\boldsymbol{n}$ is $\boldsymbol{N}$, an $M \times N$ dimensional matrix. Therefore,

$$\boldsymbol{y} = (\boldsymbol{\Psi} \otimes \boldsymbol{\Theta}^T)\boldsymbol{s} + \boldsymbol{n} = vec(\boldsymbol{\Psi S \Theta} + \boldsymbol{N}) \tag{6}$$

Where vec shows that the matrix is arranged into column vector in row priority order. Then,

$$Y = \Psi S \Theta + N \tag{7}$$

Where Y is $M \times N$ dimensional, and the matrix form of vector $\boldsymbol{y}$.

3 2D Multitasking CS Algorithm Based on SVD

3.1 2D Off-Grid Algorithm Based on Taylor Expansion

The first order approximation of Taylor expansion can be used to solve the basis mismatch problem in two-dimensional DOA estimation by applying the above signal model with separable steering matrix.

Discrete the spatial domain; suppose the angle of a target signal is $\left(\hat{\theta}, \hat{\varphi}\right)$, and the angle of its nearest grid is $\left(\theta_p, \varphi_q\right)$. δ_{θ_p} and δ_{φ_q} represent the estimation bias of the pitch dimension and azimuth dimension. Here, $\delta_\theta = [\delta_{\theta_1}, \delta_{\theta_2}, \cdots, \delta_{\theta_P}]^T \delta_\varphi = \left[\delta_{\varphi_1}, \delta_{\varphi_2}, \cdots, \delta_{\varphi_Q}\right]^T$.

Discrete steering vector as follows:

$$a\left(\hat{\varphi}, \hat{\theta}\right) = \psi(\hat{\varphi}) \otimes \phi\left(\hat{\theta}\right) \tag{8}$$

Where $\otimes$ is Kronecker product.

$$a\left(\hat{\varphi}, \hat{\theta}\right) = \left[e^{-j2\pi f_0 \tau_{(1,1)}\left(\hat{\varphi},\hat{\theta}\right)} \quad e^{-j2\pi f_0 \tau_{(1,2)}\left(\hat{\varphi},\hat{\theta}\right)} \quad \cdots \quad e^{-j2\pi f_0 \tau_{(m,n)}\left(\hat{\varphi},\hat{\theta}\right)} \quad \cdots \quad e^{-j2\pi f_0 \tau_{(M,N)}\left(\hat{\varphi},\hat{\theta}\right)}\right]^T \tag{9}$$

$$\tau_{(m,n)}\left(\hat{\varphi}, \hat{\theta}\right) = \frac{(n-1)d\sin\hat{\theta} + (m-1)d\sin\hat{\varphi}}{c} \tag{10}$$

$\psi(\hat{\varphi})$ is the steering vector of pitch dimension, expressed as follows.

$$\psi(\hat{\varphi}) = \left[e^{-j2\pi f_0 \tau_1(\hat{\varphi})} \quad e^{-j2\pi f_0 \tau_2(\hat{\varphi})} \quad \cdots \quad e^{-j2\pi f_0 \tau_m(\hat{\varphi})} \quad \cdots \quad e^{-j2\pi f_0 \tau_M(\hat{\varphi})}\right]^T \tag{11}$$

$\phi\left(\hat{\theta}\right)$ is the steering vector of azimuth dimension, expressed as follows.

$$\phi\left(\hat{\theta}\right) = \left[e^{-j2\pi f_0 \tau_1\left(\hat{\theta}\right)} \quad e^{-j2\pi f_0 \tau_2\left(\hat{\theta}\right)} \quad \cdots \quad e^{-j2\pi f_0 \tau_n\left(\hat{\theta}\right)} \quad \cdots \quad e^{-j2\pi f_0 \tau_N\left(\hat{\theta}\right)}\right]^T \tag{12}$$

Express the steering vectors of pitch and azimuth dimensions of the target signal separately with first order approximation of Taylor expansion as follows.

$$a\left(\hat{\varphi}, \hat{\theta}\right) = \left(\psi(\varphi_p) + b(\varphi_p)(\hat{\varphi} - \varphi_p)\right) \otimes \left(\phi(\theta_q) + c(\theta_q)\left(\hat{\theta} - \theta_q\right)\right) \tag{13}$$

Where $b(\varphi_p)$ and $c(\theta_q)$ are first derivative vectors of $\psi(\varphi_p)$ and $\phi(\theta_q)$.

Put the above into matrix form, and the signal model of 2D DOA estimation can be expressed as follows.

$$\boldsymbol{Y} = (\boldsymbol{\Psi} + \boldsymbol{B}\boldsymbol{\Delta}\delta_\varphi)\boldsymbol{S}(\boldsymbol{\Theta} + \boldsymbol{\Delta}\delta_\theta\boldsymbol{C}) + \boldsymbol{N} \tag{14}$$

Where $\boldsymbol{\Delta}\delta_\varphi = diag(\delta_\varphi)$ and $\boldsymbol{\Delta}\delta_\theta = diag(\delta_\theta)$, $\boldsymbol{B}$ is the matrix obtained by deriving each element in the steering vector of pitch dimension $\boldsymbol{\Psi}$, so does $\boldsymbol{C}$, and $\boldsymbol{N}$ is the noise.

In (14), there are three unknown variables $\boldsymbol{\Delta}\delta_\varphi$, $\boldsymbol{\Delta}\delta_\theta$ and $\boldsymbol{S}$. Solve the matrix $\boldsymbol{S}$, then solve $\boldsymbol{\Delta}\delta_\varphi$ and $\boldsymbol{\Delta}\delta_\theta$ with the method of alternate iteration as follows.

In the solution of signal matrix $\boldsymbol{S}$, initialize $\boldsymbol{\Delta}\delta_\varphi$ and $\boldsymbol{\Delta}\delta_\theta$ as zero matrixes, and (14) can be expressed as follows.

$$\boldsymbol{Y} = \boldsymbol{\Psi}\boldsymbol{S}\boldsymbol{\Theta} + \boldsymbol{N} \tag{15}$$

The original problem degenerates into the basic problem of separable DOA estimation, which can be regarded as a rough solution of $\boldsymbol{S}$. Due to the off-grid problem, the solution obtained must be deviated from the true value.

Then solve $\boldsymbol{\Delta}\delta_\varphi$ and $\boldsymbol{\Delta}\delta_\theta$.

In the solution of $\boldsymbol{\Delta}\delta_\varphi$, initialize $\boldsymbol{\Delta}\delta_\theta$ as unit matrix, regard $\boldsymbol{S}(\boldsymbol{\Theta} + \boldsymbol{\Delta}\delta_\theta\boldsymbol{C})$ in (14) as a fixed value, let $\boldsymbol{H} = \boldsymbol{S}(\boldsymbol{\Theta} + \boldsymbol{\Delta}\delta_\theta\boldsymbol{C})$, then

$$\boldsymbol{Y} = (\boldsymbol{\Psi} + \boldsymbol{B}\boldsymbol{\Delta}\delta_\varphi)\boldsymbol{H} + \boldsymbol{N} \tag{16}$$

The minimum deviation is required between (θ_p, φ_q), the angle of signal matrix, and the true angle of the target signal, which means δ_φ and δ_θ are restrained to the minimum. Constrain their sparsity with 2-norm, and the optimization problem can be obtained.

$$\min\|\delta_\varphi\|_2^2 + \|\boldsymbol{Y} - (\boldsymbol{\Psi} + \boldsymbol{B}\boldsymbol{\Delta}\delta_\varphi)\boldsymbol{H}\|_F^2 \tag{17}$$

Least-squares solution is applied, and each column in matrix $\boldsymbol{Y}$ in (17) meets the following equation:

$$\boldsymbol{Y}[n] = (\boldsymbol{\Psi} + \boldsymbol{B}\boldsymbol{\Delta}\delta_\varphi)\boldsymbol{H}[n] \tag{18}$$

Where $\bullet[n]$ is the n-th column, $n = 1, 2, \cdots, N$. Given the following equation

$$\boldsymbol{\Delta}\delta_\varphi\boldsymbol{H}[n] = \boldsymbol{\Delta}\boldsymbol{H}[n]\varphi \tag{19}$$

Where $\boldsymbol{\Delta}\boldsymbol{H}[n] = diag(\boldsymbol{H}[n])$. The optimization problem can be transformed into the problem of obtaining the least square solution.

$$\begin{bmatrix} \boldsymbol{Y}[1] - \boldsymbol{\Psi H}[1] \\ \boldsymbol{Y}[2] - \boldsymbol{\Psi H}[2] \\ \vdots \\ \boldsymbol{Y}[N] - \boldsymbol{\Psi H}[N] \end{bmatrix} = \begin{bmatrix} \boldsymbol{B\Delta H}[1] \\ \boldsymbol{B\Delta H}[2] \\ \vdots \\ \boldsymbol{B\Delta H}[N] \end{bmatrix} \delta_{\varphi} \tag{20}$$

The solution to δ_{φ} can be obtained with the least square method.

$$\delta_{\varphi} = (\boldsymbol{B}_{\Delta H})^{+} \boldsymbol{Y}_{\Psi H} \tag{21}$$

Where $(\bullet)^{+}$ is the generalized inverse of the matrix.

$$\begin{aligned} \boldsymbol{B}_{\Delta H} &= [\boldsymbol{B\Delta H}[1], \boldsymbol{B\Delta H}[2], \cdots, \boldsymbol{B\Delta H}[N]]^{T} \\ \boldsymbol{Y}_{\Psi H} &= [\boldsymbol{Y}[1] - \boldsymbol{\Psi H}[1], \boldsymbol{Y}[2] - \boldsymbol{\Psi H}[2], \cdots, \boldsymbol{Y}[N] - \boldsymbol{\Psi H}[N]]^{T} \end{aligned} \tag{22}$$

Substitute δ_{φ} into (14), and let $\boldsymbol{G} = (\boldsymbol{\Psi} + \boldsymbol{B}\Delta\delta_{\varphi})\boldsymbol{S}$, and solve δ_{θ} in the same way. Calculate δ_{φ} and δ_{θ} alternatively, and set the condition to finish iterative process. $\boldsymbol{S}$, $\Delta\delta_{\varphi}$ and $\Delta\delta_{\theta}$ are ultimately solved, then the true angles of the target signal can be expressed as $\hat{\theta}_p = \theta_p + \delta_{\theta_p}$ and $\hat{\varphi}_q = \varphi_q + \delta_{\varphi_q}$.

3.2 OMP Reconstruction Method Based on SVD

The two-dimensional off-grid algorithm based on the first-order Taylor expansion has some limitations. When reconstructing the sparse matrix $\boldsymbol{S}$, the noise of the received data matrix $\boldsymbol{Y}$ is zero by default, which leads to a poor accuracy of the solution to the sparse matrix and a less likely improvement in the following solution. In addition, in the case of a low SNR, more sampling snapshots can achieve certain estimation accuracy, and the algorithm is only suitable for single snapshot, which needs to be extended to multiple snapshots and solve the problem of high computational complexity.

In order to reduce the computational complexity and improve the anti-noise performance, the subspace methods are often applied in array signal processing. In this paper, singular value decomposition (SVD) is used to extract signal subspace to process array received signal matrix.

Without considering the coherence among signal sources, this paper conducts the singular value decomposition to the received data matrix $\boldsymbol{Y}$ with the purpose of removing noise from the received signal.

$$\boldsymbol{Y} = \boldsymbol{ULV}^{H} = [\boldsymbol{U}_S \boldsymbol{U}_N]\boldsymbol{LV}^{H} \tag{23}$$

Where V is an orthogonal matrix, and U is a matrix formed by arranging singular values from large to small. U_S is the signal subspace formed by singular vectors corresponding to the preceding K singular values. U_N is the noise subspace. Suppose the number of signal source K is known, take the preceding K columns of matrix U, and the $N \times K$ dimensional matrix $\boldsymbol{Y}_S = \boldsymbol{ULD}_K = \boldsymbol{YVD}_K$ formed by signal components is obtained, where $\boldsymbol{D}_K = [\boldsymbol{I}_K \quad \boldsymbol{O}]^H$. $\boldsymbol{I}_K$ is the $K \times K$ dimensional unit matrix, O is the $K \times (T - K)$ dimensional zero matrix, and the low-dimensional form of $\boldsymbol{Y}$ can be expressed as follows.

$$\boldsymbol{Y}_S = \boldsymbol{\Psi S}_S\boldsymbol{\Theta} + \boldsymbol{N}_S \tag{24}$$

Where $\boldsymbol{S}_S = \boldsymbol{SVD}_K$, and $\boldsymbol{N}_S = \boldsymbol{NVD}_K$. $\boldsymbol{S}_S$ remains the sparsity unchanged. It is the signal theoretically denoised from $\boldsymbol{S}$, then the solution to $\boldsymbol{S}$ can be simplified as the solution to $\boldsymbol{S}_S$. Compared to the common high sampling frequency under the practical condition, data dimension can decrease from T to K with the application of this algorithm, hence the calculation is significantly reduced. Simultaneously, SVD can be comprehended as a denoising process, which is helpful in DOA estimation under low SNR.

The algorithm steps are as follows:

(1) Conduct SVD to observation matrix $\boldsymbol{Y}$ to get $\boldsymbol{Y}_S$, and keep the right singular vector $\boldsymbol{U}_K$;
(2) Initialize residual matrix and index set, i.e. $\boldsymbol{R} = \boldsymbol{Y}$ and $\boldsymbol{\Lambda}_0 = \emptyset$;
(3) Take the inner product of each column in $\boldsymbol{U}_K$ and $\boldsymbol{\Phi}$, i.e. $\boldsymbol{G}^n = \left|\boldsymbol{\Phi}^H \boldsymbol{U}_{Kn-1}\right|$;
(4) Find out the row index value λ corresponding to the maximum norm of row vector q in $\boldsymbol{G}^n$, and update the index set $\boldsymbol{\Lambda}_n = \boldsymbol{\Lambda}_{n-1} \cup \{\lambda_n\}$ and its column vector set $\boldsymbol{\Psi}_n = \boldsymbol{\Psi}_{n-1} \cup \{\boldsymbol{\Phi}_{\lambda_n}\}$;
(5) Obtain its approximate solution $\boldsymbol{S}_S^n = \left(\boldsymbol{\Psi}_n^H \boldsymbol{\Psi}_n\right)^{-1} \boldsymbol{\Psi}_n^H \boldsymbol{Y}_S$ with the least square method;
(6) Update the residual error $\boldsymbol{R}_n = \boldsymbol{Y}_S - \boldsymbol{\Psi S}_S^n$, where $n = n + 1$;
(7) Judge whether the end condition of the algorithm is reached. If so, the calculation will be terminated. Otherwise, skip to Step (2) and repeat.

3.3 Multi-task Processing

The improved algorithm applies the current separable array signal model, which is only suitable for single snapshot. To solve the problem of DOA estimation under the condition of multiple snapshots, this paper introduces the idea of multitasking Bayesian compressed sensing into the algorithm.

Define snapshot as K, solve the data of the i-th snapshot to obtain δ_{φ} and δ_{θ},

$$\begin{aligned} \boldsymbol{Y}_i &= (\boldsymbol{\Psi} + \boldsymbol{B}\boldsymbol{\Delta}\delta_{\varphi i})\boldsymbol{H} + \boldsymbol{N}_i \\ \boldsymbol{Y}_i &= \boldsymbol{G}(\boldsymbol{\Theta} + \boldsymbol{\Delta}\delta_{\theta i}\boldsymbol{C}) + \boldsymbol{N}_i \end{aligned} \tag{25}$$

Conduct sparsity constraint to $\boldsymbol{\Delta}\delta_{\varphi i}$ and $\boldsymbol{\Delta}\delta_{\theta i}$ with 2-norm to get the optimized problem:

$$\begin{aligned} \min\|\delta_{\varphi i}\|_2^2 + \|\boldsymbol{Y}_i - (\boldsymbol{\Psi} + \boldsymbol{B}\boldsymbol{\Delta}\delta_{\varphi i})\boldsymbol{H}\|_F^2 \\ \min\|\delta_{\theta i}\|_2^2 + \|\boldsymbol{Y}_i - \boldsymbol{G}(\boldsymbol{\Theta} + \boldsymbol{\Delta}\delta_{\theta i}\boldsymbol{C})\|_F^2 \end{aligned} \tag{26}$$

Solve it with least-squares solution to get the analytical solution of $\delta_{\varphi i}$

$$\delta_{\varphi i} = (\boldsymbol{B}_{\Delta H})^{+}\boldsymbol{Y}_{\Psi Hi} \tag{27}$$

And the analytical solution of $\delta_{\theta i}$

$$\delta_{\theta i} = (\boldsymbol{G}_{\Delta C})^{+}\boldsymbol{Y}_{G\Theta i} \tag{28}$$

To sum up, the solving steps of 2D multitasking compressed sensing off-grid algorithm based on SVD are summarized as follows:

(1) Set a pitch dimension $\boldsymbol{\Psi}$ and an azimuth dimension $\boldsymbol{\Theta}$, and conduct derivation to get matrixes $\boldsymbol{B}$ and $\boldsymbol{C}$.
(2) Obtain spectral matrix $\boldsymbol{S}$ by applying the OMP reconstruction method to matrixes $\boldsymbol{Y}$, $\boldsymbol{\Psi}$ and $\boldsymbol{\Theta}$.
(3) Take the sampling data of the i-th snapshot $i(i = 1, \cdots, K)$, and obtain $\delta_{\theta i}$ and $\delta_{\varphi i}$ according to (26) and (27) till the termination conditions are met.
(4) Work out the incident angles with (θ_p, φ_q), $\delta_{\theta i}$, and $\delta_{\varphi i}$ in $\boldsymbol{S}$ obtained in Step 2.

4 Performance Study

To verify the effectiveness of the modified algorithm, a supposed simulation model is a 8×8 matrix URA with $d = \lambda/2$. Two independent narrow-band signals are separately incident on the array with azimuth angles 9.9° and 18.4°, and pitch angles 3.3° and 24.2°. Their ranges of spatial discrete grids are both $[0°, 90°]$. The following experiments are conducted based on the above conditions.

Experiment 1

To test the feasibility of signal model with separable array manifold under the condition of single snapshot, the DOA estimation performance of MMV-OMP, BCS and off-grid algorithms under different SNR is simulated, with SNR increasing from −15 dB to 10 dB, the number of snapshots 1, and Monte-carlo simulation 500 times. Results are shown in Fig. 4.

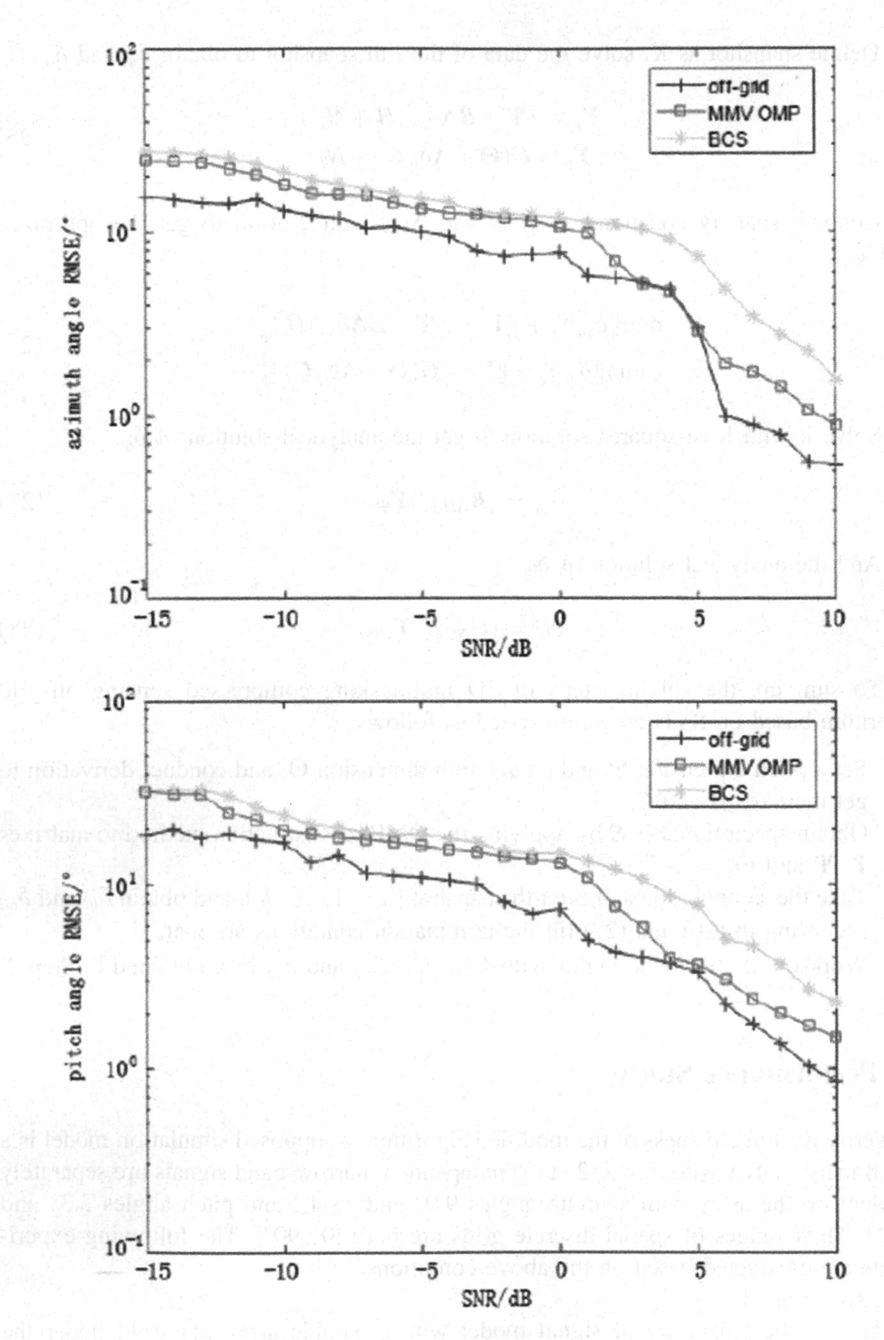

Fig. 4. Curves of mean square error

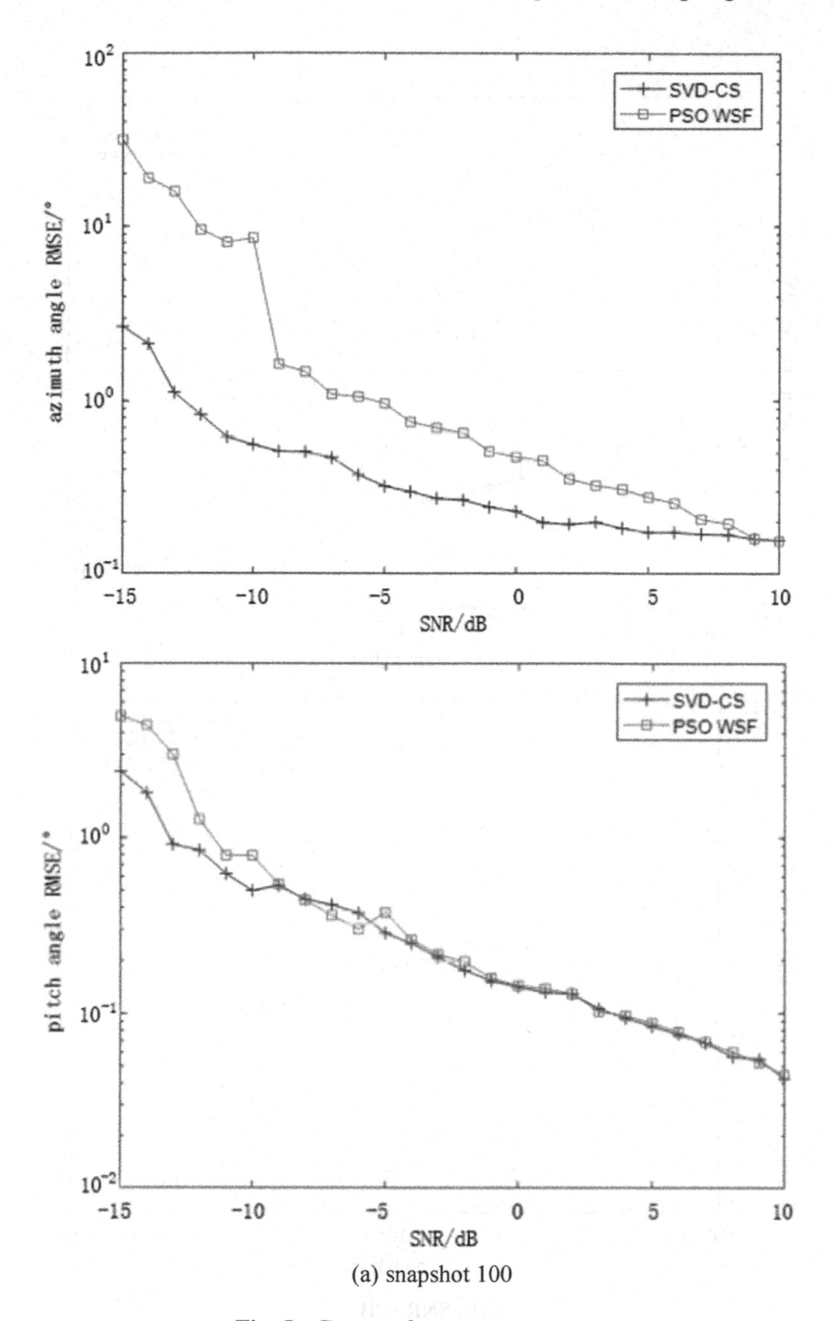

(a) snapshot 100

Fig. 5. Curves of mean square error

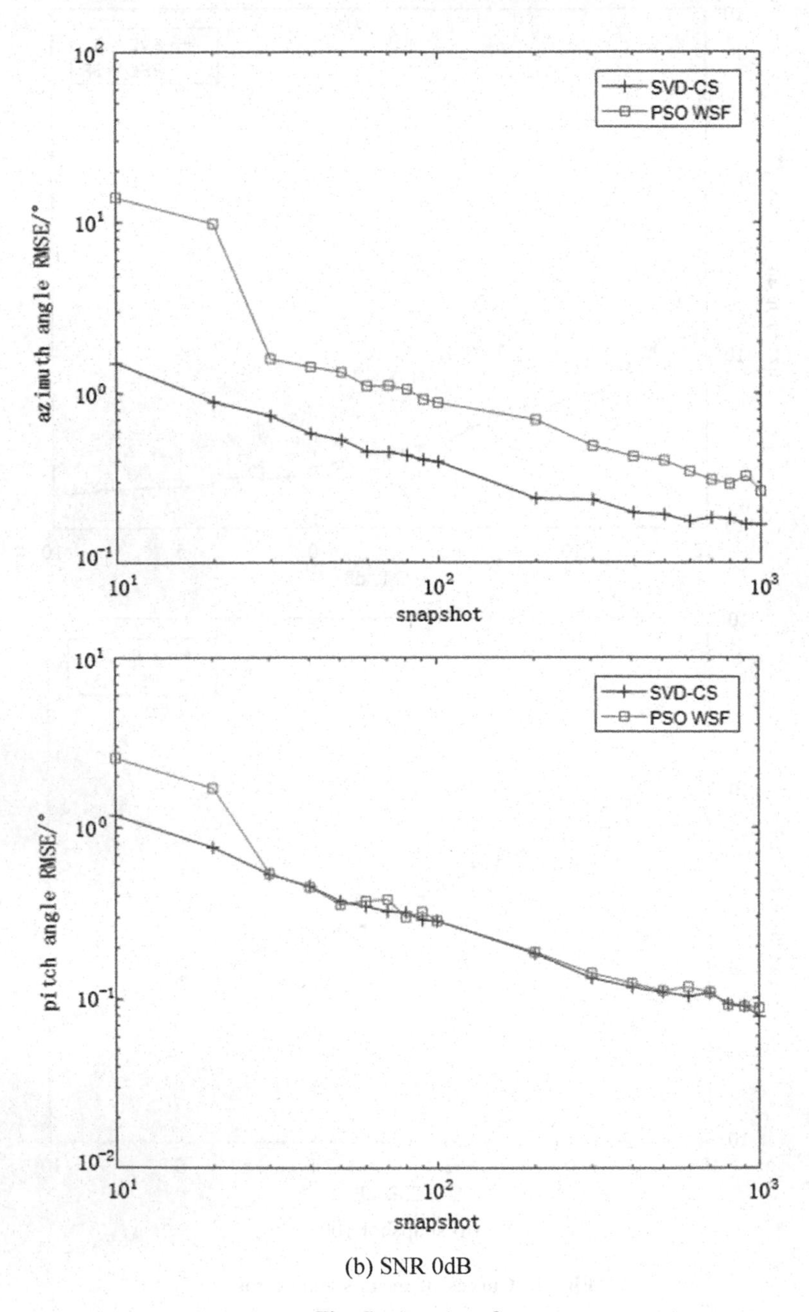

(b) SNR 0dB

Fig. 5. (*continued*)

Simulation results verify the effectiveness of the model and a higher estimation accuracy of the 2D off-grid algorithm based on the first-order Taylor expansion for the off-grid incident signal under the condition of single snapshot.

Experiment 2

To verify its effectiveness, the modified algorithm is compared with WSF algorithm based on particle swarm optimization (PSO) algorithm. The distance between two adjacent grids is 1°, with SNR increasing from −15 dB to 10 dB, snapshot from 10 to 1000, and Monte-carlo simulation 500 times. Results are shown in Fig. 5.

Simulation results show that the proposed algorithm can deal with the more snapshots problem, and improve the solution accuracy under off-grid conditions with better robustness. Compared with the traditional subspace DOA estimation algorithms, it has better accuracy and is less affected by SNR and snapshot number. It solves the off-grid problems better than the basic compressed sensing methods.

5 Concluding Remarks

This paper introduces the signal model with separable array manifold and its 2D off-grid algorithm based on the first-order Taylor expansion. The effectiveness of the algorithm in off-grid conditions is analyzed by simulation, while unsuitable in more snapshots. In order to solve this problem, a 2D multitasking compressed sensing off-grid algorithm based on SVD is proposed. Experimental results show this scheme can solve the problem of 2D off-grid DOA estimation in such cases, and its performance is significantly improved compared with the traditional subspace DOA estimation algorithms.

References

1. Gorodnitsky, I.F., Rao, B.D.: Sparse signal reconstruction from limited data using FOCUSS: a re-weighted minimum norm algorithm. IEEE Trans. Signal Process. **45**(3), 600–616 (2002)
2. Wang, S.H., Ruan, H.L.: DOA estimation of LFM signals based on compressed sensing. Comput. Simul. **36**(11), 175–179 (2019)
3. Zuo, L., Wang, J., et al.: Super-resolution DOA estimation method of passive bistatic radar based on TLS-CS. Syst. Eng. Electron. **42**(01), 61–66 (2020)
4. Zhao, H.W., Liu, B., Liu, H.: Improved PSO and its application to CS DOA estimation. Microelectron. Comput. **33**(05), 33–36+41 (2016)
5. Gretsistas, A., Plumbley, M.D.: An alternating descent algorithm for the off-grid DOA estimation problem with sparsity constraints. In: Signal Processing Conference. IEEE (2010)
6. Yang, Z., Xie, L., Zhang, C.: Off-grid direction of arrival estimation using sparse bayesian inference. IEEE Trans. Signal Process. **61**(1), 38–43 (2013)
7. Lin, B., Liu, J., Xie, M., et al.: Super-resolution DOA estimation using single snapshot via compressed sensing off the grid. In: 2014 IEEE International Conference on Signal Processing, Communications and Computing (ICSPCC). IEEE (2014)

Perceptual Quality Enhancement with Multi-scale Deep Learning for Video Transmission: A QoE Perspective

Chaoyi Han[1,2], Yiping Duan[1,2], Xiaoming Tao[1,2](✉), Rundong Gao[1], and Jianhua Lu[1,2]

[1] Department of Electronic Engineering, Tsinghua University, Beijing 100084, China
{hancy16,gaord}@mails.tsinghua.edu.cn
{yipingduan,taoxm,lhh-dee}@tsinghua.edu.cn
[2] Beijing National Research Center for Information Science and Technology (BNRist), Beijing 100084, China

Abstract. With the development of mobile Internet technologies, wireless communication is facing huge challenges under the explosive growth of multimedia data, e.g. video conferences, online education. This makes it difficult to guarantee the communication quality where communication resources (bandwidth, channel, etc.) are limited. In this paper, we propose an image enhancement method to transform blurred images into images with high perceptual quality. The proposed method serves as a post-processing part for communication systems and is incorporated into the receiver. Specifically, we learn the prior of high quality images using a collected dataset. We train a neural network to accomplish this task and adopt a multi-scale perceptual loss as the objective, which is more consistent with the quality of experience (QoE). To validate the proposed method, we train our model on a large dataset with both blurred images and high quality images. Experimental results show that, using a pre-collected dataset with high quality images, the proposed approach can effectively restore the blurred images.

Keywords: Quality of experience (QoE) · Perceptual image enhancement · Wireless communications

1 Introduction

With the rapid development of mobile Internet technologies, multimedia data such as images and videos are becoming the mainstream. Cisco Visual Network reported that Global IP video traffic will grow four-fold from 2017 to 2022, with a compound annual growth rate (CAGR) of 29%. In particular, live Internet video has the potential to drive large amounts of traffic as it is replacing traditional broadcast services. Nowadays, live video already accounts for 5% of Internet video traffic and will reach 17% by 2022 [1]. Correspondingly, the global

S. Shi et al. (Eds.): AICON 2020, LNICST 356, pp. 26–40, 2021.
https://doi.org/10.1007/978-3-030-69066-3_3

average broadband speed will double from 2017 to 2022, from 39.0 Mbps to 75.4 Mbps, which exhibits a notable mismatch. When transmitting images/videos with limited bandwidth, image compression must be applied to significantly save the encoding bit rate [3,4]. However, commonly used image compression methods usually have artifacts such as blocking and ringing, which may servery degrade the quality of user experience (QoE). Moreover, these artifacts may reduce the accuracy of subsequent classification, recognition and other high-level tasks. Therefore, it is necessary to study the quality enhancement methods to make the blurred image high-definition [2]. The quality enhancement module follows the decoder to improve the degraded image quality that caused by limited bandwidth, channel and other transmission issues. The proposed approach can not only help to improve the quality of experience (QoE) significantly but can also be used to relieve the pressure on communication bandwidth.

Recently, there has been increasing interest in enhancing the quality of the compressed images/videos [5,6]. The quality enhancement of image aims to restore the original undistorted image as much as possible from the degraded image, and at the same time improving the perceptual quality [7,15]. According to different research methods, it can be roughly classified into two types, namely model-based degraded image quality enhancement and data-based degraded image quality enhancement. The whole quality enhancement process of the model-based method includes three parts: modeling the degradation process, estimating the degradation degree and inferring the reconstructed signal. Since restoring the original data from degraded data is an ill-conditioned problem, it is generally necessary to add certain prior constraints to solve it. In actual application, it is necessary to know the degradation process in advance, and then estimate the degradation degree so as to select the corresponding model for restoration and solution. Data-based enhancement of degraded images is expected to use real low-quality images and high-quality image data through parameterized methods [9,10]. According to whether the training data is paired, it is divided into: 1. The same image pair in the training data with low-quality and high-quality image pairs, so the whole process can be trained by supervised learning; 2. Degraded images and high-quality images that not corresponding to each other, then optimization can be performed by minimizing the distance between the reconstructed data and the distribution of high-quality images, and the consistency of the image content before and after the restoration enhancement can be ensured by the cyclic mapping method.

In this paper, we proposed a perceptual quality enhancement method with multi-scale nerual network for video transmission toward QoE. Specifically, we train an encoder-decoder model to exploit the relationship between the blurred images and the high quality image for each scale. Moreover, we present multiscale perceptual loss function that mimics conventional coarse-to-fine approaches. Experiments on the benchmark dataset show that using the loss function of the feature domain for training, the neural network has improved the subjective perceptual quality of the restored image, and even achieved better results in

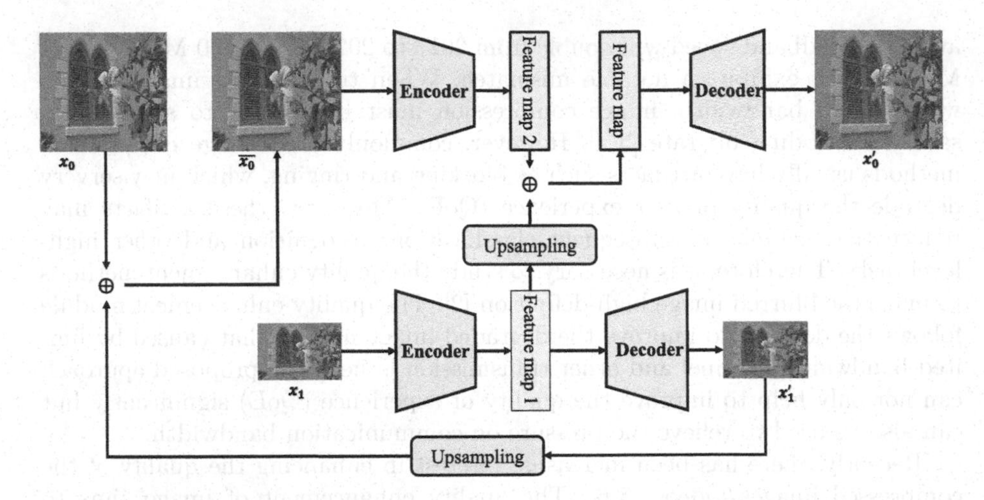

Fig. 1. The framework of the proposed approach (scale $s = 2$). x_0 is the original scale and x_1 is the downsampling scale with the stride $ds = 2$. $\bigoplus$ represents the average weighted sum.

automated evaluation metrics. The corresponding framework is shown in Fig. 1. The contributions can be summarized as follows,

- We established a QoE-oriented image quality enhancement framework. A novel optimization objectives combing mean square error and feature-level error to preserve the fidelity from pixel-level and semantic-level.
- We develop a multi-scale deep learning method to learn the relationships between the blurred image and the high quality image. The multi-scale and multi-feature learning are used to improve the performance of the proposed approach.

The rest of this paper is organized as follows. Section 2 reviews the related work about image enhancement and Sect. 3 presents the proposed perceptual quality enhancement model including the network architecture and multi-scale end-to-end optimization. Section 4 presents the experimental results on a benchmark dataset and Sect. 5 concludes this paper.

2 Related Work

Recently, extensive works have focused on enhancing the quality of images after compression [11–14]. In general, the input is a blurred image and the output is a high-definition image. Through a deep neural network, a complex nonlinear relationship between the blurred image and the high-definition image is established to improve the image quality.

The method based on deep learning needs to solve three extremely important basic problems when applied to a specific field: high-quality data, a network structure that can effectively extract features, and a loss function that can effectively evaluate the results. In terms of data, early researchers mainly used some simple methods such as downsampling to generate fuzzy images on the existing clear images to build a data set [15]. However, this simple way of generating blurred images makes the training data not in line with the real-world data distribution, which greatly restricts the effectiveness of deep learning. Therefore, some researchers have built a data set closer to the real world. This data set is called the GoPro dataset [12], which meets the data needs of training neural networks. In terms of network structure, because image deblurring is a pixel-dense task, the network is required to generate output at each pixel, and this is similar to some classic computer vision tasks in terms of output, such as image semantic segmentation tasks [16]. Therefore, researchers have migrated the classic network U-shaped network in the field of image semantic segmentation to the field of image deblurring and the U-shaped network has become almost the only network basic framework in the field of image deblurring [17–19]. In terms of loss function, researchers generally use a pixel-by-pixel two-norm loss function. However, in recent years, research work has shown that the pixel-by-pixel two-norm loss function cannot effectively describe the subjective perceptual quality of the image. This phenomenon is called the "perceptual gap", that is: higher PSNR (Peak Signal to Noise Ratio) is not necessarily more in line with human subjective perception [20–22].

Kim et al. [23] tried to solve non-uniform blind image deblurring problem. In this paper, in contrastive the restrictive assumption that the underlying scene is static and the blur is caused by only camera shake, authors address the deblurring problem of general dynamic scenes which contain multiple moving objects as well as camera shake. [11] to use deep learning technique to solve image deblurring problem. In this paper, authors address the problem of estimating and removing non-uniform blur from a single blurry image. They propose a deep learning approach to predicting the probabilistic distribution of motion blur at the patch level using a convolution neural network (CNN). In [12], researchers present a deep learning framework that mimics conventional coarse-to-fine approaches, which restores sharp images in an end-to-end manner through a multi-scale convolution neural network. To tackle the above problems, in [24], researchers present a deep hierarchical multi-patch network inspired by Spatial Pyramid Matching to deal with blurry images via a fine-to-coarse hierarchical representation. To deal with the performance saturation w.r.t depth, they propose a stacked version of their multi-patch model.

3 Perceptual Quality Enhancement Model

3.1 Network Architecture

The framework of the proposed approach with multi-scale structure is show in Fig. 1. For each scale, we adopt the encoder-decoder structure in Fig. 2. This structure accepts a blurred image as input, and outputs a high-definition image with same size as the input image. The encoder is responsible for extracting the original image features for processing, and the decoder is responsible for restoring a high-definition image based on the extracted features.

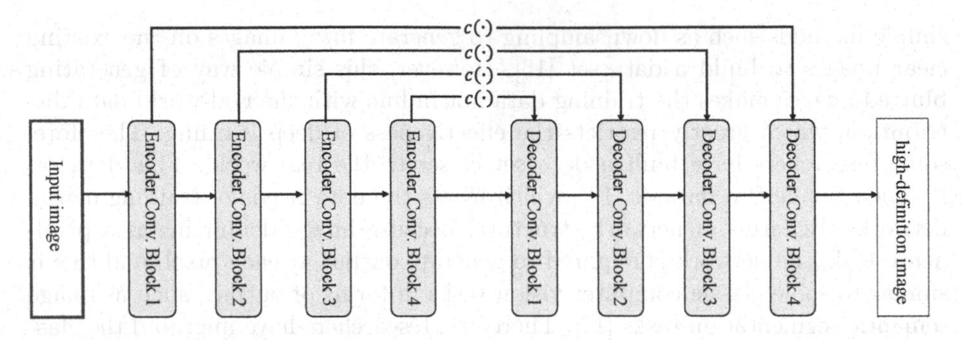

Fig. 2. The network architecture of each scale. The network includes 4 encoder convolutional block and 4 decoder convolutional block. Encoder convolution block 1 is the inverse process of decoder deconvolution block 1, and so on.

As shown in Fig. 1, the encoder-decoder structure consists of two basic convolutional blocks stacked: encoding convolutional block and decoding convolutional block. The internal structure of the two convolutional blocks is shown in Fig. 2. The encoding convolutional block is first composed of a convolution layer (the convolution kernel size is 3×3) followed by two residual convolution blocks [25]. The decoding convolutional block is first convolved by two residuals followed by a deconvolution layer [26] (the deconvolution kernel size is 4×4). The residual convolution block contains a two-layer convolution layer with ReLu activation function [27]. With such basic components, we can build an encoder-decoder structure: the encoding convolutional blocks are stacked to become an encoder, and the decoding convolutional blocks are stacked to become a decoder, and the two modules are symmetrical. The structure first reduces the size of the feature map through the multi-layer convolutional neural network on the encoder side, and increases the number of feature channels at the same time. The encoder composed of a multi-layer convolutional neural network extracts the image semantic features necessary for the deblurring task from the original input image, and then such features are input to the decoder, and the multi-layer convolutional network on the decoder side is gradually upsampled and increased The feature map size is reduced while the number of feature channels is reduced, and the processed image semantic features are decoded to generate a clear image with the same size as the input after deblurring operation. $\oplus$ splices the feature maps of the encoder and decoder as the input of the corresponding decoding convolution block. By this way, the decoder can make full use of different Hierarchical information including the low-level and high-level features (Fig. 3).

The specific structure parameters of the network we adopted are as follows. The encoder is composed of encoding convolutional block. The first encoding convolutional block converts the image from a 3-channel (RGB) original image to a 32-channel feature map with the same size. Subsequently, each of the three encoding convolutional blocks doubles the number of input feature channels, and at the same time reduces the size of the feature map by one time. Because of

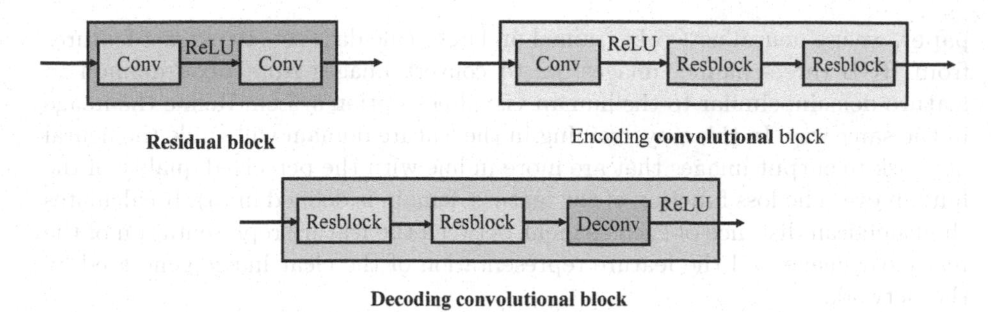

Fig. 3. The left is the residual block. The middle is the encoder convolution. The right is the decoder decovolution.

the symmetry of the encoder-decoder structure, the decoder is also composed of four decoded convolutional blocks. Each of the first three decoded convolutional blocks reduces the number of feature channels of the input feature map by a factor of two, while the feature map size doubled. The last decoder convolution block converts the input feature map into a 3-channel restored image as the final output.

3.2 Perceptual Quality Optimization

The current convolutional neural network based image quality enhancement employ PSNR or SSIM as the optimization target. However, the PSNR of the image does not consider the quality of experience (QoE) of users. Perceptual quality is an objective measure of QoE. Therefore, we define the optimization target from pixel-level and semantic-level as,

$$L_{loss} = L_{pixel} + L_{feature} \tag{1}$$

where m and n represents the height and width of the image, respectively. S is the real clear image and O is the clear image learned by the network. In (1), the optimization target includes two parts of the loss function in the pixel domain and the feature domain. The loss function of the pixel domain is defined in (2). It directly calculates the Euclidean distance of each pixel between the real clear image and the clear image learned by the network.

$$L_{pixel} = \frac{1}{m \times n} \sum_{i=1}^{m} \sum_{j=1}^{n} (S(i,j) - O(i,j))^2 \tag{2}$$

where n and m respectively represent the length and width of the image, S represents the real clear image, and O represents the clear image generated by the network.

Some researchers began to search for new loss functions to guide the neural network to produce images that fit the human eye's perceptual quality. In this paper, we use neural networks trained on large-scale data sets to extract features from RGB three-channel images, or to convert images from pixel domain to feature domain, similar to the human visual perception system Refine the image in the same way. In this way, training in the feature domain will guide the neural network to output images that are more in line with the perceived quality of the human eye. The loss function of the feature domain is defined in (3). It calculates the Euclidean distance of each element between the feature representation of the real clear image and the feature representation of the clear image generated by the network.

$$L_{feature} = \frac{1}{m \times n} \sum_{i=1}^{m} \sum_{j=1}^{n} (f(S(i,j)) - f(O(i,j)))^2 \tag{3}$$

We choose the pre-trained VGG16 [6] on ImageNet dataset as the feature extraction function. The network is a neural network structure developed by the Google DeepMind research team and the Oxford University Computer Vision Laboratory. The neural network is formed by stacking a 3×3 convolutional layer and a 2×2 maximum pooling layer. In order to use multiple feature maps, the feature loss can be written as,

$$L_{feature} = \frac{1}{m \times n} \sum_{i=1}^{m} \sum_{j=1}^{n} \sum_{k=1}^{C} \alpha_k (f(S(i,j)) - f(O(i,j)))^2 \tag{4}$$

where α_k is the weights and C is the number of the feature maps.

3.3 Multi-scale Deep Learning Model

Intuitively, images always contain different features at different scales. The image will show more texture details at large resolutions and the overall structure of the image will be more compact at a small resolution. Therefore, under large and small resolutions, different levels of information can be captured effectively. In this way, multi-scale algorithms can fully extract the features of different levels of the image and increase the accuracy of image feature description.

The multi-scale structure based on the encoder-decoder is shown in Fig. 1. The encoder-decoder network is a fully convolutional network, and the convolutional layer has no dimensional assumptions on the input of the image. Figure 1 shows the two-scale network architectures. We can see that the two encoder-decoders are completely the same in structure. The overall network structure is from bottom to top, and the resolution of the processed image is from small to large. In addition, in order to accelerate convergence and strengthen the interaction between different scales, residual connections are also introduced in the intermediate feature maps of the encoder and decoder of the two scales.

$$L_{feature} = \frac{1}{m \times n} \sum_{i=1}^{m} \sum_{j=1}^{n} \sum_{k=1}^{C} \alpha_k (f(S(i,j)) - f(O(i,j)))^2 \tag{5}$$

4 Experiments

Recently, a researcher proposed that the blurred frames of long exposure time can be approximated by aggregating several consecutive short exposure time frames in the video recorded by high-speed cameras, and released a public data set called GoPRO data through this method [12]. The researchers used the professional sports camera GoPRo Hero 4 Black to record. When the camera continuously receives light during the exposure process, a blurred image is generated. The fuzzy generation process can be modeled as,

$$B = g(\frac{1}{T}\int_t^T s(t)dt) \approx g(\frac{1}{M}\sum_{i=1}^{M} s[i]) \tag{6}$$

where $s(t)$ represents the clear image corresponding to time t, and T represents the exposure time. Correspondingly, M represents the number of sampled video frames, and $s[i]$ represents the $i-th$ clear image signal sampled during the exposure time. g represents the camera response function, which converts the signal received by the camera into an image signal that we can observe.

The researchers used the GoPro data set to shoot $240FPS$ videos, and then gathered 7 to 13 consecutive frames to obtain different degrees of blurred images, using the middle frame in the blurred frame segment as the target clear image. The training data of the GoPRo dataset can simulate complex camera shake and object movement scenarios, which fits the real application scene very well, and the amount of training data is larger than the previous dataset, which can greatly satisfy the neural network's training data. Therefore, this dataset has also become the most important evaluation benchmark dataset for image deblurring methods based on deep learning. The data set contains 3214 clear-fuzzy image pairs. We use 2103 data for training, and the remaining 1111 data are used as the validation set.

4.1 Settings

In the training phase, the adaptive momentum estimation optimization algorithm (Adam) [8] is used to optimize the neural network, and the Adam algorithm is widely used in deep learning training. The model is trained for a total of 3000 rounds, the initial learning rate is set to 0.0001, and after each 1000 rounds of training, it is reduced to 0.3 times the current learning rate. According to the experimental results, 3000 rounds of training can make the model fully converge. In each iteration, we sample two blurred images and randomly crop the image area of 256×256 size as a batch (and in the test, the input is the original size, that is, 720×1280), because the original input is 720×1280 resolution is very large. If the original resolution of 720×1280 is used as input in the training phase, the video memory requirement cannot be met, and the experimental results show that a 256×256 cropped image block contains enough information to make the neural network Learn the mapping relationship from blurred images to clear images. The input image is divided by 255 and normalized to the range [0, 1], then 0.5 is subtracted, and normalized to the range [−0.5, −0.5]. All trainable parameters of the model are initialized using Xavier [9] initialization method. All experiments are performed on an NVIDIA Titan X.

4.2 Performance of the Proposed Approach

Objective Evaluation. This section compares several previous algorithms on the benchmark data set. Because the training data of the GoPro data set contains general camera shake blur and object movement blur, the model trained on such a data set has the ability to deal with non-uniform blur, so it is compared with many traditional algorithms under the assumption of uniform blur. it's meaningless. The algorithm proposed by Whyte [12] can be used as a representative of the classic non-uniform deblurring algorithm. In addition, the comparison algorithm also includes the algorithm proposed by Nah [3], the algorithm proposed by Tao [10] and the algorithm proposed by Sun [2]. Nah introduced multi-scale structure into the field of image deblurring for the first time. On the basis of Nah, a recurrent neural network was added, and Sun used a convolutional neural network to estimate the local fuzzy kernel, and then applied random fields and traditional deconvolution algorithms to restore the entire clear image.

Table 1. Performance of different approaches.

Approaches	Sun [11]	Nah [12]	Tao [31]	Ours
PSNR	24.64	29.08	30.10	30.43
SSIM	0.8429	0.9135	0.9323	0.9031
Time	20 min	3.09 s	1.6 s	0.26 s

The automated evaluation indicators are shown in Table 1. It can be seen from Table 1 that this model is not inferior to the comparison algorithm in the automated evaluation index (the PSNR is even better than the comparison algorithm). In addition, one advantage of this model compared to the comparison algorithm is that it takes less time to process a picture (0.26 s vs 1.6 s), which makes this model more practical in some scenarios that are extremely sensitive to time loss (such as video Stream processing, etc.).

Subjective Evaluation. The visual results of the proposed approach is shown in Fig. 4 with three images randomly sampled from the test dataset. The first row is the original blur image. The second row is the quality enhancement result by Whyte et al. It can be observed that Whyte algorithm failed to restore the photos with good sharpness, and the visual effect is very poor, such as the first picture. The black streaks seriously affect the sensory quality. Sun's algorithm is also unable to restore effective clear images. It can be seen that the third row has almost no effect on removing blur compared to the first row, because Sun's model is trained on artificially synthesized data sets, and Compared with real fuzzy data, artificially synthesized data is too simple to represent the real-world data distribution well. Nah's algorithm can output a certain quality of clear images. Nah's algorithm can output a certain quality of clear images, but there is still a considerable degree of blur compared to the proposed model, because the number of layers of the proposed model is deeper. Tao's model can produce clearer restored images. The results of the proposed model are similar to those of Tao. But the overall look and feel of the results of the proposed model is sharper, such as the text in the blue box area in the second row. In the output result of the proposed model, the text in this area is clearer than the result of Tao. Moreover, the proposed approach takes less time to process an image, which makes this model more suitable for actual production scenarios, because in actual production and life, the application scenarios for image deblurring are often some Scenarios that require rapid response, such as monitoring equipment or video stream processing. In these scenarios, the image processing speed and image quality are equally important. Compared with Tao's model, the delay of this model is reduced by more than 1 s.

4.3 Ablation Study

Performance of Different Scales. In order to analyze the effect of multi-scale structure, models of different scale levels are trained. The evaluation results of these models are shown in Table 2. Considering the trade-off between the limitations of video memory and training time and the performance improvement brought by increasing the scale level, the network is only stacked to three scales at most.

Table 2. Performance of different scales.

Scale-level	1	2	3
PSNR	30.37	30.45	30.43
SSIM	0.9018	0.9031	0.9030

It can be seen from Table 2 that the introduction of multi-scale has improved the performance of the neural network. Among the three scale parameter settings, the neural network with a scale parameter of 2 has the best performance. Compared with a basic encoder-decoder network, That is, for a network with a scale level of 1, the PSNR increased by 0.08 dB, and the SSIM increased by 0.0013. However, the network performance of scale level 3 has not been further improved, but has slightly decreased. This phenomenon is more difficult to explain. In the specific experiment, at the beginning, we guessed that the stacking of scales caused the network to deepen and the trainability problem, so we tried to use the multi-scale loss function proposed in the Nah algorithm, that

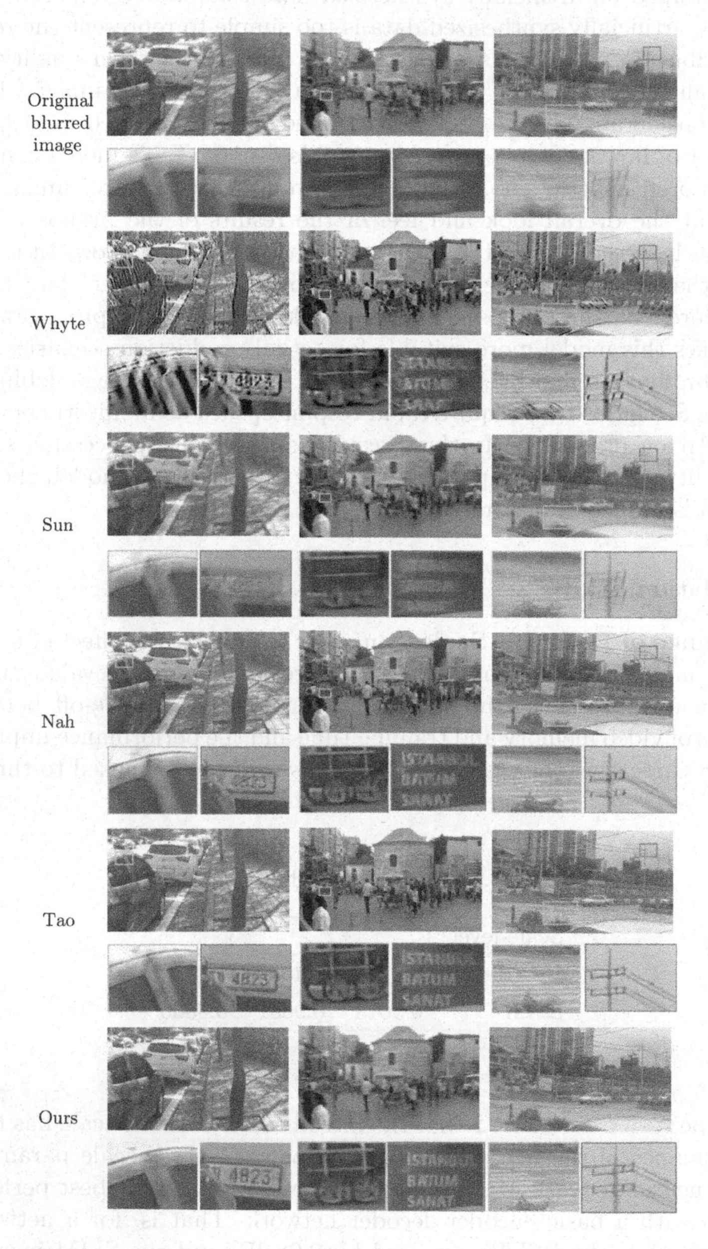

Fig. 4. Visualization of image quality enhancement by different methods. The first row is the original blurred image. The second row is the result of Whyte's method. The third row is the result of Sun's method. The fourth row is the result of Nah's method. The fifth row is the result of Tao's method. The sixth row is the result of our proposed method.

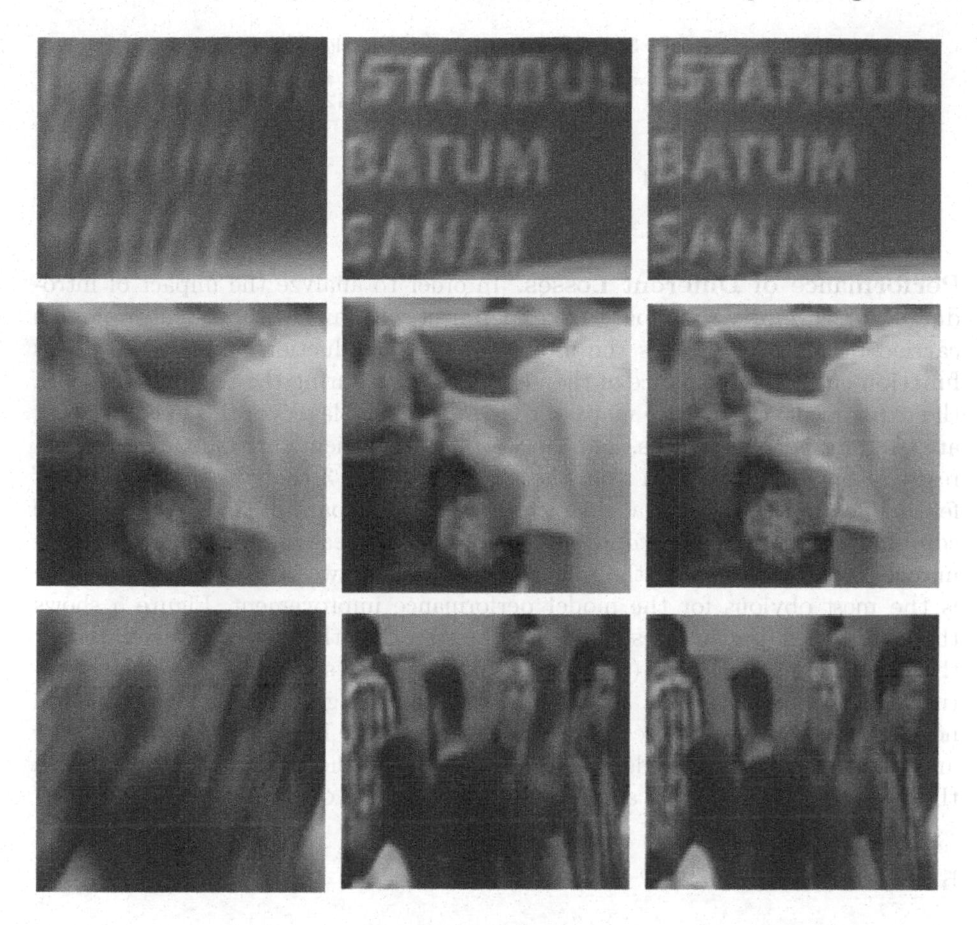

Fig. 5. Visualization of image quality enhancement with different losses. The first column is the original blurred image. The second column is the result with MSE loss. The third column is the result with MSE loss and perceptual loss.

is, the output result of each scale is subjected to the loss function The calculation (ground truth is directly obtained by downsampling the clear image of the original resolution), but the result of the training strategy training in this way has been described in the previous section. The training loss of the model drops very quickly, but in the validation set The PSNR and SSIM indicators are far lower than those of models trained without multi-scale loss function. Therefore, we believe that the introduction of a multi-scale loss function may cause the loss function of the model to be dominated by the low-resolution image deblurring results, that is to say, it is overfitted to the low-resolution output results. This is obviously not the result we expected, so the multi-scale loss function is not used.

Table 3. Performance of different losses.

Loss function	MSE	MSE+VGG9	MSE+VGG23	MSE+VGG_9_23
PSNR	30.16	30.30	30.21	30.45
SSIM	0.8991	0.9009	0.9007	0.9031

Performance of Different Losses. In order to analyze the impact of introducing feature loss, a comparative experiment with and without feature loss is carried out. Table 3 shows the impact of the introduction of the feature loss function on the performance of the neural network during the training phase. In the experiment, the feature maps of the $9th$ and $23rd$ layers of the VGG network are used for perceptual loss, and the weighted coefficients are 0.002 and 0.0015, respectively. It can be seen from the results in Table 3 that the introduction of feature loss, whether it is a separate $9th$ layer, a separate $23rd$ layer, and the combination of $9th$ and $23rd$ layers, has brought a great improvement in performance. It can be seen that the feature loss of adding two layers at the same time is the most obvious for the model performance improvement. Figure 5 shows three examples of visual results. It can be seen that the networks generated by the two training strategies (introducing feature loss vs without introducing feature loss) can remove blur better, and the deblurring generated by the neural network that introduces the feature loss function The image will become clearer and more realistic at the edges (such as the edge of the clothes of the person on the left) and texture (such as the front window glass of the car in the picture).

5 Conclusions

For the purpose of image deblurring, this paper adds a simple loss function to the original pixel-wise two-norm loss function and performs multi-scale expansion on the network structure to improve the subjective perception quality and objective quality of the restored image. We established a QoE-oriented image quality enhancement framework and adopted a novel optimization objective that combines mean square error and feature-level error to preserve the fidelity from pixel-level and semantic-level. We developed a multi-scale deep learning method to learn the relationships between the blurred image and the high quality image. The multi-scale and multi-feature learning are used to improve the performance of the proposed approach. Experiments on the benchmark data set for image deblurring show that when using feature domain loss function for training, the multi-scale image processing network has improved the subjective perceptual quality of restored images, and achieved better results in objective quality.

Acknowledgement. This work was supported by the National Key R&D Program of China (2018YFB1 800804) and the National Natural Science Foundation of China (NSFC 61925105, 61801260).

References

1. LNCS Homepage. https://www.cisco.com/c/en/us/solutions/collateral/executive-perspectives/annual-internet-report/white-paper-c11-741490.html. Accessed 4 July 2020
2. Fergus, R., Singh, B., Hertzmann, A., et al.: Removing camera shake from a single photograph. ACM Trans. Graph. **25**(3), 787–794 (2006)
3. Kim, Y., Cho, S., Lee, J., Jeong, S., Choi, J., Do, J.: Towards the perceptual quality enhancement of low bit-rate compressed images. In: 2020 IEEE/CVF Conference on Computer Vision and Pattern Recognition Workshops (CVPRW), Seattle, WA, USA, pp. 565–569. IEEE (2020)
4. Lee, J., et al.: FBRNN: feedback recurrent neural network for extreme image super-resolution. In: 2020 IEEE/CVF Conference on Computer Vision and Pattern Recognition Workshops (CVPRW), Seattle, WA, USA, pp. 565–569. IEEE (2020)
5. Ding, D., Wang, W., Tong, J., Gao, X., Liu, Z., Fang, Y.: Biprediction-based video quality enhancement via learning. IEEE Trans. Cybern. (2020, in press)
6. Guan, Z., Xing, Q., Xu, M., Yang, R., Li, T., Wang, Z.: MFQE 2.0: a new approach for multi-frame quality enhancement on compressed video. IEEE Trans. Pattern Anal. Mach. Intell. (2020, in press)
7. Singh, G., Mittal, A.: Various image enhancement techniques a critiacal review. Int. J. Innov. Sci. Res. **10**(2), 267–274 (2014)
8. Rani, S., Jindal, S., Kaur, B.: A brief review on image restoration techniques. Int. J. Comput. Appl. **150**(12), 30–33 (2016)
9. Chen, Y., Wang, Y., Kao, M., Chuang, Y.: Deep photo enhancer: unpaired learning for image enhancement from photographs with GANs. In: 2018 IEEE/CVF Conference on Computer Vision and Pattern Recognition (CVPR), Salt Lake City, UT, USA, pp. 6306–6314. IEEE (2018)
10. Park, J., Lee, J., Yoo, D., Kweon, I.: Distort-and-recover: color enhancement using deep reinforcement learning. In: 2018 IEEE/CVF Conference on Computer Vision and Pattern Recognition (CVPR), Seattle, WA, USA, pp. 5928–5936 569. IEEE (2018)
11. Sun, J., Cao, W., Xu, Z., Ponce, J.: Learning a convolutional neural network for non-uniform motion blur removal. In: 2015 IEEE Conference on Computer Vision and Pattern Recognition (CVPR), Boston, MA, USA, pp. 769–777. IEEE (2015)
12. Nah, S., Kim, T., Lee, K.: Deep multi-scale convolutional neural network for dynamic scene deblurring. In: 2017 IEEE Conference on Computer Vision and Pattern Recognition (CVPR), Honolulu, HI, USA, pp. 257–265. IEEE (2017)
13. Johnson, J., Alahi, A., Fei-Fei, L.: Perceptual losses for real-time style transfer and super-resolution. In: Leibe, B., Matas, J., Sebe, N., Welling, M. (eds.) ECCV 2016. LNCS, vol. 9906, pp. 694–711. Springer, Cham (2016). https://doi.org/10.1007/978-3-319-46475-6_43
14. Ledig, C., et al.: Photo-realistic single image super-resolution using a generative adversarial network. In: 2017 IEEE Conference on Computer Vision and Pattern Recognition (CVPR), Honolulu, HI, USA, pp. 257–265. IEEE (2017)
15. Dong, W., Zhang, L., Shi, G., Wu, X.: Image deblurring and super-resolution by adaptive sparse domain selection and adaptive regularization. IEEE Trans. Image Process. **20**(7), 1838–1857 (2011)
16. Han, C., Duan, Y., Tao, X., Lu, J.: Dense convolutional networks for semantic segmentation. IEEE Access **7**, 43369–43382 (2019)

17. Whyte, O., Sivic, J., Zisserman, A., Ponce, J.: Non-uniform deblurring for shaken images. In: 2010 IEEE Conference on Computer Vision and Pattern Recognition (CVPR), San Francisco, CA, USA, pp. 491–498. IEEE (2010)
18. Jia, C., Zhang, X., Zhang, J., Wang, S., Ma, S.: Deep convolutional network based image quality enhancement for low bit rate image compression. In: 2016 Visual Communications and Image Processing (VCIP), Chengdu, China, pp. 1–4. IEEE (2016)
19. Yu, J., Gao, X., Tao, D., Li, X., Zhang, K.: A unified learning framework for single image super-resolution. IEEE Trans. Neural Netw. Learn. Syst. **25**(4), 780–792 (2014)
20. Hameed, A., Dai, R., Balas, B.: A decision-tree-based perceptual video quality prediction model and its application in FEC for wireless multimedia communications. IEEE Trans. Multimedia **18**(4), 764–774 (2016)
21. Ahar, A., Barri, A., Schelken, P.: From sparse coding significance to perceptual quality: a new approach for image quality assessment. IEEE Trans. Image Process. **27**(2), 879–893 (2018)
22. Tao, X., Duan, Y., Xu, M., Meng, Z., Lu, J.: Learning QoE of mobile video transmission with deep neural network: a data-driven approach. IEEE J. Sel. Areas Commun. **37**(6), 1337–1348 (2019)
23. Kim, T., Ahn, B., Lee, K.: Dynamic scene deblurring. In: 2013 IEEE International Conference on Computer Vision, Sydney, NSW, Australia, pp. 3160–3167. IEEE (2013)
24. Zhang, H., Dai, Y., Li, H., Koniusz, P.: Deep stacked hierarchical multi-patch network for image deblurring. In: 2019 IEEE/CVF Conference on Computer Vision and Pattern Recognition (CVPR), Long Beach, CA, USA, USA, pp. 5971–5979. IEEE (2019)
25. He, K., Zhang, X., Ren, S., Sun, J.: Deep residual learning for image recognition. In: 2016 IEEE Conference on Computer Vision and Pattern Recognition (CVPR), Las Vegas, NV, USA, pp. 770–778. IEEE (2016)
26. Zeiler, M., Taylor, G., Fergus, R.: Adaptive deconvolutional networks for mid and high level feature learning. In: 2011 International Conference on Computer Vision, Barcelona, Spain, pp. 2018–2025. IEEE (2011)
27. Glorot, X., Bordes, A., Bengio, Y.: Deep sparse rectifier neural networks. J. Mach. Learn. Res. (11), 315–323 (2011)
28. Sergey, L., Christian, S.: Batch Normalization: Accelerating Deep Network Training by Reducing Internal Covariate Shift (2015). http://arxiv.org/abs/1502.03167
29. Ba, J., Kiros, J., Hinton, G.: Layer Normalization (2016). https://arxiv.org/abs/1607.06450
30. Wu, Y., He, K.: Group normalization. Int. J. Comput. Vision **128**(3), 742–755 (2020)
31. Tao, X., Gao, H., Shen, X., Wang, J., Jia, J.: Scale-recurrent network for deep image deblurring. In: 2018 IEEE Conference on Computer Vision and Pattern Recognition (CVPR), Salt Lake City, UT, USA, pp. 8174–8182. IEEE (2018)

Indoor Map Construction Method Based on Geomagnetic Signals and Smartphones

Min Zhao, Danyang Qin(✉), Ruolin Guo, and Xinxin Wang

Heilongjiang University, Harbin 150080, People's Republic of China
qindanyang@hlju.edu.cn

Abstract. Indoor map construction techniques based on geomagnetic signals can achieve better effects for constructing indoor maps, because indoor geomagnetic signals are uniquely representative. An Indoor Map Construction Method based on Geomagnetic Signals and Smartphones (IMC-GSS) is proposed. The magnetic trajectory data are collected through smartphones and crowdsourcing technology, the Dynamic Time Warping (DTW) is utilized to cluster the obtained magnetic trajectory data, and the trajectory fusion technology based on the affinity propagation algorithm is applied to fuse the trajectories belonging to the same cluster in the magnetic trajectory domain to obtain a relatively accurate indoor path. The experimental results show that constructed indoor fingerprint map is reliable as well as effective.

Keywords: Indoor map construction · Geomagnetic signals · Smartphones · Affinity propagation algorithm · Hierarchical clustering

1 Introduction

With the rapid development and widespread popularization of the mobile internet technology, the locating service industry has continued to expand. Satellite positioning [1] and base station positioning [2] can only meet the requirements of individuals in the outdoor environment. However, more and more individuals in modern life are in crowded places such as indoor shopping malls, stations, airports and work units [3], which makes the demand for location-based services more urgent. However, the traditional method of hiring professionals to make indoor maps layer by layer is expensive and time-consuming, and cannot be applied to large-scale indoor coverage. In addition, occasional changes of indoor floor plans pose challenges to updating indoor maps within a reasonable time. The use of stable geomagnetic signals and smartphones for map construction can bring great convenience to people.

This work is supported by the National Natural Science Foundation of China (61771186), University Nursing Program for Young Scholars with Creative Talents in Heilongjiang Province (UNPYSCT-2017125), Distinguished Young Scholars Fund of Heilongjiang University, and postdoctoral Research Foundation of Heilongjiang Province (LBH-Q15121), Outstanding Youth Project of Provincial Natural Science Foundation of China in 2020 (YQ2020F012). Heilongjiang University Graduate Innovative Research Project (YJSCX2020-061HLJU).

S. Shi et al. (Eds.): AICON 2020, LNICST 356, pp. 41–50, 2021.
https://doi.org/10.1007/978-3-030-69066-3_4

An indoor map construction method based on geomagnetic signals and smartphones is proposed, aiming at the problems that the existing indoor maps constructing technologies are expensive as well as time-consuming. The geomagnetic signals can be served as an indicator for indoor locating because of the uniqueness [4]. Based on geomagnetic signals, this paper uses data from the compass, accelerometer and gyroscope sensors of smartphones, combines dead reckoning methods, geomagnetic signals observation models and trajectory fusion techniques based on affinity propagation algorithms to realize indoor maps construction.

2 Related Work

In recent years, with the increasing demand for indoor locating, methods for location research are constantly emerging. There are several methods for constructing indoor maps that use different technologies. Li et al. [5] proposed an automatic radio map construction system based on crowdsourcing Pedestrian Dead Reckoning (PDR). The system processed some opportunistic PDR traces (such as a user walking through a building), and generated partial road paths by panning, rotating, and zooming the traces based on opportunistic GPS locations and doorways as landmarks. Then, turn errors were processed and PDR trajectories were merged based on the similarity of Wi-Fi fingerprints. Finally, road coverage was further expanded based on the merged PDR trajectory. Sun et al. [6] proposed an RSS-based indoor map construction algorithm using only Wi-Fi fingerprint information, which converted the indoor map construction problem into the classification problem of reference points in the fingerprint database. The algorithm used a hierarchical classification system consisting of a single AP-based basic classifier and a multi-AP-based combined classifier to ensure the accuracy of the classification results. An accurate line segment feature map was obtained by identifying the dividing line between two categories from a hierarchical classification system. Chen et al. [7] proposed a dynamic method for constructing indoor floor plans using existing motion sensing functions. This method abstracted the unknown indoor map into a matrix and combined three mobile sensing technologies (Accelerometer supporting dead reckoning, Bluetooth RSSI detection and Wi-Fi RSSI detection) using curve-fitting fusion technology. The floor plan reconstruction was extended from one room to the entire building by using shadow rate and anchor point analysis. The techniques and ideas used in these methods provide inspiration for the research in this paper.

Based on the above research, this paper aims to design an indoor map localization method based on geomagnetic signals and smartphones. This method is based on stable geomagnetic field data, and performs indoor map constructing using the built-in accelerator sensor, gyroscope sensor, and compass in the smartphone.

3 IMC-GSS Algorithm

Aiming at the problems of the expensive and time-consuming of making indoor maps, this paper proposes an indoor map construction method based on geomagnetic signals and smartphones (IMC-GSS). The overall structure of proposed method is shown in

Fig. 1, which can be divided into four parts: pedestrian trajectory atomization, hierarchical clustering, trajectory fusion, map construction.

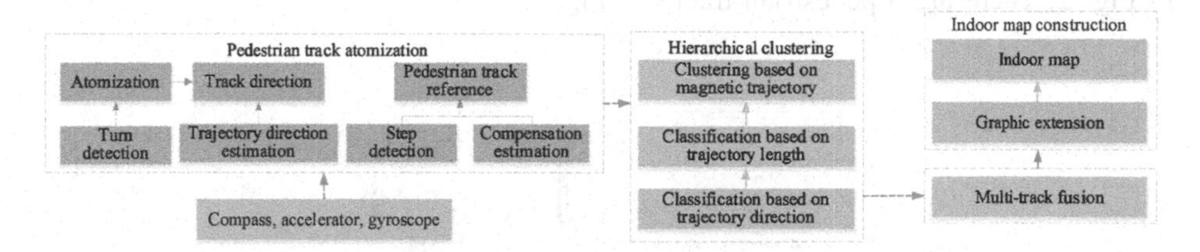

Fig. 1. Architecture of the map building system

3.1 Pedestrian Track Atomization

The complex indoor environment is characterized by a combination of large long trajectories generated by various pedestrian walking paths These trajectories only partially overlap, it is difficult to cluster them as a whole and fuse them into precise indoor paths. So, a turn detection method (i.e., trajectory segmentation) is used to solve this problem, dividing the long trajectory into short straight-line segments.

The turning point of the indoor path is taken as the natural segmentation point for dividing the long indoor trajectory into regular short trajectories, which will make all short trajectories be straight lines with no inflection points in the middle. A turn detection method is used to accurately detect the turning points on the trajectory, which uses both an accelerator and a gyroscope, as shown in Eq. (1):

$$\begin{cases} \theta = \sum_{i=1}^{m} w_v^i \Delta t \\ w_v^i = \left(w_x^i, w_y^i, w_z^i\right) \cdot \left(\overline{a}_x, \overline{a}_y, \overline{a}_z\right) \Big/ \sqrt{\overline{a}_x^2 + \overline{a}_y^2 + \overline{a}_z^2} \\ \overline{a}_x = \frac{1}{n} \sum_{j=1}^{n} a_x^j, \overline{a}_y = \frac{1}{n} \sum_{j=1}^{n} a_y^j, \overline{a}_z = \frac{1}{n} \sum_{j=1}^{n} a_z^j \end{cases} \tag{1}$$

where n is the sliding window size corresponding to the even steps. w_v^i represents the vertical speed. $\left(w_x^i, w_y^i, w_z^i\right)$ represents the gyroscope observation vector. $\left(a_x^i, a_y^i, a_z^i\right)$ represents the accelerator observation vector. w_v^i is the dot product of $\left(w_x^i, w_y^i, w_z^i\right)$ and $\left(\overline{a}_x, \overline{a}_y, \overline{a}_z\right) \Big/ \sqrt{\overline{a}_x^2 + \overline{a}_y^2 + \overline{a}_z^2}$ (unit vector of gravitational acceleration). A moving average method of acceleration data is used to offset acceleration fluctuations caused by pedestrians walking to obtain the unit vector of gravitational acceleration (The average vector of the accelerator $\left(\overline{a}_x, \overline{a}_y, \overline{a}_z\right)$ approximates the gravity acceleration vector, which points directly at the center of gravity of the earth.). In a pre-defined sliding window, the cumulative rotation angle θ is used to determine whether a turning event occurs when a pedestrian walk on a certain path. m represents the corresponding window size for the time period.

The turn detection algorithm recognizes the turning activity of pedestrians as shown in Fig. 2. There are 5 pedestrian tracks in Fig. 2.

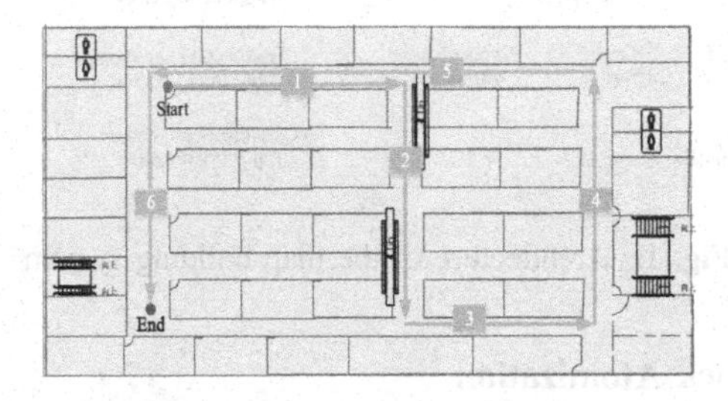

Fig. 2. Trajectory path diagram

3.2 Hierarchical Clustering

Due to the influence of the inertial sensor and compass sensor noise, the indoor trajectory obtained by walking by only one pedestrian will produce step and heading errors. The accuracy of indoor trajectory and indoor plane planning can be improved by identifying and merging multiple atomic trajectories corresponding to the same indoor path. Reliably identifying and correctly correlating multiple inferred trajectories on the same indoor path is the core of the indoor map construction algorithm used in this paper. A hierarchical clustering algorithm is used to improve the clustering accuracy and speed of the algorithm, which mainly includes three classifiers: Classifier based on trajectory direction, classifier based on trajectory length, and classifier based on magnetic sequence (as shown in Fig. 1).

Classification Based on Trajectory Direction. The atomic trajectory direction is defined as the average compass direction of all detected steps on the atomic trajectory. The direction of atomic trajectory will fluctuate in the direction of the indoor compass and deviate from the actual direction, because affected by the geomagnetic anomaly caused by ferromagnetic building materials and the shaking during walking [8]. The average direction of the compass along this atomic trajectory is took to increase the accuracy of the atomic trajectory direction. Figure 3 shows a comparison of step orientations and average orientation each time when walking on the same straight path. In the figure, the step orientation 1 indicates the step orientation of the first round of walking on the same straight path, and the average orientation 1 indicates the average step orientation of the first round of walking on the same straight path. It can be found that the average orientation remains stable when repeatedly walking on the same indoor atomic trajectory, but the orientation of each step fluctuates greatly due to the non-uniform magnetic anomaly in the indoor environment.

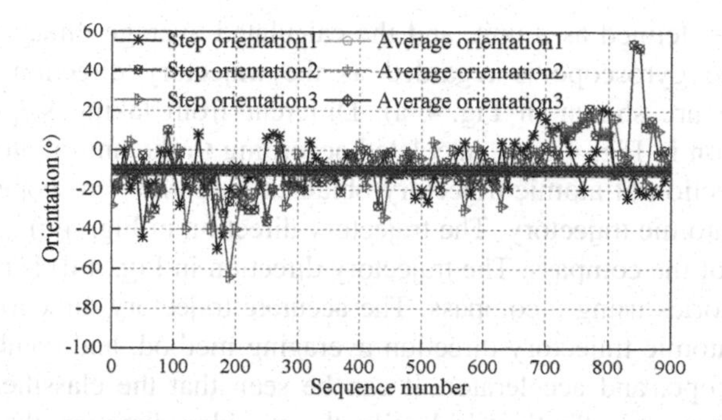

Fig. 3. Comparison of step orientations and average orientation along same atomic trajectory

The classifier based on the trajectory direction is characterized by the average direction of each atomic trajectory, and clusters all atomic trajectories into several clusters. The clustering results are basically consistent with the actual directions of indoor trajectories. Multiple walking experiments along a square path are conducted to further evaluate the effectiveness of proposed method. The estimated trajectories of different direction estimation methods are shown in Fig. 4.

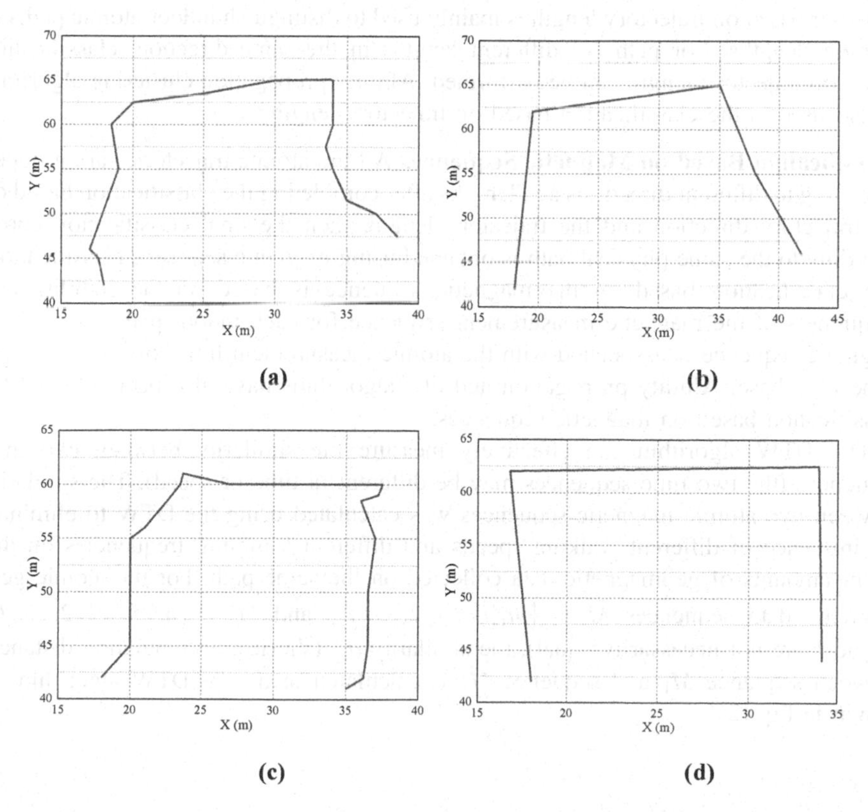

Fig. 4. Trajectory inference using (a) Gyroscope, (b) Atomic trajectory based on gyroscope, (c) Compass and (d) Atomic trajectory based on compass

First, a step is defined as a unit, and the calculated average direction of each step measured by the gyroscope is regarded as the trajectory direction, the trajectory inference results are shown in Fig. 4 a). Different from taking steps as units, the trajectory direction in Fig. 4 b) is calculated according to the unit of atomic trajectory. The average direction of atomic trajectory obtained from the gyroscope is regarded as the direction of atomic trajectory. The trajectory direction in Fig. 4 c) is inferred based on the step unit of the compass. The trajectory direction in Fig. 4 d) is inferred in units of atomic trajectories using a compass. The accurate trajectory direction estimates are obtained using atomic trajectory direction averaging method, and combining with the compass, gyroscope, and accelerator. It can be seen that the classifier based on the trajectory direction can effectively identify the corridor direction that is parallel or perpendicular to the building direction.

Classification Based on Trajectory Length. After completing the clustering based on the trajectory direction and implementing the classification for all atomic trajectories in different directions, the atomic trajectory is further grouped using a classifier based on the trajectory length (trajectory length is an estimate of the distance for a user walking on an atomic trajectory).

There are many different indoor paths in an indoor environment, and many different indoor paths in these paths have the same direction and different path lengths. The classifier based on trajectory length is mainly used to distinguish indoor atomic paths of different lengths. For paths of different lengths in the same direction, classification based on trajectory length can be performed. Affinity propagation clustering algorithm is also used in the classification based on trajectory length.

Classification Based on Magnetic Sequence. All the atomic trajectories are grouped according to different directions and lengths after completing the classification based on the trajectory direction and the trajectory length, then the final classification corresponding to the same physical path is obtained using magnetic sequence classification. The classification based on the magnetic sequence is based on the stability and uniqueness of the magnetic measurement sequence for each indoor path. The atomic magnetic sequence is associated with the atomic measurement trajectory based on the same time base. Affinity propagation and dtw algorithms have also been adopted for classification based on magnetic sequences.

The DTW algorithm can effectively measure the similarity between two time sequences (the two time sequences may be different in time or speed). The similarity between two atomic magnetic sequences was calculated using the DTW to eliminate the influence of different walking speeds and different sampling frequencies on different amounts of geomagnetic data collected on the same path. For the atomic geomagnetic data sequences $M_1 = \{m_1^i | i = 1, 2, \ldots, n\}$ and $M_2 = \{m_2^j | j = 1, 2, \ldots, k\}$ (n and k are not necessarily equal), the similarity $d_{i,j}$ (shortest accumulation distance) between sequence M_1 and sequence M_2 is calculated using the DTW algorithm, as shown in Eq. (2).

$$d_{i,j} = \min\left(d_{i-1}, d_{i,j-1}, d_{i-1,j-1}\right) + dist_{i,j} \quad (2)$$

where $dist_{i,j}$ is the Euclidean distance between sequence m_1^i and sequence m_2^j, and $i \in (1, n), j \in (1, k)$. $d_{n,k}$ represents the similarity of two atomic geomagnetic data sequences. In addition, a fast DTW algorithm with a time complexity of $O(n)$ is used in this paper [9].

3.3 Multi-track Fusion

Hierarchical clustering operation divides all crowdsourcing trajectories into different groups, which respectively correspond to indoor linear trajectory segments. A multi-track fusion scheme is adopted to address the problem that a single motion trajectory has a lot of noise due to sensor error accumulation. Multi-track fusion scheme fuses all atomic trajectories belonging to the same group to get the final trajectory.

All points in the non-clustered representative magnetic sequence in the same cluster are mapped to the cluster representative members after the DTW-based magnetic sequence clustering is completed. All mapping positions of the non-clustered representative magnetic sequence are averaged with the corresponding positions of the cluster centers to obtain the path position estimates.

Indoor path estimation is obtained by fusing all positions in the cluster center as shown in Fig. 5. Two different types of smartphones (OPPO R9s and Huawei Nova Youth) are used by different users to collect six atomic trajectories with different trajectory lengths and trajectory directions. The results show that the six original atomic trajectories deviate from the actual path. But the fusion trajectory after DTW-based multi-track fusion is similar to the actual observed trajectory.

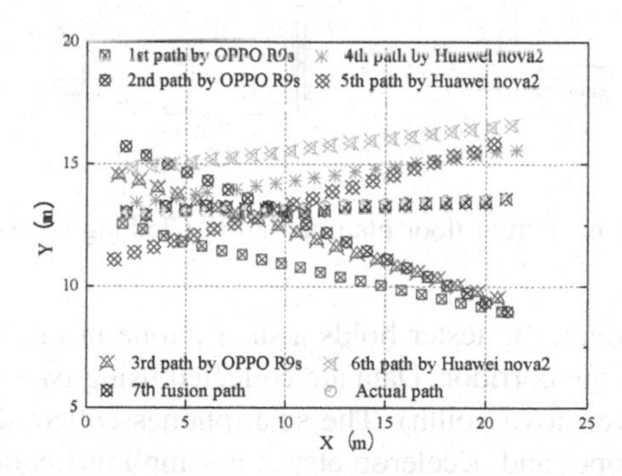

Fig. 5. Comparison of atomic trajectories

3.4 Map Construction Based on Graphic Extension Algorithm

The following expansion process is performed to systematically generate a path graph with path width information. First each fusion trajectory is drawn on the canvas, then the fusion trajectory is expanded to 0.5 m wide according to the distribution of related atomic trajectories (that is, from lines to shapes). The pixel occupied by the atomic trajectory is proportionally weighted according to its distance from the fusion trajectory (that is, the closer the pixel is, the weight will be the higher).

Because the multiplicity of atomic trajectories, the extended atomic trajectories will overlap, the weights of the overlapping pixels will be added. The path expansion process continues until all atomic trajectories are traversed, and finally a complete path graph is formed. A shrinking process is used to delete those external pixels whose weight is less than a certain threshold to eliminate the large error effect of some atomic trajectories. In view of the fact that some atomic trajectories may experience more frequently than other atomic trajectories, a dynamic weight threshold is adopted, and the dynamic weight threshold of each pixel is set to a certain proportion of the maximum weight of the pixel neighborhood. The threshold proportion is set to 20%. Finally, the isolated pixels are removed and the edges of the reduced path map are smoothed to obtain the final path map.

4 Simulation and Results Analysis

4.1 Experimental Setup

The verification experiment about the map construction is performed on the second floor of Harbin Clothing Market. The experiment area and the indoor layout are shown in Fig. 6.

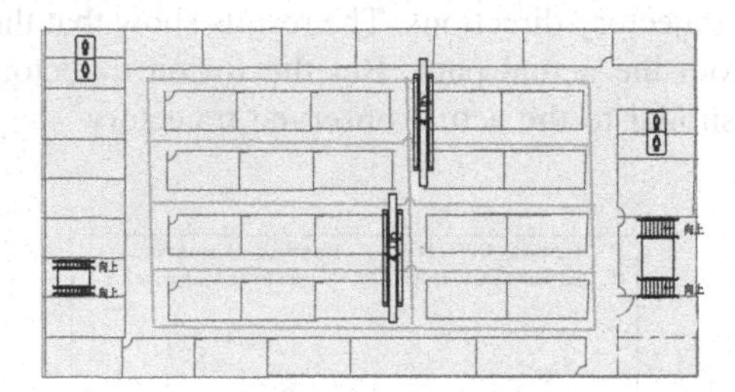

Fig. 6. Partial floor plan of Harbin Clothing Market

During the experiment, the tester holds a smartphone in his hand, places it on his chest, and walks along the corridor. Data are collected using two Android smartphones (OPPO R9s and Huawei nova youth). The smartphones collect data generated by the magnetometer, gyroscope, and accelerometer at a sampling frequency of 2Hz. During the experiment, testers walk at a constant speed and collect data on 8 different paths. The 10 test points on each path are uniformly selected and their actual positions are recorded. When the testers pass these test points, they trigger the relevant application to obtain the position data of the corresponding test points. Finally, according to the cumulative distribution function, map construction accuracy analysis is performed on the selected test points.

4.2 Evaluation of Performance

The comparison of map construction results is shown in Fig. 7. The two figures are the result of constructing the orange path of Fig. 2. Figure 7 (a) shows the trajectory constructed using the step trajectory fusion algorithm (that is, the original direction of each step is used as the path direction). Figure 7 (b) shows a plan view constructed by the proposed algorithm. Due to the direction fluctuation caused by environmental factors, the direction trajectory derived by the step trajectory fusion algorithm has a large deviation from the actual trajectory, but the proposed algorithm shows very good performance.

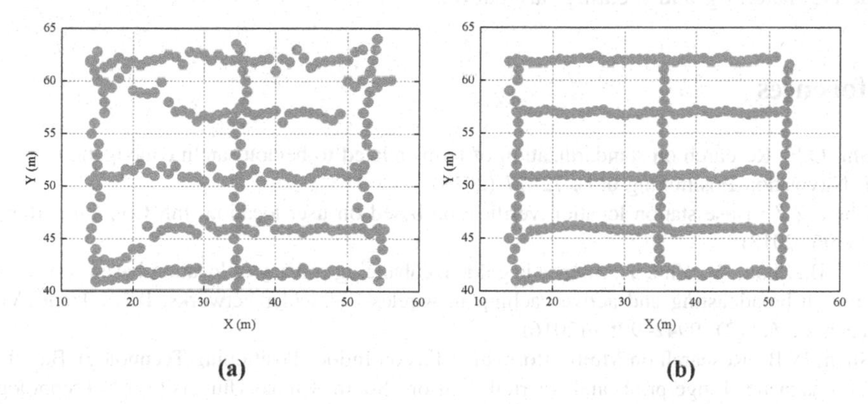

Fig. 7. Comparison of map constructing results of (a) Step trajectory fusion algorithm and (b) Proposed algorithm

In addition, a comparative analysis of the CDF of map construction errors is also performed. Figure 8 shows the CDF comparison results of the proposed algorithm and the step trajectory fusion algorithm. It can be seen from the figure that when the step trajectory fusion algorithm achieves a 1m map construction error, the confidence is only 40%. However, the confidence is about 90% when the proposed algorithm achieves a 1 m map construction error, which shows good performance.

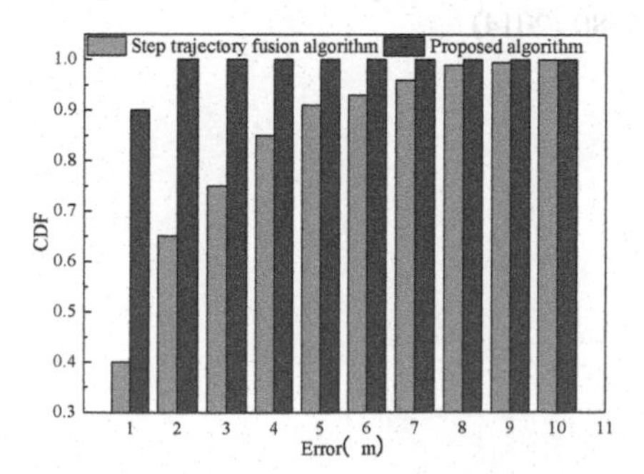

Fig. 8. CDF of map construction error

5 Conclusion

This paper proposes an indoor map construction method that relies on the geomagnetic signals and uses the synergy of multiple sensors in a smartphone. The method uses the turn detection algorithm to atomize the pedestrian trajectory, uses affinity propagation algorithm and DTW to cluster crowdsourcing trajectories, merges atomic trajectories that belong to the same cluster, and expands the linear trajectory into a trajectory with a certain width by graphic expansion algorithm to complete the construction of the indoor map. The experimental results show that the proposed algorithm successfully constructs a high-precision indoor map, which provided guarantee for subsequent map database matching and locating navigation.

References

1. Shi, Q.F.: Research on standardization of terms related to beidou satellite navigation system. J. Navigation Positioning **6**(4), 72–77 (2018)
2. Chen, Z.Z.: Base station location verification based on user location. Inf. Commun. **185**(5), 93–95 (2018)
3. Liu, B., Zhou, W., Zhu, T.: Silence is golden: enhancing privacy of location-based services by content broadcasting and active caching in wireless vehicular networks. IEEE Trans. Veh. Technol. **65**(12), 9942–9953 (2016)
4. Shen, W.B.: Research on Multi-information Fusion Indoor Positioning Technology Based on Geomagnetic Fingerprint and Inertial Sensor. South China University of Technology, Guangzhou (2017)
5. Li, Z., Zhao, X.H., Liang, H.: Automatic construction of radio maps by crowdsourcing PDR traces for indoor positioning. In: 2018 IEEE International Conference on Communications (ICC) (2018).
6. Sun, T.: RSS-based map construction for indoor localization. In: 2018 International Conference on Indoor Positioning and Indoor Navigation (IPIN) (2018).
7. Qiu, C., Mutka, M.W.: iFrame: dynamic indoor map construction through automatic mobile sensing. In: 2016 IEEE International Conference on Pervasive Computing and Communications (PerCom) (2016).
8. Zhang, Y., Xiong, Y., Wang, Y.: An adaptive dual-window step detection method for a waist-worn inertial navigation system. J. Navigation **69**(3), 659–672 (2016)
9. Salvador, S., Chan, P.: Toward accurate dynamic time warping in linear time and space. Intell. Data Anal. **11**(5), 561–580 (2014)

An Improved Generation Method of Adversarial Example to Deceive NLP Deep Learning Classifiers

Fangzhou Yuan[1,2], Tianyi Zhang[1], Xin Liang[1,2], Peihang Li[1,2], Hongzheng Wang[1,2], and Mingfeng Lu[1,2(✉)]

[1] School of Information and Electronics, Beijing Institute of Technology, Beijing 100081, China
yuanfz_shaddock@foxmail.com, lumingfeng@bit.edu.cn
[2] Beijing Key Laboratory of Fractional Signals and Systems, Beijing 100081, China

Abstract. Deep learning has been developed rapidly and widely used over the last decade. However, the concepts of adversarial example and adversarial attack are proposed, that is, adding some perturbations to the input of a deep learning model could easily change the prediction result. Deep learning-based NLP (natural language processing) classification algorithms also have this vulnerability. DeepWordBug algorithm is an advanced algorithm for generating adversarial examples, which can effectively deceive common NLP classification models. However, this algorithm needs to modify too many words to cheat NLP classification models, which limits its applications. In response to the shortcomings of DeepWordBug algorithm, this paper proposes an improving method to improve DeepWordBug. Drawing on the idea of Textfooler algorithm, the improved DeepWordBug adopts the method of dynamically adjusting the number of modified words, limits the maximum number of modified words. The new algorithm greatly reduces the number of words that need to be modified while ensuring the accuracy of NLP classification models as around 30%. It also ensures better practicality while maintaining transferability.

Keywords: Adversarial example · Deep learning · NLP · DeepWordBug algorithm

1 Introduction

Deep learning has been developed rapidly in the past ten years. It has promoted the breakthroughs in many fields, such as computer vision, natural language processing (NLP), and speech processing. Accordingly, more and more deep learning models have been put into practical applications. However, there are serious vulnerabilities in these models, which bring great risks to the practical application of various deep learning technologies.

In 2014, Szegedy et al. [1] found that after adding some disturbances to the input of the deep learning model, the prediction results of the model can be easily changed. Subsequent studies name this type of disturbance as adversarial perturbation, and the

S. Shi et al. (Eds.): AICON 2020, LNICST 356, pp. 51–62, 2021.
https://doi.org/10.1007/978-3-030-69066-3_5

input after the disturbance is called adversarial examples, the process of inputting the adversarial examples to mislead the model is called adversarial attack.

Goodfellow et al. [2] attributed this vulnerability to the linear characteristics of the deep learning model. Even if the deep network is not a linear network, some non-linear activation functions and network structures, such as ReLU activation functions, LSTM network structures, etc., are deliberately used in a nearly linear manner to make the model easy to optimize, otherwise a lot of time will be spent on model debugging when training the network. Equation (1) uses a dot product operation to simulate this situation: under linear condition, when a small disturbance η is added to the model input X, after the disturbance input X' is input to the model, the change of model output $W^T\eta$ will be produced, where W represents a weight vector. In the deep learning model, the dimensionality of W is very high, so its output change $W^T\eta$ will also be large. This leads the model to make a false judgment.

$$W^TX' = W^TX + W^T\eta \tag{1}$$

Since the concept of adversarial examples was proposed, some white box attack methods [3–5] based on FGSM (Fast Gradient Sign Method) [2] were proposed. After that, researchers have shifted their interest to black box attack, some black box attack methods [6–10]were proposed.

NLP is an important research direction in the field of deep learning and artificial intelligence. The aim of the research is making use of computers to process and understand human languages to achieve effective communication between humans and computers. NLP is mainly used for question-answering systems, sentiment analysis, machine translation, etc. NLP classification algorithms based on deep learning are currently used in many fields such as spam classification, public opinion monitoring, product analysis, etc. Correspondingly, the generation method of adversarial example used to attack the NLP classifiers can also be used in many aspects. For example, the processing of advertising mail can deceive the spam classification system, and the processing of current affairs news can deceive the public opinion monitoring system.

DeepWordBug [6] is a black box adversarial attack algorithm, which is proposed by Gao et al. A scoring function is used for the first time to rank the importance of each word in the input, and then the most important words are selected to make letter-level modification. The extent of this change is controlled by the edit distance before and after the modification, so that the modification is not easily noticed by human observers. But when the number of modified words is too large, human observers will still perceive that this adversarial example is a modified text.

Then a new algorithm, Textfooler [7], appeared in 2019, which makes word-level changes. In this algorithm, a strategy of dynamically changing the number of words that needs to be modified is applied, and good results have been achieved.

This paper focuses on the weakness of the DeepWordBug algorithm that is not practical, the strategy of Textfooler is added in, and an improved method is proposed. Improved DeepWordBug could dynamically change the number of modified words, and limit the maximum number of modified words, and it has better practicality.

2 Performance and Problem Analysis of DeepWordBug Adversarial Example Generation Algorithm

2.1 Introduction of DeepWordBug

DeepWordBug is a black box attack algorithm using character-level changes, which can be carried out in a scene without model information. The following are the specific implementation steps of DeepWordBug algorithm.

Step 1: The first step of DeepWordBug algorithm is to calculate the importance score for each word x_i in the text example X, and rank the words in X in descending order according to the importance score. The scoring function in DeepWordBug uses Combined Score (CS), which is composed of the weighted sum of Temporal Head Score (THS) and Temporal Tail Score (TTS). The calculation formula is shown in formula (2) to formula (4).

$$THS(x_i) = F_y(x_1, x_2, \ldots, x_{i-1}, x_i) - F_y(x_1, x_2, \ldots, x_{i-1}) \tag{2}$$

$$TTS(x_i) = F_y(x_i, x_{i+1}, \ldots, x_n) - F_y(x_{i+1}, x_{i+2}, \ldots, x_n) \tag{3}$$

$$CS(x_i) = THS(x_i) + \lambda(TTS(x_i)) \tag{4}$$

THS is the difference between the output confidence after inputting from the first word to the i-th word, and the output confidence after inputting from the first word to the I − 1th word, shown in formula (2). The larger the result of this scoring function is, the more the output confidence reduced after x_i is removed. It means x_i is more important to the classification result. However, this scoring function does not consider the contribution of the words after the i-th word to the classification result, so TTS is added.

TTS is the difference between the output confidence after inputting from the i-th word to the last word, and the output confidence after inputting from the i + 1-th word to the last word, shown in formula (3).The larger the result of this scoring function is, the more the output confidence reduced after x_i is removed. It means x_i is more important to the classification result. Opposite to THS above, this scoring function does not consider the contribution of the words before the i-th word to the classification result.

THS and TTS only consider part of the input, while ignoring the impact of the other part on the classification results. CS adds the results of the two scoring functions, both including the influence of the i-th word and the words before the i-th word in THS, and the influence of the i-th word and the words after the i-th word in TTS, shown in formula (4), λ is a hyperparameter. Some experiment results have shown that using CS is better than using THS or TTS.

Step 2: Important words will be modified in this step. The first m words in the ranked word sequence are selected to make character-level changes, where m is a hyperparameter. The modification method is also a hyperparameter, and the available methods include insert, delete, swap, and substitute.

The insert method is to randomly select a position first, and then insert a random letter at that position. The delete method is to delete a letter in a random position. The swap method is to select two random letters in the word to swap. The substitute method is to select a random position in the word, delete the letter in this position and replace it with another different random letter. Table 1 is an example of these four modification methods.

Table 1. Modification methods of DeepWordBug

Original word	Modification method	Modified word
Sequence	Insert	Sequnence
	Delete	Sequnce
	Swap	Seqeunce
	Substitute	Sequerce

2.2 Performance of DeepWordBug

In this part, we choose two datasets to test DeepWordBug adversarial example generation algorithm, namely Yelp review dataset and Ag news dataset. We use two commonly used text classification algorithms, LSTM [11] and TextCNN [12], to classify these two datasets. After that, we use adversarial example generated by DeepWordBug to attack the classification model, and analyze its performance and problems.

Yelp review dataset is a two-category dataset of a text collection. The average length of each text example is 146 words. It divides the reviews of restaurants, venues, movies, etc. on Yelp website into two categories: positive reviews and negative reviews. There are 560,000 texts in the training set and 38,000 texts in the test set. Ag news dataset is a multi-category dataset. The average length of each text is only 42 words. It divides the sentences extracted from the news into four categories: global news, sports news, business news, and technology news. There are 120,000 texts in the training set and 7,600 texts in the test set.

The two datasets are very different in length and classification category. The purpose of this choice is to test whether good results can be obtained from DeepWordBug algorithm, in short text or long text, multi-classification or two-class classification.

Table 2 shows the classification accuracy of LSTM model and TextCNN model. It can be seen from the experimental results that the classification accuracy of LSTM is slightly higher than the TextCNN model on two datasets, and the classification accuracy of the two models on Yelp comment dataset is slightly higher than that on Ag news dataset. In general, the two models have achieved an accuracy of more than 90% on both datasets, and the effect is good.

Table 2. The classification accuracy of classification models on two datasets

	AG dataset	Yelp dataset
LSTM model	92.57%	96.52%
TextCNN model	91.76%	95.92%

First of all, we perform the DeepWordBug algorithm on the two datasets. Table 3 shows the experimental result of the accuracy experiment on the two models and two datasets. In comparison, the accuracy of Ag news dataset decreases more, and the accuracy of Yelp dataset decreases less. This is because the number of modified words m = 30, the average length of Ag news dataset is 42 words, for Ag dataset, the proportion of changes is relatively large. The average length of Yelp comment dataset is 146 words, so the proportion of changes is small. So the accuracy is different between the two datasets.

Table 3. The accuracy on two models and two datasets

DeepWordBug (m = 30, insert)	AG dataset	Yelp dataset
LSTM model	20.0%	48.2%
TextCNN model	32.7%	56.3%

Then we analyze the impact of number of modified words m on the accuracy of adversarial example classification. The experimental results using the TextCNN model and Yelp dataset as examples are shown in Fig. 1. From this experimental result, it can be found that as m increases, the classification accuracy of the adversarial examples decreases, but the decrease becomes more and more gentle. This is because in the process of increasing m, the modified words added later have lower importance rankings. So their changes will contribute less to the decline in the classification accuracy than the higher ones, so the curve becomes more and more flat.

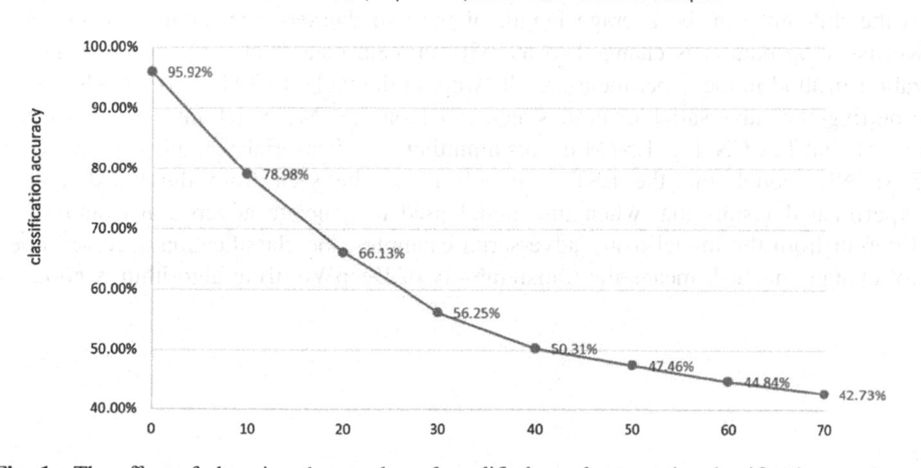

Fig. 1. The effect of changing the number of modified words m on the classification accuracy

Table 4 shows the adversarial example generated by DeepWordBug algorithm when m = 30. The classification model is TextCNN, the dataset is Yelp, and the modification method is swap. The red words are the modified words. In this example, the actual number of modified words m = 27, which is less than m. This is because in the swap operation, if two identical letters in a word are swapped, the original word will be obtained.

Table 4. An example of adversarial examples generated by DeepWordBug.

DeepWordBug (m=30, swap)	prediction	text
original text	positive review	when i went to fry's today it was a great experience. looked like they just went under a re-model and i think it makes the place look much more clean and welcoming. also, the staff there was very nice! since it was under remodel i had no clue where everything was and the front end manager helped me out! very sweet nice people working there!
adversarial example	negative review	when i wnet to fry's otday ti was a graet experience. looked ilke they just went under a er-model adn i think ti makes the place look much more clena nad wlecoming. alos, the staff tehre was veyr ncie! sicne it was undre remodel i had no clue wheer everytihng was and hte front edn manager hleped em out! evry sewet ince people wokring there!

Table 5 shows the transferability experiment results on the two datasets. According to the difference in the average length of the two datasets, the number of modified words in Ag dataset is changed to m = 30, in Yelp dataset m = 50, and the modification method in the experiment are all swap. In the table, LSTM- > TextCNN means inputting the adversarial examples generated on LSTM model into the TextCNN model, and TextCNN- > LSTM means inputting the adversarial examples generated on TextCNN model into the LSTM model. It can be seen from the transferability experimental results that when the model used to generate adversarial examples is different from the model using adversarial examples, the classification accuracy does not change much. It means the transferability of DeepWordBug algorithm is good.

Table 5. Transferability experiment results on two datasets

DeepWordBug	Ag dataset		Yelp dataset	
	Original accuracy	Accuracy after being attacked	Original accuracy	Accuracy after being attacked
LSTM	92.6%	20.0%	96.5%	36.1%
LSTM -> TextCNN	91.8%	25.7%	95.9%	48.3%
TextCNN	91.8%	32.7%	95.9%	47.5%
TextCNN -> LSTM	92.6%	36.6%	96.5%	50.5%

2.3 Problem Analysis

From the above experimental results, it can be seen that the accuracy of the adversarial example generated by DeepWordBug algorithm is still not very satisfactory. Especially on Yelp dataset, the accuracy is more than 35%. We found that DeepWordBug algorithm has the following problems:

First, in DeepWordBug algorithm, the number of modified words m is a fixed hyperparameter. In the experiment, if we fix m = 30, all adversarial examples will change 30 words on the basis of the original example. As we can see in Table 5, too many changed words will make the observer easily perceive the change of the text content. At this time, even if the adversarial example can deceive the classifier, it is not practical.

Second, in DeepWordBug algorithm, the modified m words are the top m words in the importance score ranking. However, modifying the top m words in the importance score ranking does not ensure that the effect of modifying these words is better than modifying any other m words. Because when calculating the importance score of each word, we only modified this word in the text, that is to say, we only measured the contribution of this word to the classification of the entire text, but when modifying m words, we cannot assume that the contribution of the m words to the entire text classification is equal to the simple addition of each words contribution to the text classification. Therefore, it is likely that there are other combinations of m words that can play a greater role in changing the classification result than the previous m words.

3 Improvement Strategy

3.1 Improvement of DeepWordBug

During the research, we found that for some sentences, only changing one or two words can cause changes in the classification results when using TextFooler algorithm. Based on the above analysis, we decide to add TextFooler algorithm to DeepWordBug to increase the effectiveness. This strategy is used to dynamically change the number of modified words, and only select the words with better effects to modify after the changes, and stop editing when the classification result can already be changed. In other words, even for the top-ranked words, we have to check whether the confidence of the original prediction category of the sentence decreased after the modification. If the confidence does not decrease but rises, we can also give up changing the top-ranked words, instead, we can choose to change the lower-ranked words which can reduce the confidence of the original prediction category. Its purpose is to reduce the number of modified words while reducing the classification accuracy of adversarial examples, and hope to maintain the transferability of adversarial examples generated by DeepWordBug algorithm. The specific process of the improvement plan is as follows (Fig. 2):

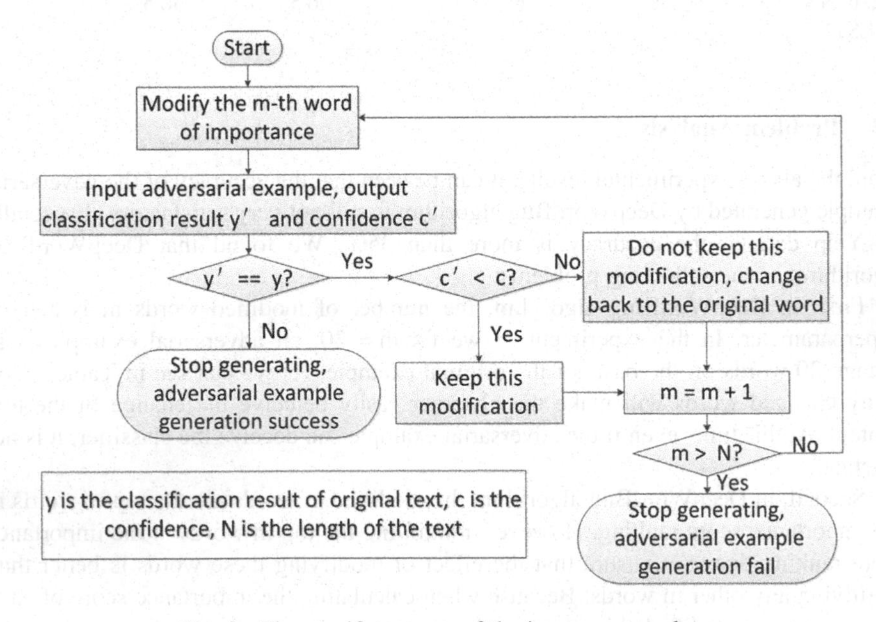

Fig. 2. The specific process of the improvement plan.

After adding the effectiveness check to DeepWordBug algorithm, the experimental results on Ag news dataset and Yelp comment dataset are shown in Table 6. The word modification methods used in the experiment before and after the improvement are both insert. Before the improvement, set m = 30 on Ag dataset and m = 50 on Yelp dataset.

Table 6. Experimental results and comparison of the improved algorithm

Dataset	Ag dataset		Yelp dataset	
Parameters	Insert, m = 30 before the improvement, dynamic m after the improvement		Insert, m = 50 before the improvement, dynamic m after the improvement	
Classification model	LSTM model	TextCNN model	LSTM model	TextCNN model
Original accuracy	92.6%	91.8%	96.5%	95.9%

(continued)

Table 6. (*continued*)

Dataset	Ag dataset		Yelp dataset	
Accuracy of adversarial examples generated by DeepWordBug	20.0%	32.7%	36.1%	47.5%
Accuracy of adversarial examples generated by improved DeepWordBug	6.5%	11.3%	5.4%	15.4%
Average number of modified words	11.6	15.3	30.9	35.1
Average number of Total words	42	42	146	146
Average change rate	27.6%	36.4%	21.2%	24.0%
Minimum number of changed words	1	1	1	1
Maximum number of changed words	116	107	210	447

According to the experimental results, it can be seen that from the comparison of the effectiveness of the adversarial examples, the classification accuracy of adversarial examples generated by the improved DeepWordBug algorithm has dropped significantly, from 20–50% to about 10%. From the comparison of the practicality of adversarial examples, the average number of changed words is less than the fixed number of modified words m on both datasets, and the accuracy of adversarial examples is lower.

However, from the "average number of modified words", "minimum number of changed words" and "maximum number of changed words" in Table 6, it can be seen that after adding the improvement strategy, although the average number of modified words is reduced, modifying a small number of words can change the classification result in some sentences, but the number of changed words is still very large in some sentences. The maximum number of modified words on Ag dataset reached 116 and 107, and the maximum number of modified words on Yelp dataset reached 210 and 447. This is because there is no restriction on the maximum number of modified words. For any sentence, if the classification result is always not changed, you can continue to change the next word in importance ranking until all words are modified.

3.2 Further Improvement

So we improved this method again and added the limit of the maximum number of modified words. For Ag dataset, we limited the maximum number of modified words to 30, and for Yelp dataset, we limited the maximum number of modified words to 50. As before, the difference between the experimental settings on the two datasets is still based on the difference in the average number of words in the two datasets. After adding the limit, the experimental results are shown in Table 7. For the convenience of comparison, the table also shows the experimental results of the algorithm before the improvement and the algorithm after the improvement without adding the limit of the maximum number of modified words.

Table 8 shows the experimental results when the maximum number of modified words is 30, the model is TextCNN, and the modification method is swap. The red words are the modified words. In addition, we also did a transferability test on the improved algorithm again. The experimental results are shown in Table 9.

Table 7. Experimental results and comparison of the improved algorithm

Dataset	Ag dataset		Yelp dataset	
Parameters	Insert, m = 30 before the improvement, dynamic m after the first improvement, limit m not more than 30 after the second improvement		Insert, m = 50 before the improvement, dynamic m after the first improvement, limit m not more than 50 after the second improvement	
Classification model	LSTM model	TextCNN model	LSTM model	TextCNN model
Original accuracy	92.6%	91.8%	96.5%	95.9%
Accuracy of adversarial examples generated by DeepWordBug	20.0%	32.7%	36.1%	47.5%
Accuracy of adversarial examples generated by improved DeepWordBug	6.5%	11.3%	5.4%	15.4%
Accuracy of adversarial examples generated by further improved DeepWordBug	11.3%	20.0%	24.5%	36.3%
Average number of modified words of improved algorithm	11.6	15.3	30.9	35.1
Average number of modified words of further improved algorithm	10.1	12.9	18.0	18.3
Average number of total words	42	42	146	146

Table 8. An example of adversarial examples generated by improved DeepWordBug

DeepWordBug	prediction	text
original text	positive review	when i went to fry's today it was a great experience. looked like they just went under a re-model and i think it makes the place look much more clean and welcoming. also, the staff there was very nice! since it was under remodel i had no clue where everything was and the front end manager helped me out! very sweet nice people working there!
adversarial example (improved DeepWordBug)	negative review	when i went to fry's today it was a greta experience. looked like they just went under a re-model and i think it makes the place look much more clean and welcoming. also, the staff there was very nice! since it was under remodel i had no clue where everything was and the front end manager helepd me out! very sweet ncie people working there!
adversarial example (original DeepWordBug)	negative review	when i wnet to fry's otday ti was a graet experience. looked ilke they just went under a er-model adn i think ti makes the place look much more clena nad wlecoming. alos, the staff tehre was veyr ncie! sicne it was undre remodel i had no clue wheer everytihng was and hte front edn manager hleped em out! evry sewet ince people wokring there!

Table 9. Transferability experiment results of improved DeepWordBug on two datasets

DeepWordBug (improved twice)	Ag dataset		Yelp dataset	
	Original accuracy	Accuracy after being attacked	Original accuracy	Accuracy after being attacked
LSTM	92.6%	11.3%	96.5%	24.5%
LSTM -> TextCNN	91.8%	58.0%	95.9%	63.9%
TextCNN	91.8%	20.0%	95.9%	36.3%
TextCNN -> LSTM	92.6%	54.8%	96.5%	68.9%

From the experimental results in Table 7, it can be seen that even when the limited maximum number of modified words is equal to the fixed number of modified words of the original algorithm, the two models of the two datasets can achieve lower accuracy than the original algorithm, and the average number of changed words is obviously less than the original algorithm.

From the adversarial examples in Table 8, it can also be seen that the improved algorithm can change the important words more accurately, and stop the changes immediately when the classification results change. So the number of changed words in the adversarial examples is small, and the effect is remarkable.

For example, the changed words in this example are "great", "help" and "nice". These three words are the words that could clearly reflect the emotional tendency of the reviewer. After the modification, they become an unknown word vector and can no longer be recognized by the model, and they cannot make great contribution to the classification result, so the classification result can be changed while maintaining high practicability.

In addition, from the transferability results in Table 9, it can be seen that the transferability of improved DeepWordBug is worse than that of the original algorithm. This is because for different classification models, the same word has different influence on classification results. When the number of modified words is large, there must be more words that have an impact on the prediction result in another classification model, so the transferability of the algorithm will be good, otherwise, less modified words will greatly affect the transferability of the algorithm. However, in practical applications, we can use this algorithm only for white box attack, and the weakness in transferability can be avoided.

4 Conclusion

In this paper, we first introduced an algorithm to generate adversarial example to deceive NLP deep learning classifiers: DeepWordBug, and illustrated some shortcomings in the practicality of the algorithm through experiments. Then, we used the idea of TextFooler algorithm for reference and improved DeepWordBug algorithm. Using the improved DeepWordBug to generate adversarial examples to attack the NLP classification models can maintain the classification accuracy at about 30%, and the number of modified words is greatly reduced. The experimental results show that on the two datasets and two classification models, our improved DeepWordBug algorithm has a significant improvement in practicability while ensuring the success rate of the attack.

Acknowledgements. This work is supported by Beijing Natural Science Foundation (L191004).

References

1. Krizhevsky, A., Sutskever, I., Hinton, G.E.: ImageNet classification with deep convolutional neural networks. In: Advances in Neural Information Processing Systems, pp. 1097–1105 (2012)
2. Goodfellow, I.J., Shlens, J., Szegedy, C.: Explaining and harnessing adversarial examples. arXiv preprint arXiv:1412.6572 (2014)
3. Papernot, N., McDaniel, P., Swami, A., et al.: Crafting adversarial input sequences for recurrent neural networks. In: 2016 IEEE Military Communications Conference, MILCOM 2016, pp. 49–54. IEEE (2016)
4. Samanta, S., Mehta, S.: Towards crafting text adversarial samples. arXiv preprint arXiv: 1707.02812 (2017)
5. Ebrahimi, J., Rao, A., Lowd, D., et al.: HotFlip: white-box adversarial examples for text classification. arXiv preprint arXiv:1712.06751 (2017)
6. Gao, J., Lanchantin, J., Soffa, M.L., et al.: Black-box generation of adversarial text sequences to evade deep learning classifiers. In: 2018 IEEE Security and Privacy Workshops (SPW), pp. 50–56. IEEE (2018)
7. Li, J., Ji, S., Du, T., et al.: TextBugger: generating adversarial text against real-world applications. arXiv preprint arXiv:1812.05271 (2018)
8. Jin, D., Jin, Z., Zhou, J.T., Szolovits, P.: Is BERT really robust? Natural language attack on text classification and entailment. arXiv:1907.11932 (2019)
9. Liang, B., Li, H., Su, M., et al.: Deep text classification can be fooled. arXiv preprint arXiv: 1704.08006 (2017)
10. Alzantot, M., Sharma, Y., Elgohary, A., et al.: Generating natural language adversarial examples. arXiv preprint arXiv:1804.07998 (2018)
11. Hochreiter, S., Schmidhuber, J.: Long short-term memory. Neural Comput. **9**(8), 1735–1780 (1997)
12. Kim, Y.: Convolutional neural networks for sentence classification. arXiv preprint arXiv: 1408.5882 (2014)

Encryption Analysis of Different Measurement Matrices Based on Compressed Sensing

Mengna Shi, Shiyu Guo, Chao Li, Yanqi Zhou, and Erfu Wang(✉)

Key Lab of Electronic and Communication Engineering,
Heilongjiang University, No. 74 Xuefu Road, Harbin, China
efwang_612@163.com

Abstract. The randomness of the traditional measurement matrix in compressed sensing is too strong to be implemented on hardware, and when compressed sensing is used for image encryption, the measurement matrix transmitted as a key will consume time and storage space. Combined with the sensitivity of the chaotic system to the initial value, this paper uses Logistic-Chebyshev chaotic map to obtain random sequences with fewer parameters and construct measurement matrix. To test the measurement performance of the chaotic matrix, compare it with the Gaussian measurement matrix and the Bernoulli measurement matrix in the same compression encryption scheme. Pixel scrambling operation is carried out on the compressed image to complete the final encryption step, and the encrypted image is obtained. The reconstruction algorithm adopts the orthogonal matching tracking method to restore the image. The experimental simulation results show that the chaotic matrix has more advantages than the other two random matrices in image quality, and the encryption and decryption time is shorter.

Keywords: Compressed sensing · Measurement matrix · Chaotic system · Encryption

1 Introduction

In the scope of computer networks, information transmission in the channel is vulnerable to attack. Therefore, information security becomes the focus. When the information demand increases, the processing efficiency in the process of information acquisition is required higher [1, 2]. In 2006, the compressed sensing theory published by Candes and Donoho et al. broke the Nyquist conventional signal sampling technology and realized signal compression through uncorrelated observation of sparse signals, where the observed measurement matrix had to satisfy the condition of restricted isometry property (RIP) [3]. In recent years, compressed sensing is often applied to encryption [4]. Chaotic system is sensitive to the initial value, and the pseudo-random sequence can be generated with fewer parameters to construct the measurement matrix. This paper compared the reconstruction image quality of chaotic

This work was supported by the Natural Science Foundation of Heilongjiang Province, China (no. LH2019F048).

S. Shi et al. (Eds.): AICON 2020, LNICST 356, pp. 63–69, 2021.
https://doi.org/10.1007/978-3-030-69066-3_6

matrix, Gaussian matrix, and Bernoulli matrix as measurement matrices in the same compressed encryption scheme. The simulation results show that the chaotic matrix is feasible as a measurement matrix in compression and encryption.

2 Theoretical Basis

2.1 Compressed Sensing Model

Compressed sensing, as a new signal sampling theory, utilizes the sparse characteristics of signals to obtain discrete samples of signals with random sampling under the condition that the sampling rate is much lower than Nyquist sampling rate, and restores the signal through a nonlinear reconstruction algorithm. Different from traditional compression, compressed sensing is the simultaneous sampling and compression, that is, the sampling value is the compressed value, which greatly reduces the number of transmission, transmission time and storage space.

If the N-dimensional signal x can be expanded under a certain set of sparse basis $\{\psi_i\}_{i=1}^N$, that is:

$$x = \sum_{i=1}^{N} s_i \psi_i, \tag{1}$$

where the expansion coefficient $s_i = \langle x, \psi_i \rangle = \psi_i^T x$ can be written in matrix form to obtain:

$$x = \psi s, \tag{2}$$

where ψ is an $N \times N$-dimensional sparse basis matrix, s is an N-dimensional sparse coefficient vector, and the number of non-zero coefficients in s is much smaller than N. Use a matrix ϕ that is not related to the sparse basis to project the signal x. The mathematical expression is as follows:

$$y = \phi x, \tag{3}$$

where ϕ is an $M \times N$-dimensional matrix ($M \ll N$), and y is an M-dimensional observation vector. Then the complete compression process of signal x is

$$y = \phi \psi s, \tag{4}$$

and the essence of compressed sensing is to reduce the dimensionality of the signal, from N dimension to M dimension.

The Compressed sensing reconstruction of the signal x is achieved by solving underdetermined set of equations, which is a l_0 norm minimization problem. The l_0

norm solution is an NP hard problem, so it is often converted to a l_1 norm minimization solution, denoted by

$$\min\|s\|_1 s.t. y = \phi\psi s. \tag{5}$$

Commonly used reconstruction methods include base pursuit method, gradient projection method, orthogonal matching pursuit method, etc. In this paper, orthogonal matching pursuit (OMP) method is used [5–7].

2.2 Chaotic Mapping

Chaos is a kind of nonlinear dynamic system, which is generated by control parameters within a certain range, and is sensitive to initial conditions. The sequence generated by the chaotic system is unpredictable. One-dimensional Logistic mapping is defined as follows

$$x_{n+1} = \mu x_n(1 - x_n). \tag{6}$$

When $\mu \in (3.5699456, 4]$, $x_n \in (0, 1)$, the system is chaotic state [8]. One-dimensional Chebyshev mapping is defined as follows

$$x_{n+1} = \cos(t.\arccos x_n). \tag{7}$$

Where t is the order of Chebyshev. When $x_n \in (-1, 1)$, $t \geq 2$, this system is chaotic [9]. The Logistic-Chebyshev chaotic system is defined as follows

$$x_{n+1} = \mathrm{mod}((ax_n(1 - x_n) + ((4 - a)/4) * \cos(b.\arccos x_n)), 1). \tag{8}$$

When $a \in (0, 4]$, $b \in N$, $x_n \in (0, 1)$, the system is in chaos.

2.3 Measurement Matrix

The quality of signal reconstruction in compressed sensing depends on the measurement matrix. When the measurement matrix satisfies the RIP condition, M measurements and measurement matrices can be used to recover the original signal. Common random class matrices satisfy the RIP characteristics, but the strong randomness of these matrices is limited by hardware conditions in practical application, and in the decryption process, the whole measurement matrix as key transmission and storage will waste resources [10]. Chaotic systems use fewer parameters to obtain random sequences, and consequently, when the chaotic matrix is used for compressed sensing, it overcomes the drawback of the large transmission volume of the traditional random matrix. In this paper, Gaussian matrix, Bernoulli matrix and chaotic matrix generated by Logistic-Chebyshev system are tested and compared for image compression and encryption.

3 Encryption Scheme

In compression encryption, we uniformly use discrete wavelet transform to sparse the plain image P with the size of $N \times N$, and use different measurement matrices of size $M \times N$ for observation [11]. To simplify the encryption step, we scrambling the compressed image to obtain the cipher image C [12]. The scrambling steps are as follows.

Step 1: After the Logistic system discards the first 1000 iteration values, the chaotic sequence L with length $M * N$ is generated.

Step 2: sort L to obtain index sequence L'.

Step 3: permutation the pixels of compressed image according to the position of L' to obtain encrypted image C with the size of $M \times N$.

The overall structure of encryption and decryption is shown in Fig. 1. Image decryption is the inverse step of encryption, and the reconstruction algorithm is OMP.

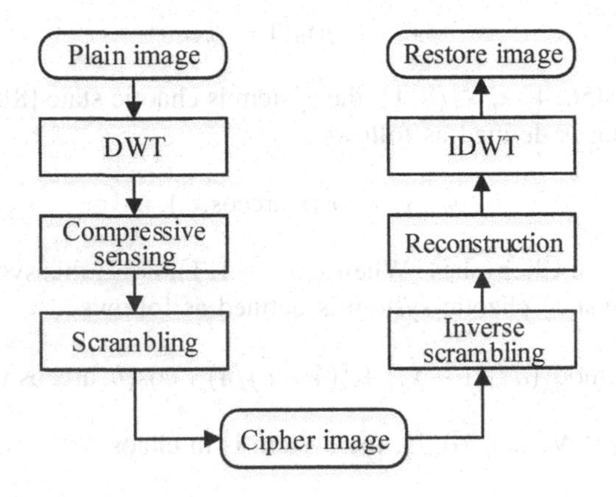

Fig. 1. The overall structure of encryption and decryption.

4 Simulation Results and Analysis

We use Matlab R2016a to verify the performance of the proposed chaotic matrix, the simulation experiments are carried out in a desktop computers with Windows7 operating system, 3.30 GHz CPU and 4 GB RAM. Lena with the size of 256×256 was selected as the test image, and 190×256 sized the measurement matrix. The Logistic chaotic system parameters used for scrambling are set as $\mu = 3.99$, $x_0 = 0.8326$, The Logistic-Chebyshev chaotic system used to construct the measurement matrix, its parameters set to $a = 3.99$, $b = 20$, $x_0 = 0.23$. Encryption and decryption are carried out according to the method in Sect. 3. Figure 2 is a simulation diagram of encryption and decryption of test images compressed with different measurement matrices. It can be seen from Fig. 2 that the reconstruction of the chaotic matrix is as clear as the other two random matrices.

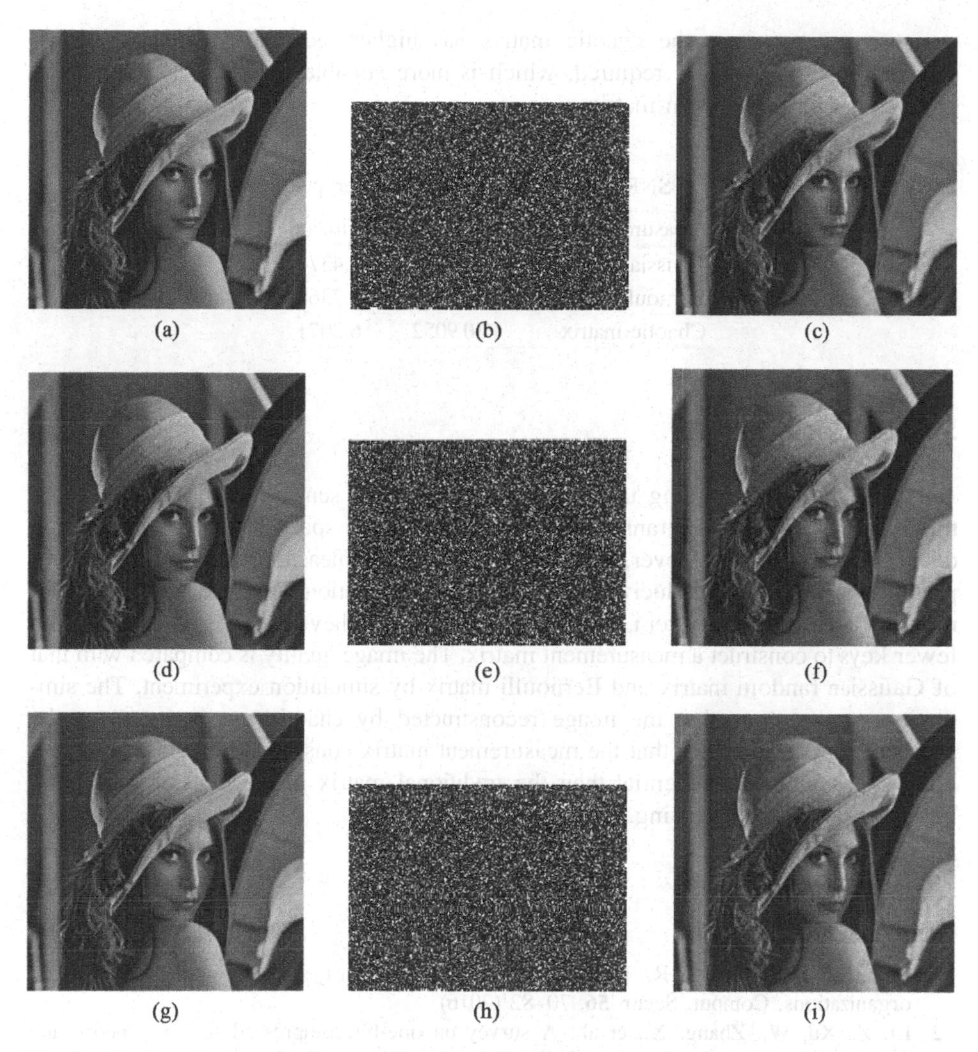

Fig. 2. (a) Test image; (b) Encrypted image of Gaussian matrix measurement; (c) Restore image; (d) Test image; (e) Encrypted image of Bernoulli matrix measurement; (f) Restore image; (g) Test image; (h) Encrypted image of chaotic matrix measurement; (i) Restore image.

Peak signal-to-noise ratio (PSNR) is often used to evaluate the quality of image compression reconstruction, which can be obtained from the definition

$$PSNR = 10\lg \frac{255 \times 255}{\frac{1}{MN}\sum_{i=1}^{M}\sum_{i=1}^{N}\left(P(i,j) - C(i,j)\right)^{2}}, \tag{9}$$

where P is a test image, C is an encrypted image, and M and N are the height and width of the encrypted image respectively [13]. Table 1 lists the PSNR of different decrypted images and encryption and decryption time. The results show that for the three measurement matrices, the chaotic matrix has higher reconstruction performance, shorter time and less key required, which is more suitable for practical application compared with the random matrix.

Table 1. PSNR and time comparison of decrypted images.

Measurement matrix	PSNR (dB)	Time(s)
Gaussian matrix	30.6039	6.4374
Bernoulli's matrix	30.6448	6.7368
Chaotic matrix	30.9052	6.3071

5 Conclusion

Applying the new sampling technology of compressed sensing to the field of cryptography will reduce the transmission time and storage space in the encryption and decryption process. However, the traditional random measurement matrix in compressed sensing is not conducive to hardware implementation. With the natural pseudo-randomness of chaotic system, and the Logistic-Chevbyshev system is utilized through fewer keys to construct a measurement matrix. The image quality is compared with that of Gaussian random matrix and Bernoulli matrix by simulation experiment. The simulation results show that the image reconstructed by chaotic matrix improves the accuracy, which indicates that the measurement matrix constructed by chaotic system has better application potential than the traditional matrix in the field of encryption based on compressed sensing.

References

1. Safa, N.S., Von Solms, R., Furnell, S.: Information security policy compliance model in organizations. Comput. Secur. **56**, 70–82 (2016)
2. Li, Z., Xu, W., Zhang, X., et al.: A survey on one-bit compressed sensing: theory and applications. Front. Comput. Sci. **12**(2), 217–230 (2018)
3. Candès, E.J., Wakin, M.B.: An introduction to compressive sampling. IEEE Sig. Process. Mag. **25**(2), 21–30 (2008)
4. Cambareri, V., Mangia, M., Pareschi, F., et al.: Low-complexity multiclass encryption by compressed sensing. IEEE Trans. Sig. Process. **63**(9), 2183–2195 (2015)
5. Ujan, S., Ghorshi, S., Pourebrahim, M, et al.: On the use of compressive sensing for image enhancement. In: 2016 UKSim-AMSS. In: 18th International Conference on Computer Modelling and Simulation (UKSim), Cambridge, UK, pp. 167–171. IEEE (2016)
6. Liu, J.K., Du, X.L.: A gradient projection method for the sparse signal reconstruction in compressive sensing. Appl. Anal. **97**(12), 2122–2131 (2018)
7. Wen, J., Zhou, Z., Wang, J., et al.: A sharp condition for exact support recovery with orthogonal matching pursuit. IEEE Trans. Sig. Process. **65**(6), 1370–1382 (2016)
8. Kong, X., Bi, H., Lu, D., et al.: Construction of a class of logistic chaotic measurement matrices for compressed sensing. Pattern Recogn. Image Anal. **29**(3), 493–502 (2019)

9. Zhu, S., Zhu, C., Wang, W.: A novel image compression-encryption scheme based on chaos and compression sensing. IEEE Access **6**, 67095–67107 (2018)
10. Candes, E.J., Tao, T.: Near-optimal signal recovery from random projections: Universal encoding strategies? IEEE Trans. Inf. Theor. **52**(12), 5406–5425 (2006)
11. Yao, S., Chen, L., Zhong, Y.: An encryption system for color image based on compressive sensing. Opt. Laser Technol. **120**, 105703 (2019)
12. Wang, X., Gao, S.: Application of matrix semi-tensor product in chaotic image encryption. J. Franklin Inst. **356**(18), 11638–11667 (2019)
13. Yuan, X., Zhang, L., Chen, J., et al.: Multiple-image encryption scheme based on ghost imaging of Hadamard matrix and spatial multiplexing. Appl. Phys. B **125**(9), 174 (2019)

Indoor Visual Positioning Based on Image Retrieval in Dense Connected Convolutional Network

Xiaomeng Guo, Danyang Qin(✉), and Yan Yang

Heilongjiang University, Harbin 150080, People's Republic of China
qindanyang@hlju.edu.cn

Abstract. As now available methods or systems based on image retrieval and visual researchs are implemented in an indoor environment, their retrieval accuracy and real-time positioning still have their own limitations. For this reason, this paper designs a visual indoor positioning system based on densely connected convolutional network image retrieval. Combine visual positioning with DenseNet-based image retrieval method. The problem of excessively deep network layers caused by the original convolutional neural network in pursuit of high retrieval accuracy is improved. Under the advantage of ensuring the high accuracy of image retrieval based on depth features, the problem of low real-time positioning caused by the long training time of the convolutional network model is improved. The simulation results show the feasibility of the positioning method in indoor environment, and the comparison experiment verifies the improvement of accuracy and speed as well as the reliability of the method.

Keywords: DenseNet · Feature extraction · Image retrieval · Indoor positioning

1 Introduction

With the increasingly prominent advantages of visual indoor positioning, it has become one of the hot spots in the research field of indoor positioning due to its low cost, high accuracy, and high real-time positioning [1]. In recent years, regarding indoor visual positioning attempts, Kohoutek et al. used the CityGML, an internal architectural model with digital spatial semantics, to determine the location and direction of the distance imaging camera [2] at the highest level of detail (LoD 4). Hile and Borriello compared the floor plan with current cell phone images. Kitanov et al. The image taken by the camera installed on the robot is used to detect the image line and compare it with the 3D vector model [3]. The computer vision algorithm summarized by Schlaile et al.

This work is supported by the National Natural Science Foundation of China (61771186), University Nursing Program for Young Scholars with Creative Talents in Heilongjiang Province (UNPYSCT-2017125), Distinguished Young Scholars Fund of Heilongjiang University, and postdoctoral Research Foundation of Heilongjiang Province (LBH-Q15121), Outstanding Youth Project of Provincial Natural Science Foundation of China in 2020 (YQ2020F012).

S. Shi et al. (Eds.): AICON 2020, LNICST 356, pp. 70–79, 2021.
https://doi.org/10.1007/978-3-030-69066-3_7

It also depends on feature detection in the image sequence. Muffert et al. Determine the track of the omnidirectional camera according to the relative direction of the continuous image. Due to the limited indoor space, there is a high probability that the scene and the indoor layout tend to be highly similar [4]. At this time, the collected images and the extracted vector features will also be difficult to distinguish. In this case, it is easy to cause problems such as inaccurate estimation of camera parameters and mismatch of pose information. Therefore, high-precision feature information is very necessary for indoor positioning.

This paper proposes an indoor visual positioning scheme based on densely connected convolutional neural network [5] image retrieval. Through the extraction of image depth features, it combines the advantages of visual positioning and deep learning positioning, while ensuring high-precision positioning while ensuring it improves the real-time and robustness of indoor positioning [6].

2 System Description

The main though of this method is to utilize the collected indoor scene images to determine the user's location through static objects. The system flowchart is shown in Fig. 1. The proposed visual indoor positioning system incorporate three portions: static image collection, image retrieval based on depth features, and user position estimation.

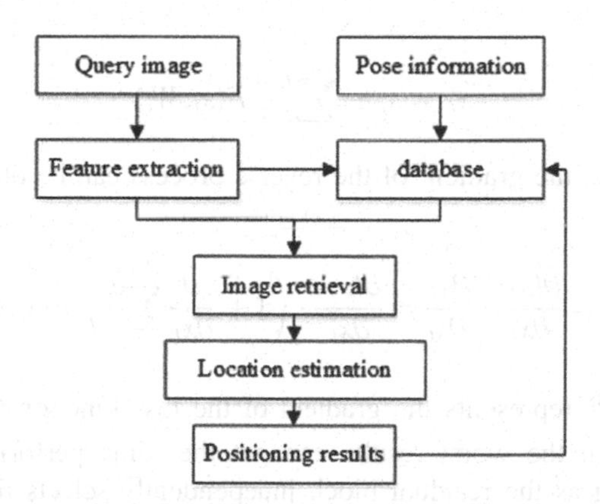

Fig. 1. Visual positioning system block diagram.

In the offline data set preparation stage, a camera is used to capture user RGB images. It contains the pose information of the captured image. The pre-trained DenseNet model extracts depth features from all collected images and sends them to the server. In the image retrieval stage, deep feature extraction is performed on images through densely connected convolutional networks, and distances are calculated for image features using metric learning methods such as Euclidean distance, and the distances of the images are sorted to obtain the primary search results, and then based on the context information of the image data and The manifold structure reorders the image retrieval results to output the final retrieval results. The last part is the user position estimation stage, the two most similar images obtained from the image retrieval stage and the query image pose. Fuse the pose prediction with the relative distance and angle of the user. The final revised estimate and sensor deviation will be resent to the user to complete the result recall. The last two stages are completed online.

2.1 Resnet Structure

The main reason that the residual network works is that the residual block can easily learn the identity function. Therefore, it can be ensured that network performance will not be affected, and efficiency can even be improved in many cases. First, the residual element can be expressed as Formula 1 and 2:

$$y_l = h(x_l) + F(x_l, W_l) \tag{1}$$

$$x_{l+1} = f(y_l) \tag{2}$$

Where x_l and x_{l+1} represent the input and output of the first residual unit respectively, and each first residual unit usually contains a multilayer structure. F is the residual function, which means the residual of learning, and $h(x_l) = x_l$ means the identity mapping, and F is the ReLU activation function. According to the above formula, the feature vector of similarity is obtained, and the calculation process is shown in Formula 3:

$$x_L = x_l + \sum_{i=1}^{L-1} F(x_i, W_i) \tag{3}$$

Using the chain rule, the gradient of the reverse process can be obtained as shown in Formula 4:

$$\frac{\partial loss}{\partial x_l} = \frac{\partial loss}{\partial x_L} \cdot \frac{\partial x_L}{\partial x_l} = \frac{\partial loss}{\partial x_L} \cdot \left(1 + \frac{\partial}{\partial x_L} \sum_{i=l}^{L-1} F(x_i, W_i)\right) \tag{4}$$

The first factor $\frac{\partial loss}{\partial x_L}$ represents the gradient of the loss function to L. This structure guarantees that even the worst results can get the same performance as the plain network, so as long as the residual block independently selects the working mechanism, the overall performance of the network can be improved, and the burden on the network itself is not increased. Figure 2 shows the residual structure [7]. In this case, the deep network should have at least the same performance as the shallow network without degradation.

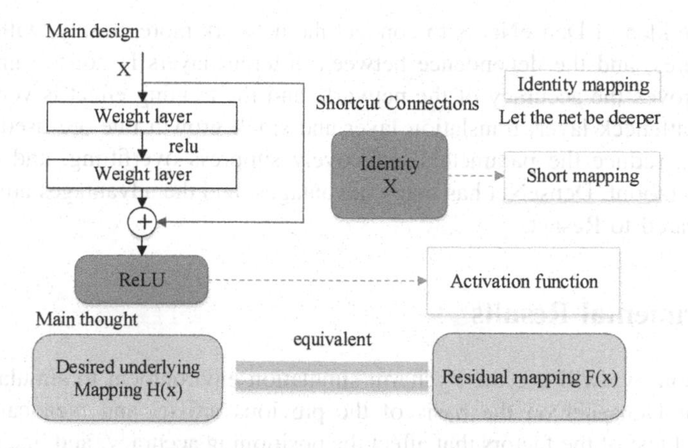

Fig. 2. Residual block principle.

2.2 DenseNet Network Structure

Compared with Resnet, Densenet's design philosophy is as follows: a tighter network mechanism is created, that is, all layers are connected to each other, which means that each layer will accept the output of all previous layers as its additional input, which is the most obvious. The effect is to realize the repeated use of features, so the efficiency of the network is also significantly improved. The nonlinear transformation of DenseNet is shown in Eq. 5:

$$x_l = H_l([x_0, x_1, \ldots, x_{l-1}]) \tag{5}$$

As shown in Fig. 3, when performing convolution operations on different layers, in order to ensure that the feature size is the same, the Densenet is divided into multiple Dense blocks, and the feature size in each Dense block remains the same. In addition, in order to ensure The connection between layers and the realization of downsampling function, set up transition layers between different dense blocks [8]. The problem of gradient loss is alleviated by tight connections, feature propagation is strengthened, feature maps are fully utilized, and the amount of parameters is greatly reduced.

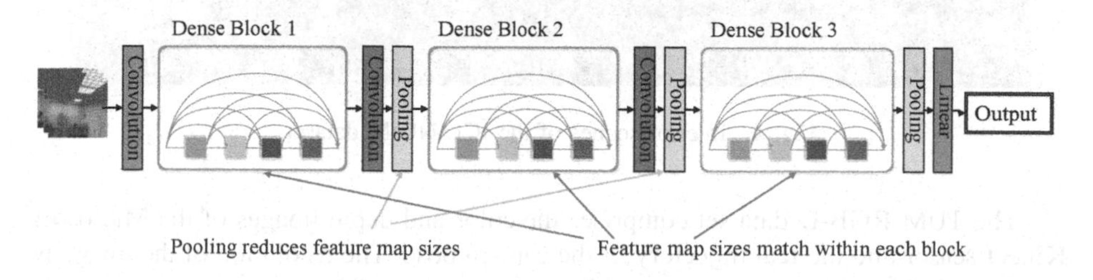

Fig. 3. Densenet network structure.

The core idea of DenseNet is to connect the network more closely, with relatively robust features, and the dependence between different layers is not too great, which further improves the accuracy of the network, and the training effect is very good. In addition, bottleneck layer, translation layer and small growth rate are used to narrow the network, reduce the parameters, effectively suppress overfitting, and reduce the calculation amount. DenseNet has many advantages, and the advantages are very clear when compared to Resnet.

3 Experimental Results

In this section, we will use the Pycharm simulation environment to simulate the performance of Densenet on the basis of the previous article, and then carry out the experimental test of the factors that affect the positioning accuracy, and finally evaluate the accuracy of image retrieval and indoor positioning. The research is based on densely connected convolutional network image retrieval. The visual positioning achieves the trade-off effect between positioning accuracy and real-time performance. The deep learning library uses Keras, TensorFlow, OpenCV, etc.

3.1 Dataset Preparation

In order to compare with existing indoor positioning methods, we decided to choose classic open source data sets ICL-NUIM and TUM RGB-D, and select appropriate indoor scenearios from the data set as the test scenes of this experiment.

The ICL-NUIM dataset includes RGB-D images of camera trajectories from two indoor scenes. The images are collected by a handheld Kinect RGB-D camera, and Kintinuous is used to obtain the ground truth of the trajectory. The images are taken at a resolution of 640 × 480 (Fig. 4).

Fig. 4. Interior scenes of the ICL-NUIM dataset.

The TUM RGB-D data set comprises the color and depth images of the Microsoft Kinect sensor and the real trajectory of the camera pose. The resolution of the image is 640 × 480. The data set consists of 89 sequences from different camera movements (Fig. 5).

Fig. 5. Interior scenes of the TUM RGB-D dataset.

3.2 Feature Extraction

Feature extraction refers to the extraction of more advanced features from the original pixels, these features can capture the difference between each category. This feature extraction uses an unsupervised method, and no image category labels are used when extracting information from pixels [9]. Commonly used traditional features include GIST, HOG, SIFT, LBP, etc. After feature extraction, use these features of the image and their corresponding category labels to train a classification model. Commonly used classification models include SVM, LR, random forest and decision tree.

Figure 6 shows the visualization of the depth feature vector (512 dimensions) extracted from the final convolutional layer of the Densenet-B and Densenet-C networks. The leftmost represents the vector of the input query image; the middle image represents the vector of the image with the highest score retrieved. Vector; The image on the right shows the vector of the feature vector image of the second matching image with the highest score in the same scene [10].

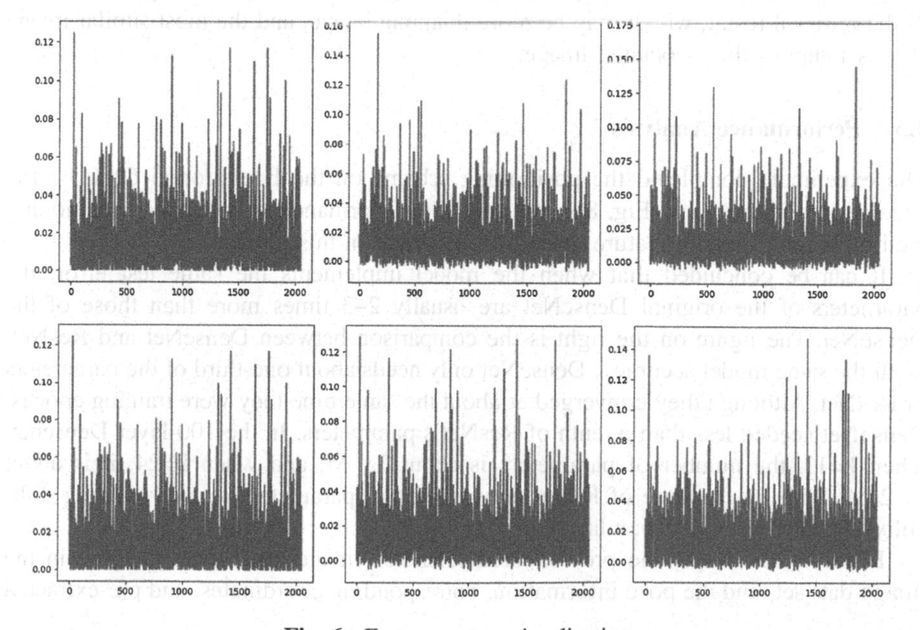

Fig. 6. Feature vector visualization.

For the CNN-based image retrieval method [11], it is mainly based on the pre-training network to extract the depth feature normalized and aggregated feature vectors to optimize the retrieval accuracy rate. Therefore, it is necessary to define an appropriate visual feature similarity measurement method, which undoubtedly has a great influence on the effect of image retrieval [12]. The accuracy of image feature similarity comparison lies in the depth feature extracted by the convolutional neural network. The public data set used in this paper is suitable for 224 × 224 input. Since each training is a random sample pair, the training takes longer and the generalization is better, but it should not be too long. Iteration 3 W round, batch_size = 128.

Euclidean distance is used to count the similarity between the feature vector vc of the retrieved image (as shown in Formula 6) and the feature vector vi of the query image (as shown in Formula 7). The smaller the distance, the higher the similarity. The distance between the two is shown in Formula 8:

$$\mathrm{V}_c = \left[V_c^0, V_c^1, \ldots V_c^{512}\right] \tag{6}$$

$$\mathrm{V}_i = \left[V_i^0, V_i^1, \ldots V_i^{512}\right] \tag{7}$$

$$\mathrm{D}(\mathrm{V}_c, \mathrm{V}_i) = \sqrt{\sum\nolimits_{c,i=1}^{512} (\mathrm{V}_c - \mathrm{V}_i)^2} \tag{8}$$

After feature extraction, image features are aggregated into fixed-length vectors. According to the current research status, similar to the general framework of most retrieval, we used the DenseNet model to extract the depth features of the input image of 224 × 224, and then mapped to the public space to obtain a unified representation and the same fixed-length feature vector. Then, the similarity is calculated according to the distance measurement method. In the final stage, the top ranked image is extracted as the retrieval result, which may be more than one image, and the most similar image [13] is found as the associated image.

3.3 Performance Analysis

The experiment completed the positioning scheme on the Intel Corei5 6200U CPU @2.3 GHz. Figure 7 and Fig. 8 summarize the performance of the indoor positioning method based on depth feature extraction proposed in this paper.

It can be concluded that when the model implements the same test error, the parameters of the original DenseNet are usually 2–3 times more than those of the DenseNet. The figure on the right is the comparison between DenseNet and ResNet. With the same model accuracy, DenseNet only needs about one-third of the parameters of ResNet. Although they converged at about the same time they were training epochs, DenseNet needed less than a tenth of ResNet's parameters. In the 100-layer Densenet, when k=12, the number of parameters is about 7.0 M, and when k=24, it is about 27.2 M. With the increase of K and network depth, parameter two and training difficulty also increase correspondingly.

In the offline stage, the pre-collected reference images are directly stored in the image data set, and the pose information, corresponding coordinates, and pre-extracted

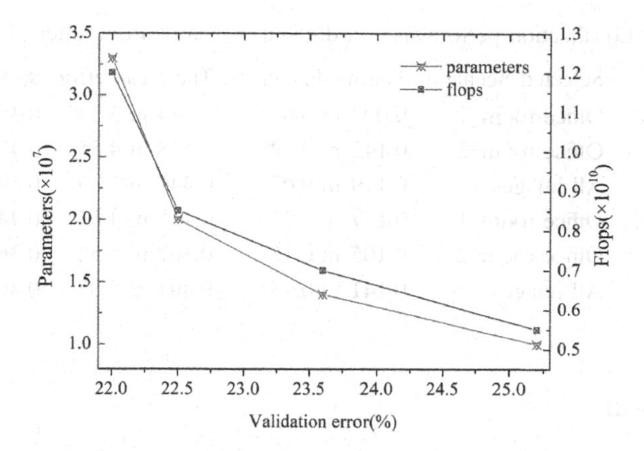

Fig. 7. Densenet network performance.

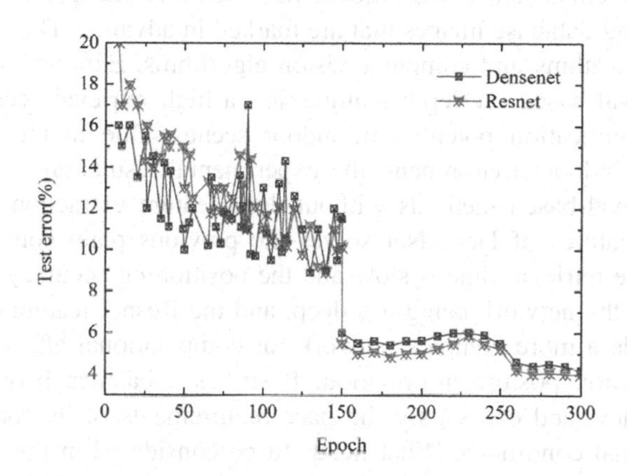

Fig. 8. Comparison of dense net and resnet test errors.

depth feature vectors contained in these images are also recorded and stored. In the online phase, the query images in the test set are input into the pre-trained densenet, and the depth features extracted from the query images can be compared and analyzed with the previous features to obtain the feature vector with the highest similarity, and then output the most similar image. It also contains its location information, etc., and the coordinate information is fed back to the user. So in order to save query time, we need to pre-train the dense net network and collect coordinate and vector information. At the same recognition rate, Densenet's parameter complexity is about half that of ResNet.

For the ICL-NUIM dataset, the mean errors of the direction and translation of the method proposed in this paper are 3.89° and 0.446 m, and the median errors are 0.142° and 0.12 m. For the TUM RGB-D data set, the mean error was 5.07°, 0.401 m, and the median error was 0.18° 0.141 m. For most of the indoor scenes [14] in the two data sets, our method has an average positioning success rate of more than 90% (Table 1).

Table 1. Localization performance in different scenarios from different datasets.

Dataset	Selected Scenario	The median error	The mean error	90% Accuracy
ICL-NUIM	Office room_1	0.097 m 0.03°	0.354 m 3.24°	0.354 m 1.37°
	Office room_2	0.142 m 0.12°	0.538 m 4.55°	0.436 m 0.89°
	All images	0.119 m 0.07°	0.446 m 3.89°	0.395 m 1.13°
TUM RGB-D	Office room_1	0.177 m 0.22°	0.435 m 4.62°	0.445 m 0.93°
	Office room_2	0.105 m 0.15°	0.367 m 5.52°	0.364 m 1.70°
	All images	0.141 m 0.18°	0.401 m 5.07°	0.405 m 1.31°

4 Conclusion

This paper proposes a system of assisting indoor visual positioning by image retrieval based on densely connected convolutional nets. Identify the given query image by retrieving matching database images that are marked in advance. The system combines deep learning algorithms and computer vision algorithms. Experimental results show that image retrieval based on depth features has a high retrieval accuracy and has a wide range of application potential in indoor scenes. Due to the complexity and instability of the indoor environment, the experimental results are significantly more robust than retrieval-based methods without deep feature extraction. Image retrieval based on deep features of DenseNet solves the previous plain convolutional neural network well. The retrieval time is slow and the positioning accuracy is not improved obviously due to the network being too deep, and the Resnet feature extraction is not obvious. We trade a more complex network for computational efficiency, and finally obtain high-precision posture information. It strikes a balance between positioning speed and accuracy, and can satisfy the pace requirements of indoor moving pedestrians under normal conditions. What needs to be considered in the future is a more efficient and robust method to be suitable for complex large-scale indoor scenes.

References

1. Li, F., et al.: A reliable and accurate indoor localization method using phone inertial sensors. In: Proceedings of the 2012 ACM Conference on Ubiquitous Computing, pp. 421–430 (2012)
2. Ali, H.M., Omran, A.H.: Floor identification using smart phone barometer sensor for indoor positioning. Int. J. Eng. Sci. Res. Technol. **4**(2), 384–391 (2015)
3. Carrillo, D., Moreno, V., Beda, B., Skarmeta, A.F.: MagicFinger: 3D magnetic fingerprints for indoor location. Sensors **15**(7), 17168–17194 (2015)
4. Liu, J., Chen, R., Pei, L., Guinness, R., Kuusniemi, H.: A hybrid smartphone indoor positioning solution for mobile LBS. Sensors **2012**(12), 17208–17233 (2012)
5. Khan, S., Hayat, M., Bennamoun, M., Sohel, F., Togneri, R.: A discriminative representation of convolutional features for indoor scene recognition. IEEE Trans. Image Process. **25**, 3372–3383 (2016)
6. Xiao, A., Chen, R., Li, D., Chen, Y., Wu, D.: Indoor positioning system based on static objects in large indoor scenes by using smartphone cameras. Sensors **18**, 2229 (2018)

7. He, K.M., Zhang, X.Y., Ren, S.Q., et al.: Deep residual learning for image recognition. In: IEEE Conference on Computer Vision and Pattern Recognition, pp. 770–778 (2016)
8. Huang, G, Liu, Z, Weinberger, K.Q, et al.: Densely connected convolutional networks. arXiv preprint arXiv:1608.06993 (2016)
9. Gong, Y., Wang, L., Guo, R., Lazebnik, S.: Multi-scale orderless pooling of deep convolutional activation features. In: Proceedings of the European Conference on Computer Vision, 6–12 September 2014, Zurich, Switzerland, pp. 392–407 (2014)
10. Tolias, G., Avrithis, Y.: Image search with selective match kernels: aggregation across single and multiple images. Int. J. Comput. Vis. **116**, 262 (2016)
11. Sattler, T., Leibe, B., Kobbelt, L.: Efficient & effective prioritized matching for large-scale image-based localization. IEEE Trans. Pattern Anal. Mach. Intell. **2017**(39), 1744–1756 (2017)
12. Zakaria, L., Iaroslav, M., Surya, K., Juho, K. Camera relocalization by computing pairwise relative poses using convolutional neural network. In: Proceedings of the IEEE International Conference on Computer Vision Workshop, Venice, Italy, October, pp. 22–29, 920–929 (2017)
13. Gordo, A., Almazán, J., Revaud, J., Larlus, D.: Deep image retrieval: learning global representations for image search. In: Proceedings of the European Conference on Computer Vision, Amsterdam, pp. 241–257. The Netherlands, 11–14 October (2016)
14. Liang, J.Z., Corso, N., Turner, E., Zakhor, A.: Image based localization in indoor environments. In: proceedings of the Fourth International Conference on Computing for Geospatial Research and Application, San Jose, CA, USA, 22–24 July, pp. 71–75 (2013)

Coin Recognition Based on Physical Detection and Template Matching

Jie Wang[1], Long Bai[1], Jiayi Yang[1], and Mingfeng Lu[2](✉)

[1] School of Optics and Photonics,
Beijing Institute of Technology, Beijing 100081, China
[2] Beijing Key Laboratory of Fractional Signals and Systems,
School of Information and Electronics,
Beijing Institute of Technology, Beijing 100081, China
lumingfeng@bit.edu.cn

Abstract. At present, coin circulation automation technology has been widely used in many aspects, so it is necessary to install coin recognition and detection devices in related equipment to prevent coin confusion. However, many current coin detection methods have high requirements for hardware,which increases costs and makes it difficult to install and use equipment in narrow spaces. In this paper, we propose a coin recognition method with low hardware requirements and high accuracy. The design is to take a picture of the coin, detect the image and match the template to distinguish three different coins in the fourth edition of RMB. Compared with other detection methods using the eddy current method, it is lighter and easier to assemble, and can be easily embedded in narrow places.

Keywords: Coin recognition · Image processing · Template matching

1 Introduction

Facing the increasing coin circulation, the market needs a mature and reliable coin identification device. Therefore, the efficient detection of coins has become a very meaningful work, and a large number of coin detection methods have been proposed, some coin sorting devices also appeared.

The coin sorting machine [1] is a system that recognizes and counts coins passing at high speed, and at the same time eliminates counterfeit coins and residual coins. It is the basis of many coin processing tools such as sorting machines, counting machines, packaging machines, and destruction machines. Due to different national conditions and currency systems, it is unrealistic to develop a unified coin sorting system for all countries. Therefore, in this field, researchers have done a lot of work [1, 3, 16–23]. The developed products are roughly divided into three grades, low-end, mid-range and high-end. The low-end sorting speed [2] is below 1000 pieces/min, the mid-range is about 1000–1500 pieces/min, and the high-end is above 1500 pieces/min. There are

J. Wang and L. Bai—Contributed equally to the paper.

S. Shi et al. (Eds.): AICON 2020, LNICST 356, pp. 80–88, 2021.
https://doi.org/10.1007/978-3-030-69066-3_8

two main categories of sorting methods used, one is sorting based on physical technology, and the other is sorting based on performance indicators.

In the fourth set of coins currently used in China, the materials of one-yuan, five-dime and one-dime are nickel-plated steel core, copper-zinc alloy, and aluminum-magnesium alloy. They are special alloys specially used for coinage. There are three thresholds for the machine to distinguish coins: size, material, and weight. The inside of the machine generally has a limiting device composed of a high-frequency oscillation circuit [3], which can directly screen coins. The principle of screening is that when a coin enters the magnetic field generated by the high-frequency oscillation circuit through the coin slot, the difference between the metal material and the size of the coin affects the inductance, which leads to the change of the oscillation frequency, and then the detected frequency change, Compare with the set value, after confirming a certain coin type, the frequency signal is changed into a voltage signal and outputted to complete the identification of the metal coin. The weight recognition is carried out by the speed of the coin sliding to a certain set position in the machine.

At present, the commonly used coin denomination recognition methods include image methods and eddy current sensor methods. Modi S et al. [22] used Hough transform, pulsed averaging technology to extract the features of the coin image, and then sent it as input to a trained neural network for judgment. Gupta V et al. [23] established a coin recognition system based on image subtraction technology. By subtracting the input image and the database image, the value obtained was compared with the threshold value to complete the coin recognition. The School of Engineering of Soochow University [16] adopted a compensation measure to connect two identical eddy current sensors in series and install them on both sides of the coin rail to improve the detection accuracy. The University of Electronic Science and Technology of China [21] designed a dual-channel eddy current sensor composed of high-frequency reflection and low-frequency transmission to detect different parameters of coins, and realize high-speed and reliable detection of coins through FPGA. Shanghai Jiao Tong University [17], Beijing Architecture University [18], Hangzhou Institute of Electronic Technology [20], Tianjin University [19] and many other institutions have done in-depth research on how to correctly identify coins, and the eddy current method is generally used in the mechanism.

In this paper, we propose a coin recognition method based on digital image processing and machine learning technology, which can quickly and accurately verify coins, having broad application prospects.

In our work, we find that we can take advantage of the characteristic that five-dime coins are yellow to detect the number of five-dime coins in advance. In this way, the follow-up work will be greatly simplified, because then we only need to complete the two-classification problem of one-yuan and five-dime coins.

2 Approach

This device is mainly composed of two basic structures: imaging system and coin sorting system (image processing system). The imaging system is responsible for imaging the front and back sides of the coin, and the coin classification system is responsible for classifying various coins judged after imaging.

2.1 Equipment Selection in the Imaging System

In the imaging system, we need an industrial camera to shoot the target, and considering the lack of lighting, we need a light source for supplementary lighting.

For industrial cameras [5], we need to consider factors such as its resolution, frame rate, focal length, working distance, magnification and field of view. After the resolution and frame rate are determined, we need to combine specific application scenarios to determine the magnification and other factors, so we mainly consider the resolution and frame rate here. The first is the resolution. For the range of 30 mm diameter, the theoretical resolution calculation formula is:

Resolution (theoretical) = (high field of view/precision) * (width of field of view/precision) = (30/0.04) * (30/0.04) = 750 * 750

Taking into account the distortion of the camera's edge field of view and the stability requirements of the system, generally one-pixel unit is not used to correspond to a measurement accuracy value. Generally, a multiple of 2 or higher is selected, so that the single-directional resolution of the camera is 1500, so the camera resolution 1500 * 1500 = 2.25 million, so the camera pixel needs to be greater than 2.25 million. The second is the frame rate. The frame rate of a conventional industrial camera is basically greater than 30FPS, which means that it can take 1,800 photos per minute, which basically meets the needs.

In light source selection, we choose the LED light source [4]. It is the current mainstream machine vision light source, because it is powered by DC, has no stroboscopic, and has a very long life and high brightness. It can also be flexibly designed as a line light source with different structures, such as direct light, with condenser lens, backlight, coaxial and a diffuse reflection line light source similar to a bowl.

2.2 Coin Classification System (Image Processing System)

The flow of the whole system is shown in Fig. 1. For one-yuan and one-dime coins, the size of a one-yuan coin in the image is very different from that of a one-dime coin. The circle radius can be detected and compared in the process of coin recognition. The five-dime coin is very different from the one-yuan and one-dime coins in color. When detecting coins, the yellow component can be extracted and threshold segmentation can be performed to extract the number of five-dime coins.

First, we extract HSV color components [6] from the image. In the RGB model, the extraction of single-channel components is far from the effect that we need to extract the yellow component for threshold segmentation, so we use other color models to try. After many experiments, the extraction of the color component of the second channel in the HSV model is of high definition and can meet our requirements, so the HSV component is selected.

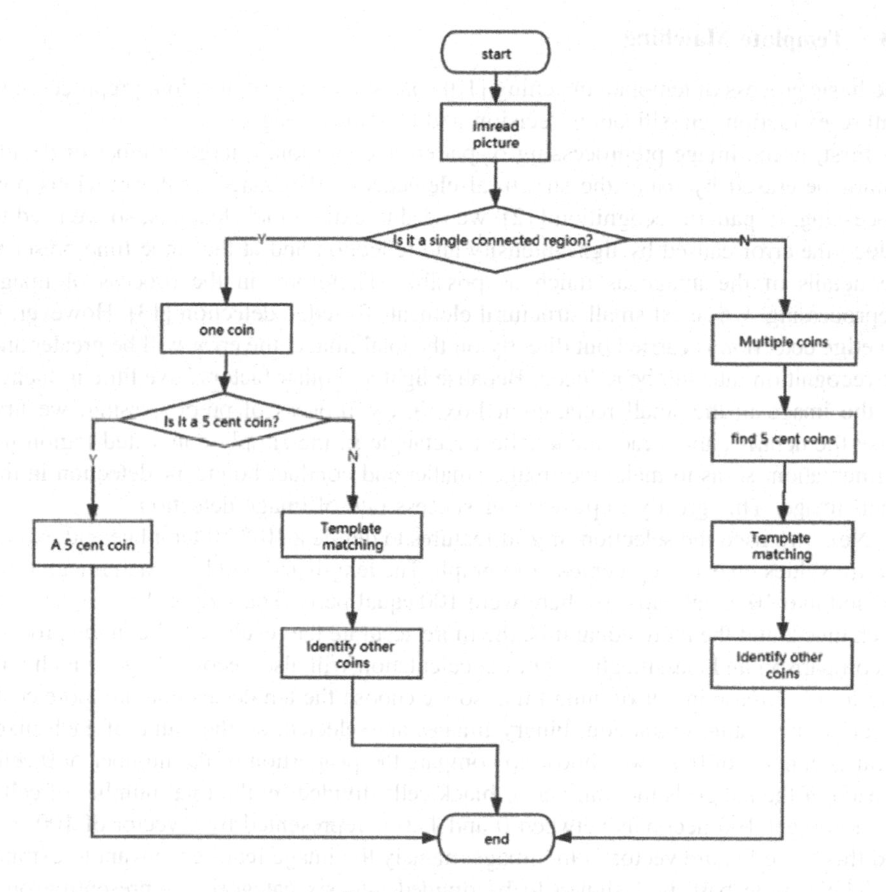

Fig. 1. Coin identification flow chart

Then, we perform image binarization on the image from which the yellow component has been extracted, and use image opening and closing operations, edge detection [8, 9, 13], inverted color changes to fill holes, and median filtering [7] in the preprocessing to make the image clear, Calculate the number of penny coins in the image to be tested by determining the number of connected regions.

Then, we change the gray scale and binarize the image again, and preprocess the image, use the above method to make the image clear and no noise, and then calculate the number of connected regions, that is, the total number of coins, and use the sub-function to find The center and radius of the function, using the value of the radius to extract the number of one yuan and the last of a corner, get the number of one corner based on the number of one yuan and five corners and the total number.

At the same time, if there are only multiple or one one-yuan or one-dime coin in the image, it cannot be recognized by the above method, so we select one of the connected regions of the image for template matching, and use Bayesian classifier to classify.

2.3 Template Matching

The basic process of template matching [10] consists of data acquisition, preprocessing, feature extraction, classification decision and classifier design.

First, in the image preprocessing of pattern recognition, a large number of details cannot be erased by using the structural elements of the image as the previous preprocessing. In pattern recognition [12], we need to extract their features, so we need to reduce the error caused by light intensity and reflection and at the same time preserve the details of the image as much as possible. Therefore, in the process of image preprocessing, we select small structural elements for edge detection [13]. However, if the edge detection is carried out directly on the total image, the error will be greater and the recognition rate will be reduced. Because light and other factors have little influence on the image in the small rectangular box, in the process of preprocessing, we first erase the details and extract the smallest rectangle in the simply connected region for segmentation, so as to make the image smaller and conduct boundary detection in the small image. This greatly improves our success rate of image detection.

Next, we used the selection of grid features to define a 10 * 10 template and extract feature values on each segmented coin graph. The length and width of each image were divided into 10 equal parts, so there were 100 equal parts. The size of the template can be changed, and the more equal it is, the more accurate the results will be in the process of comparison and classification, but the calculation will also become huge, which will lead to an increase in our running time, so we choose the ten decals that are more cost-effective. In feature extraction, binary images are selected, so the value of each pixel point is either 1 or 0, so we choose to compare the proportion of the number of 0 cells in each of the subgrids the number of black cells divided by the total number of cells. Then you get 100 decimals between 0 and 1. It is represented by a vector of 100 * 1, and this is the feature vector of the image, namely the image feature we want to extract.

The sample base is designed to be divided into six categories, representing one-yuan, five-dime, one-dime coins and their front and back respectively. We designed nine learning samples for each coin, each using a one-yuan, five-dime, one-dime coin and its front and back in different situations. When the features of each sample are extracted, all the sample data is stored in a template file, and each coin USES a structure to store the information.

Finally, we use Bayes formula to design the classifier. Bayes formula [11] is as follows:

$$P(A|B) = P(A)\frac{P(A|B)}{P(B)}$$

The so-called prior probability is the probability distribution based on the existing knowledge and experience, that is, the probability of the occurrence of this event, which is mostly uniform distribution. The posterior probability is the probability obtained after some actual observation statistics.

When the characteristic condition is the state of the object X to be tested, the posterior probability of ω is the largest among the different coin categories, that is, X belongs to the category ω.

Prior probability $P(\omega_i)$ is approximately calculated by the number of samples and the total number of those. N_i is the number of samples of coin i, and N is the total number of samples:

$$P(w_i) \approx \frac{N_i}{N}, i = 0, 1, 2, \ldots, 9$$

First, we compute the $P_j(\omega_i)$, and then we compute the conditional probability $P(X|\omega_i)$.

$$P_j(w_i) = \left(\sum_{k=0, X \in w_i}^{N_i} x_{ij} + 1 \right) / (N_i + 2)$$

Where i = 0, 1, 2 ..., 9, j = 0, 1, 2 ..., 99. The threshold value is set as 0.05. When the value of the 100 eigencomponents is greater than 0.05, the eigenvalue is considered to be 1; otherwise, it is 0. Pj(ωi) is an estimate of the probability that the jth component of X under test is $1(x_j = 1)$ when sample $X(x_0, x_1, x_2, \ldots, x_{99})$ is in the ω_i category.. From this we can calculate:

$$P(x_j = 1 | X \in x_l) = P_j(w_i)$$
$$P(x_j = 0 | X \in x_i) = 1 - P_j(w_i)$$

Where i = 0,1,2..., 9, j = 0,1,2..., 99. It is assumed that X no matter what kind of coins, all of these classes sample characteristic, the first component is the first 1/100 grid probability value is 0, then we can assume that the eigenvalue of X must also be 0, and the probability of the inverse is 1.. Bayes emphasizes that the sample is fixed, the frequency is random, and the posterior probability is affected by the sample. Therefore, the conditional probability of the object X to be tested is:

$$P(X|w_i) = P[X = (x_0, x_1, x_2, \ldots, x_{99}) | X \in w_i] = \prod_{j=0}^{99} P(x_j = a | X \in w_i)$$

Where i = 0,1,2... 9, a = 0 or 1.

Applying Bayes formula to calculate the posterior probability, namely the possibility that X to be tested belongs to i, we can get

$$P(w_i|X) = \frac{P(w_i)P(X|w_i)}{P(w_0)P(X|w_0) + P(w_1)P(X|w_1) + \ldots + P(w_9)P(X|w_9)}$$

Where i = 0,1,2... 9. The magnitude of P(X) is the sum of all the different categories of X in the denominator.

The category that computes the maximum value of the posterior probability is the category to which the coin belongs.

3 Results

Fig. 2. Part of test images [Fengxiaode1778 2020] (The images are 1, 2, 3, 4, 5, 6 from left to right and top to bottom)

We collected some pictures of one-yuan, five-dime and one-dime coins in the fourth edition of RMB from the Internet to test the software part of this method. At the same time, we collected some coins by ourselves and took pictures with the equipment meeting the imaging requirements of this paper. These photos were used to test the software part of this method.

The results show that our method has a good effect on the differentiation of one-yuan, five-dime and one-dime in the fourth edition of RMB. At the same time, when the coin is in a dark environment, LED lighting can also make the image become clear, to meet the needs of coin recognition.

The results can be seen in Fig. 2 and Table 1.

Table 1. Part of test result

	Test1	Test2	Test3	Test4	Test5	Test6
One-yuan coins	1	0	0	1	2	1
Five-dime coins	0	1	1	0	0	1
One-dime coins	0	0	1	1	0	1

4 Conclusion

Prior work concentrated only on physical methods by eddy current sensors or light detectors, having poor identification of counterfeit coins, either a high cost.

In this study, we propose a simple coin recognition method based on digital image processing and machine learning technology, which has a lower cost. At the same time, this method shows its extremely high precision in the pictures we choose or take. If we can have more data sets to train, we will get better results.

However, we still have some problems that need to be improved. First of all, for the detection of overlapping coins, we sometimes cannot realize the correct detection; Secondly, our program takes a long time to run and needs to be improved.

In the following work, we can use a token-based watershed segmentation [14] to split the overlapping coins. At the same time, if there are overlapping coins, we can extract a section of arc separately, and then use this section of arc to fit the circle equation, so as to calculate the number of coins.

Acknowledgement. This paper is sponsored by Beijing Natural Science Foundation (L191004).

References

1. Kavale, A., Shukla, S., Bramhe, P.: Coin counting and sorting machine. In: 2019 9th International Conference on Emerging Trends in Engineering and Technology-Signal and Information Processing (ICETET-SIP-19), pp. 1–4. IEEE (2019)
2. Qing, X., Wang, Z.: Coins, coin materials, coin circulation. Mach. Manuf. Res. **5**, 30–31 (2000)
3. Zhang, N., Cao, J., Wang, S., et al.: Design and calculation of planar eddy current coil in coin identification. IEEE Trans. Appl. Superconduct. **26**(7), 1–4 (2016)
4. Shailesh, K.R., Kini, S.G., Kurian, C.P.: Summary of LED down light testing and its implications. In: 2016 10th International Conference on Intelligent Systems and Control (ISCO), pp. 1–5. IEEE (2016)
5. Liao, W., Tai, X., Li, G., et al.: An industrial camera for color image processing with mixed lighting conditions. In: 2010 The 2nd International Conference on Computer and Automation Engineering (ICCAE), vol. 5, pp. 381–385. IEEE (2010)
6. Su, C.H., Chiu, H.S., Hsieh, T.M.: An efficient image retrieval based on HSV color space. In: 2011 International Conference on Electrical and Control Engineering, pp. 5746–5749. IEEE (2011)
7. George, G., Oommen, R.M., Shelly, S., et al.: A survey on various median filtering techniques for removal of impulse noise from digital image. In: 2018 Conference on Emerging Devices and Smart Systems (ICEDSS), pp. 235–238. IEEE (2018)
8. Perumal, E., Arulandhu, P.: Multilevel morphological fuzzy edge detection for color images (MMFED). In: 2017 International Conference on Electrical, Electronics, Communication, Computer, and Optimization Techniques (ICEECCOT), pp. 269–273. IEEE (2017)
9. Yuan, L., Xu, X.: Adaptive image edge detection algorithm based on canny operator. In: 2015 4th International Conference on Advanced Information Technology and Sensor Application (AITS), pp. 28–31. IEEE (2015)

10. Brunelli, R.: Template Matching Techniques in Computer Vision: Theory And Practice. Wiley, Hoboken (2009)
11. Kass, R.E., Raftery, A.E.: Bayes factors. J. Am. Stat. Assoc. **90**(430), 773–795 (1995)
12. Tou, J.T., Gonzalez, R.C.: Pattern Recognition Principles (1974)
13. Lee, J., Haralick, R., Shapiro, L.: Morphologic edge detection. IEEE J. Robot. Autom. **3**(2), 142–156 (1987)
14. Gonzalez, R.C., Woods, R.E., Eddins, S.L.: Digital Image Processing Using MATLAB. Pearson Education India (2004)
15. Wang, Y.: Coin Recognition and Detection Device. Central South University (2009)
16. Xiao, S.: Application of eddy current sensor in RMB coin recognition system. J. Soochow Univ. (Nat. Sci. Edn.) **12**(1), 60–63 (1996)
17. Zhu, X.: Design and Realization of High Accuracy Coin Identification Device. Shanghai Jiao Tong University (2008)
18. Wang, M., He, R., Ma, C., Wang, H.: Research on coin classification algorithm based on image processing. Image Signal Process. **7**(4), 227–235 (2018). https://doi.org/10.12677/JISP.2018.74026
19. Fu, L.: Coin Recognition System Based on Machine Learning. Tianjin University (2009)
20. Bai, J., Xu, Z.: A practical microcomputer coin identification device. Electron. Technol. **08** (1997)
21. Mo, L.: Research on FPGA-Based Coin Recognizer. University of Electronic Science and Technology (2009)
22. Modi, S., Bawa, D.: Automated coin recognition system using ANN. arXiv preprint arXiv:1312.6615 (2013)
23. Gupta, V., Puri, R., Verma, M.: Prompt Indian coin recognition with rotation invariance using image subtraction technique. In: 2011 International Conference on Devices and Communications (ICDeCom), pp. 1–5. IEEE (2011)
24. Fengxiaode1778. Image-based coin detection and counting (2020). https://download.csdn.net/download/ajianlee/12257137?utm_medium=distribute.pc_relevant_download.none

Generative Adversarial Network for Generating Time-Frequency Images

Weigang Zhu[1,2(✉)], Kun Li[2], Wei Qu[2], Bakun Zhu[2], and Hongyu Zhao[1]

[1] State Key Laboratory of Complex Electromagnetic Environment Effects on Electronics and Information System, Luoyang 471003, China
wg_zhu@outlook.com

[2] Space Engineering University, Beijing 101416, China
yi_yun_hou@163.com

Abstract. To deal with the problem of de-noising and enhancement of radar signal time-frequency images, a method of secondary generating time-frequency images by generative adversarial network is proposed. Firstly, time-frequency analysis is used to generate the time-frequency image of the radar signal as the original data set 1. Then, after learning the data set 1 by using the generative adversarial network, a new data set 2 is generated, and the data set 2 has de-noising and enhancement effects relative to data set 1. Finally, the validity of the data set 2 generated by the time-frequency image singular value feature is checked. Experiments on the time-frequency images of five common radar signals are carried out. The results show that the method is effective in time-frequency image de-noising and increasing sample diversity.

Keywords: Radar emitter identification · Time-frequency image · GAN · Sample diversity · SVD

1 Introduction

Radar emitter identification refers to the process of analyzing radar system, usage and working status from the radar signal characteristic parameters obtained by processing, and is the key link of radar reconnaissance. The traditional radar emitter identification extracts the inter-pulse five parameters of the radar signal: pulse width (PW), carrier frequency (CF), pulse amplitude (PA), time of arrival (TOA), angle of arrival (AOA) forms the identified feature vector. However, with the increasing complexity of the electromagnetic environment and the continuous development of radar technology, various new radars with complex systems have emerged, such as multi-function phased array radars with functions of search, tracking, guidance, etc. As a result, traditional methods relying on inter-pulse parameter identification fail [1], and more and more researchers are turning their attention to the extraction of intra-pulse features of radar emitter signals. As a carrier of intra-pulse features, time-frequency images have more abundant radar information and robustness than inter-pulse features [2].

S. Shi et al. (Eds.): AICON 2020, LNICST 356, pp. 89–98, 2021.
https://doi.org/10.1007/978-3-030-69066-3_9

Preprocessing the time-frequency image can reduce the interference and redundant information and enhance the signal characteristics, which is of great significance for improving the recognition rate of the back-end classifier. The traditional method of preprocessing the time-frequency image is to use image pre-processing methods such as grayscale, standardization, binarization, cropping [3] to achieve the effect of denoising and weakening the cross-term interference of time-frequency distribution, and the processing process is cumbersome. At the same time, due to the extremely short interception time of the radar signal on the battlefield, it will result in: (1) the number of reconnaissance samples is limited enough to support the training of the radar emitter identification model; (2) the insufficient diversity of reconnaissance samples leads to the radar emitter identification model. Poor ability or training is easy to over-fit. Some commonly used digital image data enhancement methods, such as rotation, translation, etc. will destroy the time-frequency relationship of radar radiation source signals. Obviously, these methods are poorly processed for time-frequency images. In recent years, the generative adversarial network (GAN) has been successfully used in the image field because it can learn the true distribution of samples [4], eliminate interference, generate diverse samples. Therefore, this paper proposes a time-frequency image generation method based on generating an anti-network, which is effective in denoising time-frequency images [5], reducing cross-term interference, and increasing sample diversity.

2 Time-Frequency Analysis Generates Time-Frequency Images

Time-frequency analysis is a powerful tool for dealing with non-stationary signals such as radar signals. It is essentially a combination of a two-dimensional joint function of time and frequency. For the radar signal x(t), the mathematical expression of the bord-jondan time-frequency distribution is as follows:

$$C_x(t,f) = \iint \varphi(t - t', \tau)x^*\left(t' - \frac{\tau}{2}\right)x\left(t' + \frac{\tau}{2}\right)e^{-j2\pi f\tau}dt'd\tau \tag{1}$$

$$\varphi(t,\tau) = \begin{cases} \frac{1}{|\tau|}, & \left|\frac{t}{\tau}\right| < \frac{1}{2} \\ 0, & \left|\frac{t}{\tau}\right| > \frac{1}{2} \end{cases} \tag{2}$$

Among them, $\varphi(t, \tau)$ is a kernel function [6]. Because it has better time-frequency aggregation while suppressing cross-term interference, this paper chooses bord-jondan distribution as the time-frequency analysis method of radar signal for five classics. The radar signal is simulated. As shown in Fig. 1, it can be seen that under the condition of

−5 dB SNR, the time-frequency diagrams of the five radar signals have certain identifiable time-frequency structural characteristics, but the signal-to-noise ratio is too low. And the bord-jondan distribution does not suppress the cross term incompletely, and still causes abnormal disturbances such as burrs, edge characteristics, and structural fractures in the time-frequency diagram of the radar signal, especially when the time-frequency structure characteristics of the radar signal are prominent. The influence of interference will also increase. Conversely, weakening the influence of interference will also obscure the time-frequency structure characteristics of the signal. When the signal-to-noise ratio is reduced to −10 dB (Fig. 2), the signal time-frequency structure has versely, weakening the influence of interference will also obscure the time-frequency structure characteristics of the signal. When the signal-to-noise ratio is reduced to -10 dB (Fig. 2), the signal time-frequency structure has been completely destroyed by noise and cannot be recognized by the naked eye.

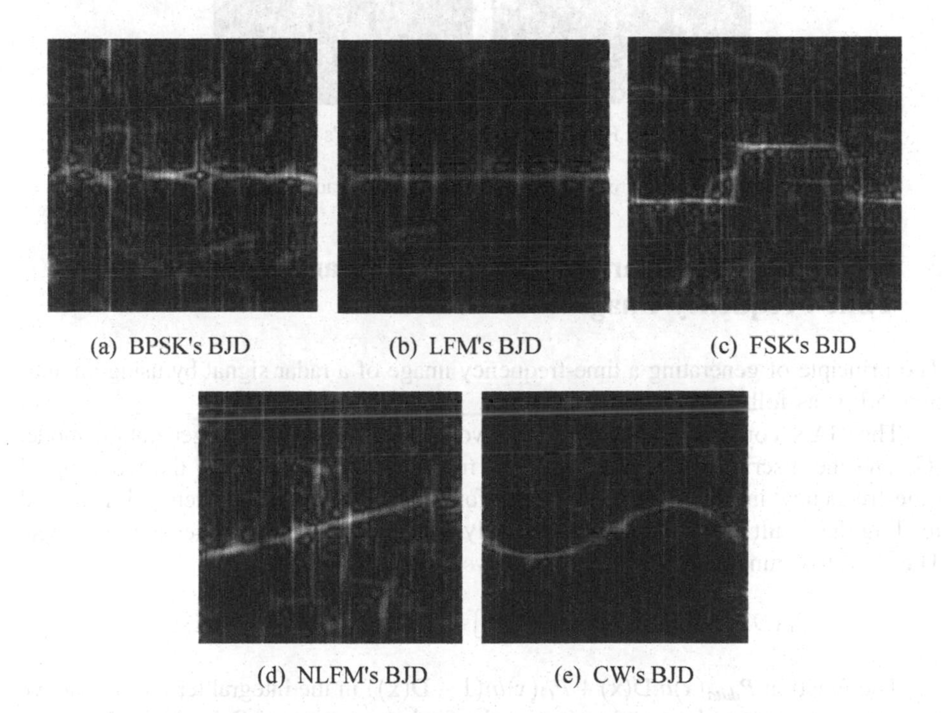

Fig. 1. Bord-Jondan time-frequency distribution of five radar signals (SNR = −5 dB)

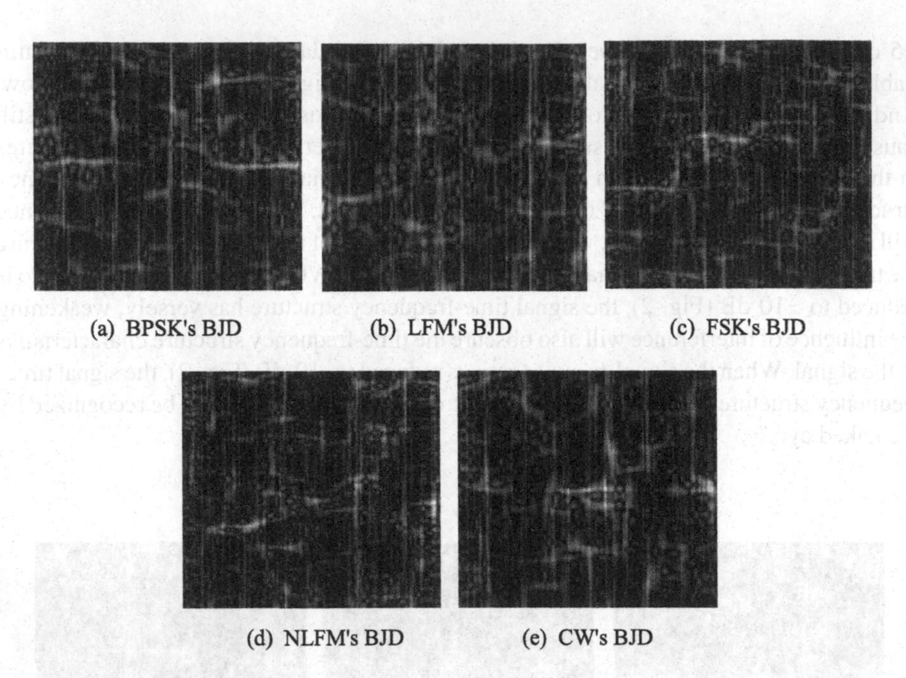

Fig. 2. Bord-Jondan time-frequency distribution of five radar signals (SNR = −5 dB)

3 Generative Adversarial Network Secondary Generation Time-Frequency Image

The principle of generating a time-frequency image of a radar signal by using an anti-network is as follows:

The GAN consists of two neural networks [7], which are the generation model (G) and the discriminant model (D). G is responsible for generating the radar signal time-frequency image. D is responsible for judging the time-frequency image and feeding the result back to G to continuously improve the quality of generating image. The objective function is defined as follows:

$$V(D, G) = E_{X \sim P_{data}(x)}[ln\mathrm{D}(\mathrm{x})] + E_{Z \sim P_Z(Z)}[ln(1 - \mathrm{D}(\mathrm{G}(\mathrm{z})))] \tag{3}$$

The function $P_{data}(x)ln\mathrm{D}(\mathrm{x}) + P_G(x)ln(1 - \mathrm{D}(\mathrm{x}))$ in the integral term of the above formula is derived to be equal to 0, and the final expression of D is obtained as:

$$\begin{aligned} V(D, G) &= E_{X \sim P_{data}(x)}[ln\mathrm{D}(\mathrm{x})] + E_{Z \sim P_Z(Z)}[ln(1 - \mathrm{D}(\mathrm{G}(\mathrm{z})))] \\ &= E_{X \sim P_{data}(x)}[ln\mathrm{D}(\mathrm{x})] + E_{X \sim P_G(x)}[ln(1 - \mathrm{D}(\mathrm{x}))] \\ &= \int [P_{data}(x)ln\mathrm{D}(\mathrm{x}) + P_G(x)ln(1 - \mathrm{D}(\mathrm{x}))]dx \end{aligned} \tag{4}$$

The function $P_{data}(x)ln\mathrm{D}(\mathrm{x}) + P_G(x)ln(1 - \mathrm{D}(\mathrm{x}))$ in the integral term of the above formula is derived to be equal to 0, and the final expression of D is obtained as:

$$D = \frac{P_{data}(x)}{P_{data}(x) + P_G(x)} \tag{5}$$

Next, substitute the obtained D into V(D, G) and continue to find the expression of G with the smallest V(D, G):

$$V(D, G) = \int P_{data}(x) ln \frac{P_{data}(x)}{P_{data}(x) + P_G(x)} dx + \int P_G(x) ln \frac{P_G(x)}{P_G(x) + P_{data}(x)} dx \tag{6}$$

Compare the Jensen-Shannon divergence formula:

$$JSD(P \parallel Q) = \int p(x) ln \frac{p(x)}{p(x) + q(x)} dx + \int q(x) ln \frac{q(x)}{q(x) + p(x)} dx \tag{7}$$

The expression of V(D, G) is written as:

$$V(D, G) = -ln(4) + 2 \times JSD(P_{data}(x) \parallel P_G(x)) \tag{8}$$

Since the JS divergence is non-negative, in order to minimize V(D, G), when $P_{data}(x) = P_G(x)$, the minimum is obtained, that is, the distribution of the time-frequency image of the final G-generated $P_G(x)$ and the real time-frequency The distribution of images $P_{data}(x)$ is consistent.

In the actual situation, the distribution of the radar signal $P_{data}(x)$ is unknown, and only the limited time-frequency image X of the sample is obtained, and the noise distribution $P_Z(Z)$ is known, so that the existing limited time-frequency image can be passed. The training of X on GAN causes G to map the known noise Z to the unknown X, expand the degree of freedom of generating time-frequency images, increase the diversity of the generated time-frequency images and suppress the image noise that does not belong to the $P_{data}(x)$ distribution (unknown The generation of noise). The original GAN has the defects of unstable training and slow convergence [8, 9]. The Deep Convolutional Generative Adversarial Network (DCGAN) [10] has been improved on GAN and has a significant effect on high quality image generation.

4 Validity Test Using Singular Values

Singular Value Decomposition (SVD) is an extension of the feature decomposition of the matrix matrix and is widely used in the field of image and signal processing [11]. Since the singular value can well characterize the image features, this paper extracts the singular value feature of the time-frequency image as the criterion for verifying the validity of the time-frequency image generated by DCGAN. The SVD is defined as follows:

The matrix A $\in R^{m \times n}$, if there is an orthogonal matrix U $= [u_1, u_2, \ldots, u_m] \in R^{m \times m}$, V $= [v_1, v_2, \ldots, v_n] \in R^{n \times n}$, let A $= U\Sigma V^T$, $\Sigma = \text{diag}[\sigma_1, \sigma_2, \ldots, \sigma_p]$, p $= \min(m, n)$, $\sigma_1 \geq \sigma_2 \geq \ldots \geq \sigma_p \geq 0$, then say A $= U\Sigma V^T$ is the singular value decomposition of A, $\sigma_1, \sigma_2, \ldots, \sigma_p$ is the singular value of A, and is the square root of AA^T or $A^T A$ eigenvalues $\lambda_1, \lambda_2, \ldots, \lambda_p$, That is, $\sigma_i = \sqrt{\lambda_i}$.

The change of singular value can represent the change of time-frequency image. The diversity of singular value can represent the diversity of samples. The change of position of time-frequency image cannot increase the diversity of time-frequency image samples. Only by using DCGAN to learn the true distribution of samples, it is possible to generate more time-frequency images with different characteristics. At the same time, the singular value can reflect the composition of the time-frequency image energy. The closer to the front, the singular value reflects the energy of the signal [12, 13], and the closer the singular value reflects the energy of the noise. Therefore, the distribution of singular values can be used to judge the denoising ability of DCGAN for time-frequency images.

5 Experiment and Analysis

In order to verify the validity of the method for generating time-frequency images, five kinds of radar signals are used in the experiment: binary phase shift keying (BPSK), linear frequency modulation (LFM), frequency shift keying (FSK), nonlinear frequency modulation (NLFM) and conventional radar (CW). Among them, BPSK uses 7-bit Barker code, LFM bandwidth is 17 MHz, pulse width is 10 μs, NLFM uses sine wave frequency modulation, CW carrier frequency is 25 MHz, sampling frequency is 100 MHz, and Gaussian white noise is added. A time-frequency analysis method is generated every 1 dB between the signal-to-noise ratio of −10 dB to −5 dB, and a total of 600 time-frequency images are used as the training set of DCGAN.

Simulation 1: Time-frequency image generation experiment based on DCGAN.

The experimental software environment: operating system Win10 64-bit, based on the open source deep learning framework Tensorflow1.5-GPU build model, VS2015 + CUDA8 + CUDNN7 provides support for GPU computing and improves the speed of graphics parallel computing. Hardware environment: CPU: Intel i7-7700K @ 4.20 GHz, GPU: GTX TITAN X. Training parameters: Mini-Batch is trained by 64 random gradient descent algorithm, and the Adam optimizer performs hyper parameter tuning with a learning rate of 0.0002, momentum β_1 of 0.5, and Epoch of 200. The experimental results are shown in Fig. 3 (only one generated sample is listed for each radar signal).

It can be seen from Fig. 3 that compared with the original image (Fig. 1), the generated time-frequency image has better time-frequency aggregation, and the phenomena such as burr, edge characteristic blur and structural fracture are all relieved to some extent, especially for the lower signal-to-noise ratio (Fig. 2). This effect is more obvious, and the ability to reconstruct the characteristics of the radar signal is stronger. However, DCGAN also has a common problem in the GAN model, namely the collapse mode. After a certain training phase, the generated model loses the ability to learn new features of the sample, and the discriminant model will only judge with the same standard, the repeated samples are continuously generated, and the sample features learned are incomplete. As shown in Fig. 4, DCGAN believes that the importance of the background is greater than the importance of the sample, so the background contour is more prominent, and the time-frequency structural features of the sample are diluted, losing the ability to generate new samples.

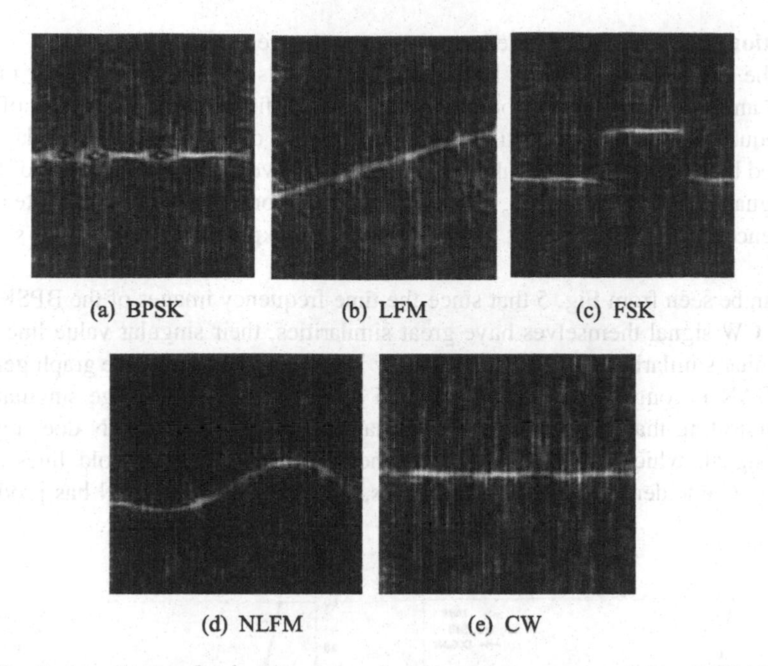

Fig. 3. Five kinds of radar signal time-frequency images generated by DCGAN

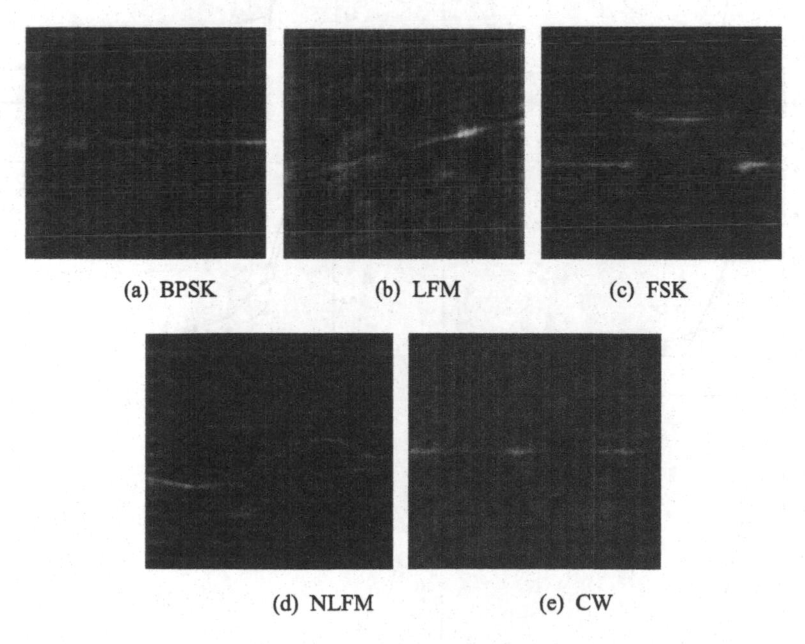

Fig. 4. Collapse mode

Simulation 2: Validity test based on singular value feature

In the experiment, the five kinds of radar signals with signal-to-noise ratios of −10 dB and −5 dB are taken from the average of the first 10 singular values of the 10 time-frequency images, and the time-frequency images of the five kinds of radar signals generated by DCGAN are also taken. Comparing the average singular values of the five radar signals for −10 dB, −5 dB, and DCGAN conditions, which can eliminate random interference and make the result more credible. The experimental results are shown in Fig. 5.

It can be seen from Fig. 5 that since the time-frequency images of the BPSK signal and the CW signal themselves have great similarities, their singular value line graphs have similar similarities. The time-frequency image singular value line graph generated by DCGAN is roughly the same as the −5 dB time-frequency image singular value trend, indicating that the time-frequency image generated by DCGAN does represent such a signal, which is effective, but at the same time, the two fold lines are not complete. Coincidence, there are differences, indicating that DCGAN has produced a

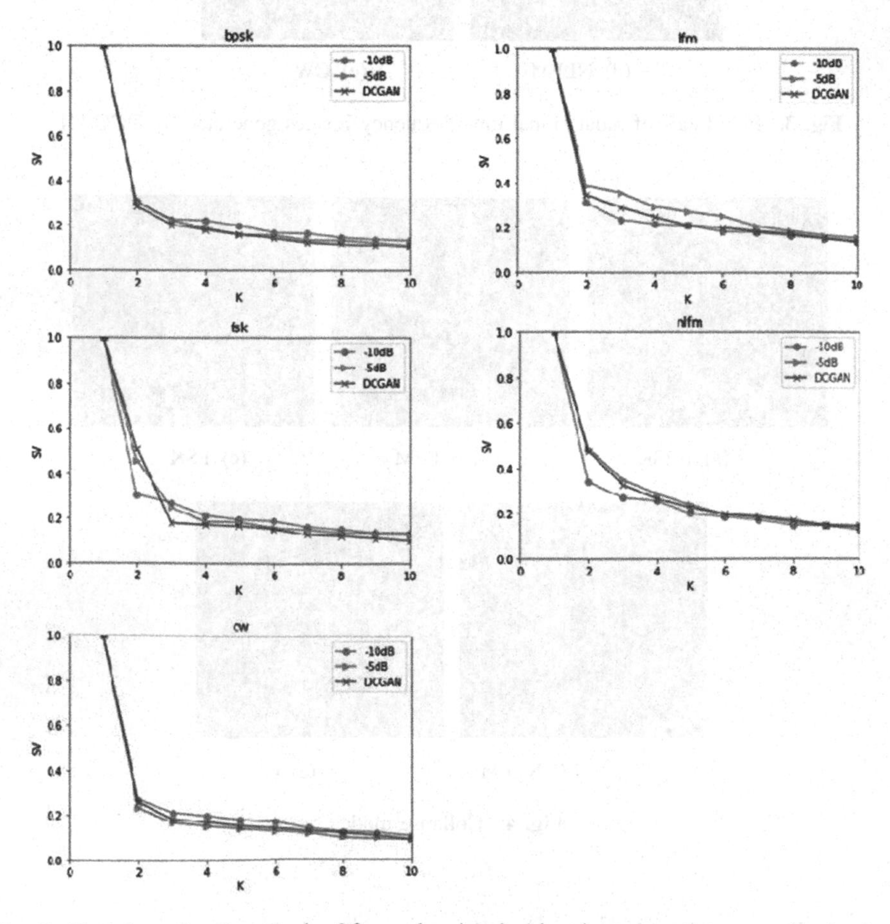

Fig. 5. Singular value line graph of five radar signals (singular value takes normalized value)

new sample after learning the true distribution of the original sample, expanding the diversity of training samples. Among the singular values, the large value reflects the signal energy, and the small value reflects the noise energy. Therefore, the suppression effect of DCGAN on the noise is reflected in the singular value line graph. The DCGAN of the first half of the X-axis is slightly closer. Above, the DCGAN's fold line in the second half of the X-axis is slightly below, that is, the signal has more energy and less noise.

6 Conclusion

Based on the in-depth analysis of the radar signal time-frequency image and the generative adversarial network, this paper proposes a method for denoising and enhancing the radar signal time-frequency image by generating the time-frequency image with generative adversarial network and using the image singular value feature to verify the validity. Different from traditional time-frequency image de-noising and enhancement methods, the generative adversarial network can learn the real distribution of samples and generate new effective samples, expand the freedom of generating samples, increase the diversity of original samples. Simulation experiments prove the effectiveness of the proposed method, but how to solve the problem of collapse mode, finding more robust time-frequency image feature representation methods and using the time-frequency image generated by this method for radar emitter signal recognition experiments One step that needs to be studied.

Acknowledgment. This work was funded by No. 2020Z0203B of the State Key Laboratory of Complex Electromagnetic Environment Effects on Electronics and Information System.

References

1. Chen, W., Tao, J.: Summary of new system radar and their countermeasure techniques. Shipboard Electron. Countermeasur. **33**(4), 9–14 (2010)
2. Djurovic, I.: QML-RANSAC instantaneous frequency estimator for overlapping multicomponent signals in the time-frequency plane. IEEE Signal Process. Lett. **25**(3), 447–451 (2018)
3. Bai, H., Zhao, Y., Hu, D.: Radar emitter identification based on image feature of Choi-Williams time-frequency distribution. J. Data Acquist. Process. **27**(4), 480–485 (2012)
4. Zhang, Y.: A survey on generative adversarial networks. Electron. Des. Eng. **26**(5), 34–38 (2018)
5. Moghadasian, S.S., Fatemi-Behbahani, E.: A structure for representation of localized time-frequency distributions in presence of noise. Signal Process. **148**, 9–19 (2018)
6. Czarnecki, K., Fourer, D., Auger, F.: A fast time-frequency multi-window analysis using a tuning directional kernel. Signal Process. **147**, 110–119 (2018)
7. Goodfellow, I.J.: Generative adversarial nets. arXiv preprint arXiv:1406.2661 (2014)
8. Salimans, T., Goodfellow, I.: Improved Techniques for Training GANs. arXiv preprint arXiv:1606.03498 (2016)

9. Arjovsky, M.: Towards Principled Methods for Training Generative Adversarial Network. arXiv preprint arXiv:1701.04862 (2017)
10. Radford, A., Metz, L.: Unsupervised representative learning with Deep Convolutional Generative Adversarial Network. arXiv preprint arXiv:1511.06434 (2016)
11. Guo, Q., Nan, P., Zhang, X., Zhao, Y., Wan, J.: Recognition of radar emitter signals based on SVD and AF main ridge slice. J. Commun. Netw. **17**(5), 491–498 (2015)
12. Miao, J., Cheng, G., Cai, Y.: Approximate joint singular value decomposition algorithm based on givens-like rotation. IEEE Signal Process. Lett. **25**(5), 620–624 (2018)
13. Deb, T., Anjan, K., Mukherjee, A.: Singular value decomposition applied to associative memory of hopfield neural network. Mater. Today Process. **5**(1), 2222–2228 (2018)

Research on Weak Signal Detection Method Based on Duffing Oscillator in Narrowband Noise

Qiuyue Li[1(✉)] and Shuo Shi[2]

[1] China Agricultural University, Beijing, China
lqyue@cau.edu.cn
[2] Harbin Institute of Technology, Harbin, China
crcss@hit.edu.cn

Abstract. One of the most important issues in communication is how to effectively detect signals. Being able to correctly detect the required signal is the basis for the partner to correctly implement the signal reception, and is also the basis for non-cooperative parties to implement information countermeasures and signal interference. The nonlinear signal detection method makes full use of the characteristics of the nonlinear system, and can detect the low SNR signal by converting the change of the signal into the change of the system state. Chaos theory is one of the nonlinear signal detection algorithms, and Duffing oscillator is the most typical. In this paper, the basic theory of Duffing oscillator is studied firstly, and the weak signal detection method based on Duffing oscillator is analyzed. In order to achieve signal frequency detection under narrowband noise conditions, narrowband noise is introduced into the Duffing oscillator to create a new Duffing oscillator model. Then analyzes the model by Melnikov equation, and the state form of the Duffing oscillator different from the traditional theory is obtained, that is, the probability period state of the oscillator under narrowband noise. A new method for weak signal detection using Duffing oscillator under narrow-band noise conditions is proposed. The state of the oscillator is judged by the period state time ratio (PSTR) method. Subsequently, using MATLAB to establish a weak signal detection platform based on PSTR method, the feasibility of detecting weak signals based on PSTR method under narrow-band noise conditions is verified.

Keywords: Signal detection · Duffing oscillator · Narrowband noise

1 Introduction

Signal detection is an important part of signal processing. With the changes in signal detection requirements and the advancement of signal detection theory and technology, researchers have proposed different signal detection algorithms. The increasing update of detection methods also makes the detection performance of the algorithm continuously improved and the scope of application is widened. With the proposed new signal

S. Shi et al. (Eds.): AICON 2020, LNICST 356, pp. 99–111, 2021.
https://doi.org/10.1007/978-3-030-69066-3_10

detection method, a lot of signal detection problem is constantly overcome. At this stage, the goal of signal detection methods is how to use less prior knowledge to achieve detection of lower signal-to-noise ratios and more signals. According to the different signal processing systems, the signal detection methods are divided into linear signal detection and nonlinear signal detection (Fig. 1).

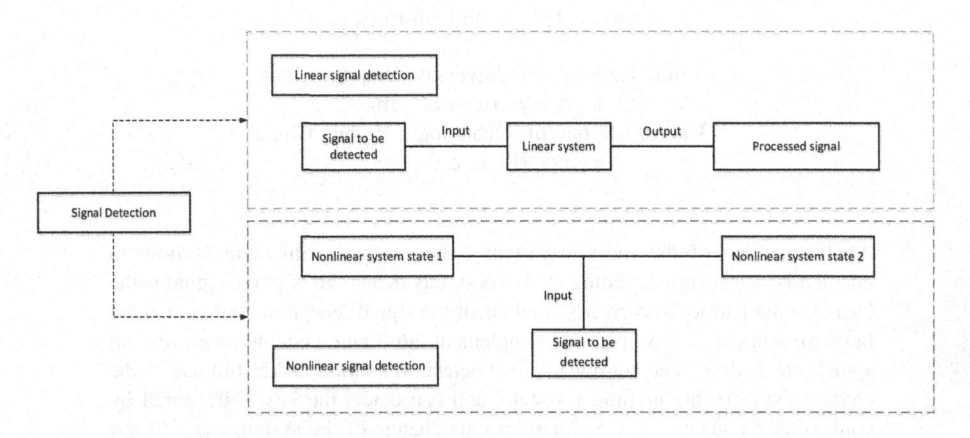

Fig. 1. Signal detection ideas and classification

The linear method is the most mature method in the current technology, but since the signal and the noise experience the same linear change, the linear method cannot detect the signal with low SNR. The nonlinear system uses a new idea to achieve signal detection. The nonlinear method inputs the signal to be tested into the nonlinear system. Due to the system characteristics of the nonlinear system, the useful signal in the input signal can be separated from the noise. The essence of nonlinear theory is that when the signal under test contains a signal that needs to be detected, the nonlinear system will change from one state to another, transforming the detection of the signal into a change in the state of the system. This shift of the signal achieved by amplifying, enabling the detection of low SNR signals. Among them, the oscillator in chaos theory has the characteristics of initial value sensitivity and noise immunity, which can transform weak signal changes into system state changes and realize the amplification of weak signals. At the same time, since the oscillator is immune to noise, the oscillator state is not affected by noise. It is possible to detect low SNR signals that are submerged under noise.

At present, the use of oscillators for signal detection is the direct use of the basic characteristics of the oscillator and theoretical analysis results, in theory to achieve the detection of weak communication signals. Many scholars and researchers have carried out a large number of analysis and simulation to prove the accuracy of the theoretical results. However, the existing simulation and analysis results are obtained under Gaussian white noise conditions, and can only be used for theoretical analysis. In the actual signal, the noise of the signal to be tested cannot be Gaussian white noise.

In this paper, the Duffing oscillator is used as the basic model of weak signal detection, and the narrowband Gaussian noise is used as the noise model to study the influence of narrowband noise on the state of the oscillator. Using the Melnikov equation to theoretically analyze and simulate the state of the oscillator under narrowband noise, and a new method based on oscillator for signal detection under narrowband noise is proposed. Then establish a signal detection model and analyze the weak signal detection performance of the new method.

2 Related Research Work

In the 1960s and 1970s, the idea of chaotic systems was proposed. In 1979, when Holmes studied chaotic motion, he found that there are special attractors in nonlinear oscillator differential equations. Therefore Duffing oscillators with fixed parameters will produce statistical properties similar to random processes [1]. In 1992, Birx D. L. and Pipenberg S. J. first realized weak signal detection with Gaussian white noise with a signal-to-noise ratio of −12 dB. All of their methods were Duffing oscillator combined with neural network method [2]. In 1995, SIMON H. and XIAO B. L first combined chaotic systems with neural networks to detect noise-containing signals in radar detection projects [3]. In 2001, Liu W. Y. and Zhu W. Q. used Duffing oscillator as the research model, and introduced bounded noise into the oscillator to obtain the stochastic process in the mean square sense of the oscillator Melnikov under noise excitation [4]. In 2011, Yi W. S. introduced the noise of non-Gaussian white noise into the Duffing oscillator and theoretically analyzed the Melnikov process of the oscillator at this time [5]. In 2012, Haykin S proposed a signal detection model based on phase space reconstruction [6]. In 2014, Zapateiro M et al. realized the frequency estimation of signals based on Duffing oscillators [7]. In 2015, Luo Z et al. proposed a signal frequency measurement method based on extended Duffing oscillators. It can improve the accuracy of frequency measurement and realize the fundamental frequency detection of power system based on extended Duffing [8]. In the same year, Zeng Z Z et al. proposed a new method for detecting weak pulse signals based on extended oscillators, which can further expand the detection range and application field of weak signals [9]. In 2016, Li N. proposed a method for adaptively adjusting the oscillator threshold parameters [10]. In 2017, Han D. Y. et al. proposed a weak signal feature extraction algorithm based on double-coupled duffing oscillator stochastic resonance to achieve effective extraction of weak signal fault feature information [11]. As one of the typical chaotic systems, Duffing oscillators have made great progress in signal detection and have lower signal-to-noise ratio thresholds.

3 Duffing Oscillator Weak Signal Detection Theory

3.1 Basic Theory of Duffing Oscillator

In 1918, Duffing used the standardized Duffing equation to describe forced motion with nonlinear restoring forces in the form of:

$$\ddot{x} + k\dot{x} + f(x) = g(x) \tag{1}$$

Where A is a nonlinear restoring force, including a nonlinear function of the x cube term, B is the oscillator cycle strategy and is a periodic function.

Based on the Duffing equation, Holmes proposed the Holmes-type Duffing equation in 1978:

$$\ddot{x} + k\dot{x} - x + x^3 = \gamma \cos(\omega t) \tag{2}$$

In Eq. (2 and 3), k is the damping ratio, $\gamma \cos(wt)$ is the periodic strategy, x is the state value of the system, and w is the angular frequency of the periodic strategy. When w takes 1 rad/s, the Duffing equation can be written as:

$$\ddot{x} + k\dot{x} - x + x^3 = \gamma \cos(t) \tag{3}$$

Its state equation is:

$$\begin{cases} \dot{x} = y \\ \dot{y} = -ky + x - x^2 + \gamma \cos(t) \end{cases} \tag{4}$$

By analyzing the state equation of the Duffing oscillator, we can see that when the system cycle power amplitude is in a small range, the oscillator will finally perform a reciprocating cycle around the saddle point. Continue to increase the cycle power of the system, the system will present the state of the same orbit. Continue to increase the system cycle power amplitude until $\gamma = 0.4$, the system has a phase trajectory of periodic bifurcation, and chaotic state. When the system power amplitude $\gamma = 0.8260$, the oscillator will be in a critical state; continue to increase the oscillator power to $\gamma = 0.83$, the system will appear a large-scale cycle state, as shown in Fig. 2.

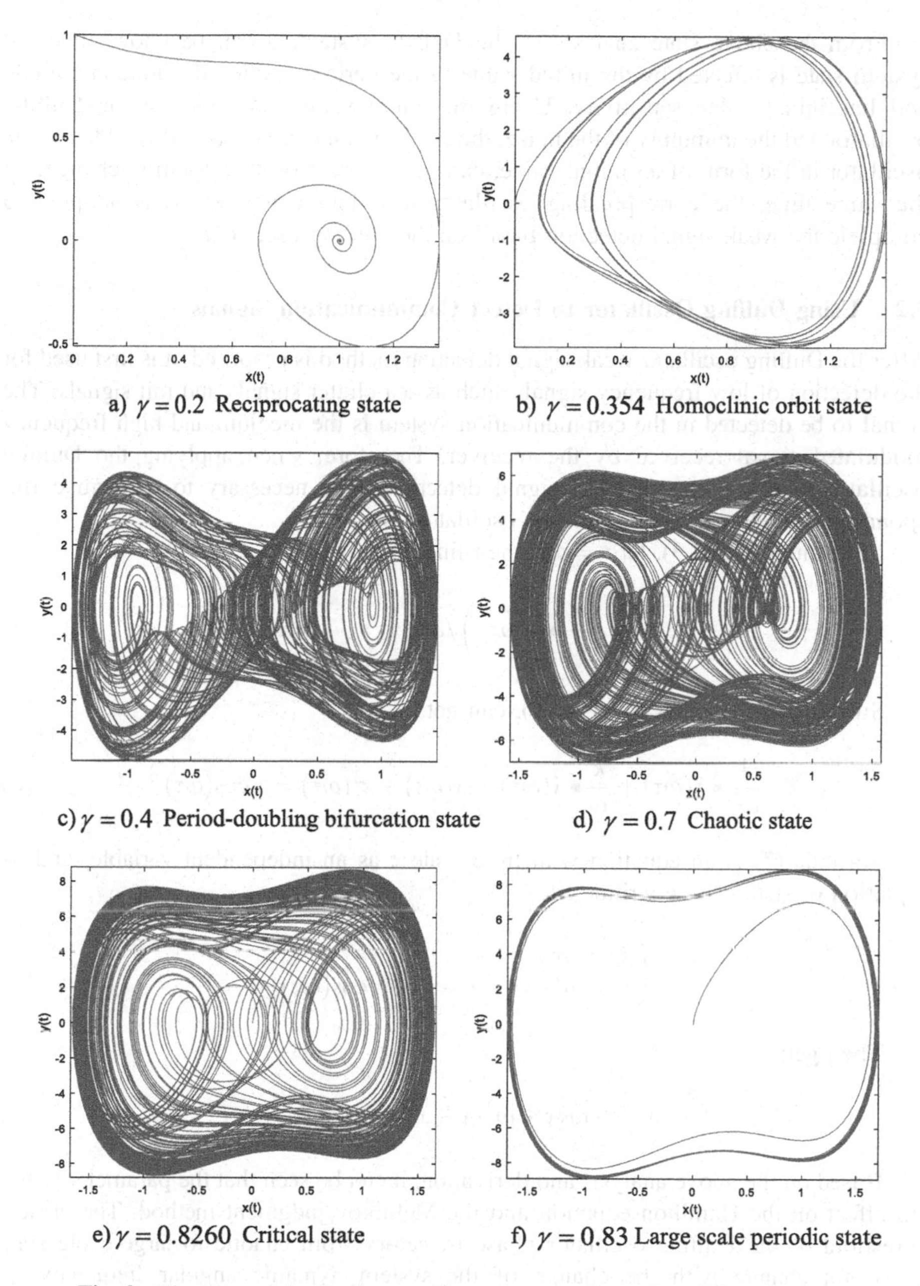

a) $\gamma = 0.2$ Reciprocating state

b) $\gamma = 0.354$ Homoclinic orbit state

c) $\gamma = 0.4$ Period-doubling bifurcation state

d) $\gamma = 0.7$ Chaotic state

e) $\gamma = 0.8260$ Critical state

f) $\gamma = 0.83$ Large scale periodic state

Fig. 2. The state of the Duffing oscillator system when γ is a different value

From the above state analysis of the Duffing system, it can be known that the system state is affected by the initial value of the periodic policy dynamic amplitude and has initial value sensitivity. Using the initial value sensitivity of the Duffing oscillator and the immunity of the noise, the weak signal to be measured is added to the oscillator in the form of an initial value, causing the state of the system to change. At the same time, the corresponding oscillator state judgment method is adopted to complete the weak signal detection based on the Duffing oscillator.

3.2 Using Duffing Oscillator to Detect Communication Signals

After the Duffing oscillator weak signal detection method is proposed, it is first used for the detection of low frequency signals such as sea clutter signals and rail signals. The signal to be detected in the communication system is the medium and high frequency modulated signal received by the receiver. Therefore, when applying the Duffing oscillator to the communication signal detection, it is necessary to perform corresponding scale transformation on the oscillator.

In formula (2 and 3), let $t = \omega\tau$, then find:

$$\ddot{x}(t) = d\left(\frac{1}{\omega} \bullet \dot{x}(\omega\tau)\right)/d(\omega\tau) = \frac{1}{\omega^2} \bullet \ddot{x}(\omega\tau) \tag{5}$$

Substituting formulas (4), (5), (6), can get:

$$\frac{1}{\omega^2} \bullet \ddot{x}(\omega\tau) + \frac{k}{\omega} \bullet \dot{x}(\omega\tau) - x(\omega\tau) + x^3(\omega\tau) = \gamma\cos(\omega\tau) \tag{6}$$

Formula (7) is an equation with time scale τ as an independent variable, and its equation of state can be written as:

$$\begin{cases} \dot{x} = \omega y \\ \dot{y} = \omega(-ky + x - x^3 + \gamma\cos(\omega\tau)) \end{cases} \tag{7}$$

Then get:

$$\ddot{x} = -\omega k\dot{x} + \omega^2(x - x^3 + \gamma\cos(\omega\tau)) \tag{8}$$

Based on the above analysis and derivation, it can be seen that the parameter w has no effect on the Hamilton equation and the Melnikov judgment method. The critical threshold of the Duffing oscillator phase trajectory from chaotic to large-cycle state does not change with the change of the system dynamic angular frequency w. Therefore, even if the frequency of the signal to be tested changes, it is only necessary to adjust the value of w to adapt to different signal frequencies to be tested, without having to solve the system critical threshold again.

4 Weak Signal Frequency Detection Method Under Narrowband Noise

In a communication system, the receiver performs band pass filtering on the received modulated signal. When performing signal detection, the signal to be tested is a mixed signal of the communication signal and the narrowband noise after being filtered by the receiver. The Gaussian white noise passing through the ideal rectangular band pass filter is called band pass white noise in communication. If the band B of the band pass filter is larger than fc at this time, it is called narrow band Gaussian white noise. The expression of narrowband noise is as follows:

$$n(t) = n_c(t)\cos(w_c t) - n_s(t)\sin(w_c t) \tag{9}$$

Narrowband Gaussian noise can be expressed as envelope and phase:

$$\begin{aligned} n(t) &= R(t)\cos(w_c t + \varphi(t)) \\ R(t) &= \sqrt{n_c^2(t) + n_s^2(t)} \\ \varphi(t) &= \tan^{-1}(n_s(t)/n_c(t)) \end{aligned} \tag{10}$$

The verification model for the influence of the narrowband Gaussian white noise on the system state can be established as shown in Fig. 3. The narrowband noise in the system is the narrowband Gaussian white noise after Gaussian white noise passes through a band pass filter. Set the band pass filter to FIR filter with a filter pass-band of 45–55 MHz and a filter bandwidth of 10 MHz. The Duffing oscillator period power frequency is 50 MHz, and the filter impulse response is shown in Fig. 4.

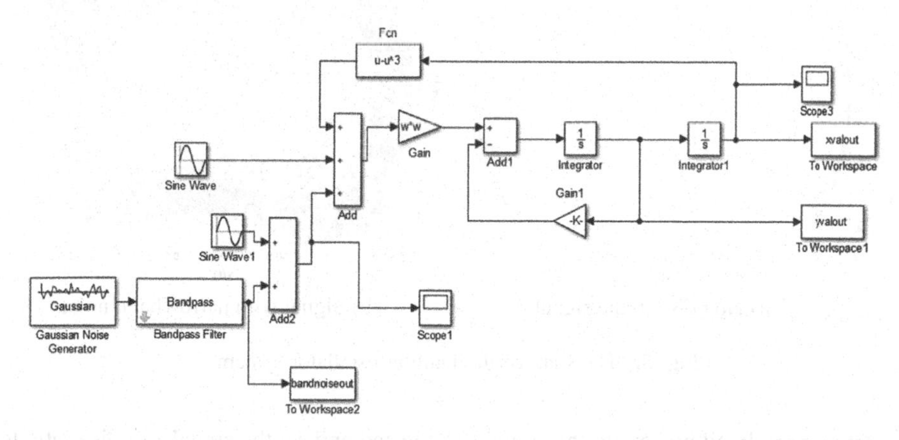

Fig. 3. Verification model for the influence of narrow-band noise on Duffing oscillator

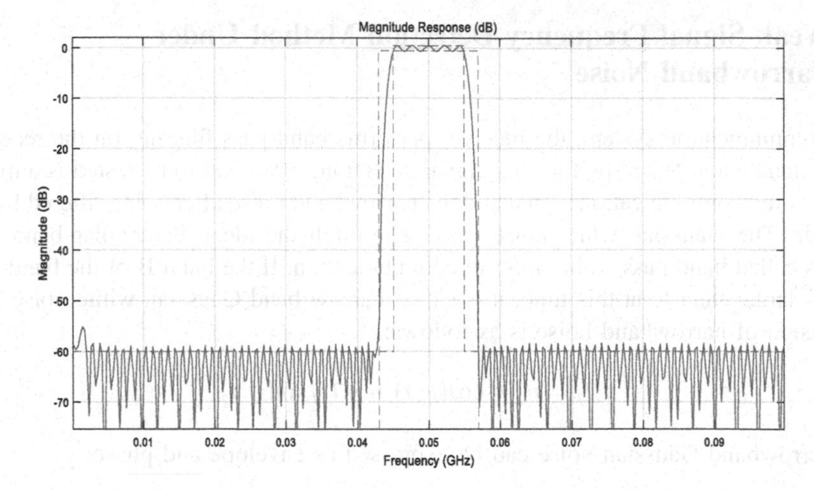

Fig. 4. Impact response function of FIR bandpass filter

Set the system strategy power amplitude to $\gamma = 0.8260$, and the power frequency to $f_0 = 50$ MHz. At this time, the oscillator is in a critical state. Assume that the signal to be tested contains a sinusoidal signal with a frequency of $f_0 = 50$ MHz, and set the damping ratio to 0.5. The simulation results are shown in Fig. 5, and the system is in a large-scale cycle state.

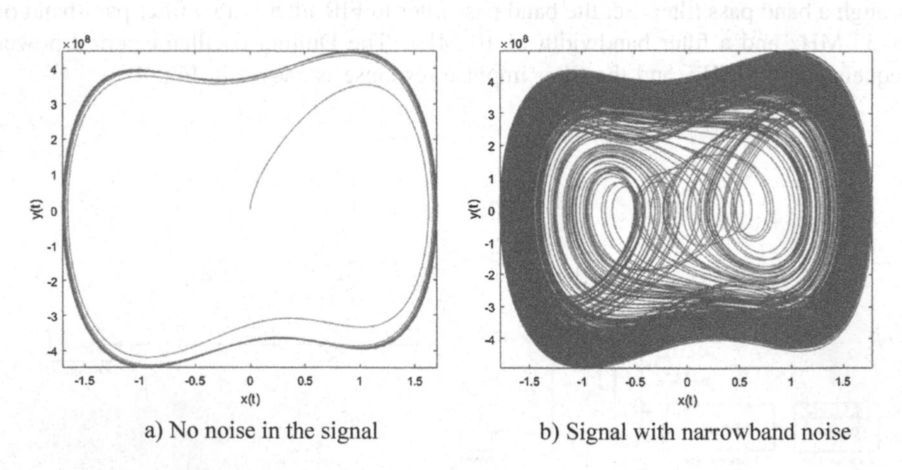

a) No noise in the signal

b) Signal with narrowband noise

Fig. 5. The state of the Duffing oscillator system

Add narrowband noise into the signal to be tested and set the signal-to-noise ratio to −10 dB. The simulation results are shown in Fig. 6. The current state of the oscillator that entered the periodic state will be almost identical to the chaotic state and difficult to distinguish.

It can be seen that the narrow-band noise will cause the oscillator that should enter the cycle to enter chaos. At this time, the oscillator state judgment according to the conventional method cannot obtain the correct result. Therefore, in the case of narrowband noise, it is necessary to find a new oscillator state judging method.

When there is a periodic force and damping ratio in the oscillator, the state equation of the Duffing oscillator is as shown in formula (5), and the Melnikov method is used to judge the oscillator state of the oscillator equation. The Melnikov equation corresponding to formula (5) is:

$$\begin{aligned} M(t_0) &= \int_{-\infty}^{\infty} \gamma \cos(t+t_0) y_0(t) dt - \int_{-\infty}^{\infty} k \times y_0^2(t) \\ &= -\sqrt{2}\pi\gamma \sec h\left(\frac{\pi}{2}\right) \sin(t_0) - \frac{4}{3}k \end{aligned} \tag{11}$$

It follows that when $\left|\sqrt{2}\pi\gamma \sec h(\pi/2)\right| = -4/3k$, the oscillator is in a critical state, and can get the critical amplitude γ_d. When $\left|\sqrt{2}\pi\gamma \sec h(\pi/2)\right| > -4/3k$, there is t_0 such that $M(t_0) = 0$, the oscillator will be able to enter the periodic state.

When noise is added into the oscillator, the oscillator equation becomes:

$$\begin{cases} \dot{x} = y \\ \dot{y} = -ky + x - x^3 + \gamma \cos(t) + n(t) \end{cases} \tag{12}$$

The corresponding Melnikov function can be expressed as:

$$\begin{aligned} M(t_0) &= \int_{-\infty}^{+\infty} (\gamma \cos(t+t_0) + n(t+t_0)) y_0(t) dt - \int_{-\infty}^{+\infty} k \cdot y_0^2(t) dt \\ &= -\sqrt{2}\pi\,\gamma \sec h\left(\frac{\pi}{2}\right) sin(t_0) - \int_{-\infty}^{+\infty} n(t+t_0) y_0(t) dt - \frac{4}{3}k \\ &= -\sqrt{2}\pi\,\gamma \sec h\left(\frac{\pi}{2}\right) sin(t_0) - \int_{-\infty}^{+\infty} n(t+t_0) \frac{e^t - e^{-t}}{(e^t + e^{-t})^2} dt - \frac{4}{3}k \\ &= M_1(t_0) + M_2(t_0) + M_3 \end{aligned} \tag{13}$$

The first part M_1 is a function of γ and t_0, which can be regarded as a sine function whose amplitude change satisfies $|M_1(t_0)| = \sqrt{2}\pi\gamma \sec h(\pi/2)$, and its amplitude function is a one-time increasing function about γ. The third part M_3 is a positive number. As can be seen from the expression of the second part M_2, due to the existence of noise n(t), M_2 also becomes a random variable related to the noise distribution. Since the noise has a value after zero time, M_2 can be written as:

$$M_2(t_0) = -\int_{-t_0}^{+\infty} n(t+t_0) \frac{e^t - e^{-t}}{(e^t + e^{-t})^2} dt = \int_{-t_0}^{+\infty} n(t+t_0) f(t) dt \tag{14}$$

$$f(t) = \frac{e^t - e^{-t}}{(e^t + e^{-t})^2}$$

It can be seen that t has the same sign as $f(t)$. So when $t \in [-t_0, 0]$, $M_2(t_0) > 0$ M_2 will prevent the oscillator from entering the cycle state. When $t \in [0, +\infty]$ M_2 will promote the oscillator into the cycle state.

So, M_2 becomes a random process related to noise. For a certain period of the engine power amplitude γ, $M(t_0)$ becomes a function of t_0, by calculating whether t_0 exists so that $M(t_0) = 0$ is established, and whether the oscillator can be in a periodic state. At this time, the probability that the oscillator appears in the periodic state can be expressed as:

$$P_{cycle} = \lim_{N \to \infty} \left[\oint_{\sigma} f_{n(t)}(n) dn \right] \tag{15}$$

Where σ is the set of all n(t) that $M(t_0) = 0$ can have solutions, and $f_{n(t)}(n)$ is the N-dimensional joint probability density of noise. In order to solve the equation, the Melnikov equation is written as:

$$\begin{aligned} M(t_0) &= M_1(t_0) + M_2(t_0) + M_3 = 0 \\ &\Leftrightarrow M_2(t_0) = -M_1(t_0) - M_3 \end{aligned} \tag{16}$$

Since $\lim_{t \to \infty} f(t) = 0$ and $\lim_{t \to \infty} M_2(t_0) = 0$, then the formula (17) can be solved equivalent to the existence of t_0 such that the formula (18) holds:

$$M_2(t_0) > \min(-M_1(t_0) - M_3) = -|M_1(t_0)| - M_3 = M(\gamma) \tag{17}$$

Each P_{cycle} corresponding to t_0 is actually the probability of $M_2(t_0, n(t)) > M(\gamma)$ which can be expressed as:

$$P_{cycle}(t_0) = \lim_{N \to \infty} \int_{M(\gamma)}^{\infty} p(x) dx \tag{18}$$

$$\begin{aligned} P_{cycle} &= \int_{-\infty}^{+\infty} P_{cycle}(t_0) dt_0 = \lim_{M \to \infty} \frac{\int_{-M}^{+M} P_{cycle}(t_0) dt_0}{2M} \\ &= \lim_{M \to \infty} \lim_{N \to \infty} \frac{\int_{M(\gamma)}^{\infty} \int_{-M}^{+M} p(x) dx dt_0}{2M} \end{aligned} \tag{19}$$

$$\begin{aligned}\frac{dP_{cycle}}{d\gamma} &= \lim_{M\to\infty}\lim_{N\to\infty}\frac{d\int_{M(\gamma)}^{\infty}\int_{-M}^{+M}p(x)dxdt_0}{2M\times d\gamma} \\ &= \lim_{M\to\infty}\frac{1}{2M}\lim_{N\to\infty}\int_{-M}^{+M}\frac{d\int_{M(\gamma)}^{\infty}p(x)dx}{d\gamma}dt_0 \\ &= \lim_{M\to\infty}\frac{1}{2M}\lim_{N\to\infty}\left[-\int_{-M}^{+M}p(M(\gamma))\times\frac{dM(\gamma)}{d\gamma}dt_0\right] \\ &= \frac{\sqrt{2}}{2}\pi\sec h\left(\frac{\pi}{2}\right)\lim_{M\to\infty}\frac{1}{M}\lim_{N\to\infty}\int_{-M}^{+M}p(M(\gamma))dt_0 > 0\end{aligned} \tag{20}$$

Therefore, under the action of narrowband noise, the Duffing oscillator will enter the periodic state according to a certain probability. It can be seen from formula (21) that the probability of entering the periodic state is an increasing function of the oscillator's power amplitude γ.

According to the previous analysis and derivation, when the noise contained in the signal to be tested is not Gaussian white noise, the Duffing oscillator no longer has complete noise immunity. The narrow-band noise mixed in the signal to be tested causes the oscillator that should enter the chaotic state to appear in a periodic state, showing a state in which the period and chaos alternately appear. This is similar to the existing intermittent chaotic state, but the difference is that the oscillator is in a periodic state or a chaotic state is subject to a certain probability distribution.

Based on the derivation and analysis, this paper proposes a method for judging the state of the oscillator by counting the period state time rate (PSTR) of the Duffing oscillator. The definition of PSTR is as follows:

$$PSTR = \frac{T_{period}}{T_{total}} \tag{21}$$

T_{period} represents the proportion of the time of the cycle state in a simulation time, and T_{total} is the total simulation time. Express it as a discrete sampling point as follows:

$$PSTR = \frac{n_{period}}{n_{total}} \tag{22}$$

Using MATLAB/Simulink to establish a statistical model of the oscillator PSTR under narrowband noise. In order to verify the correctness of the proposed method, the PSTR value of the Duffing oscillator under different cycle dynamics is simulated and analyzed. The simulation results are shown in Fig. 6.

As can be seen from Fig. 6. Under the influence of narrow-band noise, the oscillator does not exhibit stable periodic and chaotic states under classical theory. At this

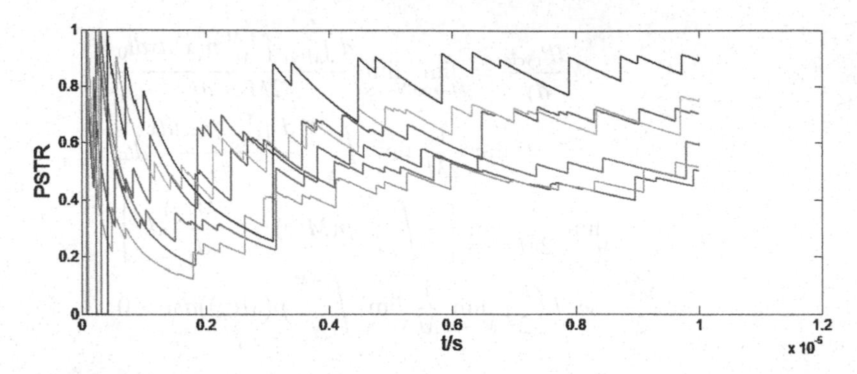

Fig. 6. PSTR of the Duffing oscillator with different initial values of the dynamic amplitude

time, the Duffing oscillator will appear alternating between the period and the chaotic state, and the probability of the periodic state appearing as the PSTR value is different. As the simulation time increases, the PSTR of the oscillator with different periodic power amplitudes will tend to a stable value. And the PSTR at the time of stabilization increases with the increase of the periodic power amplitude, which is the same as the analysis result.

5 Conclusion

As a typical chaotic system, Duffing oscillator is one of the most researching non-linear detection methods for low SNR signal detection. In this paper, the narrow-band Gaussian white noise is used as the noise of the signal to be tested, and the Duffing oscillator model under narrow-band noise conditions is established. Through the Melnikov function analysis, the conclusion that the periodic state of the Duffing oscillator exhibits a probability distribution under narrow-band noise is obtained. Based on the analysis, an oscillator state judgment method based on Duffing oscillator period state time ratio (PSTR) is proposed. The weak signal detection platform is built by MATLAB. The correctness of the method is verified by simulation analysis.

References

1. Holmes, P.: A nonlinear oscillator with a strange attractor. Philos. Trans. R. Soc. Lond. **292** (A292), 419–448 (1979)
2. Birx, D.L., Pipenberg, S.J.: Chaotic oscillators and complex mapping feed forward networks (CMFFNs) for signal detection in noisy environments. In: International Joint Conference on Neural Networks, vol. 2, pp. 881–888. IEEE (2002)
3. Haykin, S., Li, X.B.: Detection of signals in chaos. Proc. IEEE **83**(1), 95–122 (1995)
4. Liu, W.Y., Zhu, W.Q., Huang, Z.L.: Effect of bounded noise on chaotic motion of duffing oscillator under parametric excitation. Chaos Solitons Fractals **12**(3), 527–537 (2001)

5. Yi, W., Song, J.: The analysis of Melnikov process of duffing oscillator in non-Gaussian color noise environment. In: International Congress on Image and Signal Processing, pp. 306–309. IEEE (2011)
6. Haykin, S., Li, X.B.: Detection of signals in chaos. Proc. IEEE **83**(1), 95–122 (2012)
7. Zapateiro, M., Vidal, Y., Acho, L.: A secure communication scheme based on chaotic Duffing oscillators and frequency estimation for the transmission of binary-coded messages. Commun. Nonlinear Sci. Numer. Simul. **19**(4), 991–1003 (2014)
8. Luo, Z., Zeng, Z.: A highly accurate frequency-measuring method based on extended Duffing oscillator. Autom. Electric Power Syst. **39**(16), 81–85 (2015)
9. Zeng, Z.Z., Zhou, Y.H.K., Hu, K.: Study on partial discharge signals detection by extended Duffing oscillator. Acta Physica Sinica **64**(7), 1–8 (2015)
10. Li, N., Li, X., Liu, C.: A detection method of Duffing oscillator based on adaptive adjusting parameter. In: Ocean Acoustics, pp. 1–5. IEEE (2016)
11. Han, D., Sun, Y., Shi, P.: Research on weak signal feature extraction method based on double coupled Duffing oscillator stochastic resonance. In: 11th International Workshop on Structural Health Monitoring, pp. 2738–2744 (2017)

AI in Ubiquitous Mobile Wireless Communications

Research on an Intelligent Routing Strategy for Industrial Internet of Things

Xu Zhang(✉)

Department of Control Science and Engineering, Harbin Institute of Technology, Harbin 150080, China
zhangxuhit@126.com

Abstract. The Industrial Internet of Things is considered to be an important cornerstone of future industrial development and has broad development prospects. It is being deployed to the society on a large scale. Sensor nodes in the Internet of Things will inevitably face many challenges. Due to the limitations of the existing sensor nodes, the traditional energy measurement methods for wireless sensor networks have been difficult to meet. Therefore, this article proposes a new routing protocol for low-power and low-power networks (RPL). The routing measurement method EEM meets the requirements of low-power lossy networks for link quality and energy consumption, and then uses the modified simulation software to test the EEM routing measurement, and performs performance verification on the packet loss rate and network load. The experimental results show that EEM retains the requirements of ETX routing metrics for link quality, realizes the awareness of node energy consumption, and optimizes network load.

Keywords: Wireless sensor network · RPL · Routing metrics · Industrial internet of things

1 Introduction

The RPL routing protocol is an IPv6-based distance vector protocol for low-power lossy network design and development. In a low-power lossy network, the environment faced by the links between nodes is often harsh, and the information processing capabilities and energy of the nodes themselves are limited, and the limited energy and processing capabilities of the nodes need to be self-organized. networking. In the RPL routing protocol, nodes broadcast control messages to each other to establish a DODAG structure, collect information from sink nodes intermittently, and are compatible with IPv6. These features make the RPL protocol a perfect fit for the development needs of the future industrial Internet of Things. And because of the objective function and the reference of routing metrics, the RPL protocol has extremely high modularity and plasticity, so it has great research value in the field of industrial Internet of Things.

S. Shi et al. (Eds.): AICON 2020, LNICST 356, pp. 115–125, 2021.
https://doi.org/10.1007/978-3-030-69066-3_11

2 Research Background

At present, the RPL protocol mainly uses two objective functions to complete the network topology construction, namely OF0 and MRHOF. OF0 is the metric selection based on the number of hops from the node to the root node. The smaller the number of hops from the candidate parent node to the root node, the higher the probability that it will be the parent node. OF0 is the simplest basic objective function and is used in small-scale deployment networks. Often performance is excellent, but this routing strategy cannot consider the link quality between nodes. When the communication quality between nodes is poor, it will cause a large number of backhauls between nodes and cause energy consumption problems, which will compress the network life cycle. A large number of practices use the MRHOF objective function based on the ETX routing metric, and the objective function can take a variety of routing metrics, including Hops, Latency, etc., among which ETX represents the quality of the communication link between nodes. Therefore, based on the in-depth study of ETX in the RPL protocol, this article considers ETX alone as a routing metric that cannot meet the needs of the industrial Internet of Things model, and then introduces a routing strategy EEM (ETX and Energy) based on link communication quality and energy balance. balanced Metrics), and used the API provided by the objective function in the RPL protocol to implement and verify the strategy.

3 Industrial IoT Model

The industrial IoT system architecture is mainly divided into three levels, namely the perception layer, the transmission layer and the application layer. This paper focuses on the routing strategy of the sensor network in the transmission layer of the Industrial Internet of Things, so this article uses the Industrial Internet of Things model shown in Fig. 1 to study the routing strategy.

In Fig. 1, the root node with a stable power supply is used as a border router to allocate the IP addresses of other child nodes, and centrally process and transfer the information collected by the sensor nodes to the information processing center. The sub-nodes composed of sensors respectively include mobile nodes and fixed nodes, and their communication distances are limited and because they are powered by batteries, the energy of the sub-nodes is limited. The child nodes have a certain storage space and limited transmission distance, and have the ability to collect information and send information. The control messages and trickle mechanism introduced in Sect. 2 are used between the nodes to build and maintain the topology structure. The control messages contain the information of the node. Energy information, Rank value, etc., and finally the information collected by the child nodes are transmitted to the root node in a multi-hop manner. In this model, the information collected by the node needs to be gathered at the root node.

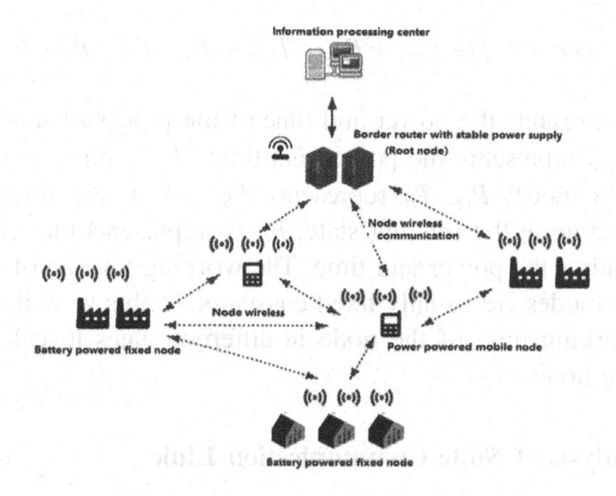

Fig. 1. Industrial Internet of Things research model

3.1 Model Structure Analysis

In this paper, the Packet Loss Ratio (Packet Loss Ratio) indicator is used to measure the cognitive function of the node on the link quality. Thc usc of this index mainly considers that the packet loss rate is closely related to the link quality. If the link quality between nodes is poor and the routing strategy does not realize the cognitive function of the link quality, it will inevitably lead to an increase in the packet loss rate., and a large number of retransmissions will lead to an increase in the cost of control messages and an increase in the duty cycle of node monitoring and sending, which makes the node's excessive energy consumption and death time earlier, which is fatal for the sensor network. This article defines the packet loss rate as follows:

$$PLR = \frac{\sum_{n} \frac{P_{received}(i)}{P_{send}(i)}}{N_{total}} \tag{1}$$

$P_{received}(i)$ represents the number of data packets sent by the node that the root node has successfully received, $P_{send}(i)$ represents the number of data packets sent by the node to the root node, and N_{total} represents the number of child nodes in the entire network topology.

Aiming at the problem of node energy consumption, this paper proposes a model of computing node energy consumption:

$$E_{consumption} = \sum_{M,S} P_{M,S} \cdot T_{M,S} \tag{2}$$

Where $P_{M,S}$ represents the power of the M module in the S state, and $T_{M,S}$ represents the working time of the M module in the S state.

Through the above analysis, calculate the energy of the processor and wireless communication modules that consume a lot of energy:

$$Power = P_{cpu} \cdot T_{cpu} + P_{lpm} \cdot T_{lpm} + P_{rt} \cdot T_{rt} + P_{rl} \cdot T_{rl} \quad (3)$$

Where P_{cpu}, T_{cpu} represents the power and time of the processor module in full-speed operation; P_{lpm}, T_{lpm} represents the power and time of the processor module in low-power consumption mode; P_{rt}, T_{rt} represents the power and time of the wireless communication module in the sending state; P_{rl},T_{rl} represents the wireless communication module Monitor the power and time. The working voltage of the node and the current in different modes are usually fixed constants. In this way, it is only necessary to measure the working time of the node in different states to calculate the energy consumption of the node.

3.2 Quality Analysis of Node Communication Link

The purpose of the introduction of ETX is to measure the reliability of the link. In practical applications, nodes use the MAC layer to receive transmission or ACK messages, collect neighbor node information, determine and calculate the ETX value between nodes. ETX calculation methods are diverse. One of the most widely used is the EWMA (Exponentially-Weighed Moving Average Function) algorithm, whose formula is as follows:

$$new\_ETX = \alpha \cdot recorded\_ETX + (1 - \alpha) \cdot packet\_ETX \quad (4)$$

If the node does not receive neighbor node information at the MAC layer, *packet_ETX* will be set to the maximum value, which can be adjusted for different network conditions. On the contrary, if the node receives neighbor node information, the maximum value will be decremented. *packet_ETX* can reflect the link quality between nodes. It is characterized as the link metric value of the neighbor node, and the new ETX value is assigned to *link_metric* after the calculation is completed.

4 Research on Intelligent Routing Strategy

MRHOF adopts the ETX measurement while considering the hysteresis phenomenon. Although the introduction of the ETX concept realizes the node's cognitive function of the link quality between nodes, a single cognitive function is difficult to adapt to the complex environment. Therefore, this article A routing metric EEM that integrates two cognitive functions is proposed:

$$EEM = Algorithm(ETX, Node\_Energy) \quad (5)$$

In the above formula, ETX characterizes the link communication quality between nodes, which is calculated by formula (4), and Node_Energy represents the total historical energy consumption value of the candidate parent node, which is calculated according to formula (3). The specific principle of EEM metric is shown in Fig. 2. When a node selects a parent node, it will first give priority to the parent node's ETX value. If the candidate parent node's ETX is within the allowable range, it will continue to consider the historical total energy consumption of the candidate parent node. If a node's historical energy consumption is higher If it is smaller, the node will be selected as the parent node first. If the ETX value difference between the candidate parent nodes is too large, the node with the best link communication quality will be selected first as the parent node. In this way, EEM integrates ETX and node energy consumption measurement, and realizes node energy load balancing while ensuring the communication link.

4.1 Routing Metric EEM

Introduce the EEM routing metrics into the MRHOF objective function. The MRHOF objective function mainly provides four program interfaces, which can be used to optimize MRHOF. The four program interfaces are the best directed acyclic graph (Best Dag), the best parent (Best Parent), the calculation of the Rank value (Calculate Rank), and the update of control information (Update Metric Container). The functions of the above four program interfaces are:

(1) Optimal directed acyclic graph. The API selects the DAG with the best candidate parent node based on OF and joins it. Different OFs will consider different selection strategies, such as whether the DAG is connected to an external network, DAG rank value and version number, and the order of DIO messages.
(2) The best parent node. The API selects the best parent node based on OF. In MRHOF, compare the ETX value of the candidate parent node. If the ETX value is within the allowable threshold, the parent node will not be changed. Otherwise, a new parent node will be selected based on the routing metric. The purpose of this is to avoid network conditions. The resulting ETX gap leads to frequent replacement of parent nodes by nodes, which leads to changes in the network topology and brings huge costs. If the two ETX values exceed the threshold, the node with the smallest ETX is selected as the default next hop according to the ETX information of the parent node.
(3) Calculation of Rank value
This interface calculates the rank value of a node based on OF, and the rank value between parent and child nodes is an increasing relationship. The calculation formula is:

$$Rank_N = Rank_P + Rank\_Increase \tag{6}$$

In MRHOF, *Rank_Increase* is the *link_metric* value of the parent node, and the value of *link_metric* is the same as the ETX value between nodes, that is, the node Rank value reflects the information of the node's ETX value.
(4) Update of control information
In the DIO control message of the RPL protocol, the interface appends the updated routing metric information of the node to the DIO control message, and broadcasts it in the network along with the DIO message. The updated routing metric message must conform to the DIO control message format, such as the node's ETX Value or energy information occupies the option field in the DIO

control message. Other nodes select the parent node by extracting DIO information of neighbor nodes.

4.2 Optimization Strategy

In order to ensure that the optimized objective function has the route establishment process and maintenance mechanism of the RPL protocol, and avoid the generation of loops, this article retains the original optimal directed acyclic graph of the RPL protocol and the method of calculating the Rank value during the implementation of the EEM., It is mainly updated for the two interfaces of the optimal parent node, namely the control information update. The updated part is as follows:

(1) Optimal parent node
The pseudo code for selecting the optimal parent node is shown in the following table:

Table 1. SCM routing metrics Best Parent pseudo code

Algorithm 1	Best Parent
Input:	Alternative node P1, alternative node P2
Output:	Best Parent
1:	**if** P1->ETX – P2->ETX <mindiff && P1->ETX – P2->ETX>-mindiff **then**
2:	**If** P1->energyest < P2->energest **then**
3:	return P1
4:	**else** return P2
5:	**end if**
6:	**end if**
7:	**if** P1->ETX <P2->ETX **then**
8:	**return** P1
9:	**else return** P2
10:	**end if**

the above pseudo code, mindiff represents the preset ETX threshold. If the candidate node is within the threshold, the optimal parent node is selected according to the node energy. energest represents the total energy consumption of the node, which is equivalent to Node_Energy in formula (5). Under the premise of ensuring the quality of the communication link between nodes, the less historical energy consumption of the candidate parent node, the higher its priority.

(2) Update of control information
The pseudo code of the DIO control message update part is shown in Table 2:
In the following pseudo code, if the rank value of the root node of the DAG graph contained in the original DIO control message is 0, which means the root node, the energy information part of the new DIO is set to 0. This approach is to increase the priority of the root node, If there is a root node in the candidate parent node list of a

node, the node will preferentially select the root node as its own parent node, otherwise it will record the energy value of the local node and update the value to the new DIO and forward it.

Table 2. SCM routing metric Update Metric Container pseudo code

Algorithm 2	Update metric container
Inout:	oldDIO
Output:	newDIO
1:	**if** oldDIO->dag->root->rank == 0 **then**
2:	newDIO->mc.energest-> ==0
3:	**else**
4:	power = cpu + lpm + rl + rt
5:	newDIO-> mc.energest-> == power
6:	**end if**
7:	return newDIO

After completing the above two parts, use EEM routing metrics to re-plan the network deployment in Fig. 2. The simulation results are shown in Fig. 3:

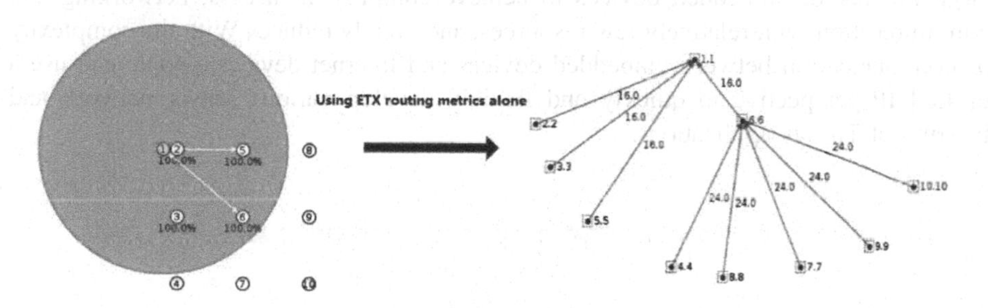

Fig. 2. ETX routing metrics

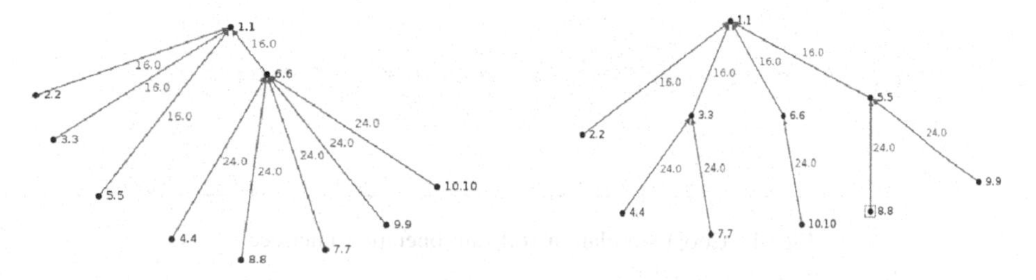

(a) Network topology based on ETX (b) Network topology based on EEM

Fig. 3. Comparison of ETX and EEM routing metric topology structures

The EEM routing metric avoids the problem of multi-node selection of node 6. It can be found that nodes 4, 7, 8, 9, 10 avoid the centralized use of 6 as its parent node, 4, 7 selects node 3 as the parent node, and 8 and 9 select Node 5 is the parent node, while node 10 retains 6 as the parent node. By using the EEM metric, the selection of the child nodes to the parent node is more balanced, and the entire topology is more dispersed. Through the analysis of the topology, it can be qualitative It is believed that EEM has completed the load balancing task of nodes in the network under the premise of ensuring the link quality.

5 Experimental Simulation

The Contiki operating system and Cooja simulation software are used to simulate and verify the routing metric and perform performance analysis to investigate the difference in packet loss rate and energy load compared with the original routing metric.

5.1 Simulation Environment

The Contiki operating system is an open source multitasking operating system. The interface is shown in Fig. 4. The uIP protocol stack implemented inside the Contiki operating system supports the TCP/IPv6 communication protocol, which enables a large number of embedded devices to achieve complete IP access, networking and communication with relatively few resources, and greatly reduces With the complexity of communication between embedded devices and Internet devices, people can use a unified IP perspective to quickly and flexibly develop various sensor network and Internet of Things applications.

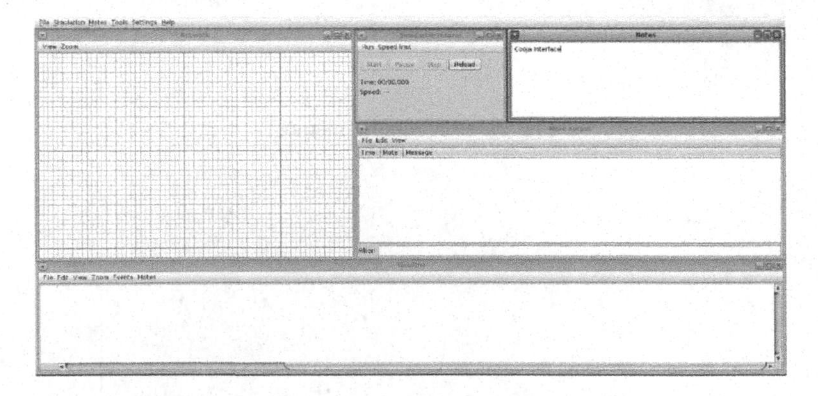

Fig. 4. Cooja simulation software operation interface

The routing strategy proposed in this paper needs to have a cognitive function on the link quality of neighbor nodes. In order to verify whether the EEM routing metric has the same link quality perception effect as ETX, this paper considers the performance of both from the perspective of packet loss rate. The experiment is based on two deployment scenarios, linear deployment and random deployment, corresponding to Fig. 4 (c) and (d). By comparing the packet loss rate data of ETX and EEM routing metrics under different packet flows, it is verified whether EEM retains the cognitive function in ETX. In the experiment, each child node sends data packets to the root node independently and regularly. If the root node successfully accepts and returns a successful acceptance message, this process is counted as a successful packet reception, otherwise it is counted as a packet loss. In the experiment, by changing the child node timer setting The number changes the node data packet flow. The simulation results are shown in Fig. 5.

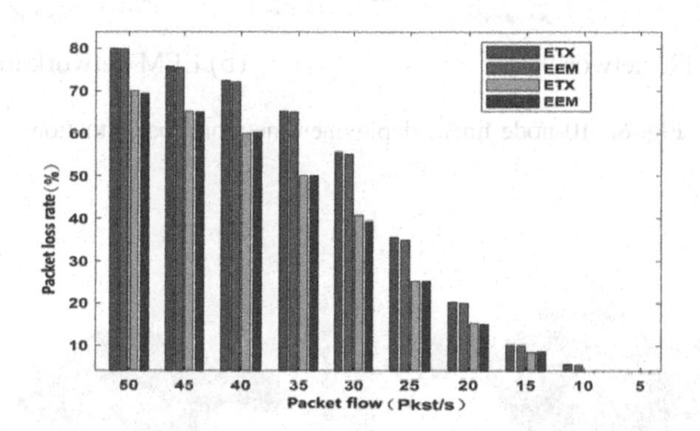

Fig. 5. Comparison of ETX and EEM packet loss rate

It can be found that with the increase of data packet traffic, the packet loss rate based on ETX and EEM routing metrics increases. When the data packet traffic is lower than 10Pkst/s, both perform well in data packet transmission. When the data packet traffic is greater than 15Pkst At/s, the packet loss rate increases significantly, and in general, the randomly deployed network structure is better than linear deployment in terms of packet loss rate. The packct loss rate performance of the two routing metrics is similar under different packet flow and topology. Based on the above results, a preliminary conclusion can be drawn that in terms of packet loss rate performance, ETX and EEM achieve the same performance effect.

EEM is more "dispersed" for the establishment of DODAG topology, and the average number of child nodes of the parent node is less than the ETX routing metric. Therefore, it is necessary to continue to verify the energy cognitive function of SCM and compare its performance on network load balancing capabilities.

Figure 6 and Fig. 7 reflect the network load situation using different routing metrics in the two deployment scenarios. There are obvious poles in the network based on the ETX standard, and the energy consumption of nodes in the entire network is not evenly distributed. Under the EEM standard, the load situation of the entire network is relatively balanced, there is no or no obvious extreme point, and the energy consumption of nodes in the entire network is relatively flat. Through experiments, it can be concluded that in terms of network load balancing, EEM has completed the expected energy cognition function, optimized network energy distribution, and realized the task of load balancing.

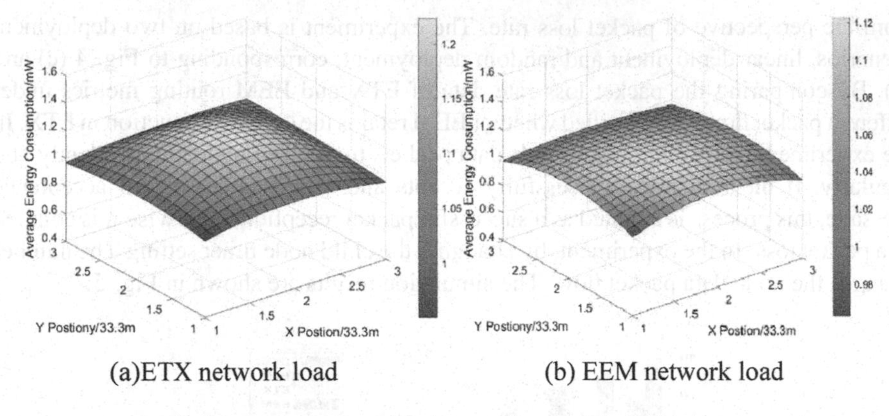

Fig. 6. 10-node linear deployment network load situation

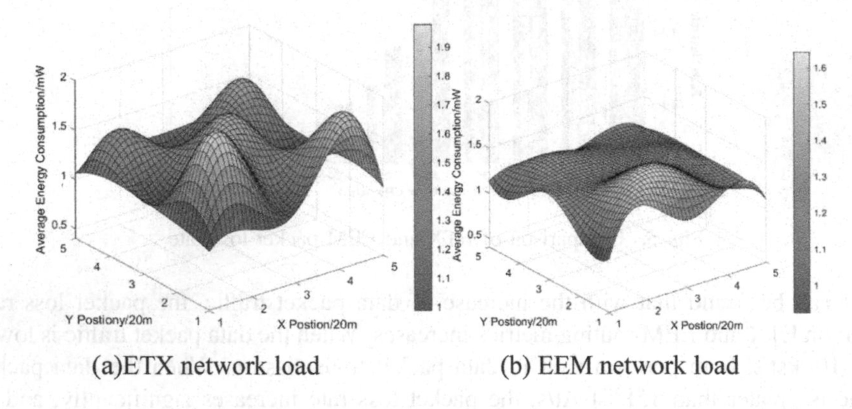

Fig. 7. 26-node linear deployment network load situation

6 Conclusion

This paper studies the routing strategy of the Industrial Internet of Things, and expands on the basis of the widely used RPL routing protocol in the field of Industrial Internet of Things. It discusses the node attributes that need to be considered for the future deployment of the Internet of Things to the society, and proposes a node For the dual functions of link quality and energy perception, a new routing metric quasi-EEM combining ETX and energy perception is designed based on this. Combined with the research platform Contiki and simulation software Cooja, the routing metrics are the packet loss rate and node energy consumption. And the network load capacity was evaluated experimentally, which verified that EEM realized the perception of node energy while retaining the cognitive function of ETX, optimized node energy consumption, and balanced network load.

References

1. Jara, A.J., Varakliotis, S., Skarmeta, A.F., et al.: Extending the internet of things to the future internet through IPv6 support. Mob. Inf. Syst. **10**(1), 3–17 (2014)
2. Deering, S., Hinden, R.: RFC2460: internet protocol, version 6 (IPv6) specification. RFC **17** (6), 1860–1864 (1995)
3. Vasseur, J.P., Dunkels, A.: Interconnecting Smart Objects with IP, pp. 1–2. Morgan Kaufmann, Burlington (2010)
4. Netze H.: Transmission of IPv6 Packets over IEEE 802.15.4 networks. In: 2012 6th International Conference on Signal Processing and Communication Systems (ICSPCS), pp. 1–6. IEEE, 2007
5. Standard I.: Wireless medium access control (MAC) and physical layer (PHY) specifications for low-rate wireless personal area networks (LR-WPANs). In: IEEE Std, pp. 13–15 (2010)
6. Kushalnagar, N., Montenegro, G., Schumacher, C.: IPv6 over low-power wireless personal area networks(6LoWPANs): overview, assumptions, problem statement, and goals, pp. 7–10. Heise Zeitschriften Verlag (2007)
7. Shelby, Z., Bormann, C.: 6LoWPAN: The Wireless Embedded Internet, pp. 15–18. Wiley, Chichester (2009)
8. Ko, J.G., Terzis, A., Dawson-Haggerty, S., Levis, P., et al.: Connecting low-power and lossy networks to the Internet. IEEE Commun. Mag. **49**(4), 96–101 (2011)
9. Winter, T., Thubert, P., Brandt, A., et al.: RPL: IPv6 routing protocol for low-power and lossy networks. IETFRFC **6550**, 2–3 (2012)

Joint Equalization and Raptor Decoding for Underwater Acoustic Communication

Miao Ke, Zhiyong Liu(✉), and Xuerong Luo

School of Information Science and Engineering, Harbin Institute of Technology, Weihai 264209, People's Republic of China
lzyhit@hit.edu.cn

Abstract. To improve the link reliability and solve the problem of long feedback delay, a joint equalization and Raptor decoding (JERD) algorithm is proposed for underwater acoustic communication. Compared with the existing approaches, the Raptor code is adopted. The Raptor code is consisted of LDPC code generated by Mackey-1A and weakened LT code, and Raptor decoding adopts the global-iteration algorithm. The detector is iteratively adapted by switching soft information between the equalization and Raptor decoding at the Turbo processing stage. Simulation results validate the feasibility and show the advantages of the proposed algorithm against the existing approaches.

Keywords: Underwater acoustic communication · Raptor codes · Joint equalization and Raptor decoding

1 Introduction

Recently, in the requirement of marine development and ocean exploration, underwater acoustic channel (UAC) has attracted increasing attention. The environment of UAC is extremely complex affected by various factors in the harsh underwater environment. The studies of UAC are confronted with the problem of the low signal-to-noise ratio and the time-space-varying channel parameters.

Considering the long feedback delay of the traditional mechanism such as automatic repeat request (ARQ) [1], a new class of sparse graph channel codes known as Fountain Codes (FCs) has been used for information transmission in the UAC. Fountain Codes can generate an infinite stream of encoded symbols from a given source message. Because the rate of the codes is not fixed a-priori [2], the Fountain Codes are rateless. There are two important classes of the Fountain Codes known as Luby Transform (LT) and Raptor codes. It has been shown that Raptor codes, which are constructed by serially concatenating LT codes with high-rate low-density parity-check (LDPC) codes, and outperform LT codes in the complexity of encoding and decoding process. Therefore, the Raptor codes can help to recover the input symbols that LT codes cannot recover [3]. Meanwhile, with the advent of turbo equalization [4, 5], there has been a wide interest in the application of turbo detection schemes for UAC communications. Compared with conventional one-time equalization, turbo equalization has a much more powerful detection capability [6]. Thus, to improve the performance of the receiver of the communication system using LT codes, an LT-Turbo

S. Shi et al. (Eds.): AICON 2020, LNICST 356, pp. 126–135, 2021.
https://doi.org/10.1007/978-3-030-69066-3_12

equalization method has been proposed, in which the adaptive linear equalization and the LT decoding are jointly optimized in the iterative process [7]. To our best knowledge, there is still lack of researches on the joint equalization and Raptor decoding algorithm.

In this paper, we propose a joint equalization and Raptor decoding (JERD) algorithm. The proposed scheme can jointly realize the adaptive equalization and the Raptor decoding. The rest of the paper is organized as follows. In Sect. 2, we review the soft decoding algorithm of Raptor codes known as global iterative Belief Propagation (BP) decoding. In Sect. 3, Joint Equalization and Raptor Decoding is introduced, where the adaptive equalization and the decoding of Raptor codes are jointly optimized in an iterative process. In Sect. 4, the performance results obtained by computer simulations of fixed-rate Raptor codes in the UAC are presented. And Sect. 5 presents the conclusions.

2 System Model

The Raptor codes in this paper concatenate weakened LT codes with LDPC codes as pre-codes that can patch the gaps in the LT code. The system model of the transmitter is shown in Fig. 1.

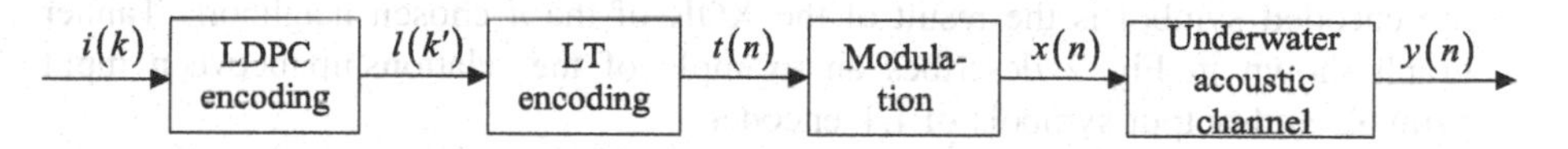

Fig. 1. The system model of the transmitter with Raptor codes.

The Tanner graph of the Raptor codes is shown in Fig. 2, it can be seen that the graph of Raptor codes consists of two component bi-partite tanner graphs (the LDPC codes graph and LT codes graph). The LDPC codes graph consists of source nodes and intermediate nodes (variable nodes), while the LT codes graph consists of intermediate nodes and encoded nodes.

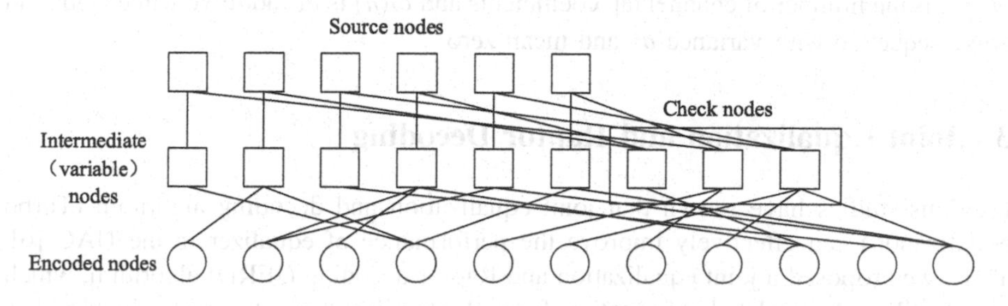

Fig. 2. The Tanner graph of Raptor codes.

2.1 LDPC Encoding

The LDPC codes can be described in terms of a sparse parity check matrix **H**, which can satisfy $\mathbf{Hi} = 0$ for all source codes i. If each column of **H** has the same weight, which means each column has the same number of non-zero elements, and the weight per row is also uniform, this class of LDPC codes is known as regular LPDC codes, otherwise, it is irregular LDPC codes. In this paper, **H** is created by Mackay-1A as followed [8]:

A K' by K matrix (K' rows, K columns) is created at random with the same weight per column, and weight per row as uniform as possible. The overlap between any two columns is no greater than 1 (The overlap between two columns means their inner product).

According to **H**, symbols can be encoded for a given rate.

2.2 LT Encoding

The length of symbols after LDPC encoding is denoted by K' and the length of encoded symbols is denoted by N. The method of LT encoding is as followed:

1. Choose the degree of each encoding symbol randomly according to a degree distribution $\rho(d)$, $d = 1, 2, 3, \ldots, K'$.
2. Select d different input symbols randomly as neighbors of the encoded symbol and the encoded symbol is the result of the XOR of the d chosen neighbors. Tanner graph shown in Fig. 2 describes an example of the relationship between input symbols and output symbols of LT encoder.

Then the output of LT encoder is modulated with Binary Phase Shift Keying (BPSK). The sampled output of a multipath channel can be expressed as followed:

$$y(n) = \sum_{i=0}^{M} h_i x(n-i) + \omega(n) \tag{1}$$

where the $x(n)$ is a BPSK symbol sequence, and $h_0, h_1, \ldots, h_M$ are the tap coefficients of multipath channel impulse response, in this paper, the Bellhop model is adopted. $M + 1$ is the number of channel tap coefficients and $\omega(n)$ is the additive white Gaussian noise sequence with variance σ^2 and mean zero.

3 Joint Equalization and Raptor Decoding

Previous studies have shown that joint equalization and decoding algorithm (Turbo equalization) can effectively improve the performance of equalizer in the UAC [6]. Thus, we proposed a joint equalization and Raptor decoding (JERD) algorithm, which could utilize the updated information from Raptor decoders. As shown in Fig. 3, a feedback loop is added between the equalizer and the Raptor decoder. The performance of detector can be improved by the Turbo iterative process.

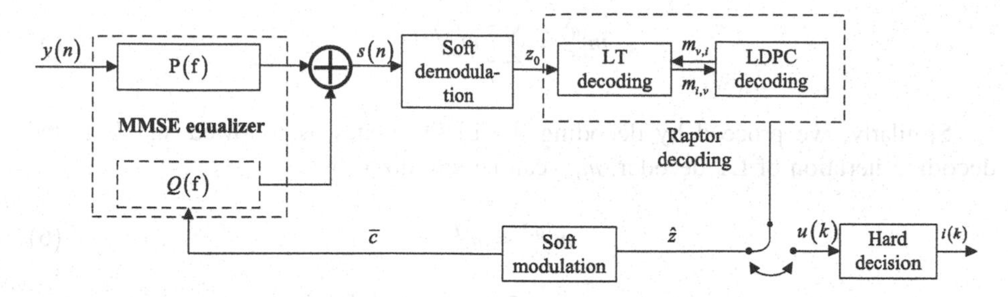

Fig. 3. The structure of JERD.

BP decoding is a common decoding algorithm for Raptor codes, in the iteration process, the message of information passes back and forth between different kinds of nodes. In this paper, we use a global-iteration algorithm that utilizes feedback between the LDPC and LT decoder in the Raptor code [9, 10].

The message of information in the decoder is denoted by m. After equalization and soft modulation, the message is received by the LT decoder. In the LT decoder, $m_{o,i}$ is the message sent from the encoded nodes o to the intermediate nodes i, and $m_{i,o}$ is the message sent from the intermediate nodes i to the encoded nodes o. In addition, the message sent from intermediate nodes i to its corresponding variable nodes v is denoted by $m_{i,v}$, from the variable nodes to the intermediate nodes oppositely by $m_{v,i}$. q is the global decoding iteration, while l_1 is the decoding iteration of LT decoder.

It has been proved that if $u = u_1 \oplus u_2$, and the log-likelihood ratio of u is defined as $L(u)$, then we can get the relation of them as [11]:

$$\begin{aligned} L(u) =& L(u_1 \oplus u_2) \\ =& 2\tanh^{-1}\left(\tanh\left(\frac{L(x_1)}{2}\right)\tanh\left(\frac{L(x_2)}{2}\right)\right) \end{aligned} \tag{2}$$

According to the encoding of LT codes shown in Fig. 2 and (2), the LT codes are decoded as followed:

$$m_{o,i}^{(q,l_1)} = 2\tanh^{-1}\left[\tanh\left(\frac{z_0}{2}\right)\prod_{i'\neq i}\tanh\left(\frac{m_{i',o}^{(q,l_1)}}{2}\right)\right] \tag{3}$$

$$m_{i,o}^{(q,l_1+1)} = \sum_{o'\neq o} m_{o',i}^{(q,l_1)} + m_{v,i}^{(q-1)} \tag{4}$$

where p_1 denotes iteration number of the LT decoder, l_1 represents the iterative number of LT decoding in progress, q expresses the qth global decoding iteration. The soft outputs on the intermediate bits of LT decoder are provided as the channel LLRs of the variable nodes of the pre-code decoder (since the intermediate bits are the variable nodes of the pre-code), as followed:

$$m_{i,v}^{(q)} = \sum_{o} m_{o,i}^{(q,p_1)} \tag{5}$$

Similarly, we proceed by decoding the LDPC codes as followed and l_2 is the decoding iteration of LT decoder, $m_{v,c}$ can be initialized by

$$m_{v,c}^{(q,1)} = m_{i,v}^{(q)} \tag{6}$$

$$m_{c,v}^{(q,l_2)} = 2\tanh^{-1}\left[\prod_{v' \neq v} \tanh\left(\frac{m_{v',c}^{(q,l_2)}}{2}\right)\right] \tag{7}$$

$$m_{v,c}^{(q,l_2+1)} = \sum_{c' \neq c} m_{c',v}^{(q,l_2)} + m_{i,v}^{(q)} \tag{8}$$

Via p_2 iterations of the LDPC decoder, the extrinsic information of the variable nodes in the LDPC decoder is provided as the a-priori information for intermediate bits in the LT decoder in the next global iteration as:

$$m_{v,i}^{(q)} = \sum_{c} m_{c,v}^{(q,p_2)} \tag{9}$$

After g iterations of the global decoder, the messages to variable nodes corresponding to the information bits are:

$$d(s) = \sum_{c} m_{c,v}^{(g,p_2)} + m_{i,v}^{(g)} \tag{10}$$

Then, a hard decision on every information bit is made to recover the source message. The soft information of the input data can be updated by the process of Raptor decoding as [7]:

$$\hat{z} = 2\tan^{-1}\left[\prod_{i}\left(\frac{m_{i,o}^{(g,p_1)}}{2}\right)\right] + z_0 \tag{11}$$

Then, we try to estimate the interfering symbols to cancel the residual ISI by modulating the soft information of those symbols from the last time of Raptor decoding iteration. Thus, the symbols are estimated as followed [12]:

$$\bar{c} = E[c|\,\hat{z}] = \tanh\left(\frac{\hat{z}}{2}\right) \tag{12}$$

To minimize the mean-squared error (MSE) between the equalizer output and the transmitted sequence, after all the information of the symbols from the decoder is obtained, the coefficients of equalizer can be updated as followed:

$$s_n = \mathbf{P_n Y_n} - \mathbf{Q_n \bar{C}_n} \tag{13}$$

$$\mathbf{P_{n+1}} = \mathbf{P_n} - \mu \mathbf{Y_n} \times (s_n - \bar{s}_n) \tag{14}$$

$$\mathbf{Q_{n+1}} = \mathbf{Q_n} + \mu \mathbf{\bar{C}_n} \times (s_n - \bar{s}_n) \tag{15}$$

where $\mathbf{P}_n$ and $\mathbf{Q}_n$ are the coefficients of the equalization, $\mathbf{Y_n}$ is a sequence received by the equalizer from the channel and $\mathbf{\bar{C}_n}$ is the estimates of symbols provided by Raptor decoder. The normalized output of the MMSE linear equalizer is denoted by s_n, while $\bar{s}_n$ is the hard-decision symbols of the output information.

In the process of Turbo equalization, the equalizer can utilize the information from the decoder to adaptively track the channel to mitigate the Inter-Symbol Inference (ISI). Meanwhile, as the iteration of Turbo equalization proceeds, the updated output information of equalizer can improve the performance of Raptor decoder.

4 Simulation Results and Discussion

In this section, we evaluate the performance of Raptor codes of 1/2 in the UAC, which is generated by the Bellhop model. The setting parameters of Bellhop model are given in Table 1. For each of the code rates, 950 information bits are encoded by a rate 0.95 LDPC code of left degree three to produce 1000 intermediate bits. In the global Raptor decoder, we perform 2 global iterations, with 7 LT decoding iterations and 3 LDPC decoding iterations per global iteration.

Table 1. Parameters Setting of Bellhop Model

Parameter	Value
Sound frequency	15 kHz
Communication distance	800 m
Sea depth	45 m
The depth of source and receiver	10 m
Sound speed	1460–1480 m/s
The number of beams	10
The launch angle of ray	−15°–+15°

The effect of iteration on the BER performance for Raptor codes with different signal to noise ratio (SNR) is shown in Fig. 4. It can be seen from Fig. 4 that with the increase of iteration number, the BER gradually decreases, and eventually tends to be flat. For SNR = 5 dB, the effect of iteration number is not obvious, the BER is only improved a little. When the SNR is set to 6 dB and 7 dB, a noticeable correlation between the iteration number and the BER performance has appeared. As the iteration number increases, the BER performance is improved significantly in the thirteen iterations and then tends to be stable. Thus, in the following simulation, we set the iteration number as thirteen.

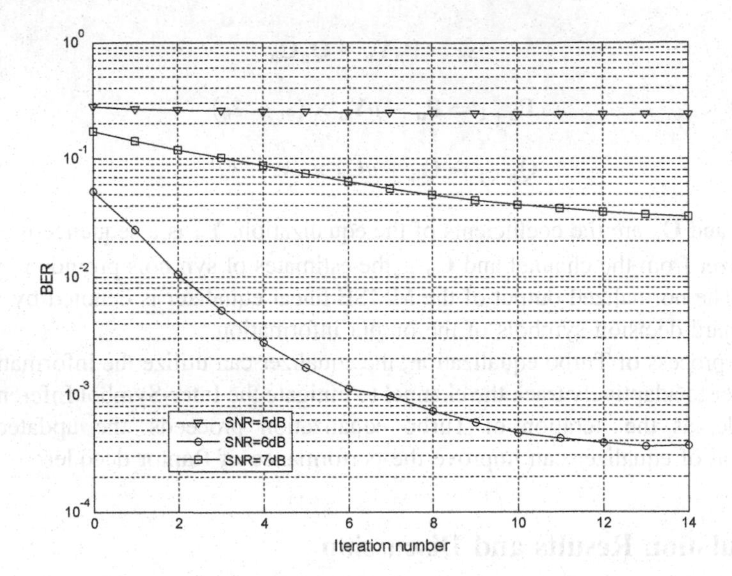

Fig. 4. The BER performance of the JERD in different iterations

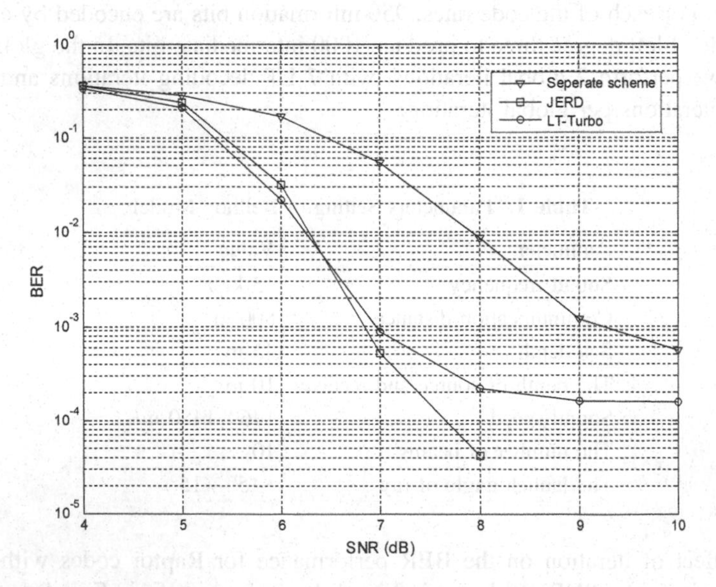

Fig. 5. The BER performance comparison of different schemes

Figure 5 gives the BER comparison of different schemes including the LT-Turbo [7], the separate scheme and the proposed JERD. In the separate scheme, the iteration process is not be used, the equalization and Raptor decoding are separately implemented. It is observed that the JERD achieves better BER performance than the separate scheme (at BER = 10^{-3}, about 2.5 dB gain). This is because the iteration process can improve the correctness of feedback information. It is also observed that when the SNR is greater than 7 dB, the BER performance of JERD is significantly better than that of LT-Turbo. This is because that the Raptor decoding outperforms the LT decoding, in the iterative process, the feedback soft information is more accurate, with the iterative number increasing, better BER performance can be achieved.

Figure 6 shows the relationship between the decoding success rate (That means the ratio of the number of complete recovery to the total number of experiments) and code redundancy. To illustrate the relationship between the decoding success rate and redundancy of the joint equalization decoding scheme based on rateless code, in the simulation, the curve in Fig. 6 is obtained by averaging on 1000 independent experiments.

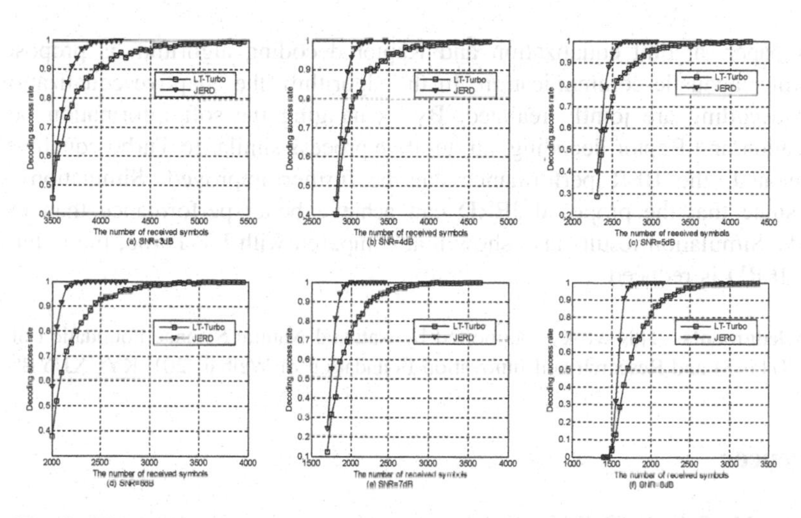

Fig. 6. The relationship between decoding success probability and redundancy

It can be seen from Fig. 6 that under different signal-to-noise ratios, for successfully decoding, the number of symbols required by JERD is less than that by LT-Turbo, this means that the redundancy required for successful decoding with JERD is smaller. It can also be observed from the Fig. 6 that as the signal-to-noise ratio increases, the redundancy required for successful decoding gradually decreases. When the decoding success rate is higher than 99%, the specific number of the received symbols is shown in Table 2.

In Table 2, the number of received symbols is given when the decoding success rate reaches 99%. In the simulation, the SNR varies from 3 dB to 8 dB. It can be seen that compared to the LT-Turbo, JERD uses fewer encoded symbols to recover the original information, which means that the redundancy rate of JERD is obviously lower.

Table 2. Simulation results for the JERD and LT-Turbo

SNR (dB)	The number of original symbols	JERD: the number of received symbols and redundancy rate	LT-Turbo: redundancy rate
3	1000	3950, 3.95	5.05
4	1000	3150, 3.15	4.25
5	1000	2600, 2.60	3.55
6	1000	2250, 2.25	3.15
7	1000	2000, 2.00	2.90
8	1000	1800, 1.80	2.75

5 Conclusions

In this paper, a joint equalization and Raptor decoding algorithm is proposed for underwater acoustic communication. In the algorithm, the adaptive equalization and Raptor decoding are jointly realized. By exchanging the soft information between equalization and Raptor decoding, an iterative process similar to Turbo equalization is implemented, the BER performance can be further improved. Simulation results demonstrate that the proposed JERD can achieve better performance than existing methods. Simulation results also show that compared with LT-Turbo, the redundancy rate of JERD is reduced.

Acknowledgments. This work is supported by National Natural Science Foundation of China (No. 61871148) and Research and Innovation Foundation of Weihai (2019KYCXJJYB04).

References

1. Chitre, M., Soh, W.: Reliable point-to-point underwater acoustic data transfer: to juggle or not to juggle? IEEE J. Ocean. Eng. **40**, 93–103 (2015)
2. MacKay, D.J.C.: Fountain codes. IEEE Proc. Commun. **152**, 1062–1068 (2005)
3. Shokrollahi, A.: Raptor codes. IEEE Trans. Inf. Theory **52**, 2551–2567 (2006)
4. Rafati, A., Lou, H., Xiao, C.: Soft-decision feedback turbo equalization for LDPC-coded MIMO underwater acoustic communications. IEEE J. Ocean. Eng. **39**, 90–99 (2014)
5. Otnes, R., Tuchler, M.: Iterative channel estimation for turbo equalization of time-varying frequency-selective channels. IEEE Trans. Wireless Commun. **3**, 1918–1923 (2004)
6. Yang, Z., Zheng, Y.R.: Iterative channel estimation and turbo equalization for multiple-input–multiple-output underwater acoustic communications. IEEE J. Ocean. Eng. **41**, 232–242 (2016)
7. Tai, Y.P.H., Wang, B., Wang, H.X., Wang, J.: A novel LT-Turbo equalization for long-range deep-water coustic communication. SCIENTIA SINICA Physica, Mechanica & Astronomica **46**, 96–103 (2016)
8. MacKay, D.J.C., Neal, R.M.: Near Shannon limit performance of low density parity check codes. Electron. Lett. **33**, 457–458 (1997)
9. Sivasubramanian, B., Leib, H.: Fixed-rate raptor code performance over correlated rayleigh fading channels. In: 2007 Canadian Conference on Electrical and Computer Engineering, Vancouver, BC, pp. 912–915 (2007)

10. Sivasubramanian, B., Leib, H.: Fixed-rate raptor codes over rician fading channels. IEEE Trans. Veh. Technol. **57**, 3905–3911 (2008)
11. Hagenauer, J., Offer, E., Papke, L.: Iterative decoding of binary block and convolutional codes. IEEE Trans. Inf. Theory **42**, 429–445 (1996)
12. Lopes, R.R., Barry, J.R.: The soft-feedback equalizer for turbo equalization of highly dispersive channels. IEEE Trans. Commun. **54**, 783–788 (2006)

Design of Wireless Communication System for CNC Machine Tools

Rui E(✉)

Heilongjiang Polytechnic, Harbin, Heilongjiang 150001, China
e_rui@126.com

Abstract. This paper first introduces the development history and future development trends of computer numerical control (CNC) machine tools. In order to greatly improve the work efficiency, this paper proposes a new type of CNC machine tool system, the central control device and the machine tool are separated, which can carry out one-to-many centralized management. In addition, in order to achieve the goal of mobile management, this paper also proposes an embedded CNC handheld terminal program, which can assist the work of CNC machine tools and save labor costs.

Keywords: CNC machine tools · Zigbee star network · Embedded Linux

1 Introduction

The machine tool is an important element of the integration of advanced manufacturing technology and manufacturing information. It is an indispensable complex production tool for the development of the machine manufacturing industry and the entire industry. It is not only an element of productivity, but also an important commodity. To a certain extent, the development and innovation of machine tools reflect the main trends of processing technology. Modern CNC machine tools comprehensively apply the latest achievements in mechanical design and manufacturing process, computer automatic control technology, precision measurement and detection, information technology, artificial intelligence and other technical fields, and will develop toward the trend of high-speed, precision, compound, flexible, extreme and so on.

The world's first CNC machine tool was first developed by the Massachusetts Institute of Technology in 1952, and was immediately put into production of 100 units. Japan developed the first CNC machine tool in 1958. and China's CNC machine tool and technology also started in 1958. The development process of nearly 50 years can be divided into three stages. The first stage, from 1958 to 1979, is in a closed development stage. The second stage is in the "sixth five-year plan", "seventh Five-year Plan" and the early stage of "eighth Five-year Plan", that is, introducing technology, digesting and absorbing, and initially establishing a localization system. The third stage is in the late period of the eighth five-year Plan and the ninth five-year Plan period, that is, to carry out industrialization research and enter the stage of market competition.

Due to the rapid development of computer technology, the technology of CNC machine tools is promoted faster. Many CNC system manufacturers in the world use the

S. Shi et al. (Eds.): AICON 2020, LNICST 356, pp. 136–145, 2021.
https://doi.org/10.1007/978-3-030-69066-3_13

rich software and hardware resources of PCs to develop a new generation of CNC systems with an open architecture. The open architecture enables the CNC system to have better versatility, flexibility, adaptability, and expandability, and it has developed greatly in the direction of intelligence and networking. In recent years, many countries have researched and developed this kind of system, such as the "Next Generation Workstation/Machine Controller Architecture" NGC jointly led by the United States Scientific Manufacturing Center (NCMS) and the Air Force, and the European Community's "Open Architecture in Automation System" "OSACA", Japan's OSEC plan, etc.

The development and research results have been applied. For example, A2100 system with open architecture has been adopted by Cincinnati Milacron since 1995 in its production of machining center, CNC milling machine, CNC lathe and other products. A new generation of open architecture CNC system, its hardware, software, and the bus specification are opening. Because there are sufficient software and hardware resources available, not only thc system integration carried out by CNC system manufacturers and users is strongly supported, but it also brings great convenience to the user's secondary development, and promotes the multi-grade, multi-variety development and wide application.

The development of information technology makes machine tools develop towards the direction of digitization and intelligentization. Intelligent technology is adopted to realize the functions of reconstruction and optimization under multi-information fusion, such as intelligent decision making, process adaptive control, error compensation intelligent control, motion trajectory optimization control of complex surface machining, fault self-diagnosis, intelligent maintenance and information integration, which will greatly improve the forming and machining accuracy. The emergence of intelligent machine tools has created conditions for the future equipment manufacturing industry to realize complete automation of production. By automatically suppressing vibration, reducing thermal deformation, preventing interference, automatically adjusting the amount of lubricating oil, reducing noise, etc., the machining accuracy and efficiency of machine tools can be improved. The development and innovation of CNC system has played a significant role in the intelligence of machine tools. It can absorb a large amount of information and store, analyze, process, judge, adjust, optimize and control all kinds of information.

Product upgrading and people's demand for diversified and individualized products have made the market's demand for manufacturing systems with good flexibility and diverse processing capabilities exceed the demand for large-scale single manufacturing systems, which makes CNC machine tools move towards modularity. The development of reconfigurable and expandable flexibility also requires efficient production in a multi-variety, variable batch environment. Especially the development of automobile manufacturing industry and electronic communication equipment manufacturing industry has put forward higher requirements for production efficiency. As the automation of the manufacturing process increases, machine tools are required not only to complete the usual processing functions, but also to have functions such as automatic measurement, automatic loading and unloading, automatic tool change, automatic error compensation, automatic diagnosis, wire feeding and networking.

The intelligence of machine tool numerical control system is the general trend, and the proportion of software in the numerical control system is increasing day by day, so that the numerical control system not only controls the machine tool movement, but also provides many possibilities for the optimized design and rational use of the machine tool. In recent years, in order to better meet the "high speed, high precision, compound, intelligence, and environmental protection" requirements of machine tool users for CNC machine tools, domestic and foreign CNC machine tool companies have continuously developed various types of CNC machine tools with intelligent functions. Especially with the development and application of machine tool design technology, Internet and information technology, and artificial intelligence technology, CNC machining has gradually shown a trend of networking and intelligence, and the intelligent functions of CNC machine tools have become more and more mature.

2 Design of Wireless Control System Based on ZigBee Star Network

2.1 Brief Introduction of ZigBee

CNC machine tool is the abbreviation of Computer numerical control machine tool. It is an automatic machine tool equipped with program control system. CNC machine tool makes it more convenient to process complex, precise, small batch and various kinds of parts. It has the advantages of good flexibility and high efficiency. In the direction of modern CNC machine tool control technology, it makes electro-mechanical integration. However, there are many problems in CNC machine tools. Firstly, the central control device is no longer connected with CNC machine tools, but wireless transmission information through ZigBee network can realize remote control. Secondly, a control device can control multiple CNC machine tools at the same time, saving resources and realizing centralized management. In addition, with the help of ZigBee network, the whole system can realize two-way communication, which can not only control the machine tool, but also monitor the working state of the CNC machine tool according to the feedback information at the control end. The above three characteristics well avoid the problems of traditional machine tools.

Zigbee Communication Protocol: ZigBee is a short distance, low power wireless communication technology based on IEEE802.15.4 standard. ZigBee represents the protocol based on Internet of things standard. It has the advantages of low power consumption, low cost, moderate transmission efficiency and distance, short delay and so on. The independence of each module in the network is relatively high. After connecting with the network, the data can be transmitted through the network, and each module is uploaded step by step. The characteristics of this technology can make industrial control, intelligent agriculture and intelligent buildings get a good development. When ZigBee establishes a network, it needs to judge whether the node is connected to other networks at the same time There can only be one coordinator in the network. When the network is established, the coordinator in the network node will enter the process router as a router. All this is because ZigBee network is distributed. As a wireless short-distance transmission technology, ZigBee has a wide range of applications in the field of Internet of things because its network can provide users with wireless transmission function conveniently.

Zigbee Network Node: ZigBee network node uses CC2530 board card. CC2530 is a real system on chip (SOC) solution for 2.4 GHz IEEE 802.15.4, ZigBee and RF4CE applications. It can build powerful network nodes with very low total material cost. ZigBee network topology mainly includes star network and network network. Different network topologies correspond to different application fields. In ZigBee wireless network, different network topologies have different configuration of network nodes. The types of network nodes are coordinator, router and terminal node. Mesh mesh network topology structure of the network has a strong function, the network can communicate through multi hop; the topology can also form a very complex network; the network also has the function of self-organization and self-healing.

2.2 Design Scheme

In this system, the CC2530 on the control side plays the role of coordinator, undertakes the task of establishing ZigBee network, and multiple CC2530 of multiple CNC machine tools play the role of convergence node, looking for and joining the ZigBee network established by the coordinator. If the memory occupied by the whole system is too large, some CC2530 nodes can be added as routers and relays to expand the communication distance of the whole network. The model of design scheme is shown in Fig. 1.

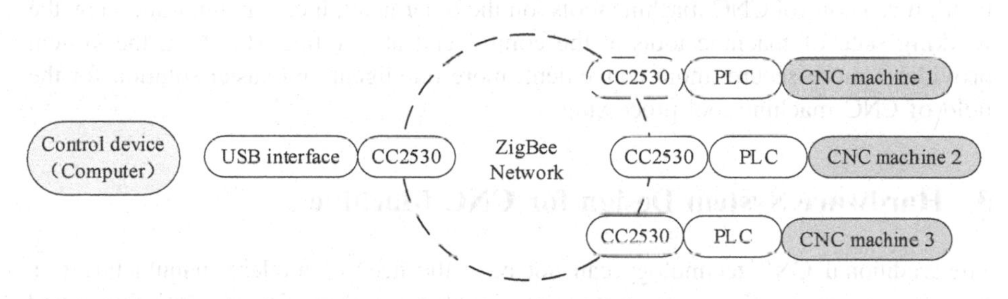

Fig. 1. Design scheme

In the control process, the command transmission process is as follows: the operator sends a command to the control device, that is, the ordinary computer. The command is written into the USB interface through the system special software or serial port assistant, and is transmitted to the CC2530 board card. The board card receives and analyzes the command, determines the controlled machine tool corresponding to the command, determines the connected CC2530 address, and calls the Z-stack protocol stack sending function The information is sent to the destination CC2530 in a wireless way. After receiving the information, the receiving board still needs to carry out some necessary processing, and then in the form of digital electrical signals sent to the CNC machine tool PLC, PLC to complete the processing task.

General Process of Zigbee Network: the initialization of ZigBee network can only be initiated by the network coordinator. Before establishing the network, it is necessary to judge that this node has not been connected with other networks. If the node has been connected with other networks, this node can only be used as a sub node of the network. Therefore, the first step is to determine the network coordinator, the second step is to scan the channel, and the third step is to set the network ID. after that, each node can join the network through the ID. The process of node joining is to find network coordinator, send association request command, wait for coordinator to process, send data request command and wait for reply.

Based on the above process, the program in CC2530 can guarantee the reliability of ZigBee network. For example, the sender adds the serial number to each message sent, and adds one to each signal sent. In this way, the receiver can judge whether there is information loss according to the serial number. If the receiver receives the information correctly, it will send a confirmation message to the sender, otherwise the sending node will resend the message.

To sum up, the system realizes the separation of operators and machine tools, and ensures the personal safety of operators to a certain extent. Through the application of ZigBee protocol, the control end and the working end of the CNC system are isolated, and the control end is separated from the harsh production environment, which reduces the requirements for the central control instrument. Because of the join of ZigBee network, one to many control mode is realized, that is, a central control computer can control multiple CNC machine tools, which is convenient for unified management and overall coordination. The system provides two-way communication link. On the one hand, it can control CNC machine tools, on the other hand, it can monitor and view the working state of machine tools at the control end at any time. In short, the system provides a more secure, more convenient, more intelligent and easier solution for the field of CNC machine tool processing.

3 Hardware System Design for CNC Machine

The traditional CNC technology can not meet the need of modern manufacturing, a new generation CNC system is towards intelligent, open, flexible and diversified development. We must seize the opportunity, take advantage of talents, develop CNC systems that adapt to the situation of modern high-end CNC systems, and promote the development of CNC industrialization and manufacturing.

At present, the handheld terminals used in CNC equipment are still in the conceptual design stage and functional testing, most of which use the 816 microcontroller as the processor, and the physical keyboard can perform some simple control operations. The hardware and function expansion are difficult and friendly. The man-machine interface is not enough. With the development of manufacturing informatization, more sophisticated and powerful handheld terminal equipment will have broad application prospects in the machinery manufacturing industry.

3.1 System Design

In order to solve these problems, this paper proposes a development plan for embedded CNC handheld terminals for workshop CNC machine tools. The plan includes two parts. The first part is the programming of the hardware development board; the second part is the programming of the mobile phone (Android terminal) For the development, the hardware development board uses a high-performance low-power 32-bit embedded microcontroller and a general-purpose operating system for embedded processors.

The hardware development board is connected to the CNC machine tool using a serial port or USB interface, and the operator sends instructions in real time (via Wi-Fi), and converts it into a format that can be recognized by the CNC machine tool. The CNC machine tool is sending work instructions.

After programming the kernel and the driver program, in terms of software design, read the user's instructions through Wi-Fi, query the commands corresponding to the semantic instructions stored in the internal storage, convert them into the format recognized by the CNC machine tool, and then pass the serial port Send to the CNC machine tool to complete the design of the hardware development board. If the scheme can be applied in practice, it can keep workers away from the machine tool, ensure safety, and repeatedly input instructions to improve work efficiency.

3.2 Hardware Selection

Based on the difficulty of system migration and further development and practical applicability, this article finally chooses OKMX6Q development board as the hardware platform.

The OKMX6Q development board uses Samsung's Cortex-A9 architecture and NXP quad-core i.MX6Q processor. It has strong internal resources and program processing capabilities, and integrates a variety of interfaces and peripherals, such as cameras, Ethernet interfaces, USB interface, LCD screen, SD card reader, etc. In addition, the OKMX6Q development board officially provides a wealth of technical information, which provides very convenient conditions for system configuration, cross-compilation environment construction, message and file transfer program writing, etc., which will lay a good foundation for further development. The OKMX6Q development board is shown in Fig. 2.

Fig. 2. The OKMX6Q development board

3.3 Uboot Startup Program

A bootloader program is required to start a Linux system, which means that a bootloader program is run after the chip is powered on. This bootloader program will first initialize DDR and other peripherals, then copy the Linux kernel from flash (NAND, NORFLASH, SD, MMC, etc.) to DDR, and finally start the Linux kernel. Of course, the actual work of the bootloader is much more complicated, but its main job is to start the Linux kernel. The relationship between the bootloader and the Linux kernel is the same as the relationship between the BIOS on the PC and Windows, and the bootloader is equivalent to the BIOS (Fig. 3).

```
/mx6ullevk/imximage-ddr512.cfg.cfgtmp board/freescale/mx6ullevk/imximage-ddr512.cfg
  ./tools/mkimage -n board/freescale/mx6ullevk/imximage-ddr512.cfg.cfgtmp -T imximage -e 0x87
800000 -d u-boot.bin u-boot.imx
Image Type:    Freescale IMX Boot Image
Image Ver:     2 (i.MX53/6/7 compatible)
Mode:          DCD
Data Size:     425984 Bytes = 416.00 kB = 0.41 MB
Load Address: 877ff420
Entry Point:   87800000
```

Fig. 3. Successfully compiled code

3.4 Operating System

Embedded Linux is an operating system that tailors the increasingly popular Linux operating system to make it run on embedded computer systems. Embedded Linux not only inherits the unlimited open source code resources on the Internet, but also has the characteristics of an embedded operating system. Embedded Linux is characterized by free copyright fees, excellent performance, easy software migration, open code, support for many application software, short application product development cycle, and rapid launch of new products because there are many open codes that can be referenced and transplanted.

3.5 Linux Kernel Tailoring and System Porting

Linux kernel tailoring. First copy the Linux3.0.35 source code compressed package to the Ubuntu home directory, and enter in the terminal:

```
# tar xjf linux-3.0.35.tar.bz3
# cd linux-3.0.35
# make distclean
#cp arch/arm/configs/imx6_s3_defconfig .config
# make imx6_defconfig
# make menuconfig
```

At the start of the graphical configuration interface, check the following parts:

Networking support → Networking options →
Networking packet filtering framework(Netfilter) → IP: Netfilter configuration →
<*> IP Userspace queueing via NETLINK(OBSOLETE),as shown in Fig. 4.

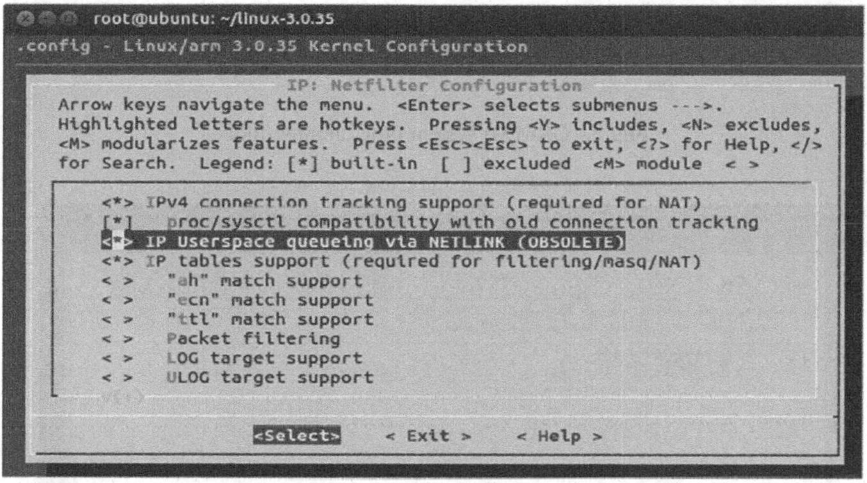

```
root@ubuntu: ~/linux-3.0.35
.config - Linux/arm 3.0.35 Kernel Configuration
                         IP: Netfilter Configuration
  Arrow keys navigate the menu.  <Enter> selects submenus --->.
  Highlighted letters are hotkeys.  Pressing <Y> includes, <N> excludes,
  <M> modularizes features.  Press <Esc><Esc> to exit, <?> for Help, </>
  for Search.  Legend: [*] built-in  [ ] excluded  <M> module  < >

      <*> IPv4 connection tracking support (required for NAT)
      [*]   proc/sysctl compatibility with old connection tracking
      <*> IP Userspace queueing via NETLINK (OBSOLETE)
      <*> IP tables support (required for filtering/masq/NAT)
      < >   "ah" match support
      < >   "ecn" match support
      < >   "ttl" match support
      < >   Packet filtering
      < >   LOG target support
      < >   ULOG target support

                   <Select>    < Exit >    < Help >
```

Fig. 4. Kernel module configuration

Finally, enter the following command:

make uImage

Complete the cross-compilation of the modified kernel and obtain the kernel file uImage.

Linux System Porting. Replace the modified and compiled kernel image uImage with the original file with the same name, together with the compiled uboot and rootfs file system, using the mfgtool3-qt-OKMX6-S-emmc.vbs system burning tool (as shown in Fig. 5), import the powered-on OKMX6Q platform together, and then reset or power on again to complete the Linux3.0.35 system migration.

Specific steps are as follows:

The first step is to install a cross compiler to build a development environment. The compiler uses arm-linux-gcc-4.5.1, which uses armv7 instructions when compiling and supports floating point operations. Unzip the compiler to the root directory of the Linux system of the computer virtual machine. Add the path of the compiler to the system environment variable.

The second step is to make the root file system. Use Busybox to make root file system with static compilation method (Fig. 6).

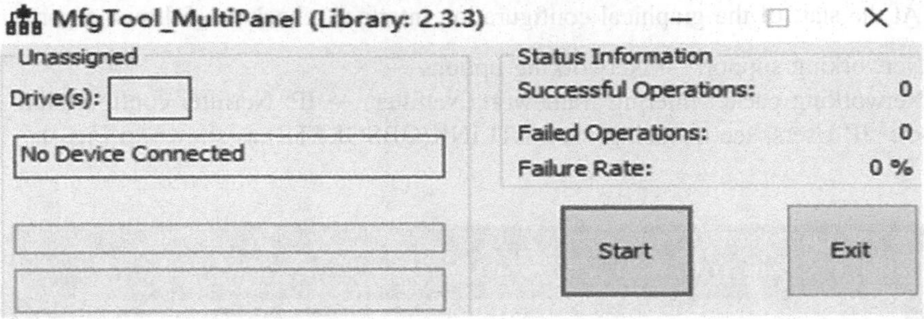

Fig. 5. OKMX6 system programming tool

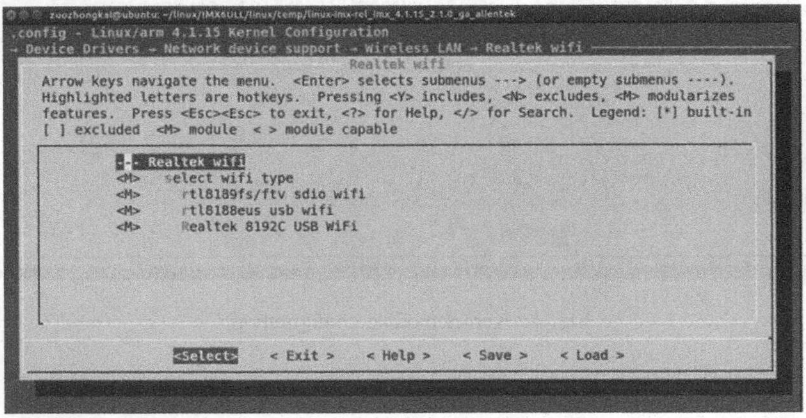

Fig. 6. WIFI configuration interface

The third step is to configure and compile the Linux kernel. Decompress the Linux kernel into the virtual machine Linux, use the make menuconfig (graphical kernel configuration interface) command, configure the root file system startup mode as nfs, use the cross compiler to compile the kernel, and output the Linux kernel image file zImage.

The fourth step is to use the dnw tool to download the Linux kernel image to the development board via USB.

3.6 Wireless Network Card Driver Transplantation and Configuration

The OKMX6Q development board currently supports two interfaces of WIFI: USB and SDIO. The chip used by USBWIFI is RTL8188EUS, and the chip used by SDIO interface is RTL8189FS, also called RTL8189FTV. Both are WIFI chips produced by Realtek. The WIFI driver does not need to be written by us, because Realtek company provides the WIFI driver source code, so we only need to add the WIFI driver source code to the Linux kernel, and then configure it through the graphical interface and choose to compile it into a module. Execute the "make menuconfig" command to open the Linux kernel configuration interface, and then select the following path to compile the rtl81xx driver into a module:

The configuration interface is the WIFI configuration interface we added. Select "rtl8189 fs/ftv sdio wifi", "rtl8188eus usb wifi" and "Realtek 8193C USB WiFi" and compile them into modules. Execute the following command to compile the module:

make modules -j13//Compile the driver module.

8188eu.ko, 8189fs.ko and 8193cu.ko are the RTL8188EUS, RTL8189FS and RTL8188CUS/8193CU driver module files, copy these three files to the ootfs/lib/modules/4.1.15 directory.

3.7 Writing Application Software

The developed application software should be able to receive the data sent by Wi-Fi, and query the corresponding machine tool commands based on the data, such as tool change, set length, etc., and send the converted commands to the PCL machine tool CNC through the serial port or infrared interface Module. When the machine tool runs, it will return a success or failure command, which is received by the hardware development board to inform the user of the operation result. If necessary, the corresponding interface program can be developed. The machine tool can be directly controlled on the development board to ensure the instruction Can accurately enter the CNC machine tool.

References

1. Shen, S., Mao, Y., Fan, Q.: The IoT concept model and system structure. J. Nanjing Univ. Posts Telecommun. Natl. Sci. Edn. **30**(004), 1–8 (2010)
2. Zhang, Z., Zhou, J., Yan, M.: Application research of Zigbee in smart home system. Ind. Control Comput. **19**(20), 7–9 (2006)
3. Dissanayake, S.D., et al.: Zigbee wireless vehicular identification and authentication system. In: 2008, 4th International Conference on Information and Automation for Sustainability, ICICAFS. IEEE (2008)
4. Chen, X.: Design of PCL control system for automatic swivel tool post of CNC lathe. Mach. Tool Electric. **3**, 38–39 (2008)
5. Gong, J., Xiong, G.: Visual C++ Communication Programming Practice. Electronic Industry Press, Beijing, 10 (2004)

Energy Efficiency Optimization for Subcarrier Allocation-Based SWIPT in OFDM Communications

Xin Liu[1,2](✉) and Yuting Guo[1]

[1] School of Information and Communication Engineering, Dalian University of Technology, Dalian 116024, China
liuxinstar1984@dlut.edu.cn, guoyuting@mail.dlut.edu.cn

[2] The 54th Research Institute of CECT, Shijiazhuang 050081, China

Abstract. Simultaneous wireless information and power transfer (SWIPT) is a promising technology to realize simultaneous information and energy transfer by utilizing radio frequency signals. It extends the life of wireless networks and is conducive to the realization of green communications. In this paper, a subcarrier allocation-based SWIPT is studied to transfer information and energy in different subcarriers of an Orthogonal Frequency Division Multiplexing (OFDM) communication system. To improve the SWIPT performance, we maximize the energy efficiency of OFDM communication system while satisfying the constraints including minimum harvested energy, target rate and transmit power budget. To obtain the optimal solution, we investigate a dual-layer iterative optimization algorithm from Lagrange dual function to solve the energy efficiency optimization problem. The simulation results show that the energy efficiency of the proposed scheme can be effectively improved.

Keywords: SWIPT · Subcarrier allocation · Power allocation · Energy efficiency

1 Introduction

For the past few years, simultaneous wireless information and power transfer (SWIPT) has been proposed as a promising method to realize green communications, due to that it can continuously provide energy to those devices supplied by finite batteries. Thus, SWIPT can lengthen the working time of energy-constrained wireless networks to some extent [1]. Since wireless signals carrying radio frequency (RF) energy can be used for transmitting information, SWIPT have the capability of performing both energy acquisition and information transmission at the same time. In [2], Varshney first proposed the notion of SWIPT and characterized the fundamental tradeoff between information and energy transfer with a proposed capacity-energy function. And the work in [2] is extended to frequency-selective channels in [3]. However, these studies are based

S. Shi et al. (Eds.): AICON 2020, LNICST 356, pp. 146–154, 2021.
https://doi.org/10.1007/978-3-030-69066-3_14

on the assumption that the circuits at the receiver for energy harvesting are ideally thought to be capable of simultaneously decoding information directly from the same received signal.

In order to implement SWIPT within practical circuit constraints, two traditional practical SWIPT schemes including power splitting (PS) [4] and time switching (TS) [5] are proposed. In the TS scheme, the receiver can switch between decoding mode and energy harvesting mode in the time domain according to the instantaneous channel condition, so it is not strictly simultaneous. And in the PS scheme, the received signals are divided into two streams at the receiver through the power splitter in a certain proportion, which can be utilized to decode information and harvest energy separately.

Orthogonal Frequency Division Multiplexing (OFDM), a popular multicarrier transmission technique, can effectively achieve high rate transmission and has been widely adopted in various standards [6]. Therefore, the strengths of SWIPT can be fully used to achieve efficient wireless transmission by combining OFDM with SWIPT [7]. The multiuser OFDM system, based on two SWIPT transmission scheme including power splitting and time switching, is studied to achieve maximum sum of information rate in [8]. However, in the SWIPT OFDM system applied with power splitting scheme or time switching scheme, a power splitter or time switcher is required at the receiver, which increases the complexity of the circuits. Thus, the study in [9] proposed a resources allocation algorithm to obtain maximum energy at the receiver which does not require a splitter.

Since saving energy has been considered a matter of great urgency. In this paper, an algorithm for obtaining energy efficiency maximization will be studied to realize the effective utilization of resources. All received subcarriers will be split into two sets, and the subcarriers in the two sets will be used for energy harvesting or information decoding respectively. The receiver knows the subcarrier distribution and carries on the information transmission.

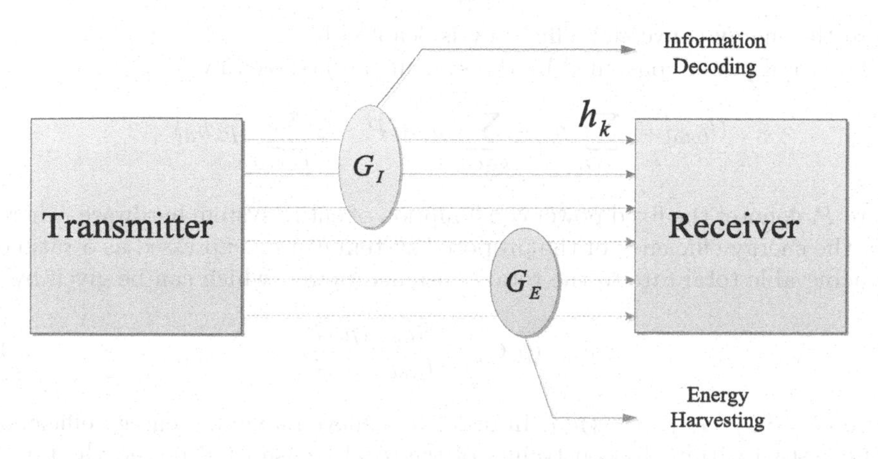

Fig. 1. System model.

2 System Model and Problem Formulation

2.1 System Model

The SWIPT enabled OFDM system, as shown in Fig. 1, is composed of a transmitter (Tx) and a receiver (Rx); each device is assumed to be equipped with one single antenna. During transmission, the entire bandwidth of the link Tx → Rx is split into K subcarriers. Then all these subcarriers are combined into a set and denoted as $K = \{1, ..., k\}$. We denote the channel power gain on the subcarrier k as h_k and assume that the transmitter knows about h_k. P is considered as the total transmission power of transmitter, and the power allocated to subcarrier k is set to p_k. At Rx, on each subcarrier noise n_k will destroy the received signal. The noise n_k is considered as an additive Gaussian white noise (AWGN) and follows a $n_k \sim CN(0, \sigma_k^2)$ distribution. Moreover, slow fading is considered in this paper that all coefficients associated with the channel conditions are assumed to be constant during one transmission period. In the transmission process, Rx uses subcarriers in G_I to decode information and subcarriers in G_E to collect energy, where $G_I \in K$, $G_E \in K$ and $G_I + G_E = K$.

2.2 Problem Formulation

For the transmission link Tx → Rx, achievable rate can express as

$$R = \sum_{k \in G_I} \ln(1 + \frac{p_k h_k}{\sigma_k^2}) \tag{1}$$

The energy harvested by Rx can be written as

$$Q = \xi \sum_{k \in G_E} (p_k h_k + \sigma_k^2) \tag{2}$$

where the energy conversion efficiency is denoted by ξ.

The total power consumed by the system can be given by

$$U_{total} = \sum_{k \in G_I} p_k + \sum_{k \in G_E} p_k + P_c - \xi \sum_{k \in G_E} (p_k h_k) \tag{3}$$

where P_c denotes the fixed power consumption of entire system hardware. Therefore, the energy efficiency of the proposed system can be expressed as a ratio of the achievable total rate to the total consumed power, which can be given by

$$E_{eff}(p, G) = \frac{R_{total}(p, G)}{U_{total}(p, G)} \tag{4}$$

where $G = \{G_I, G_E\}$, $p = \{p_k\}$. In order to achieve maximum energy efficiency of the system within the constraints of the total transmitted power, the target rate and the minimum harvested energy. Then optimization problem can be expressed as

$$\max \frac{\sum_{k\in G_I} \ln(1+\frac{p_k h_k}{\sigma_k^2})}{\sum_{k\in G_I} p_k + \sum_{k\in G_E} p_k + P_c - \xi \sum_{k\in G_E} (p_k h_k)} \tag{5a}$$

$$\text{s.t. } \xi \sum_{k\in G_E} (p_k h_k) \geq E_{min} \tag{5b}$$

$$\sum_{k\in G_I} p_k + \sum_{k\in G_E} p_k \leq P \tag{5c}$$

$$\sum_{k\in G_I} \ln(1+\frac{p_k h_k}{\sigma_k^2}) \geq R_T \tag{5d}$$

where R_T and E_{min} respectively represent target rate and minimum harvested energy requirement that need to be met. And σ_k^2 is not considered in the energy collection. It can be observed from the formulated equation that the objective function is fractional form, which is hard to be solved directly. Therefore, we can convert the fraction into a new objective function in subtractive form. The maximum achievable energy efficiency is denoted as q^*, which can be written as

$$q^* = \frac{R_{total}(p^*, G^*)}{U_{total}(p^*, G^*)} = \max \frac{R_{total}(p, G)}{U_{total}(p, G)} \tag{6}$$

We can obtain maximum achievable energy efficiency only when $\max R_{total}(p, G) - q^* U_{total}(p, G) = R_{total}(p^*, G^*) - q^* U_{total}(p^*, G^*) = 0$ is satisfied. By utilizing q, the original optimization problem is then transformed into the following form

$$\max \sum_{k\in G_I} \ln(1+\frac{p_k h_k}{\sigma_k^2}) - q(\sum_{k\in G_I} p_k + \sum_{k\in G_E} p_k + P_c - \xi \sum_{k\in G_E} (p_k h_k)) \tag{7a}$$

$$\text{s.t. } \xi \sum_{k\in G_E} (p_k h_k) \geq E_{min} \tag{7b}$$

$$\sum_{k\in G_I} p_k + \sum_{k\in G_E} p_k \leq P \tag{7c}$$

$$\sum_{k\in G_I} \ln(1+\frac{p_k h_k}{\sigma_k^2}) \geq R_T \tag{7d}$$

3 Optimal Solution

We can observe that our optimization problem is nonconvex. If subcarriers number is sufficient enough and the "time-sharing" condition can be satisfied, Lagrangian dual function and Dinkelbach method can be utilized to figure out the proposed problem. The associated Lagrangian dual function of (7a) can be written as

$$g(\beta) = \max_{\{p,G\}} L(p, G) \tag{8}$$

$$\begin{aligned} L(p,G) = &\sum_{k\in G_I} \ln(1+\frac{p_k h_k}{\sigma_k^2}) - q(\sum_{k\in G_I} p_k + \sum_{k\in G_E} p_k + P_c - \xi \sum_{k\in G_E}(p_k h_k)) \\ &+ \beta_1(\xi \sum_{k\in G_E}(p_k h_k) - E_{\min}) + \beta_2(P - \sum_{k\in G_I} p_k - \sum_{k\in G_E} p_k) \\ &+ \beta_3(\sum_{k\in G_I} \ln(1+\frac{p_k h_k}{\sigma_k^2}) - R_T) \end{aligned} \tag{9}$$

where $L(p, G)$ is shown in (9) and $\beta = (\beta_1, \beta_2, \beta_3)$ represents the dual variables vector. Then, we can give the dual optimization problem

$$\min_{\beta} g(\beta) \tag{10a}$$

$$\text{s.t.}\ \beta \geq 0 \tag{10b}$$

Since the dual function is convex, we can obtain the optimal variables $\beta = (\beta_1, \beta_2, \beta_3)$ through the subgradient-based iterative method. The subgradient can be formulated as follows

$$\begin{aligned} \Delta\beta_1 &= \eta \sum_{k\in G_E}(p_k h_k) - E_{\min} \\ \Delta\beta_2 &= P - \sum_{k\in G_I} p_k - \sum_{k\in G_E} p_k \\ \Delta\beta_3 &= \sum_{k\in G_I} \ln(1+\frac{p_k h_k}{\sigma_k^2}) - R_T \end{aligned} \tag{11}$$

The optimal β can be obtained by $\beta^{t+1} = \beta^t + v^t \Delta\beta$, where v_t denotes the step size, t is iteration times and $\Delta\beta = (\Delta\beta_1, \Delta\beta_2, \Delta\beta_3)$. During the iteration, we can find out the optimal dual variables when convergence is reached. With the following two steps, we can finally achieve the optimal $\{p, G\}$ on a given β.

3.1 Optimizing p with Fixed G

When the set G is fixed, we can calculate the partial derivative of Lagrange function (9) with $p_k, k \in G_I$ and $p_k, k \in G_E$, which can be given by

$$\frac{\partial L(p,G)}{\partial p_k} = \frac{(1+\beta_3)h_k}{\sigma_k^2 + p_k h_k} - q - \beta_2, k \in G_I \tag{12}$$

$$\frac{\partial L(p,G)}{\partial p_k} = -q + q\xi h_k + \beta_1 \xi h_k - \beta_2, k \in G_E \tag{13}$$

According to KKT condition, we can obtain the optimized p by making the partial derivative equal to zero, which is $\frac{\partial L(p,G)}{\partial p_{k,k\in G_I}} = 0$ and $\frac{\partial L(p,G)}{\partial p_{k,k\in G_E}} = 0$. Thus, the power allocated to decode information can be optimized through the following formula

$$p_k{}^* = \left(\frac{1+\beta_3}{\beta_2+q} - \frac{\sigma_k^2}{h_k}\right)^+ \tag{14}$$

where $()^+$ denotes that all negative numbers calculated by (14) will be changed to zero, while positive numbers will remain the same. And the power allocation of energy collection can also be optimized as

$$p_k* = \begin{cases} p_{\max} & \xi h_k(q+\beta_1) > q+\beta_2 \\ p_{\min} & \xi h_k(q+\beta_1) \le q+\beta_2 \end{cases} \tag{15}$$

where $p_{\min}$ is the minimum power constraint for each subcarrier while $p_{\max}$ denotes the maximum power constraint.

3.2 Obtaining the Optimal G

Substituting (14) and (15) into (9), the Lagrangian in (9) can be rewritten as (17) through algebraic transformation.

$$L(G) = \sum_{k\in GI} F_K{}^* + \sum_{k=1}^{K} (\eta p_k{}^* h_k(q+\beta_1) - p_k{}^*(q+\beta_2)) \tag{16}$$

$$- \beta_1 E_{\min} + \beta_2 P - qP_c - \beta_3 R_T \tag{17}$$

where

$$F_k^* = (1+\beta_3)\ln(1+\frac{p_k{}^* h_k}{\sigma_k^2}) - \xi p_k{}^* h_k(q+\beta_1) \tag{18}$$

We can observe from (17) that on the right hand side only the first part, F_k^*, is related to subcarrier set G_I. Therefore, the subcarrier set G_E can be optimized by finding the set which maximizes the Lagrangian function. And the optimal G_I can be written as

$$G_I^* = \arg\max_{G_I} \sum_{k\in G_I} F_k^* \tag{19}$$

It can be easy to obtain optimal G_I^*, since we can find all $k(k \in k)$ that make F_k^* positive. Then we can get optimal G_E^* which is written as

$$G_E^* = K - G_I^* \tag{20}$$

We can solve the optimization problem of subcarrier and power distribution through the above Algorithm 1, and find out the maximum optimal energy efficiency by utilizing the Dinkelbach Iterative Algorithm, which can be summarized in Algorithm 2.

Algorithm 1. The Algorithm for Optimization Problem

1: **initialize** the non-negative variables $\{\beta_1, \beta_2, \beta_3\}$.
2: **repeat**
3: Update power allocation p_k^* defined in (14) and (15).
4: Update subcarrier sets G_I^* and G_E^* according to (19) and (20).
5: Update $\{\beta_1, \beta_2, \beta_3\}$ according to (11).
6: **until** $\{\beta_1, \beta_2, \beta_3\}$ converge.

Algorithm 2. The Algorithm for Energy Effciency Optimization

1: **initialize** the stopping error ϵ and the maximum iteration times N.
2: Set $n = 0$ and $q = 0$.
3: **repeat**
4: Obtain the optimal $\{p, G\}$ according to Algorithm 1.
5: **if**
6: $R_{total}(p, G) - qU_{total}(p, G) \leq \epsilon$.
7: **return** $\{p^*, G^*\} = \{p, G\}$ and $q^* = \frac{R_{total}(p,G)}{U_{total}(p,G)}$.
8: **else**
9: Update $n = n + 1$ and $q = \frac{R_{total}(p,G)}{U_{total}(p,G)}$.
10: **end**
11: **until** $R_{total}(p, G) - qU_{total}(p, G) \leq \epsilon$.

4 Simulations Result

The performance of our proposed energy efficiency algorithm is finally demonstrated by the simulation results in terms of energy efficiency and achievable rate.

We set the number of subcarriers to 32, and the fixed power consumption P_c of the system hardware is assumed to be 0.7 mW. The channel conforms to the Rayleigh distribution, and the channel noise is considered as an additive white Gaussian noise (AWGN) random variable, which can be denoted as $n_k \sim CN(0, \sigma_k^2)$,$\sigma_k^2 = -50$ dBm. In addition, the energy conversion efficiency ξ is set to 1 for simplicity.

Figure 4 shows the allocation of power and subcarriers when $P = 3$mW, $R_T = 2$ bps/Hz and $E_{min} = 0.002\,\mu$W. Figure 2 and Fig. 3 illustrate the performance of our proposed algorithm and other algorithms.

Algorithm 1: The total transmitted power is divided into M components on average. In a fixed ratio, the power allocated on each subcarrier is split into two parts, respectively for harvesting the energy and decoding the information.

Algorithm 2: All subcarriers are randomly divided into two sets, and each set has the same number of subcarriers. The subcarrier sets are fixed and the water-filling method is utilized for power allocating. Subcarriers in different sets will be respectively used to decode information and harvest energy.

It can be clearly observed from Fig. 2 and Fig. 3 that our proposed algorithm performs better than Algorithm 1 and Algorithm 2, and Algorithm 2 is superior to Algorithm 1. We can know that in Algorithm 1, all subcarriers, regardless

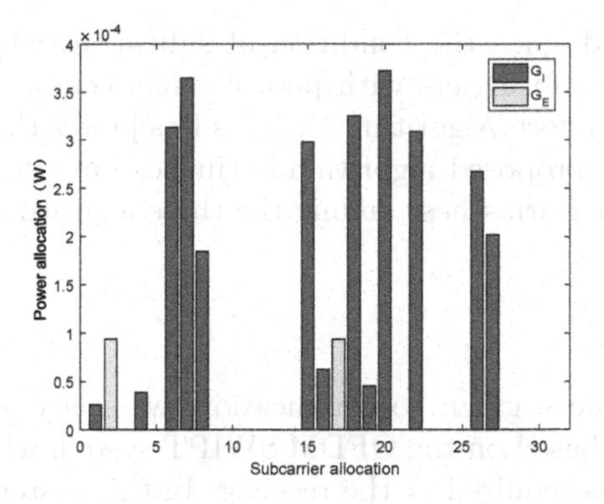

Fig. 2. Power and subcarriers allocation

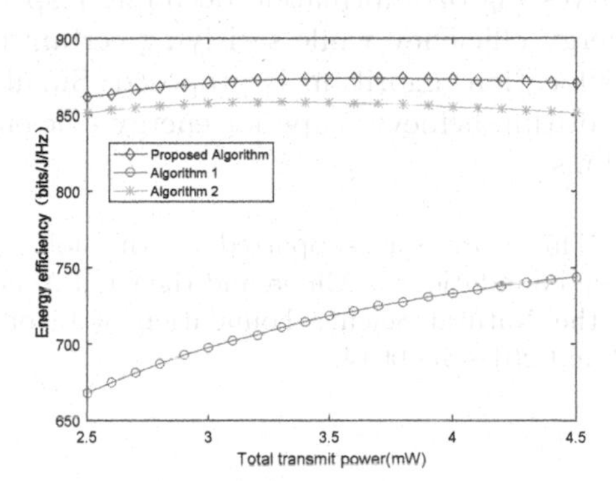

Fig. 3. Comparison of energy efficiency versus P of different algorithms

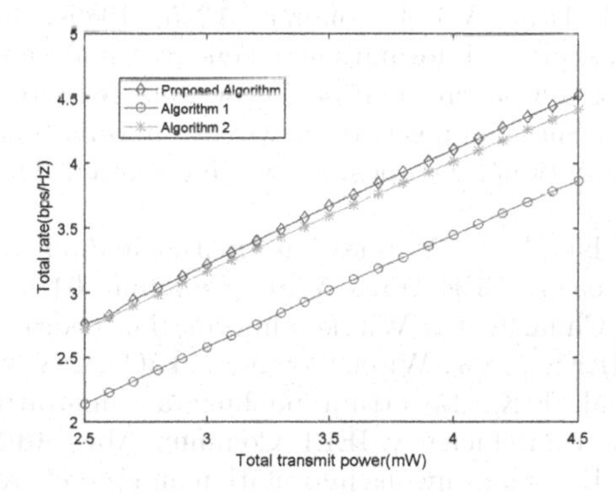

Fig. 4. Comparison of achievable rate versus P of different algorithms

of channel gain, are used for transmission, so that the power will be wasted on some poor subcarriers. In Algorithm 2, the subcarrier set is not optimal, which will inevitably cause power waste even if the power is allocated by water-filling algorithm. And since the condition of subcarries will be judged before power allocation, the subcarriers with poor channel condition will not be used for transmission. Therefore Algorithm 2 wastes less power than Algorithm 1 and performs better. Our proposed algorithm optimizes both subcarrier and power allocation and thus performs best among the three algorithms.

5 Conclusions

In this paper, to achieve green communication, we study an energy efficiency optimization scheme based on the OFDM SWIPT system where no power splitter or time switcher is required at the receiver. In this system, all received subcarriers will be split into two sets, and the subcarriers in the two sets will be utilized for energy harvesting or information decoding respectively. In order to achieve maximum energy efficiency while satisfying certain constrains, a dual-layer iterative joint allocation algorithm is proposed. Simulation results show that our proposed algorithm achieves superior energy efficiency compared with the other two algorithms.

Acknowledgements. This work was supported by the Joint Foundations of the National Natural Science Foundations of China and the Civil Aviation of China under Grant U1833102, and the Natural Science Foundation of Liaoning Province under Grants 2020-HYLH-13 and 2019-ZD-0014.

References

1. Zhang, R., Ho, C.K.: MIMO broadcasting for simultaneous wireless information and power transfer. IEEE Trans. Wirel. Commun. **12**(5), 1989–2001 (2013)
2. Varshney, L.R.: Transporting information and energy simultaneously. In: 2008 IEEE International Symposium on Information Theory, pp. 1612–1616 (2008)
3. Grover, P., Sahai, A.: Shannon meets tesla: wireless information and power transfer. In: 2010 IEEE International Symposium on Information Theory, pp. 2363–2367 (2010)
4. Liu, L., Zhang, R., Chua, K.-C.: Wireless information and power transfer: a dynamic power splitting approach. IEEE Trans. Wirel. Commun. **61**(9), 3990–4001 (2013)
5. Liu, L., Zhang, R., Chua, K.-C.: Wireless information transfer with opportunistic energy harvesting. IEEE Trans. Wirel. Commun. **12**(1), 288–300 (2013)
6. Weiss, T.A., Jondral, F.K.: Spectrum pooling: an innovative strategy for the enhancement of spectrum efficiency. IEEE Commun. Mag. **43**(3), S8–14 (2004)
7. Huang, K., Larsson, E.: Simultaneous information and power transfer for broadband wireless systems. IEEE Trans. Sig. Process. **61**(23), 5972–5986 (2013)
8. Zhou, X., Zhang, R., Ho, C.K.: Wireless information and power transfer in multiuser OFDM systems. IEEE Trans. Wireless Commun. **13**(4), 2282–2294 (2014)
9. Lu, W., Gong, Y., Wu, J., Peng, H., Hua, J.: Simultaneous wireless information and power transfer based on joint subcarrier and power allocation in OFDM systems. IEEE Access **5**, 2763–2770 (2017)

Comparative Analysis of Communication Links Between Earth-Moon and Earth-Mars

Wenjie Zhou[1(✉)], Xiaofeng Liu[1], Qing Guo[1], Xuemai Gu[2], and Rui E[3]

[1] School of Electronic and Information Engineering, Harbin Institute of Technology, Harbin 150001, Heilongjiang, China
1157553297@qq.com

[2] International Innovation Institute of HIT in Huizhou, Huizhou 516000, Guangdong, China

[3] Heilongjiang Polytechnic, Harbin 150001, Heilongjiang, China

Abstract. Deep space exploration is one of the three major aerospace activities of mankind in the new century, and deep space exploration is inseparable from the research on technologies of deep space communication. In the future, the goal of human space exploration will be extended to more and farther stars, analyzing the basic characteristics of the deep space communication link channel, and studying the specific communication problems of the nearer stars will be the necessary basis for the research of deep space communication. Focus on the characteristics of the communication channel between Earth-Moon and Earth-Mars. According to the influential parameters of different communication link, the impact of the communication link of Ka frequency band and below is simulated and analyzed to clarify each range of loss and the effect of each parameter, and the relevant channel characteristics of deep space communication link are obtained.

Keywords: Deep space communication · Channel modeling · Link budget · The lunar environment · Martian environment

1 Introduction

The communication between the probe in deep space and the earth ground station is called deep space communication. At present, deep space exploration is one of the three major human activities in the aerospace field. It integrates many high-tech technologies in the aerospace field. It is of great significance for mankind to explore deeper space, as well as to learn and use the cosmic resources. The United States is the first country to carry out deep space exploration activities, which is currently the only country that has successfully probed the sun and the eight planets of the solar system. In recent years, its exploration activities have continued to develop. The Curiosity successfully landed on the surface of Mars in 2012, and the New Horizon in 2015 flew over Pluto for the first time, NASA announced the evidence of liquid water in Mars in the same year; India successfully achieved the first Mars exploration in 2013, and launched a moon exploration project in 2019; in 2016, China's Mars exploration mission was officially established. The

S. Shi et al. (Eds.): AICON 2020, LNICST 356, pp. 155–165, 2021.
https://doi.org/10.1007/978-3-030-69066-3_15

achievements of the major aerospace nations and organizations and the planning of the new era signify that international deep space exploration has entered a new stage [1].

Deep space communication are essential for the deep space exploration missions, 1 many space agencies and researchers have been making efforts in the long distance weak signal transmissions enhancement, such as the free space optical communications, 2 large scale antenna array. 3 Because of the slowly development and great difficulties of such physical solutions, it is important for us to pay more attention on the interplanetary networking. 4 Derived from the needs of national development strategy, there might be more and more robot/manned spacecraft to explore further space in the Solar System, which would gradually construct the sparse interplanetary networks to provide end-to-end wireless communications with high quality of service [2].

Deep space exploration is inseparable from deep space communication technology. When humans extend the scope of aerospace activities to the moon and Mars, there will be many technical difficulties in the communication channel, such as low coverage and huge losses due to the distance and the impact of the cosmic environment. The deep space exploration mission marked by the exploration of moon and Mars has greatly improved the requirements for the reliability and real-time performance of the ground-air communication, the adjustment capability of the multimedia type service, the processing capacity of the transmitted data, and the throughput rate. Analyzing the basic characteristics of the link channel, and studying the specific problems of nearer astral communication will be the necessary foundation for the development of deep space communication research.

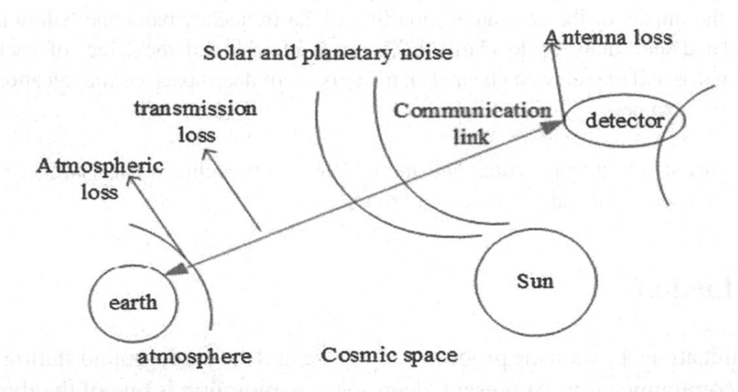

Fig. 1. The conceptual diagram of the deep space communication link

The conceptual diagram of the deep space communication link is shown in Fig. 1. Compared with the daily mobile communication, the radio signal of deep space communication has to traverse a longer distance, the path loss of the link is very serious, and the environmental impact of the surface of the star and the surface of the earth must also be considered. Here we focus on the characteristics of the communication channel between the Earth-Moon and the Earth-Mars. According to the different communication link impact parameters, the impact of the communication link in the Ka frequency band and lower frequency bands are simulated and analyzed. Clarify the value range of each loss and the effect of each parameter, and obtain the characteristics of the deep space communication link channel represented by the Earth-Moon and Earth-Mars links.

2 Similarities and Differences in the Communication Environment Between Earth-Moon and Earth-Mars

The distance between the earth and the moon is much shorter than the distance between the earth and Mars. The distance between the earth and Mars is nearly 6 * 107 km, but the distance between the earth and the moon also reaches 105 km. The communication delay between the earth and the Mars has reached several minutes or even more than ten minutes, but between the earth and the moon it is relatively small, only about one second at the maximum. At the same time, there is planctary convergence between Earth and Mars, which can be divided into superior conjunction and inferior conjunction [3]. During the superior conjunction, the communication link between Earth and Mars reached the longest. At this time, the distance between the signal transmission path of the communication link and the sun is the shortest, and the charged particles ejected by the solar wind are the most affected, which will increase the noise temperature of the communication system, causing the spectrum spread of the signal and the fluctuation of the phase and amplitude, sometimes will cause the link to be interrupted. But during the inferior conjunction, the distance between Earth and Mars is the shortest, the loss caused by space transmission is minimal, and the influence of solar activity is also small, the communication quality is the best.

In addition, the surface of moon is not covered by the atmosphere, so there will be no atmospheric fading during the transmission of radio waves. There are moon seas and land on the surface of moon, and there are no mountains, lakes and oceans, the environment is relatively simple; and compared to earth, the radius of moon is small, the curvature is bigger. The radius of Mars is about half that of the Earth, and the rotation period is similar to that of the Earth. The surface of Mars is covered by the atmosphere, but the composition is very different from the earth and relatively thin, which makes the dust difficult to fall, so it is easy to cause large-scale sandstorm and have a great impact on the quality of communications.

3 Analysis and Simulation of Mars Channel Fading Characteristics

3.1 Atmospheric Loss in the Mars Channel

Mars has an atmosphere with a composition that differs greatly from that of Earth. Table 1 compares the composition parameters of the atmosphere of Mars and the atmosphere of Earth [4]. It can be seen from the table that the water vapor content in the Martian atmosphere is extremely low, most of which is carbon dioxide. Therefore, when the communication link signal is transmitted through the Martian atmosphere, the fading is mainly caused by the absorption of radio wave energy by water vapor and oxygen molecules. Since the calculation method is similar to the earth's atmospheric loss, here we can refer to the model of the earth's atmospheric loss, and refer to the ITU-R recommendation for the atmospheric loss calculation method for the frequency band below 350 GHz to analyze the characteristics of the Martian atmosphere. For frequency below 54 GHz, the formula for calculating the loss coefficient (dB/km) of dry air is:

$$\gamma_0 = \left[\frac{7.2r_t^{2.8}}{f^2 + 0.34r_p^2 r_t^{1.6}} + \frac{0.62\xi_3}{(54-f)^{1.16\xi_1} + 0.83\xi_2}\right] f^2 r_p^2 \times 10^{-3} \quad (1)$$

Where f is frequency, the unit is GHz; p is pressure, the unit is hPa; t is temperature, the unit is °C; $r_p = p/1013$, $r_t = 288/(273 + t)$; ρ is water vapor density, the unit is g/m^3.

Table 1. Parameters of the atmosphere between Mars and Earth

Star	Temperature/K	Mean atmospheric pressure/Pa	Ingredient content (%)				
			O_2	N_2	H_2O	CO_2	Ar
Earth	300	101300	20.95	78.09	0.25	0.04	0.93
Mars	210	610	0.14	1.9	0.021	95.9	2.0

For frequency below 350 GHz, the calculation formula for the water vapor loss coefficient (dB/km) is:

$$\begin{aligned}\gamma_w = [&\frac{3.98\eta_1 e^{2.33(1-r_t)}}{(f-22.235)^2 + 9.42\eta_1^2} g(f,22) + \frac{11.96\eta_1 e^{0.7(1-r_t)}}{(f-183.31)^2 + 11.14\eta_1^2} \\ &+ \frac{0.081\eta_1 e^{6.44(1-r_t)}}{(f-321.226)^2 + 6.29\eta_1^2} + \frac{3.66\eta_1 e^{1.6(1-r_t)}}{(f-325.153)^2 + 9.22\eta_1^2} \\ &+ \frac{25.37\eta_1 e^{1.09(1-r_t)}}{(f-380)^2} + \frac{17.4\eta_1 e^{1.46(1-r_t)}}{(f-448)^2} \\ &+ \frac{844.6\eta_1 e^{0.17(1-r_t)}}{(f-557)^2} g(f,557) + \frac{290\eta_1 e^{0.41(1-r_t)}}{(f-752)^2} g(f,752) \\ &+ \frac{83328\eta_2 e^{0.99(1-r_t)}}{(f-1780)^2} g(f,1780)] f^2 r_t^{2.5} \rho \times 10^{-4}\end{aligned} \quad (2)$$

For each parameter appearing in formula (1) and formula (2), there are:

$$\xi_1 = \varphi(r_p, r_t, 0.0717, -1.8132, 0.0156, -1.6515) \quad (3)$$

$$\xi_2 = \varphi(r_p, r_t, 0.5146, -4.6368, -0.1921, -5.7416) \quad (4)$$

$$\xi_3 = \varphi(r_p, r_t, 0.3414, -6.5851, 0.2130, -8.5854) \tag{5}$$

$$\varphi(r_p, r_t, a, b, c, d) = r_p^a r_t^b \exp[c(1 - r_p) + d(1 - r_t)] \tag{6}$$

$$\eta_1 = 0.955 r_p r_t^{0.68} + 0.006\rho \tag{7}$$

$$\eta_2 = 0.735 r_p r_t^{0.5} + 0.0353 r_t^4 \rho \tag{8}$$

$$g(f, f_i) = 1 + \left(\frac{f - f_i}{f + f_i}\right)^2 \tag{9}$$

The atmospheric temperature of Earth stays constant at 300K and that of Martian remains stable at 210K. Referring to the parameters in Table 1, the characteristic loss coefficient of the Martian atmosphere can be calculated through the above formulas. Fig. 2 shows the relationship between atmospheric loss coefficient and frequency.

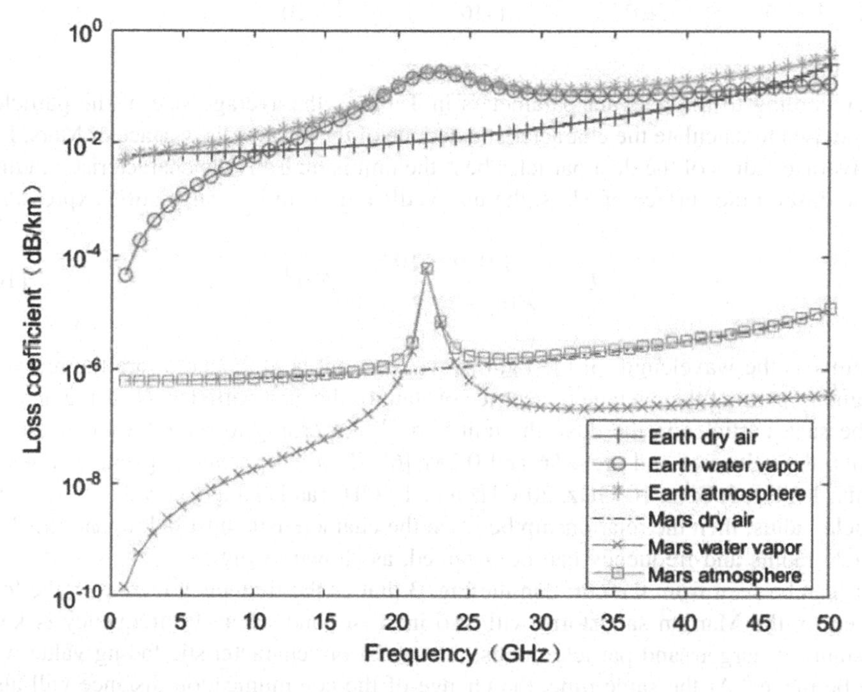

Fig. 2. Relationship between atmospheric loss coefficient and frequency on earth and Mars

It can be seen from the graph that the loss coefficient of the Martian atmosphere is small compared to the earth's atmospheric loss coefficient, and the loss caused to the communication link is basically relatively smaller. At the same time, the curve of the Martian and the earth's atmospheric fading coefficient has reached an extreme value around 22 GHz. By observing the changes of the dry air loss coefficient and the water vapor loss coefficient in the earth and Mars atmospheric fading coefficient, it can be found that around 22 GHz, the loss coefficient of dry air does not produce any special changes, and the loss coefficient curve of water vapor rises sharply to the extreme value.

3.2 Sandstorm Loss in the Mars Channel

Sandstorms are a significant feature of the weather environment on the surface of Mars. When sandstorms occur, the visibility on the surface of Mars is usually only about 4 m–10 m. Table 2 gives a comparison of the parameters of the sandstorms between the earth and Mars [5].

Table 2. Parameters of sandstorms between Earth and Mars

Star	Sand density/m^{-3}	Mass density/ (g/m^3)	Average size of sand particles/μm	Maximum size of sand particles/μm	Path length/km
Earth	10^8	$2.6 * 10^6$	30–40	80–300	10
Mars	$3 * 10^7$	$3 * 10^6$	1–10	20	10

According to the relevant parameters in Table 2, the average size of the particles can be used to calculate the characteristic decline of the dust on the surface of Mars. Let the average radius of the dust particles be r, the unit is m; Ld is the characteristic fading of the dust on the surface of Mars, the unit is dB/km, there is a calculation expression:

$$L_d = \frac{1.029 \times 10^6 \varepsilon''}{\lambda[(\varepsilon' + 2)^2 + (\varepsilon'')^2]} N_T r^3 \tag{10}$$

Where λ is the wavelength of the radio wave, the unit is m; ε' and ε'' are the real and imaginary parts of the average dielectric constant of the dust particles; N_T is the density of the sand particles in the dust, the unit is m^{-3}. According to the relevant measured results, the values of and are 4.56 and 0.251 [6]. If we take several special frequency points: 1 GHz, 5 GHz, 10 GHz, 20 GHz and 40 GHz, and change the value of the dust particle radius, then the relationship between the characteristic dust fading and the dust particle radius and frequency can be obtained, as shown in Fig. 3.

It can be seen from the curve in the Fig. 3 that as the frequency increases, the loss caused by the Martian sandstorm will also increase, and when the frequency is kept constant, for larger sand particle radius, the sandstrom characteristic fading value will also be bigger. At the same time, the change of the communication distance will also affect the value of the sandstorm fading.

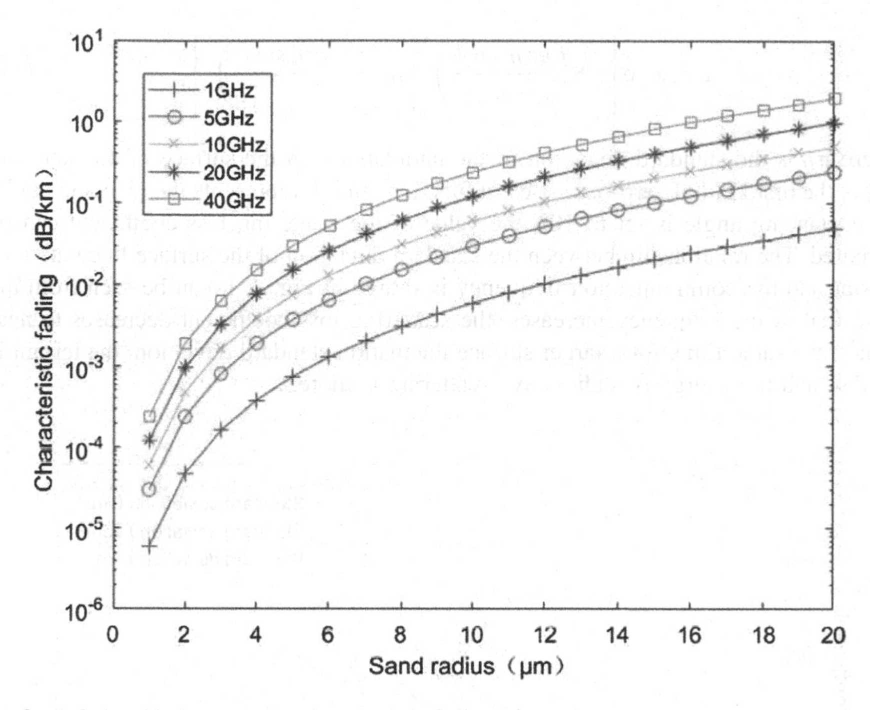

Fig. 3. Relationship between the characteristic fading of Martian sandstorm and the size of sand particles or frequency

4 The Characteristics of Radio Wave Transmission on the Surface of Moon and Mars

Compared with earth, the radius of Mars and moon has a small curvature and a large radius, and the problem of long-distance diffraction needs to be considered. At the same time, the Moon has large undulating terrain such as the moon valley, and there are obstacles of various shapes on the surface of Mars. In addition, because there is a thin atmosphere on the surface of Mars and moon has no atmosphere, so the influence of the refraction of the atmosphere during the transmission of radio waves is small or almost none. Due to the actual terrain conditions and the complex shape of obstacles, the transmission loss analysis of surface waves is divided into three stages: free transmission loss, surface reflection loss and diffraction loss. The simplified segmented model is used to study the radio wave transmission characteristics in the surface of the star [7, 8]. Here we mainly analyze the loss changes of reflection and diffraction.

When there is a signal on the reflection path on the surface of the star, for the reflected signal, the rough surface of the star will scatter part of the signal, and the reflection coefficient can be corrected by introducing the scattering loss coefficient. Let a be the scattering loss coefficient, with the expression:

$$a = \exp\left[-8\left(\frac{\pi\sigma_H \sin\psi}{\lambda}\right)^2\right] J_0\left[8\left(\frac{\pi\sigma_H \sin\psi}{\lambda}\right)^2\right] \tag{11}$$

Where σ_H is the standard deviation of the undulations on the surface of the star, and $J_O[\cdot]$ is the first kind of zero-order Bessel function, and Ψ represents the glancing angle. If the glancing angle is set to 10°, the value of the scattering loss coefficient can be estimated. The relationship between the standard deviation of the surface fluctuation of the star and the communication frequency is shown in Fig. 4. It can be seen from the curve that as the frequency increases, the scattering loss coefficient decreases to near zero; at the same time, for a larger surface fluctuation standard deviation, the terrain is not flat, and the energy of radio wave scattering is more.

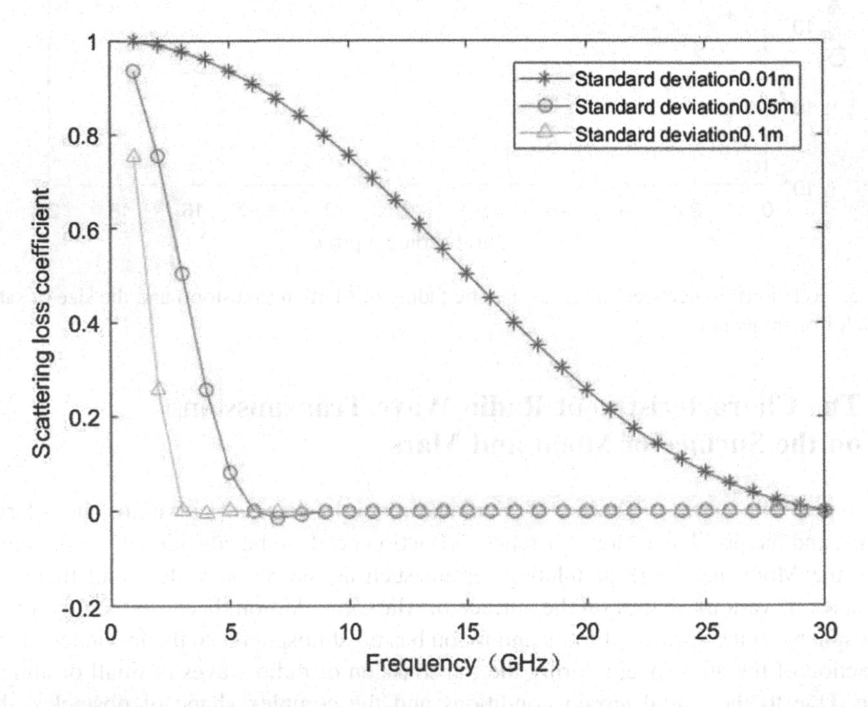

Fig. 4. Relationship between scattering loss coefficient and frequency and standard deviation of surface fluctuation

When the communication distance *d* does not exceed the sum of the maximum line-of-sight propagation distance and half of the half-shadow width, the loss of long-distance propagation and diffraction can be expressed as:

$$L = F[X(P)] + G[Y(r, P)] + G[Y(t, P)] \tag{12}$$

Where *F(X)* is the distance gain term, and *G(Y)* is the height gain term, the expressions are:

$$F[X] = \begin{cases} 11 + \log(X) - 17.6X & X \geq 1.6 \\ -20\log(X) - 5.6488X^{1.425} & X < 1.6 \end{cases} \tag{13}$$

$$G[Y] = \begin{cases} 17.6\sqrt{B-1.1} - 5\lg(B-1.1) - 8 & B > 2 \\ 20\lg(B + 0.1B^3) & 10\text{K} < B \leq 2 \\ 2 + 20\lg K + 9\lg(B/K)(\lg(B/K)+1) & K/10 < B < 10K \\ 2 + 20\lg K & B \leq K/10 \end{cases} \tag{14}$$

Among them, the calculation formula of X and Y is:

$$X = \beta\left(\frac{\pi}{\lambda r_e^2}\right)^{\frac{1}{3}} d \tag{15}$$

$$Y = 2\beta\left(\frac{\pi^2}{\lambda^2 r_e}\right)^{\frac{1}{3}} h \tag{16}$$

Where r_e is the equivalent radius; d is the transmission distance; σ is the ground conductivity; h is the corresponding height; K is the normalization factor, $B = \beta Y$, and for horizontal polarization, the parameters β and K have expressions:

$$K = \left(\frac{2\pi r_e}{\lambda}\right)^{-\frac{1}{3}} \left[(\varepsilon_r - 1)^2 + (60\lambda\sigma)^2\right]^{-\frac{1}{4}} \tag{17}$$

$$\beta = \frac{1 + 1.6 \times K^2 + 0.67 \times K^4}{1 + 4.5 \times K^2 + 1.53 \times K^4} \tag{18}$$

If the height of the transceiver is 2 m, the equivalent radius is 100 km, and the frequency of the communication wave is 1000 MHz. Taking the lunar surface as an example, under horizontal polarization conditions, the lunar surface conductivity is 3×10^{-4}, relative permittivity is 3, lunar radius is 1738 km, and in the diffraction range of 1000 m–8000 m, the long-distance propagation around the lunar surface can be obtained. The relationship between the radiation loss and the propagation distance is shown in Fig. 5.

As can be seen from Fig. 5, as the transmission distance increases, the loss of the diffraction section will gradually increase. At the same time, by comparing the curve changes of the diffraction loss at several different frequencies, it can be found that the low frequency radio waves are affected when the distance is closer. The loss is greater, and when the distance continues to increase far enough, the loss of higher frequency waves will exceed that of lower frequencies.

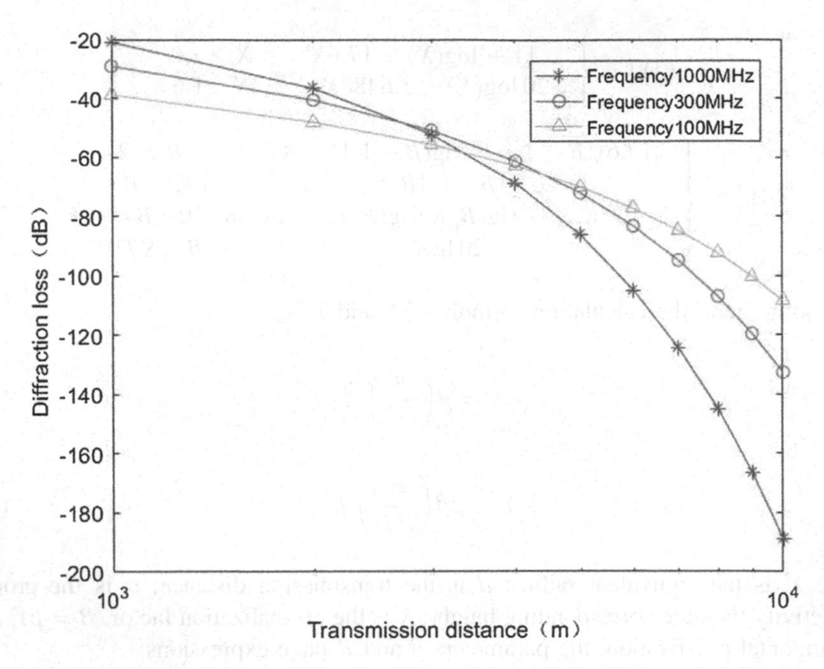

Fig. 5. Relationship between diffraction loss and transmission distance and frequency

5 Conclusion

Recently, deep space communication technology is developing rapidly, and the analysis of the channel characteristics of deep space communication links is the basis of research. This paper mainly considers the characteristics of the moon and Mars channels, gives the performance simulation results of the communication links in the Ka frequency band and lower frequency bands, and summarizes the relevant characteristics of the deep space channel. The atmospheric loss of Mars is nearly a hundred times lower than that of the earth, but similarly, there are peaks in the frequency band of 20–25 GHz; when the radius of Martian sand is large, the characteristic loss can reach nearly 1 dB/km for the radio waves of higher frequency bands; the simulation results using a simple segmented model show that uneven terrain will greatly increase the harmful scattering on the transmission path, and the diffraction loss in transmission over several kilometers can reach more than 20 dB, and at the same time, take a lower frequency can help to reduce the loss caused by scattering and diffraction.

References

1. Schlutz, J., Vange, S., Haese, M., et al.: Assessment of technology development for the ISECG global exploration roadmap. In: Global Space Exploration Conference, pp. 1–11. International Space Exploration Coordination Group, Washington DC (2012)

2. Wan, P., Zhan, Y.: A structured Solar System satellite relay constellation network topology design for Earth-Mars deep space communications. Int. J. Satell. Commun. Netw. **37**(3), 292–313 (2019)
3. Liu, Q., Mei, J., Yao, Y., Ruan, F.: Characteristic analysis and simulation of deep space channel model. J. Air Force Radar Acad. **26**(03), 181–184 (2012)
4. Hassler, D.M., Zeitlin, C., Wimmer-Schwe-Ingruber, R.F., et al.: Mar's surface radiation environment measured with the Mars Science Laboratory's Curiousity rover. Science **343**(6169), 1244–1247 (2014)
5. Du, Y., Yao, X., Fan, Y., Yan, Y., Gao, X.: Influence of Mars channel fading characteristics on communication link budget. J. Space Sci. **39**(05), 701–708 (2019)
6. ElSaid, A., Lewis, S.R., Patel, M.R., et al.: Quantifying the impact of local dust storms on martian atmosphere using the LMD/UK mars global climate model. In: The Mars Atmosphere: Modelling and Observation (2017)
7. Foore, L., Ida, N.: Path loss prediction over the lunar surface utilizing a modified longley-rice irregular terrain model. NASA TM (2007)
8. Zhu, Q., Huang, P., Chen, X., Wang, C., Wang, F., Zhou, S.: Segmented prediction model of radio wave propagation in the lunar surface environment. J. Sichuan Univ. (Eng. Sci. Ed.) **46**(02), 116–120 (2014)

Compact Miniature MIMO Array Antenna Towards Millimeter Wave

Xiangcen Liu[1(✉)], Shuai Han[1], Aili Ma[2], and Xiaogeng Hou[3]

[1] Communications Research Center, Harbin Institute of Technology, Harbin 150001, China
19S005048@stu.hit.edu.cn, hanshuai@hit.edu.cn
[2] Space Star Technology Co., Ltd., Beijing, China
aili_ma@foxmail.com
[3] China Academy of Space Technology, Beijing, China
Houxiaogeng2011@163.com

Abstract. Two types of MIMO antenna arrays are proposed, towards millimeter wave technology. The antenna resonances are at 77 GHz. This paper mainly discusses the decoupling of antenna in MIMO system. The first array antenna is microstrip feed patch MIMO antenna which unit is an elliptical patch. A decoupling branch is added in the middle to improve the isolation to less than −15 dB. The second MIMO array antenna is with a resonance frequency at 77 GHz which meets the Chebyshev distribution. We use series resonance feed, and 1 × 16 line array is used as the MIMO antenna unit. We increase the line array space to achieve an isolation of less than −20 dB in the antenna frequency band.

Keywords: MIMO antenna · 77 GHz · Isolation · Decoupling

1 Introduction

In the future, MIMO technology will also become the key technology for 5G communication. A major problem in implementing MIMO technology is the design of MIMO antennas. On the one hand, MIMO antennas will develop in the direction of miniaturization. On the other hand, in order to realize the advantages of the MIMO system, the MIMO antenna needs to have low mutual coupling [1]. Due to the limited space, shortening the antenna size and reducing the antenna distance will inevitably lead to an increase in coupling. Therefore, how to enable more antenna elements to achieve low coupling in a limited space has caused widespread academic research.

Researchers usually add decoupling structures between antenna elements, including parasitic elements, defective ground structures (DGS) and EBG structures to improve the performance of MIMO antennas. For example, an aperture coupled MIMO antenna is designed in [2], and a novel EBG structure is proposed for decoupling between two patches of the MIMO antenna. Coupling effect decreased 16.5 dB at the operating frequency of 2.45 GHz by inserting the EBG structure between the two patches. A MIMO antenna array is proposed in [3] for sub-6 GHz communication applications. The proposed antenna element is a proximity coupling fed split-ring antenna, which is

S. Shi et al. (Eds.): AICON 2020, LNICST 356, pp. 166–176, 2021.
https://doi.org/10.1007/978-3-030-69066-3_16

excited by a 50-Ω microstrip line. The researcher designed two split-rings at the outer of the antenna, a fork-like slot integrated into the middle of the ground and two L-shape-like slot at each side of the ground plane. The results proved that the MIMO antenna can provide a low coupling and a good omnidirectional radiation patterns in the operating sub-6 GHz. A compact planar wide-band MIMO antenna array with high isolation is presented in [4], which consists of four half-circular monopole radiators and four separated ground planes. Four separated grounds are designed to reduce the surface wave coupling, and four ground stubs are etched to reduce the near field coupling. And better port isolation of above 15 dB is obtained.

Therefore, miniaturization, compact layout and high isolation are the current development trends of MIMO. It has become a challenge to design high-speed links, compact, broadband and efficient antenna systems.

With the rapid development of mobile communications, millimeter wave technology has attracted widespread attention. Due to the high frequency band and few interference sources, millimeter wave technology can obviously enhance the channel anti-interference ability. The application of millimeter-wave antennas greatly reduces the size of antennas compared with microwave and realizes the miniaturization of antennas [5].

Based on previous research results, the MIMO antennas designed in this paper work are in the millimeter wave band and the resonance frequencies are all near 77 GHz. The design results in the reducing of the size of the MIMO antenna and achieves good performance of MIMO system. We had a research on MIMO antennas with different radiation units, different feeding methods, and different structures to meet the requirements of miniaturization and high isolation.

2 77 GHz Microstrip Feed Patch MIMO Antenna

2.1 The Structure of MIMO Antenna

In this section, we design a 77 GHz array antenna with microstrip feed, and an elliptical patch used as the antenna unit. We discuss the influence of the distance between the two elements on the performance of the MIMO antenna. Finally, the distance between the two elements is determined to be 0.5 λ. We design the decoupling branches to increase the isolation between MIMO units [6]. The antenna covers a frequency band of 76.4–77.6 GHz. The isolation in the antenna frequency band is less than −15 dB, and the system has a good performance.

We use microstrip feed, and the antenna unit is an oval patch. The dimensions are as follows: $a = 7.2$ mm, $b = 1.55$ mm, $c = 1.9$ mm, $d = 1.5$ mm, $e = 0.15$ mm, $f = 2.68$ mm, $g = 0.54$ mm, $h = 1.41$ mm. The distance between the two units is 0.5 λ. Blue structure is a metal radiation patch, and the yellow structure is a 0.1 mm thick FR4 dielectric board [7]. MIMO antenna has a resonance frequency of 77 GHz. The isolation within the antenna band is less than −15 dB. A decoupling branch is added between two units, and its dimensions are as follows (the unit is mm) (Fig. 2):

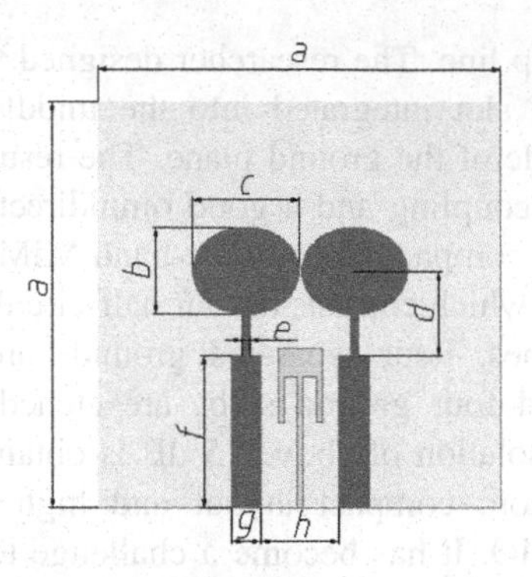

Fig. 1. MIMO antenna structure

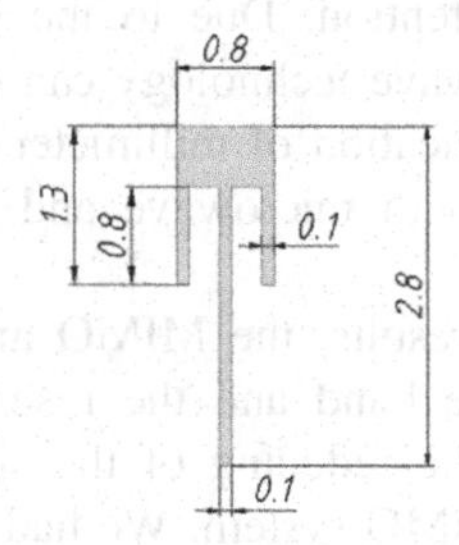

Fig. 2. Decoupling structure

2.2 Design of MIMO Antenna

As shown in Fig. 3, assuming that the unit spacing is a, here we analyze the return loss and the isolation of the MIMO antenna with a ranging from 1.948 mm (0.5 λ) to 3.986 mm (λ), which are shown in Fig. 4 and Fig. 5. The results show that the isolation increases with distance.

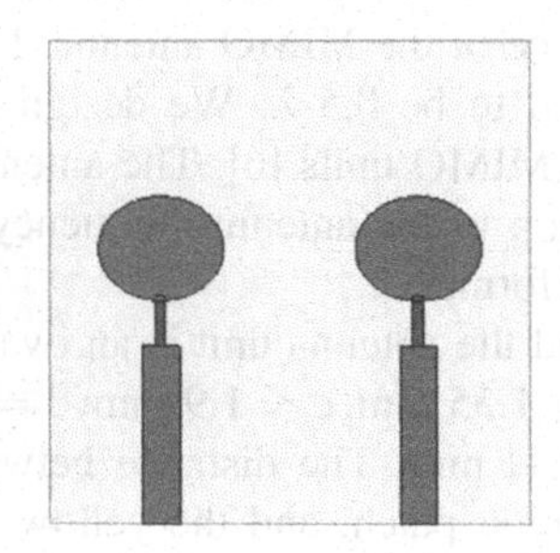

Fig. 3. MIMO antenna structure

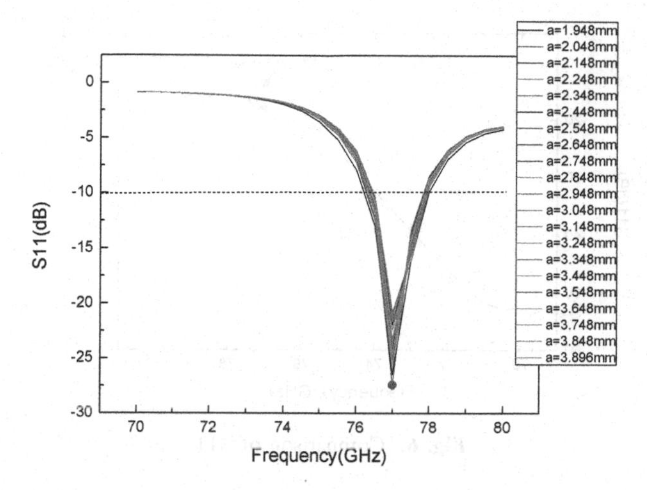

Fig. 4. The effect of the distance on S11

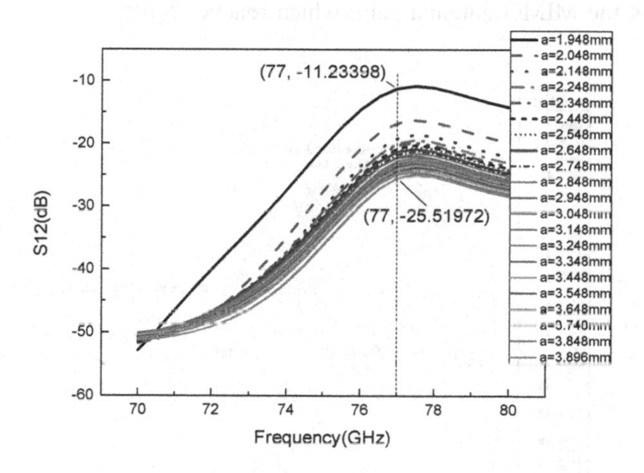

Fig. 5. The effect of the distance on S12

When a = 0.5 λ, S12 = −11.2 dB, the isolation does not meet the requirements. Increasing the distance of the MIMO antenna unit can increase the isolation, but this article uses the method of adding decoupling branch [8], as shown in Fig. 1. The comparison is shown in Fig. 6.

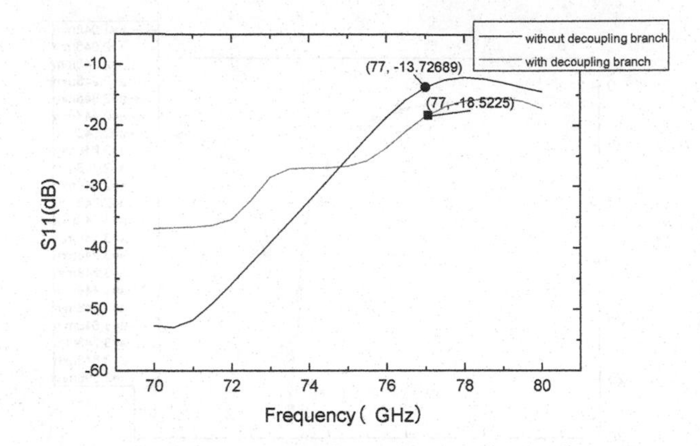

Fig. 6. Comparison of S11

It can be seen that the decoupling branch greatly improves the isolation between MIMO units, and the isolation in the antenna frequency band is less than −15 dB. Figure 7 shows the MIMO antenna gain, which reaches 5 dB.

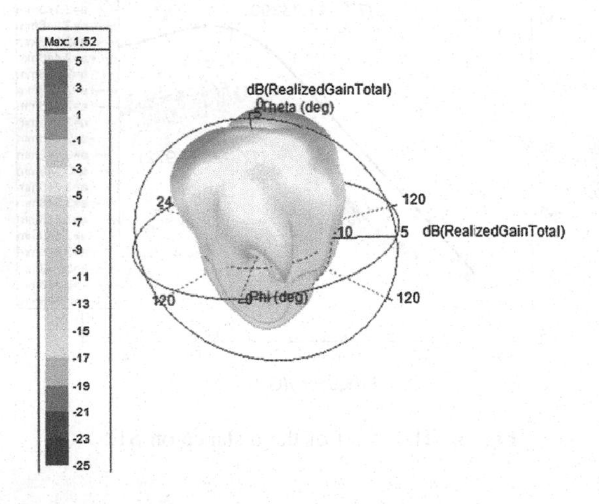

Fig. 7. 3D gain of the antenna

3 77 GHz Series-Fed MIMO Antenna for Chebyshev Distribution

3.1 The Structure of MIMO Antenna

In this section, we design a series-fed microstrip array antenna that meets Chebyshev distribution [9]. The antenna array consists of two 1 × 16 series feed units [10].

The influence of the slotting and folded T-shaped coupling branches are studied. The antenna covers a frequency band of 76.1 GHz–79.4 GHz. There are two resonance frequencies of 77 GHz and 78.5 GHz in the frequency band. The gain is 19.4 dB and the isolation is greater than 20 dB (Fig. 8).

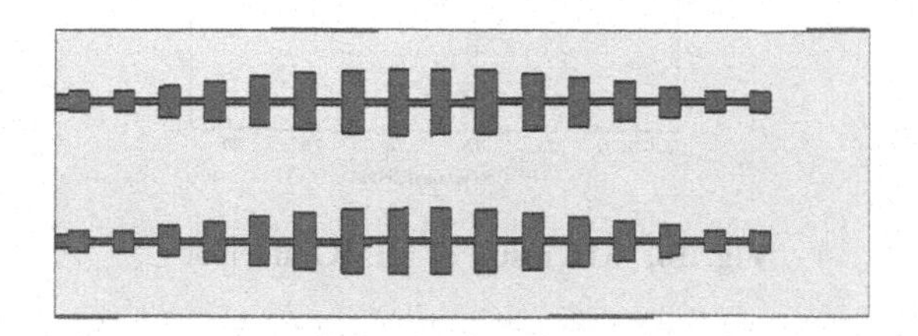

Fig. 8. Structure of MIMO antenna

A 1 × 16 array is used as the MIMO antenna unit. Blue structure is a metal radiation patch, and the green structure is a 0.254 mm thick RO3003 dielectric board. The excitation amplitude of each array element obeys Chebyshev distribution by adjusting the width of each array element to meet the Chebyshev distribution [11]. The MIMO antenna covers the frequency band of 76.1–79.4 GHz with the gain of 19.44 dB, and the isolation reaches −35.2 dB.

3.2 Design of MIMO Antenna

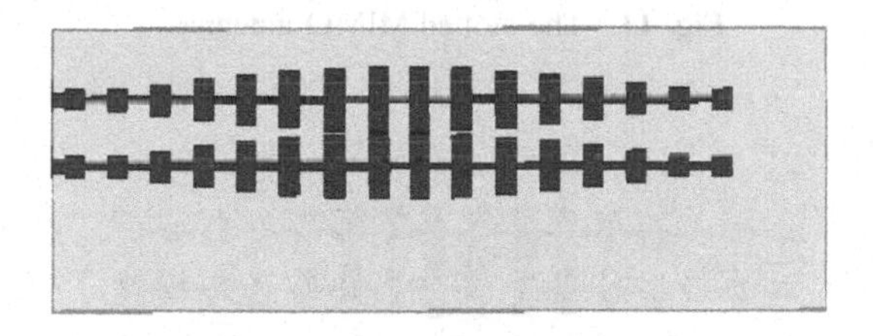

Fig. 9. MIMO antenna structure

As shown in Fig. 9, two array elements are placed in parallel on a RO3003 dielectric board with the area of 6 × 36.5 mm and thickness of 0.254 mm. Distance between two elements is 0.36 λ. S12 parameters of the MIMO antenna is shown in Fig. 10. At 77 GHz, the coupling of the MIMO antenna does not achieve the desired effect.

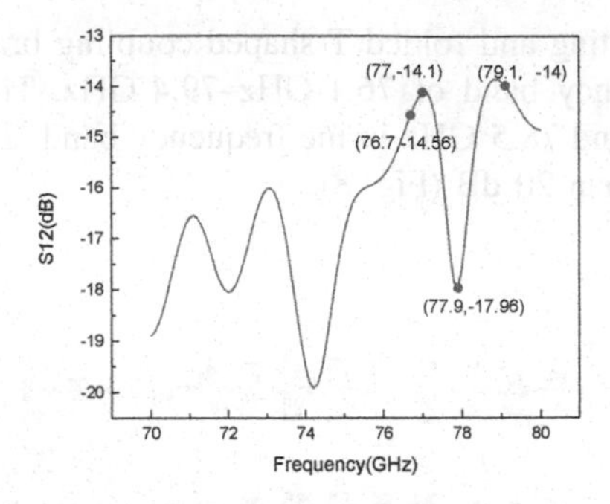

Fig. 10. S12 result of MIMO antenna

In order to reduce the coupling between the two units, the method of slotting and adding decoupling branches in the vicinity of the feed port are shown in Fig. 11 and Fig. 12 [12]. The result of S11 (as is shown in Fig. 13) is good. Figure 14 increases the resonance point of 77 GHz and increases the bandwidth by 20%.

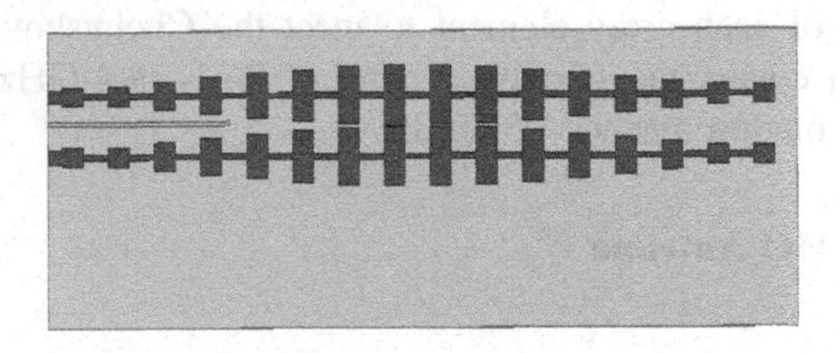

Fig. 11. The slotted MIMO antenna

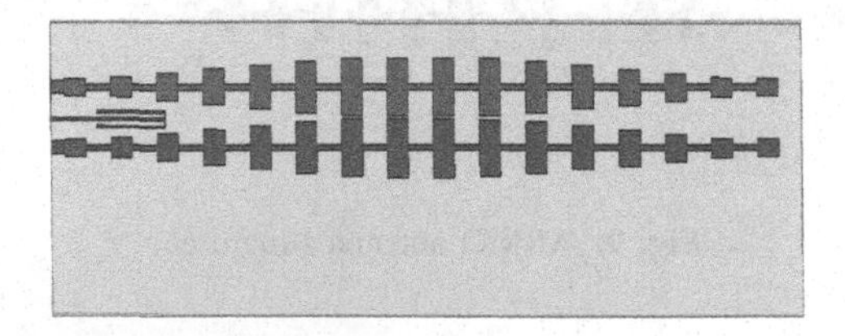

Fig. 12. MIMO antenna with decoupling branches

The S12 results are shown in Fig. 15 and Fig. 16. Figure 15 shows that slotting does not reduce the coupling. As for Fig. 16, in the 70–80 GHz frequency band, the S12 value decreases significantly near the three resonance frequencies of 70.5 GHz, 72.8 GHz, and 75.8 GHz. Because we concern the frequency of 77 GHz, the decoupling branches have not achieved the desired effect.

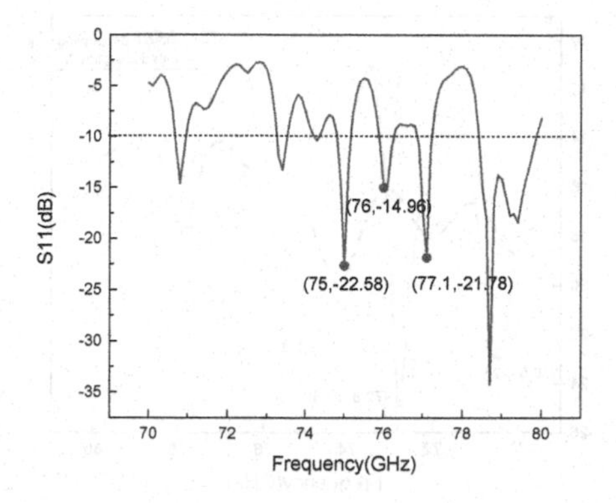

Fig. 13. S11 result of MIMO antenna after slotting

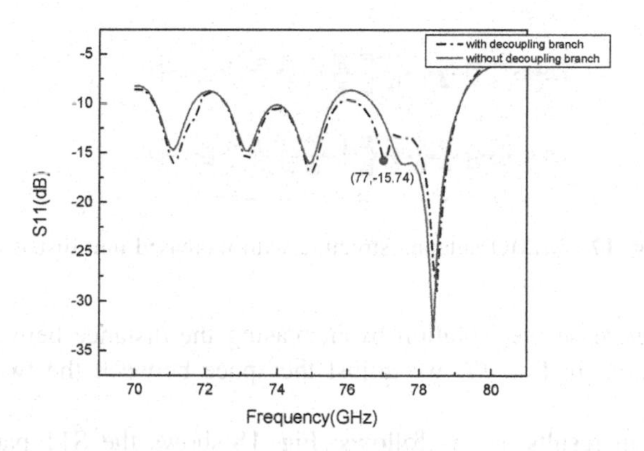

Fig. 14. Comparison of S11 before and after adding branches

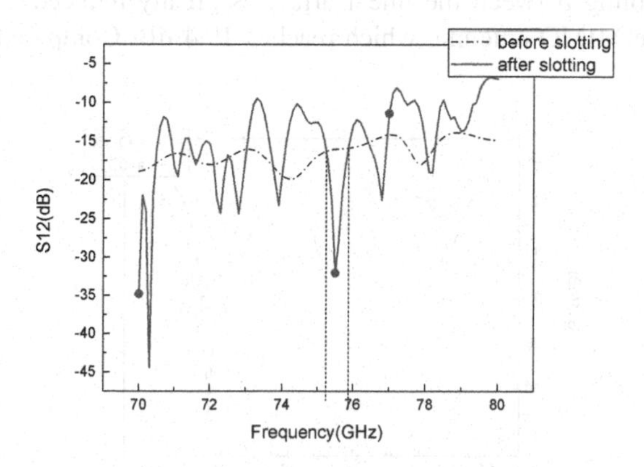

Fig. 15. Comparison of S12 before and after slotting

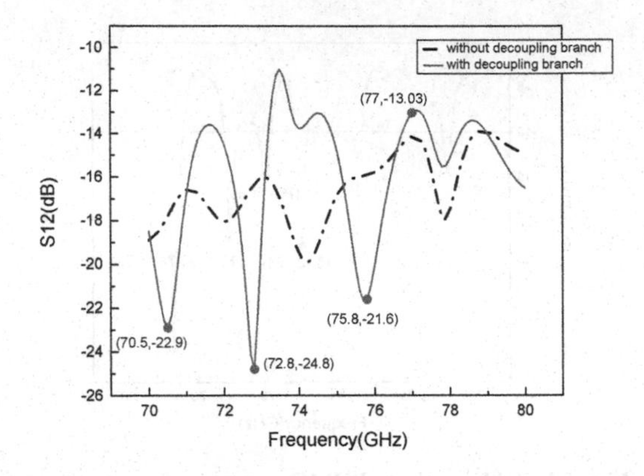

Fig. 16. Comparison of S12 before and after adding branches

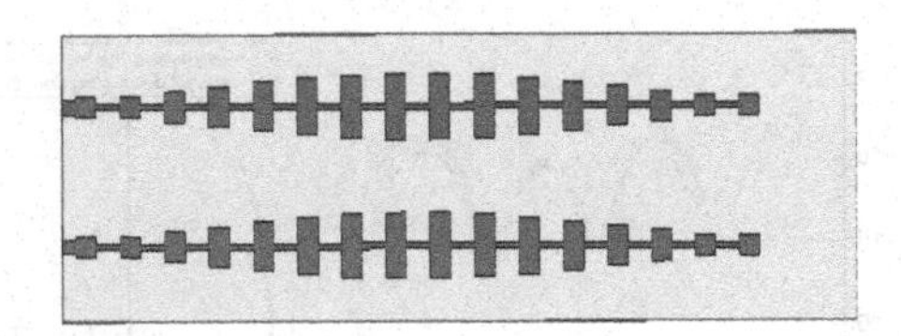

Fig. 17. MIMO antenna structure with increased unit distance

Finally, we increase the isolation by increasing the distance between the antenna elements. As shown in Fig. 17, we adjust the space between the two line arrays to 0.75 λ.

The simulation results are as follows: Fig. 18 shows the S11 parameters of the MIMO antenna with a bandwidth of 3.3 GHz. Compared with the line array spacing of 0.36 λ, the frequency band is widened. Figure 19 shows the S12 results of the MIMO antenna. The coupling between the linear arrays is greatly reduced. Figure 20 shows the 3D gain of the MIMO antenna, which reaches 19.4 dB. Compared with the linear

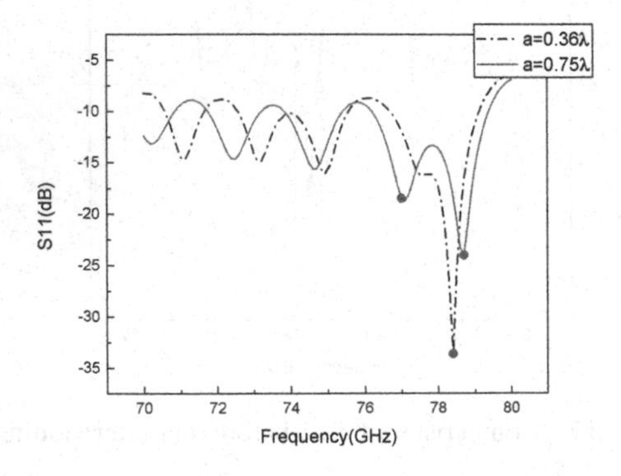

Fig. 18. Antenna S11 results with increased unit distance

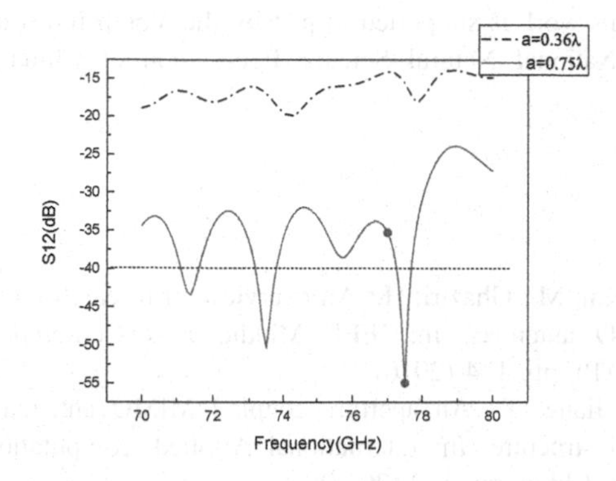

Fig. 19. Comparison of S12 before and after increasing unit distance

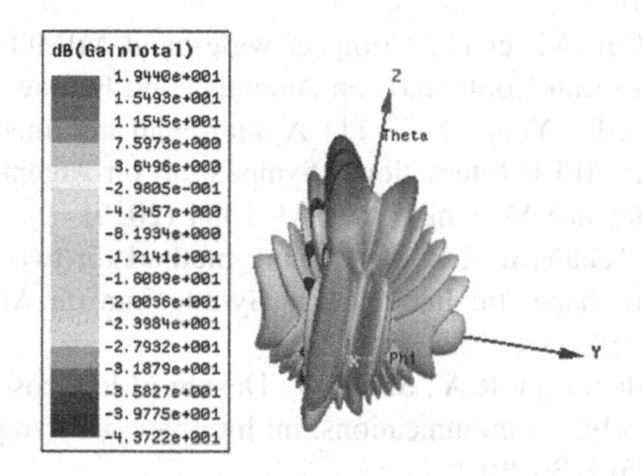

Fig. 20. 3D gain after increasing the unit distance

arrays spacing of 0.36 λ, the gain is further improved. MIMO system has a good performance.

4 Conclusion

This article focuses on the issue of millimeter-wave MIMO antennas. A MIMO array antenna with 77 GHz microstrip feed is designed. The antenna unit is an elliptical patch. We mainly discussed the influence of the unit spacing on the performance of the MIMO antenna. The distance between the two units is determined to be 0.5 λ. A decoupling branch is added in the middle to improve the isolation between MIMO units. At a result, the isolation in the antenna band is less than −15 dB, and the MIMO system has better performance. Further more, a MIMO array antenna with a resonance frequency at 77 GHz that meets the Chebyshev distribution is designed. We use series resonance feed, and 1 × 16 line array is used as the MIMO antenna unit. Slotting and decoupling branches didn't improve the isolation of the antenna.Finally we increased the line array spacing to achieve an isolation of less than −20 dB in the antenna frequency band.

Acknowledgement. This work is supported in part by the Youth Innovation Promotion Association CAS and the National Natural Science Foundation of China (No. 61831002 and 41861134010).

References

1. Yaacoub, E., Husseini, M., Ghaziri, H.: An overview of research topics and challenges for 5G massive MIMO antennas. In: IEEE Middle East Conference on Antennas and Propagation (MECAP), pp. 1–4 (2016)
2. Qin, J., Jiang, X., Jiang, T.: An aperture coupled MIMO antenna decoupling between patches using EBG structure. In: International Applied Computational Electromagnetics Society Symposium, China, pp. 1–2 (2018)
3. Mao, X., Zhu, Y., Li, Y.: A MIMO antenna array with low coupling for sub-6GHz communication applications. In: IEEE Asia-Pacific Conference on Antennas and Propagation, pp. 112–113 (2018)
4. Jiang, W., Liu, Y., Cui, Y., et al.: Compact wide-band MIMO antenna with high port isolation. In: 12th European Conference on Antennas and Propagation, pp. 1–3 (2018)
5. Han, C.-Z., Huang, G.-L., Yuan, T., et al.: A dual-band millimeter-wave antenna for 5G mobile applications. In: IEEE International Symposium on Antennas and Propagation and USNC-URSI Radio Science Meeting, pp. 1083–1084 (2019)
6. Okuda, K., Sato, H., Takahashi, M.: Decoupling method for two-element MIMO antenna using meander branch shape. In: International Symposium on Antennas and Propagation (ISAP), pp. 1–2 (2015)
7. Parchin, N.O., Abd-Alhameed, R.A., Shen, M.: Design of low cost FR4 wide-band antenna arrays for future 5G mobile communications. In: International Symposium on Antennas and Propagation (ISAP), pp.1–3 (2019)
8. Okuda, K., Sato, H., Takahashi, M.: Decoupling method using branch shape without connecting between MIMO multiple antennas. In: IEEE International Workshop on Electromagnetics (iWEM), pp. 84–85 (2014)
9. Mishra, N.K., Das, S.: Investigation of binomial and chebyshev distribution on dielectric resonator antenna array. In: International Conference on Electronic Systems, Signal Processing and Computing Technologies, pp. 434–437 (2014)
10. Chao, C.-P., Yang, S.-H., et al.: A series-fed cavity-back patch array antenna for a miniaturized 77GHz radar module. In: IEEE International Symposium on Antennas and Propagation and USNC-URSI Radio Science Meeting, pp. 657–658 (2019)
11. Enache, F., Deparateanu, D., Enache, A., et al.: Sparse array antenna design based on dolph-chebyshev and genetic algorithms. In: 8th International Conference on Electronics, Computers and Artificial Intelligence (ECAI), pp. 1–4 (2016)
12. Choi, T.H., Kim, S.S., Yoon, Y.J., et al.: Wide-band slot antenna on metal bezel for covering 28/39 GHz in 5G communication system. In: IEEE International Symposium on Antennas and Propagation and USNC-URSI Radio Science Meeting, pp. 29–30 (2019)

Adaptive Technologies of Hybrid Carrier Based on WFRFT Facing Coverage and Spectral Efficiency Balance

Ning Pan[1,2], Lin Mei[1(✉)], Linan Wang[2], Libin Jiao[2], and Bin Wang[2]

[1] Harbin Institute of Technology, Harbin, China
meilin@hit.edu.cn

[2] Science and Technology on Communication Networks Laboratory, Shijiazhuang, Hebei, People's Republic of China

Abstract. Adaptive Modulation and Coding (AMC) and power control are adopted to balance spectral efficiency and coverage facing the problem of spectrum shortage in recent years. Considering the Orthogonal Frequency Division Multiplexing (OFDM) scheme with high spectral efficiency and the Single Carrier-Frequency Domain Equalization (SC-FDE) scheme with wide coverage, this paper combines the switching of carrier schemes with AMC and power control to maximize system throughput. The proposal of hybrid carrier communication system based on Weighted-type Fractional Fourier Transform makes the integration of OFDM and SC-FDE possible, which solves the problem of smooth transition between the two schemes. This paper analyzes the relationships between coverage and spectral efficiency in the flat fading channel and gives suggestions to the user equipment of different carrier schemes. Then we have proposed the calculation strategy of power control parameters and the switching strategy of the carrier schemes and modulation and coding schemes under power control in the frequency-selective fading channel.

Keywords: Weighted-type Fractional Fourier Transform · Adaptive Modulation and Coding (AMC) · Power control · Coverage · Spectral efficiency

1 Introduction

In the early development stage of communication technology, the system can only design strategies according to the worst channel condition to ensure the quality of communication because of the time-varying fading characteristics of the wireless channels. However, if the system still uses the conservative strategies as the channel condition improves, extend resources such as spectrum, power and channel capacity will be wasted. Link-adaptive transmission technology, which includes power control, Adaptive Modulation and Coding (AMC), etc., has been proposed to solve this problem.

J. Hayes proposed the power control technology in 1968 [1]. This technology avoids the impact on other users and ensures the communication quality at the same time. The core idea of AMC technology is to dynamically change the Modulation and Coding Schemes (MCSs) at the transmitter according to the time-varying channel conditions. To maximize the spectral efficiency on the basis of the reliability of the

S. Shi et al. (Eds.): AICON 2020, LNICST 356, pp. 177–196, 2021.
https://doi.org/10.1007/978-3-030-69066-3_17

wireless transmission system, the low-order MCSs will be considered under poor channel conditions and high-order MCSs will be considered under excellent channel conditions [2, 3].

High-speed communication can be achieved easily using Multi Carrier (MC) scheme such as Orthogonal Frequency Division Multiplexing (OFDM) because of its high throughput. Nevertheless, OFDM signals have the problem of high Peak-to-Average Power Ratio (PAPR). If PAPR is high, the efficiency of Power Amplifier (PA) will be reduced because of the back-off. Thus, it is difficult for OFDM scheme to realize long-distance transmission. On the other hand, the Single Carrier (SC) signals have low PAPR and wide coverage can be realized easily.

Considering the characteristics of OFDM and SC schemes, if the two carrier schemes can be integrated under one system framework, it is possible to achieve both wide coverage and high spectral efficiency. A hybrid SC/MC system has been proposed by the research team at Tohoku University [4–7]. It is verified that the system can improve the throughput performance and extend the transmission distance.

The proposal of Hybrid Carrier (HC) communication system based on Weighted-type Fractional Fourier Transform (WFRFT) can also make the integration of the two schemes possible [8]. The system can transform the carrier scheme from SC to OFDM smoothly through parameters adjustment, which is different from the system above. The problem of incompatibility between the SC and MC schemes is solved through weighted integration of the two carrier schemes, providing the technical basis for further enhancing the flexibility and adaptive capabilities of the communication system.

Technologies such as AMC and power control are the main methods to reflect the system adaptive capabilities. In the existing research, SC or MC carrier scheme is usually combined with the technologies above, and the balance between the transmission rate and coverage is achieved through the switching of the MCSs. Based on the existing research, this paper introduces HC system based on WFRFT, utilizes the diverse characteristics provided by HC signals, and combines existing AMC and power control technologies to maximize the throughput of the system.

This paper is divided into five sections. The Sect. 2 introduces the HC system based on WFRFT. The Sect. 3 analyzes the relationships between coverage and spectral efficiency in the flat fading channel. The Sect. 4 introduces the switching strategy of the carrier schemes and MCSs under power control. In Sect. 5, the conclusions are presented.

2 HC System Based on WFRFT

2.1 HC Communication System

The system diagram of Single Carrier-Frequency Domain Equalization (SC-FDE) and OFDM is shown in Fig. 1. It can be seen that SC-FDE moves the Inverse Fast Fourier Transform (IFFT) module at the transmitter of OFDM to the receiver, so the SC system can also perform Frequency Domain Equalization (FDE).

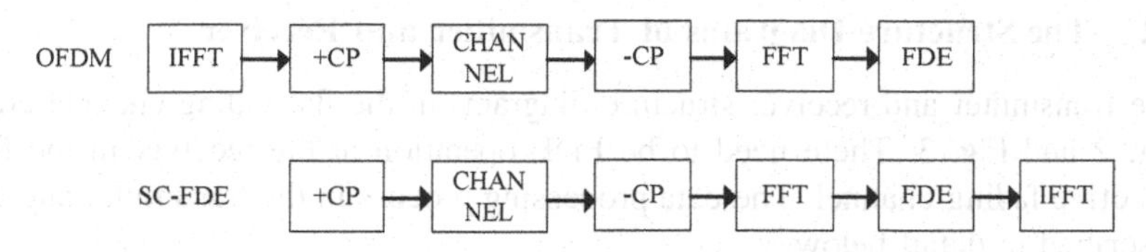

CP: Cyclic Prefix; FFT: Fast Fourier Transform

Fig. 1. System diagram of OFDM and SC-FDE

The proposal of WFRFT introduced the concept of HC. The HC is composed of SC and MC components, and the coefficient α determines the proportion of SC and MC in the HC system. The expression of α-order WFRFT is as follows

$$F_{4W}^{\alpha}[x(n)] = w_0(\alpha)x(n) + w_1(\alpha)X(n) + w_2(\alpha)x(-n) + w_3(\alpha)X(-n) \tag{1}$$

Where $x(n)$ is the data information to be processed and $X(n)$ is the Discrete Fourier Transform of $x(n)$. The expression of α in the Eq. (1) is as follows

$$w_l(\alpha) = \cos\left[\frac{(\alpha - l)\pi}{4}\right] \cos\left[\frac{2(\alpha - l)\pi}{4}\right] \exp\left[\pm i\frac{3(\alpha - l)\pi}{4}\right] \tag{2}$$

The matrix expression form of α-order WFRFT is shown below

$$\boldsymbol{W}_{\alpha} = w_0(\alpha)\boldsymbol{I} + w_1(\alpha)\boldsymbol{F} + w_2(\alpha)\boldsymbol{\Gamma} + w_3(\alpha)\Gamma\boldsymbol{F} \tag{3}$$

The matrix $\boldsymbol{I}$ is the identity matrix, and the matrix $\boldsymbol{F}$ is the matrix form of FFT, $[F]_{m,n} = \frac{1}{\sqrt{N}}\exp[-j2\pi mn/N]$, where m and n range from 1 to $N - 1$. Matric $\boldsymbol{\Gamma}$ is

$$\boldsymbol{\Gamma} = \begin{pmatrix} 1 & 0 & 0 & \cdots & 0 \\ 0 & 0 & 0 & \cdots & 1 \\ \vdots & \vdots & \vdots & \ddots & \vdots \\ 0 & 0 & 1 & \cdots & 0 \\ 0 & 1 & 0 & \cdots & 0 \end{pmatrix}_{N \times N} \tag{4}$$

This paper is based on the HC system. The carrier scheme is changed by adjusting the coefficient α. That is, the system mentioned above is MC transmission when $\alpha = 1$ and SC transmission when $\alpha = 0$. When α is between 0 and 1, it is HC transmission.

In [9–11], the PAPR characteristic of the HC signals is studied and the conclusion showed that the PAPR has a smooth transition between SC and MC as α changes. The BER performance in the frequency-selective channel also transforms smoothly between the two schemes [12].

2.2 The Structure Diagrams of Transmitter and Receiver

The transmitter and receiver structure diagrams in the flat fading channel are given in Fig. 2 and Fig. 3. There need to be FDE operation at the receiver in the frequency-selective fading channel. The data processing process in the transmitter and receiver is described in detail below.

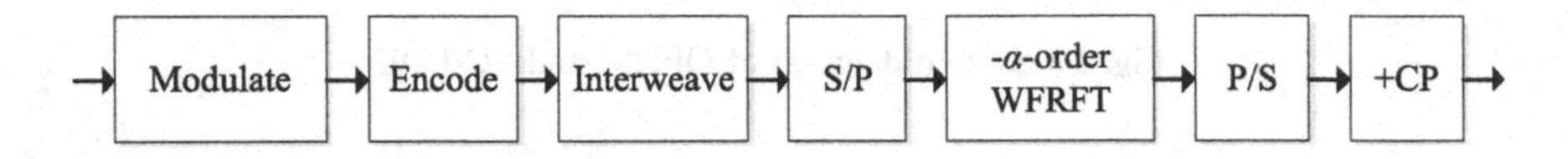

S/P: Serial/Parallel Convert; P/S: Parallel/Serial Convert

Fig. 2. Structure diagram of transmitter

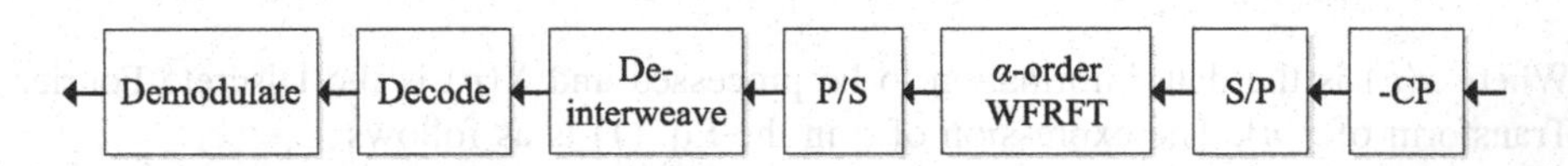

Fig. 3. Structure diagram of receiver

At the transmitter, the data need to be modulated, encoded, interweaved, S/P converted, and then sent to the -α-order WFRFT module. After the P/S conversion, CP is added and the data are transmitted over the multipath channel.

After receiving data, the receiver removes CP first, then performs S/P conversion, α-order WFRFT, and P/S conversion. Finally, the data need to be de-interweaved, decoded, and demodulated respectively.

3 The Relationships Between Coverage and Spectral Efficiency in the Flat Fading Channel

In the flat fading channel, the Block Error Rate (BLER) performance of each carrier scheme is the same, and the coverage is only determined by the Maximum Average Output Power (MATP). On the other hand, the PAPR performances of different carrier schemes are different, so the MATP values are also different. We simulated the relationships between coverage and spectral efficiency of three typical carrier schemes supposing that the signals are transmitted with MATP and gave suggestions to the User Equipment (UE) of different carrier schemes.

3.1 PAPR and MATP

In the traditional SC and OFDM carrier schemes, both the number of subcarriers and the constellation mapping methods affect the PAPR characteristics of the signals. The HC system based on WFRFT can control the PAPR of the signals through another dimension: the transformation order α. The PAPR curves of SC, HC and OFDM signals are obtained by simulating the HC system with α from 0 to 1 in steps of 0.1.

Figure 4, Fig. 5 and Fig. 6 show the Complementary Cumulative Distribution Function (CCDF) curves of PAPR for HC signals with different modulation schemes. The abscissa is the threshold of $PAPR_0$, and the ordinate is the probability that the PAPR value exceeds $PAPR_0$. The number of subcarriers is 512, and the oversampling factor is 8 [13]. Table 1 represents the PAPR of these three typical carrier schemes. When α changes from 1 to 0, the PAPR characteristic of the HC signals transforms smoothly from that of OFDM to SC.

At the transmitter, if the PA works near the saturation region, it can achieve higher power efficiency. However, when the signals have a high PAPR value, the PA can only process it by Input Back-Off (IBO) to ensure that it does not enter the nonlinear region. Among them, the IBO factor is defined as follows

$$IBO = 10\log_{10}(\frac{P_{sat}}{P_{av}}) = P_{sat}[\text{dB}] - P_{av}[\text{dB}] \quad (5)$$

where $P_{sat}[\text{dB}]$ and $P_{av}[\text{dB}]$ are the saturation and average power of signals sent to PA, respectively. In order to ensure the probability that the output signals occur non-linear distortion less than a certain threshold, the IBO cannot be lower than the corresponding $PAPR_0$.

After the power back-off, the average transmission power can be expressed as

$$P_t = P_{max} - IBO \quad (6)$$

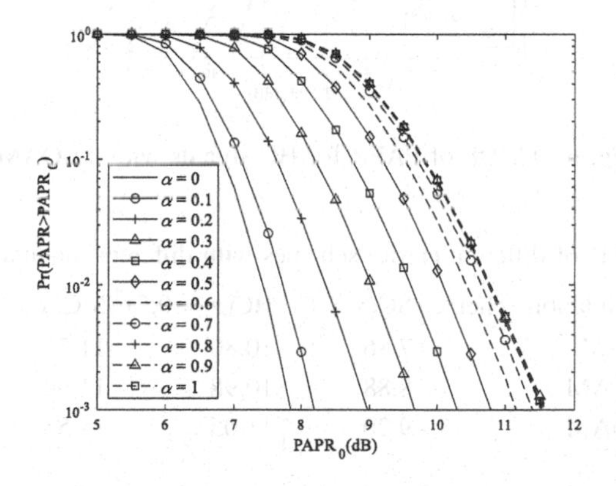

Fig. 4. CCDF of PAPR for HC signals with 4-QAM

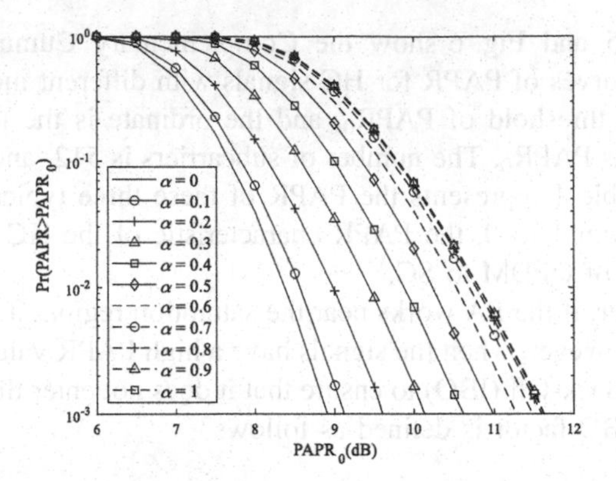

Fig. 5. CCDF of PAPR for HC signals with 16-QAM

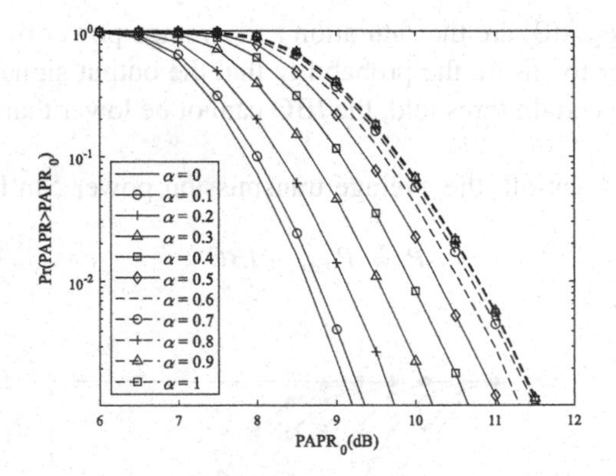

Fig. 6. CCDF of PAPR for HC signals with 64-QAM

Table 1. PAPR of different carrier schemes with different modulation schemes

Modulation scheme	SC(α = 0)	HC(α = 0.5)	MC(α = 1)
4QAM	7.86	10.80	11.57
16QAM	8.88	10.98	11.62
64QAM	9.20	11.05	11.56

where P_{max} is the maximum output power of PA without regard of IBO. This paper assumes P_{max} = 35 dBm. MATP refers to the transmission power under the condition of IBO = $PAPR_0$. For example, when the SC scheme adopts 4QAM, its $PAPR_0$ and MATP are 7.86 dB and 27.14 dBm, respectively.

Then we need to calculate the received power P_r and Signal-to-Noise Ratio (SNR) to select the optimal MCS for UE at each distance. P_r is defined as follows

$$P_r = P_t + 20\log_{10}(\frac{\lambda}{4\pi d_0}) - 10\log_{10}(\frac{d}{d_0})^{3.8} \tag{7}$$

In Eq. (7), the second term was free space electromagnetic loss and third term was propagation loss. At 2 GHz, λ and d_0 were 0.15 m and 100 m [14].

The received SNR is calculated by the received power P_r, single sideband noise power spectral density N_O of −174 dBm/Hz and system bandwidth B

$$SNR = P_r - N_0 B \tag{8}$$

3.2 The Relationships Between Coverage and Spectral Efficiency

The MCSs selected in this paper are shown in Table 2. The coding scheme is convolutional code and the constrained length is 7 for rate 1/2 and 8 for rate 1/3.

Figure 7 shows the BLER performances of different MCSs in the flat fading channel. We can find that the BLER performances reduce gradually from MCS1 to MCS6 and the abscissa in Fig. 7 is E_b/N_O.

Table 2. Modulation and coding schemes selected

MCS index	Modulation scheme	Coding rate
MCS1	4QAM	1/3
MCS2	4QAM	1/2
MCS3	16QAM	1/3
MCS4	16QAM	1/2
MCS5	64QAM	1/3
MCS6	64QAM	1/2

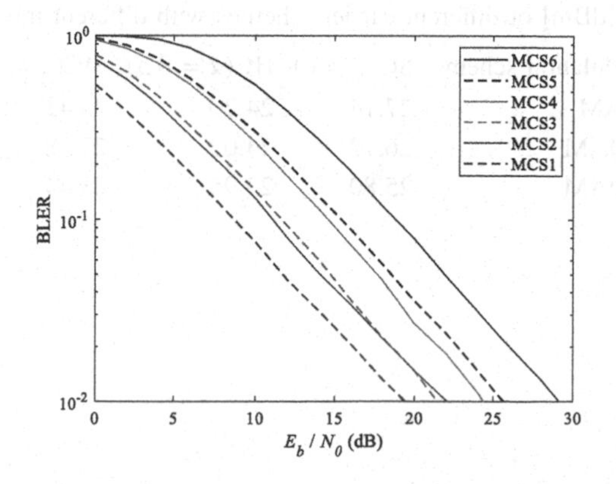

Fig. 7. Simulation of BLER performances in the flat fading channel

However, the SNR calculated from the received power is E_s/N_O, and the conversion method of E_s/N_O [dB] and E_b/N_O [dB] is shown in Eq. (9)

$$E_s/N_0[\mathrm{dB}] = E_b/N_0[\mathrm{dB}] + 10\lg R_b \tag{9}$$

In Eq. (9), the spectral efficiency R_b(bit/s/Hz) is defined as

$$R_b = \frac{N \cdot log_2 J \cdot \eta_{coding}}{(N+M)\cdot T_S \times 1/T_S} = \frac{N \cdot log_2 J \cdot \eta_{coding}}{N+M} \tag{10}$$

where T_S is the sample period, N is the symbol length, M is the CP length, η_{coding} refers to the coding rate, and $log_2 J$ refers to modulation order of J-QAM. When J = 4, 16, 64, the modulation orders are 2, 4, 6, respectively.

This paper uses BLER = 10% as the standard to delimit the switching thresholds of the MCSs [15]. The results are shown in Table 3.

Then the relationships between the spectral efficiency and coverage of the SC (α = 0), HC (α = 0.5), and MC (α = 1) carrier schemes in the flat fading channel are analyzed. Suppose that the three carrier schemes are all transmitted with their MATP, and the MATP values of different carrier schemes with different modulation schemes are shown in Table 4.

Table 3. Switching thresholds of different MCSs in the flat fading channel

MCS index	E_b/N_O[dB]	E_s/N_O[dB]
MCS1	8.72	6.96
MCS2	11.1	11.1
MCS3	11.67	12.91
MCS4	14.61	17.62
MCS5	15.43	18.44
MCS6	18.82	23.59

Table 4. MATP [dBm] of different carrier schemes with different modulation schemes

Modulation scheme	SC(α = 0)	HC(α = 0.5)	MC(α = 1)
4QAM	27.14	24.20	23.43
16QAM	26.12	24.02	23.38
64QAM	25.80	23.95	23.44

Table 5. Received SNR [dB] of different modulation schemes and MCS selected at each distance under SC scheme

Distance/m	4QAM	16QAM	64QAM	MCS index
100	42.68	41.64	41.34	MCS6
150	35.99	34.97	34.65	MCS6
200	31.24	30.22	29.90	MCS6
250	27.56	26.54	26.22	MCS6
300	24.55	23.53	23.21	MCS4
350	22.01	20.99	20.67	MCS4
400	19.80	18.78	18.46	MCS4
450	17.96	16.84	16.52	MCS3
500	16.12	15.10	14.78	MCS3
550	14.56	13.53	13.21	MCS3
600	13.11	12.09	11.77	MCS2
650	11.79	10.77	10.45	MCS2
700	10.57	9.55	9.23	MCS1
750	9.43	8.41	8.09	MCS1
800	8.36	7.34	7.02	MCS1
850	7.36	6.34	6.02	MCS1
900	6.42	5.40	5.08	None

Taking the SC scheme as an example, the received SNR under different modulation schemes and the selected MCS are shown in Table 5. At 400 m, the MCS5 also meets the SNR requirements, but the MCS4 with the same spectral efficiency is used because of its better BLER performance. In actual systems, if the SNR conditions permit, high coding rates such as 3/4 and 5/6 can be selected to achieve higher spectral efficiency.

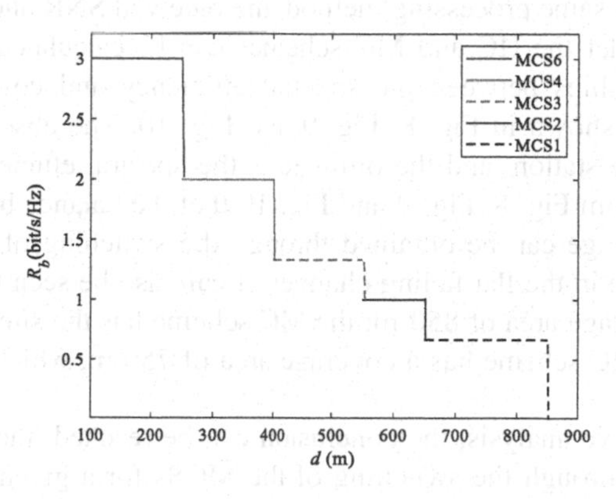

Fig. 8. Simulation of spectral efficiency and coverage under SC scheme

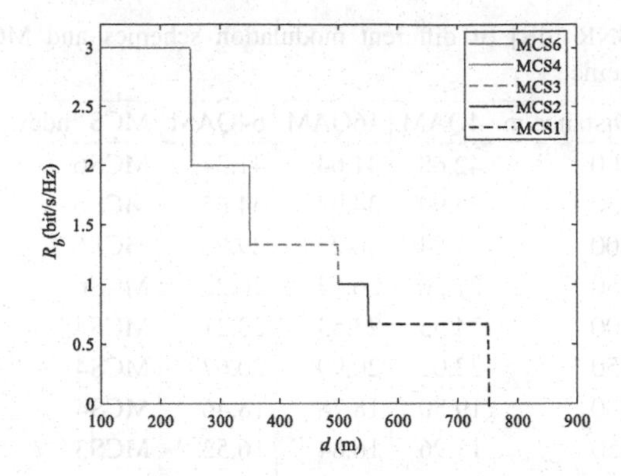

Fig. 9. Simulation of spectral efficiency and coverage under HC scheme

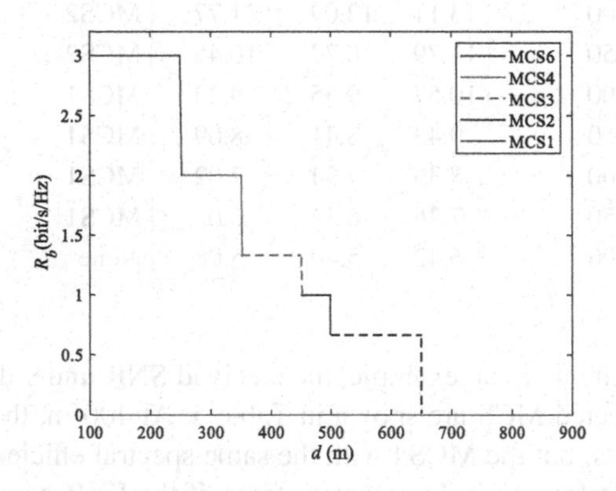

Fig. 10. Simulation of spectral efficiency and coverage under MC scheme

According to the same processing method, the received SNR and the selected MCS at each distance under the HC and MC schemes can be calculated. According to the results, the relationships between the spectral efficiency and coverage of the three carrier schemes are shown in Fig. 8, Fig. 9 and Fig. 10. The abscissa is the distance from UE to the base station, and the ordinate is the spectral efficiency.

It can be seen from Fig. 8, Fig. 9 and Fig. 10 that the balance between the spectral efficiency and coverage can be obtained through the switching of the MCSs for any given carrier scheme in the flat fading channel. It can also be seen that the SC scheme has the largest coverage area of 850 m; the MC scheme has the smallest coverage area of 650 m; and the HC scheme has a coverage area of 750 m, which is between that of SC and MC.

Through the above analysis, the conclusion can be reached: the UE can maximize system throughput through the switching of the MCSs for a given carrier scheme. In the flat fading channel, if the UE adopts the SC scheme, there is no need to change the carrier scheme, and the SC scheme transmission can obtain the best performance; if the OFDM scheme is adopted, the PAPR performance can be improved through WFRFT precoding to obtain the same coverage as the SC scheme.

4 Switching Strategy of Carrier Schemes and MCSs Under Power Control in the Frequency-Selective Fading Channel

In the frequency-selective fading channel, different carrier schemes have different BER performances. Compared with the flat fading channel, the frequency-selective fading channel can also achieve gain through the switching of the carrier schemes. We propose a new uplink selection strategy of the carrier scheme and MCS under power control in the frequency-selective channel. This strategy considers the PAPR, BLER performances of the signals, and the power control parameters.

4.1 Power Control

The power control in this paper aims at the reliability of the receiver, and the power control parameters are calculated based on the SNR switching thresholds of different MCSs in the flat fading channel. Thus, the UE can transmit signals with the lowest power on the basis of ensuring the BLER performance of the receiver. The calculation process is shown in Fig. 11 and the specific steps are as follows.

1. Starting from 100 m, select the SNR switching threshold of the MCS with the highest spectral efficiency, that is, 23.59 dB of the MCS6.
2. The transmission power is calculated by Eq. (11), which is derived from Eq. (7) and Eq. (8)

$$P_t = SNR + N_0B - \left(20\log_{10}(\frac{\lambda}{4\pi d_0}) - 10\log_{10}(\frac{d}{d_0})^{3.8}\right) \tag{11}$$

3. Judge whether the P_t is greater than the MATP of the MC scheme under this modulation scheme. If less, P_t is the power control parameter at this distance, and continue to use the SNR at the next distance until the calculated power is greater than the MATP of MC scheme under this modulation scheme. Otherwise, the SNR switching threshold of the MCS with the highest spectral efficiency in the remaining MCSs is selected at this distance.
4. After switching to a new MCS, continue to perform steps 2 and 3 until there is no MCS with lower spectral efficiency. Use P_t as the power control parameter at this distance, continue to use the SNR at the next distance and calculate P_t.
5. Judge whether P_t is greater than the MATP of the SC scheme under this modulation scheme. If less, P_t is the power control parameter at this distance, and the SNR will continue to be used at the next distance until the calculated power is greater than the MATP of SC scheme under this modulation scheme. Then end the process.

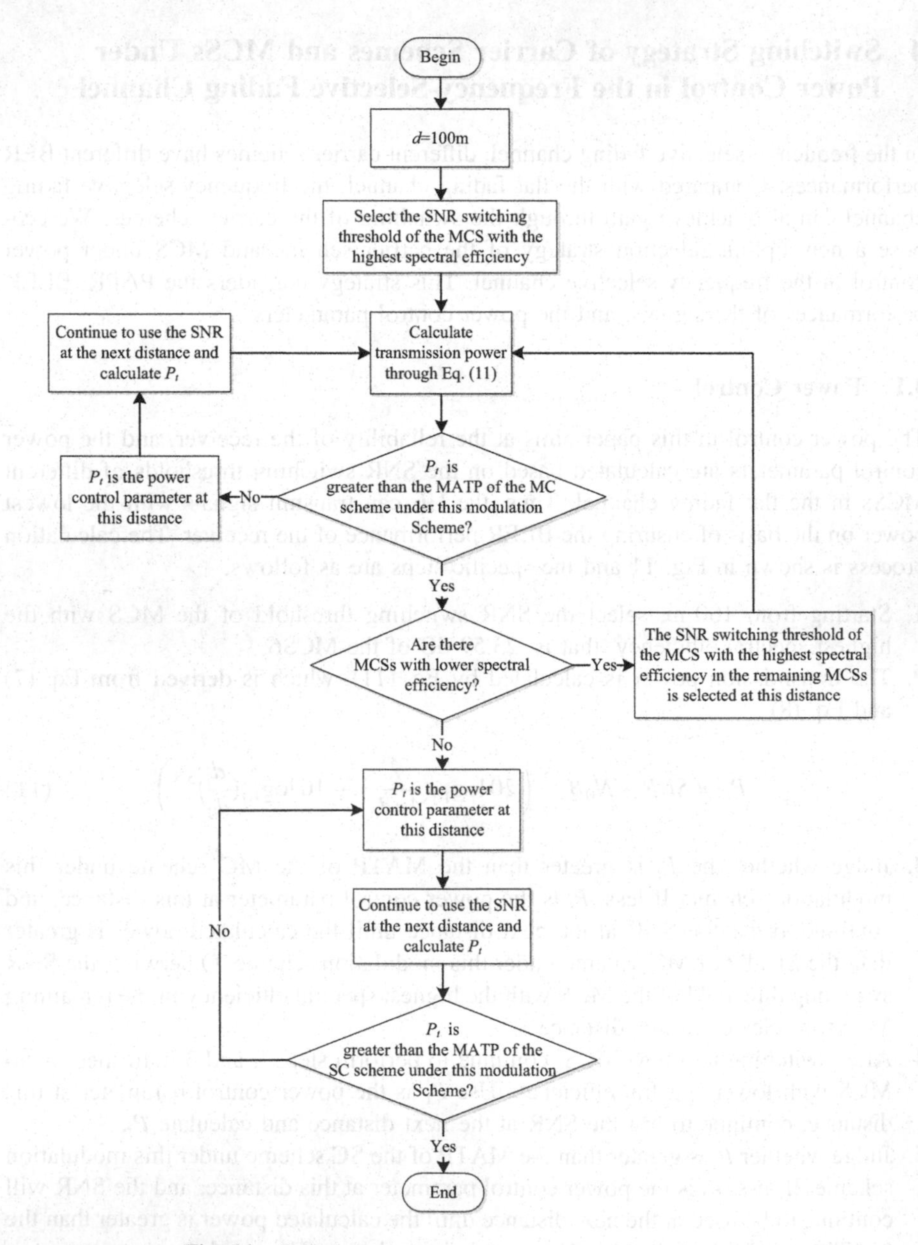

Fig. 11. Calculation process of power control parameters

Finally, the calculated power control parameters are shown in the second column of Table 6. In order to meet the requirements of the actual system better, the parameters in the second column of Table 6 are rounded as the integer multiple of 0.5, and the rounded results are shown in the third column of Table 6.

It can be seen from Table 6 that the power control parameters drop at 250–300 m, 350–400 m, 450–500 m and 500–550 m because of the switching of the MCSs.

Table 6. Power control parameters rounded or not at each distance

Distance/m	Unrounded power control parameters/dBm	Rounded power control parameters/dBm
100	8.05	8.5
150	14.74	15.0
200	19.49	19.5
250	23.17	23.5
300	20.21	20.5
350	22.75	23.0
400	20.25	20.5
450	22.19	22.5
500	22.12	22.5
550	19.55	20.0
600	20.99	21.0
650	22.31	22.5
700	23.53	24.0
750	24.67	25.0
800	25.74	26.0
850	26.74	27.0

4.2 BLER Performances of Different Carrier Schemes and MCSs

Table 7 shows the simulation conditions in the frequency-selective fading channel. Figure 12 shows the BLER performances of 4QAM under Zero Forcing (ZF) and Minimum Mean Square Error (MMSE) equalization. The simulation also takes the uncoded signals into consideration for comparison. The BLER performances of 16QAM and 64QAM are also simulated with the same parameters shown in Table 7. Then the SNR switching thresholds of all MCSs can be obtained.

In Fig. 12, we can find that the BLER performances of the MC signals under ZF and MMSE equalization are almost the same and that of the SC signals are different. The BLER performances of the HC signals are always between that of SC and MC.

Table 7. Simulation parameters of HC system in the frequency-selective fading channel

Parameter name	Value
Transmission Scheme	SC, HC(α = 0.5), MC
Decision Mode	Hard
Carrier Frequency	2 GHz
Channel Bandwidth	100 MHz
Number of Subcarriers	512
Symbol Period	10 ns
MCS Index	MCS1–MCS6
FDE Algorithms	FD-ZF, FD-MMSE
Multipath Delay	[0 100 200 300 500 700] ns
Average Path Gain	[0 − 3.6 − 7.2 − 10.8 − 25.2] dB
CP Length	80 symbols' length
Doppler Frequency Offset	0

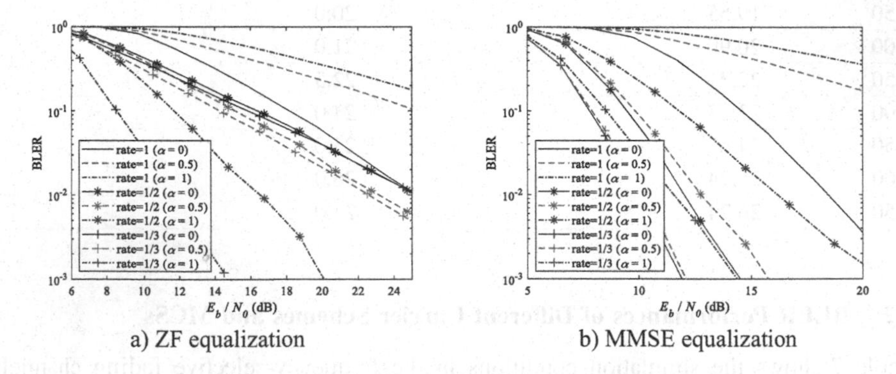

a) ZF equalization b) MMSE equalization

Fig. 12. BLER performances of 4QAM under different equalization algorithms

4.3 Switching Strategy of Carrier Schemes and MCSs Under Power Control

We propose a new uplink selection strategy of the carrier scheme and MCS based on the previous work. This strategy considers the PAPR, BLER performances of the signals, and the power control parameters. There are 18 schemes totally through all combination of the three carrier schemes and six MCSs. This strategy prefers to select the scheme with the highest spectral efficiency among the schemes that meet the requirements of the received SNR and MATP. If there is more than one scheme with the highest spectral efficiency, the scheme with the lowest SNR switching threshold will be selected, so that the system can have strong robustness. Among them, meeting the SNR requirement means that the switching SNR of the MCS needs to be less than the received SNR at this distance and meeting the MATP requirement means that the MATP of the scheme needs to be greater than the power control parameter.

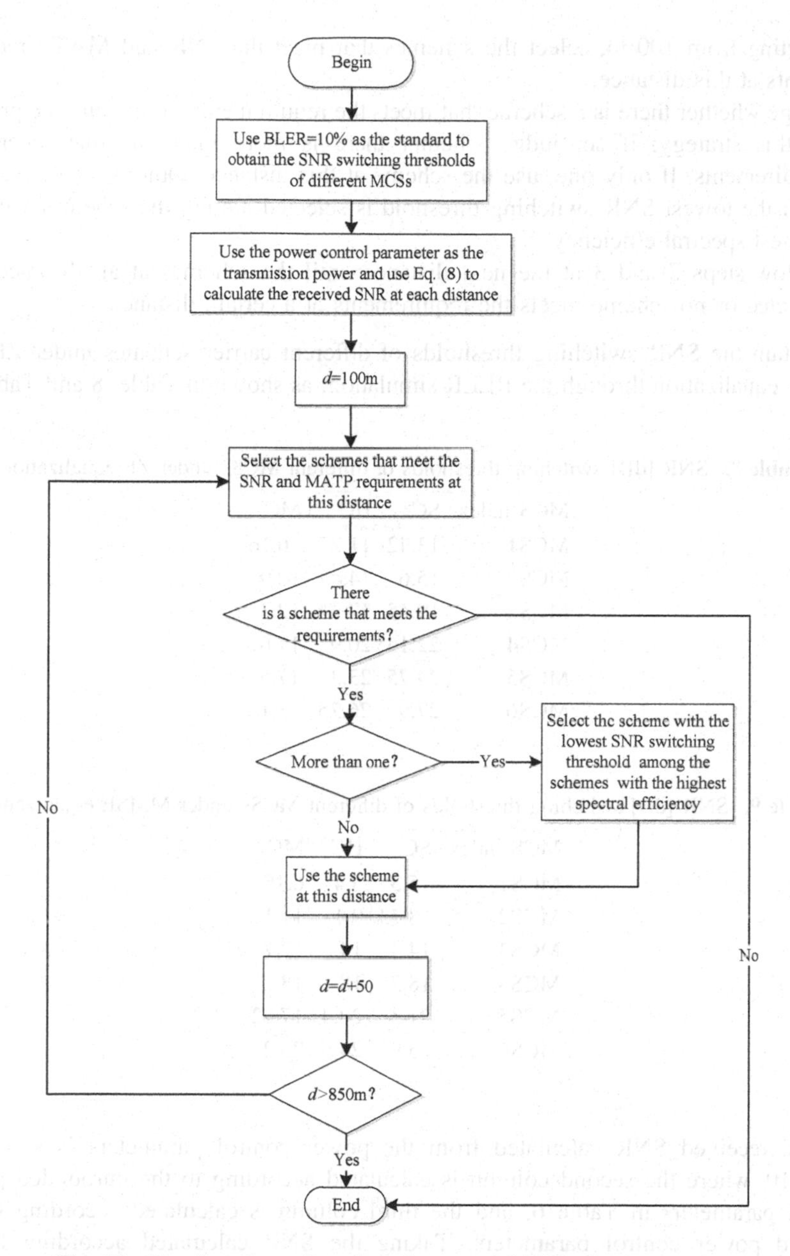

Fig. 13. The flowchart of carrier scheme and MCS selection strategy

The strategy flowchart is shown in Fig. 13, and the specific steps are as follows.

1. Use BLER = 10% as the standard to obtain the SNR switching thresholds of different MCSs under different carrier schemes; use the power control parameter as the transmission power, and use Eq. (8) to calculate the received SNR at each distance.

2. Starting from 100 m, select the schemes that meet the SNR and MATP requirements at this distance.
3. Judge whether there is a scheme that meets the requirements, if not, end the process of this strategy; if so, judge whether there is more than one that meets the requirements. If only one, use the scheme at this distance. Otherwise, the scheme with the lowest SNR switching threshold is selected among the schemes with the highest spectral efficiency.
4. Follow steps 2 and 3 at the next distance until the schemes at all distances are selected or no scheme meets the requirements at a certain distance.

We obtain the SNR switching thresholds of different carrier schemes under ZF and MMSE equalization through the BLER simulation, as shown in Table 8 and Table 9.

Table 8. SNR [dB] switching thresholds of different MCSs under ZF equalization

MCS index	SC	HC	MC
MCS1	13.42	11.75	6.16
MCS2	15.6	14.44	10.9
MCS3	19.45	17.94	12.3
MCS4	22.15	20.9	17.62
MCS5	24.75	23.3	17.5
MCS6	27.9	26.75	23.1

Table 9. SNR [dB] switching thresholds of different MCSs under MMSE equalization

MCS index	SC	HC	MC
MCS1	5.3	5.4	6.16
MCS2	8.73	9.1	11.1
MCS3	14.4	14	12.7
MCS4	18.2	18	18
MCS5	21.3	20.4	17.62
MCS6	25.9	24.9	23.2

The received SNR calculated from the power control parameters is shown in Table 10, where the second column is calculated according to the unrounded power control parameters in Table 6, and the third column is calculated according to the rounded power control parameters. Taking the SNR calculated according to the unrounded parameters as an example, the received SNR is 23.59 dB and the transmission power is 8.05 dBm at 100 m. There are 15 schemes which meet the requirements of SNR and MATP and only one scheme with the highest spectral efficiency, that is, the MC scheme with MCS6.

Table 10. Received SNR [dB] calculated by rounded power or not at each distance

Distance/m	Received SNR calculated by unrounded power	Received SNR calculated by rounded power
100	23.59	24.04
150	23.59	23.85
200	23.59	23.60
250	23.59	23.92
300	17.62	17.91
350	17.62	17.87
400	12.91	13.16
450	12.91	13.22
500	11.1	11.48
550	6.96	7.41
600	6.96	6.97
650	6.96	7.15
700	6.96	7.43
750	6.96	7.29
800	6.96	7.22
850	6.96	7.22

Finally, the switching results under ZF and MMSE equalization are shown in Table 11 and Table 12, where the second and third columns are calculated according to the second column of Table 10 and the fourth and fifth columns are calculated according to the third column.

From Table 11, the UE adopts the MC scheme at all distances under ZF equalization. As the distance increases, the spectral efficiency gradually decreases. The SC and HC schemes have not been adopted due to the high SNR switching thresholds. At the same time, a new MCS is switched at 250 m because the transmission power is rounded from 23.17 dBm to 23.5 dBm, which is greater than the MATP of the MC scheme at 64QAM.

It can be seen from Table 12 that the UE adopts the MC scheme with high-order MCSs at near distances and the SC scheme with low-order MCSs at far distances under MMSE equalization. The switching of the carrier schemes and MCSs is achieved, so that the HC system can reach a better balance between the coverage and spectral efficiency through the parameter adjustment.

From Table 11 and Table 12, it can be found that a new MCS is switched at 250 m under MMSE equalization, which is the same as ZF equalization. The difference between the two is that the UE adopts the MC scheme at all distances under ZF equalization, while the UE adopts the MC scheme with high-order MCSs at near distances and the SC scheme with low-order MCSs at far distances under MMSE equalization. The reason is that the SNR gain brought by the low PAPR performances of the SC signals is not as large as that brought by the BLER performances of the MC signals for low-order MCSs under ZF equalization. The SC scheme outperforms the MC scheme under low-order MCSs and its MATP is higher when MMSE equalization

Table 11. Carrier scheme and MCS selected at each distance under ZF equalization

Distance/m	Carrier scheme	MCS index	Carrier scheme	MCS index
100	MC	MCS6	MC	MCS6
150	MC	MCS6	MC	MCS6
200	MC	MCS6	MC	MCS6
250	MC	MCS6	MC	MCS5
300	MC	MCS5	MC	MCS5
350	MC	MCS5	MC	MCS5
400	MC	MCS3	MC	MCS3
450	MC	MCS3	MC	MCS3
500	MC	MCS2	MC	MCS2
550	MC	MCS1	MC	MCS1
600	MC	MCS1	MC	MCS1
650	MC	MCS1	MC	MCS1

Table 12. Carrier scheme and MCS selected at each distance under MMSE equalization

Distance/m	Carrier scheme	MCS index	Carrier scheme	MCS index
100	MC	MCS6	MC	MCS6
150	MC	MCS6	MC	MCS6
200	MC	MCS6	MC	MCS6
250	MC	MCS6	MC	MCS5
300	MC	MCS5	MC	MCS5
350	MC	MCS5	MC	MCS5
400	MC	MCS3	MC	MCS3
450	MC	MCS3	MC	MCS3
500	SC	MCS2	SC	MCS2
550	SC	MCS1	SC	MCS1
600	SC	MCS1	SC	MCS1
650	SC	MCS1	SC	MCS1
700	SC	MCS1	SC	MCS1
750	SC	MCS1	SC	MCS1
800	SC	MCS1	SC	MCS1
850	SC	MCS1	SC	MCS1

is adopted. Thus, it is used at long distances. The HC scheme has not been adopted under ZF and MMSE equalization, which shows that the HC scheme cannot achieve the optimal performance of the spectral efficiency and coverage in this scenario, and the system can only switch schemes between SC and MC.

5 Conclusion

We propose the adaptive technologies of HC based on WFRFT facing the balance of coverage and spectral efficiency. The relationships between the spectral efficiency and coverage of different carrier schemes in the flat fading channel are analyzed. We can find that the coverage is gradually reduced when the carrier scheme transforms from SC to MC. Therefore, if the UE adopts the SC scheme, there is no need to change the carrier scheme. If the OFDM scheme is adopted, the PAPR performance of the OFDM signals can be improved by WFRFT precoding, so that it can obtain the same coverage as the SC scheme. In the frequency-selective channel, this paper takes power control into consideration. The UE adopts the MC scheme at all distances under ZF equalization, because the SNR gain brought by the low PAPR performances of the SC signals is not as large as that brought by the BLER performances of the MC signals for the low-order MCSs. The UE adopts the MC scheme with high-order MCSs at near distances and the SC scheme with low-order MCSs at far distances under MMSE equalization, which maximizes the overall spectral efficiency of the system, and the incompatibility of SC and MC can be solved easily through HC system based on WFRFT.

Acknowledgement. This work was supported in part by The National Key Research and Development Program of China (254), in part by Science and Technology on Communication Networks Laboratory under Grant SXX19641X072.

References

1. Hayes, J.: Adaptive feedback communications. IEEE Trans. Commun. **16**(1), 29–34 (1968)
2. Wang, Y., Zhang, L., Yang, D.: Performance analysis of type III HARQ with turbo codes. In: IEEE 57th Vehicular Technology Conference (VTC 2003), pp. 2740–2744. IEEE, Jeju (2003)
3. Otsu, T., Okajima, I., Umeda, N., Yamao, Y.: Network architecture for mobile communications systems beyond IMT-2000. IEEE Personal Commun. **8**(5), 31–37 (2001)
4. Tanno, M., Kishiyama, Y., Taoka, H., Miki, N., Higuchi, K., Morimoto, A., Sawahashi, M.: Layered OFDMA radio access for IMT-advanced. In: IEEE 68th Vehicular Technology Conference (VTC 2008). IEEE, Calgary (2008)
5. Kashiwamura, I., Tomita, S., Komatsu, K., Tran, N., Oguma, H., Izuka, N., Kameda, S., Takagi, T., Tsubouchi, K.: Investigation on single-carrier and multi-carrier hybrid system for uplink. In: IEEE 20th Personal, Indoor and Mobile Radio Communications, PIMRC 2009, pp. 3188–3192. IEEE, Tokyo (2009)
6. Tomita, S., Miyake, Y., Kashiwamura, I., Komatsu, K., Tran, N., Oguma, H., Izuka, N., Kameda, S., Takagi, T., Tsubouchi, K.: Hybrid single-carrier and multi-carrier system: improving uplink throughput with optimally switching modulation. In: IEEE 21st Personal, Indoor and Mobile Radio Communications, pp. 2438–2443. IEEE, Istanbul (2010)
7. Miyake, Y., Kobayashi, K., Komatsu, K., Tanifuji, S., Oguma, H., Izuka, N., Kameda, S., Suematsu, N., Takagi, T., Tsubouchi, K.: Hybrid single-carrier and multi-carrier system: widening uplink coverage with optimally selecting SDM or Joint FDE/Antenna diversity. In: IEEE 14th International Symposium on Wireless Personal Multimedia Communications WPMC, pp. 1–5. IEEE, Brest (2011)

8. Mei, L.: Weighted Fractional Fourier Transform and its application in communication systems. Harbin Institute of Technology Ph.D. thesis, pp. 51–60 (2010)
9. Mei, L., Sha, X., Zhang, N.: PAPR of hybrid carrier scheme based on weighted-type fractional Fourier transform. In: 2011 6th International ICST Conference on Communications and Networking in China (CHINACOM), pp. 237–240. IEEE, Harbin (2011)
10. Wang, X., Mei, L., Zhang, N., Xie, W.: PAPR approximation of continuous-time WFRFT signals. In: IEEE International Conference on Communication Systems, pp. 303–307. IEEE, Macau (2014)
11. Wang, Z., Mei, L., Wang, X., Zhang, N., Wang, S.: Joint suppression of PAPR and sidelobe of hybrid carrier communication system based on WFRFT. In: 2015 IEEE/CIC International Conference on Communications in China (ICCC), pp. 1–5. IEEE, Shenzhen (2015)
12. Wang, Z., Mei, L., Sha, X., Zhang, N.: BER analysis of STBC hybrid carrier system based on WFRFT with frequency domain equalization. Sci. China Inf. Sci. **61**(8), 1–3 (2018). https://doi.org/10.1007/s11432-017-9259-0
13. Wang, X.: Research on order selection strategy and non-orthogonal multiple access technology in hybrid carrier power domain. Harbin Institute of Technology Ph.D. thesis, pp. 67–69 (2018)
14. European Cooperation in the Field of Scientific and Technical Research EURO-COST 231, Urban Transmission Loss Models for Mobile Radio in the 900 and 1,800 MHz Bands (revision 2). The Hague, Netherlands (1991)
15. 3GPP: Evolved Universal Terrestrial Radio Access (EUTRA); Physical Layer Procedures. 3rd Generation Partnership Project (3GPP), Technical Specification (TS) 36.213 (2019)

Towards Knowledge-Driven Mobility Support

Zhongda Xia and Yu Zhang(✉)

Harbin Institute of Technology, Harbin 150001, Heilongjiang, China
{xiazhongda,yuzhang}@hit.edu.cn

Abstract. Mobility refers to the ability to conduct "seamless" communication with network entities whose network location constantly changes. This paper examines the mobility support problem in IP and Named Data Networking (NDN), and identifies two dimensions in the mobility support solution space: the host dimension and data dimension. Existing host dimension solutions have exhausted the available design choices, and have not been able to achieve new breakthroughs in performance. Recognizing this limitation, this paper proposes a novel knowledge dimension. In the knowledge dimension, two knowledge-driven mobility support approaches, Topology-driven Intermediate Placement (TIP) and Trajectory-driven Reachability Update (TRU), are proposed. These approaches exploit knowledge such as network topology and movement trajectory to tweak the network and network services for better overall mobility support performance. A cross-architectural quantitative evaluation framework covering two communication scenarios and 5 quantifiable metrics is proposed to evaluate mobility support performance. Experiment results show that the knowledge-driven approaches significantly improve mobility support performance, demonstrating the potential of the knowledge-driven vision for providing better mobility support.

Keywords: Mobility · Mobility support approach · Knowledge-driven networking · Internet architecture

1 Introduction

With the proliferation of the computer networking technology, applications that thrive on the cyberspace are booming in every aspect of human activities. Along comes the ever-increasing demand on innovations that can satisfy novel needs such as supporting mobility [28], security [12,26], and autonomy [24]. Specifically, mobility refers to the ability to conduct "seamless" communication with network entities (e.g., subnets, hosts, applications or data) whose network location constantly changes. This paper examines the mobility support problem in IP and NDN, and identifies two dimensions in the mobility support solution

S. Shi et al. (Eds.): AICON 2020, LNICST 356, pp. 197–216, 2021.
https://doi.org/10.1007/978-3-030-69066-3_18

space: host dimension and data dimension. Host dimension solutions employ a certain intermediate to track the location of mobile entities; such solutions have exhausted the available design choices in the host dimension, and have not been able to achieve new breakthroughs in performance. Recognizing this limitation, this paper proposes a new design dimension, namely the knowledge dimension. In the knowledge dimension, knowledge such as network topology and movement trajectory can be used to tweak the network to both improve user experience and reduce network overhead. Under this vision, two novel knowledge-driven mobility support approaches, Topology-driven Intermediate Placement (TIP) and Trajectory-driven Reachability Update (TRU), are proposed. TIP places the intermediate at a location to shorten most forwarding paths, while TRU optimizes the way reachability information is updated to reduce signaling overhead.

To evaluate mobility support performance across network architectures, a cross-architectural quantitative evaluation framework is proposed. Under two communication scenarios, 5 quantifiable metrics reflecting user experience and network overhead are defined. The calculation formulae for these metrics are given by analyzing the key differences among mobility support approaches and architectures. By feeding network topology and movement track to the formulae, mobility support performance can be quantified via numerical simulations. Experiment results show that the proposed knowledge-driven approaches are superior according to all 5 metrics, demonstrating the potential of the knowledge-driven vision for providing better mobility support.

2 Background

2.1 IP and NDN Architecture

IP and NDN transmit application-layer information across a network by specifying a network-wide consensus including network identifier namespace, packet format, and per-hop behaviors (PHB). In IP, an IP address identifies a network interface, and each IP packet carries two IP addresses, namely source address and destination address. Each IP node maintains a Forwarding Information Base (FIB), which maps IP address (prefix) to network interface. Upon receiving an IP packet, the carried destination address is matched against the FIB using the longest prefix match (LPM) algorithm to determine the output interface. In NDN, a data name identifies a piece of named data. An Interest packet carries a data name, and is sent by data consumers as a request for data; a Data packet carries a data name and represents a piece of named data. Each NDN node maintains a FIB, a Pending Interest Table (PIT), and a Content Store (CS). The FIB maps data name (prefix) to network interface. The PIT keeps track of incoming Interests, including storing the input network interfaces. The CS caches received Data. Upon receiving an Interest, the Interest is forwarded according to the FIB if no matching Data is found in the CS and the Interest is not in the PIT, the incoming interface is then stored in the PIT. Upon receiving a Data, matching Interest is looked up in PIT, and if a match is found, Data is sent through all recorded interfaces towards consumers.

2.2 The Mobility Support Problem and Solution Space

Under the computer networking paradigm, mobility often refers to the ability to conduct "seamless" communication with network entities whose network location constantly changes. In IP, the mobility support problem concerns how to deliver IP packets to mobile entities using the same IP address. In NDN, there are two types of mobility, namely consumer mobility and producer mobility. Consumer mobility is natively supported in NDN since Data is forwarded along the reverse path of Interest. To retrieve lost Data after relocation, a mobile consumer simply retransmits the Interest. The NDN producer mobility problem is rather similar to the IP mobility problem, Interests need to be forwarded to/toward the mobile producer according to FIBs.

This paper considers IP and NDN producer mobility support problem, which share a common design space. Solutions under this common design space take a host-centric approach by maintaining the persistence and reachability of the network identifiers associated with mobile network entities. In NDN, the design space is expanded by data-centric solutions that attempt to make data constantly available by moving or disseminating data [27]. Thus, we depict the current design space as a two-dimensional space consisting of a host and a data dimension. Along the host dimension, host-centric solutions employ an intermediate to track the network location of mobile entities. Along the data dimension, solutions employ an intermediate to rendezvous requests for data with requested data, which may rely on host dimension approaches for reachability.

3 Cross-architectural Host Dimension Approaches

Host dimension solutions can be broken down into two processes: reachability information update and packet forwarding. The former updates the reachability information of a mobile entity's identifier, while the latter uses the updated information to deliver packets to the mobile entity. Considering the two processes, existing host dimension solutions may be categorized as 4 types of cross-architectural approaches, namely proxy-based, resolution-based, routing-based and trace-based approach. In the rest of this section, M represents a mobile device, S represents an immobile server that provides mobility support service, and C represents the entity that communicates with M. The term relocation refers to the event that M changes its Point-of-Attachment (PoA) to the network.

3.1 Proxy-Based Approach (Proxy)

Proxy maintains the mapping between M's network identifier and a globally reachable network identifier associated with M's current PoA, namely M's locator. In the reachability information update process, M sends a signaling packet

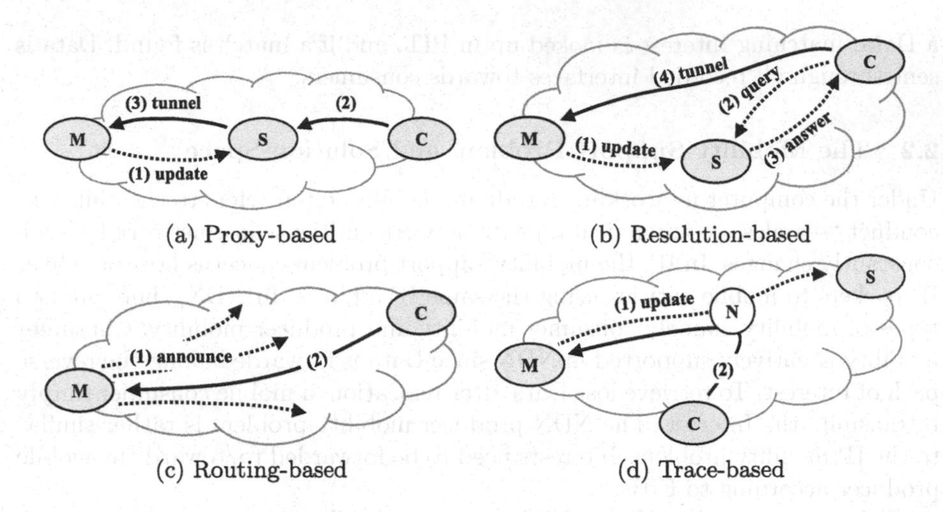

(a) Proxy-based (b) Resolution-based

(c) Routing-based (d) Trace-based

Fig. 1. Host dimension approaches

to S after a relocation (Fig. 1a: step 1); the signaling packet identifies M and contains M's current locator; S then updates the mapping accordingly. In the packet forwarding process, M's network identifier is made constantly reachable at S, so that packets destined to M always reach S first; upon receiving such packets, S will tunnel the packet to M using M's locator (Fig. 1a: step 3).

Compared with the described approach, existing work in IP [17,22,25] and NDN [4,10,13,15] differs in three major aspects: (1) S may be deployed in a distributed manner as multiple cooperative instances; (2) tunnelling may be implemented in various ways; and (3) the protocol may or may not be transparent to endpoints. Considering mobility support performance evaluation, (2) is irrelevant because tunnelling by definition produces a direct path to M; (3) is irrelevant because it is mostly an engineering choice; (1) is relevant, and this research does not consider distributed deployment.

3.2 Resolution-Based Approach (Resolution)

Like Proxy (Sect. 3.1), Resolution also maintains the mapping between M's network identifier and locator, thus shares the same reachability information update process. In the packet forwarding process, C will query S about M's locator, producing an exchange of packets between C and M (Fig. 1b: step 2 and 3). After learning M's locator, C will directly tunnel packets to M.

Resolution shares with Proxy the same differences between existing work in IP [2,9,20,29] and NDN [1,10,13], and the same considerations regarding these differences, thus not repeated here.

3.3 Routing-Based Approach (Routing)

Routing updates the FIB on each node to produce optimal forwarding paths to M's current PoA. In the reachability information update process, M broadcasts signaling packets into the network after a relocation (Fig. 1c: step 1). As a result, FIBs are updated to establish shortest forwarding paths for M's network identifier from anywhere in the network. In the packet forwarding process, packets sent by C will follow FIBs to reach M through a shortest path.

Compared with the described approach, existing solutions in IP [11,19] and NDN [3] differ in two major aspects: (1) propagation range of signaling messages; (2) route calculation. Both are relevant to evaluation, and Routing behaves as follows: for (1), signaling messages will be flooded to reach every node; for (2), shortest forwarding paths are always produced.

3.4 Trace-Based Approach (Trace)

Trace establishes forwarding paths to M on top of existing routing information. In the reachability information update process, M sends a signaling packet to S after a relocation (Fig. 1d: step 1), making each node on the shortest path between M and S record a FIB entry for M's network identifier pointing to M, namely a trace, constructing a forwarding path from S to M. In the packet forwarding process, packets destined to M will be first forwarded toward S, and eventually meet traces either at S or before (e.g., at node N as marked in Fig. 1d). After meeting a trace, the packet will follow traces to reach M.

Compared with the described approach, existing work in IP [7,18] and NDN [8,13,23,28] differs in two major aspects: (1) how traces are created, and (2) how traces are maintained. Both are relevant, and Trace behaves as follows: for (1), signaling packets from M will directly establish the complete forwarding path from S to M; for (2), traces will disappear upon relocation.

4 The Knowledge Dimension of Mobility Support

Existing solutions have exhausted the design choices in available design space, thus can only make tradeoff decisions to balance between user experience and network overhead. This paper proposes the knowledge dimension in the mobility support solution space. This new dimension offers a new set of design choices that may nurture solutions that comprehensively improve mobility support performance. In the knowledge dimension, various types of knowledge including current states (ground truth knowledge, e.g., network topology, placement of immobile entities), predictable states (inferred knowledge, e.g., movement track of mobile entities), and task-specific strategies (driver knowledge, i.e., how various information should be used to meet application needs) are exploited as the knowledge to drive the network to satisfy stringent performance requirements.

The expanded mobility support solution space is illustrated in Fig. 2. In the knowledge dimension, knowledge-driven approaches (or KD approaches for

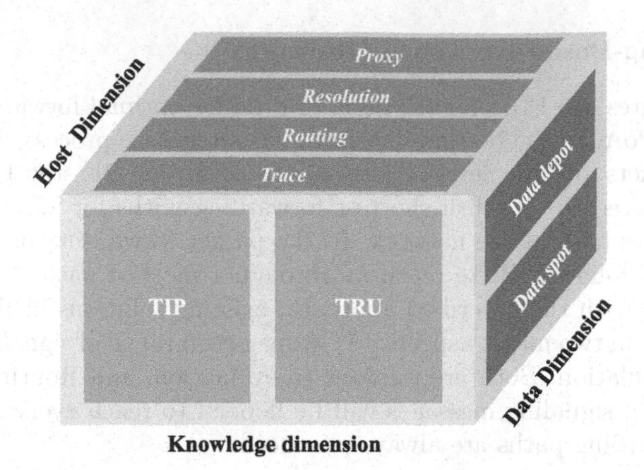

Fig. 2. The expanded mobility support solution design space

short) take knowledge as input, and output operational decisions that tweak the network and network services for better mobility support performance. This paper proposes two novel KD approaches: (1) Topology-driven Intermediate Placement (TIP), and (2) Trajectory-driven Reachability Update (TRU). Two knowledge-driven approaches can be combined with host dimension approaches to improve user experience AND reduce network overhead. In the rest of this section, we first describe the two knowledge-driven approaches, then briefly sketch out the architectural requirements to support them.

4.1 Topology-Driven Intermediate Placement (TIP)

TIP places the intermediate at an optimal network location in order to optimize forwarding paths, and may be combined with host dimension approaches that rely on an intermediate without affecting the internal mechanisms. Given a network topology and network locations (i.e., nodes in the topology) of relevant entities, an optimal network location of the intermediate is calculated using the betweeness centrality algorithm. The intermediate is then migrated to this optimal location. TIP employs the following ground truth knowledge:

- network topology graph (G): a graph representing the actual topology;
- network location set of mobile endpoints ($MLoc$): a set consisting of all potential network locations of the mobile endpoints that use the mobility support service provided by the intermediate in question;
- network location set of correspondent endpoints ($CLoc$): a set consisting of the network locations of the endpoints that communicate with the mobile endpoints in question.

The driver knowledge employed by TIP is an algorithm that outputs an optimal location. Specifically, the betweeness centrality of each node in a given topology G is calculated for a set of source nodes $CLoc$ and a set of destination nodes $MLoc$, the node with the highest value is selected as the output.

TIP then drives the network to change the location of the intermediate. Let the optimal location be B, and the intermediate's current network location be A, the migration of the intermediate from A to B may involve the following steps:

1. A collection of executables and runtime states of the service provided by the intermediate is encapsulated, and transferred from A to B;
2. routing announcements are made from B if necessary, to lead packets to B.

4.2 Trajectory-Driven Reachability Update (TRU)

TRU optimizes the reachability information update process of each host dimension approach when fed with a network topology and predicted movement track of a mobile endpoint. TRU employs the following ground truth knowledge:

- network topology graph (G): a graph representing the actual topology;
- underlying host dimension approach (*App*): which host dimension approach is already deployed and used.

TRU also employs an inferred knowledge:

- mobile endpoint's network location sequence ($Track$): a temporal sequence of network locations that may be inferred from movement pattern or trajectory that may further be based on auxiliary information about the mobile endpoint; for a mobile endpoint that relocates n times, the sequence is $Track = (t_i, P_i)(1 \leq i \leq n)$, where t_i is time when the ith relocation finishes, and P_i is the mobile endpoint's PoA after the ith relocation.

The driver knowledge employed by TRU is a centralized reachability information update mechanism that exploits the knowledge above to reduce signaling traffic. According to *App*, the mechanism exploits other knowledge as follows:

- Proxy/Resolution-based: schedule the intermediate to automatically change the stored locator to P_i at t_i;
- routing-based: calculate FIB changes at each t_i, and schedule each node to automatically update its FIB upon t_i as needed;
- trace-based approach: upon t_i, schedule each node on the shortest path between P_i and the intermediate to update its FIB to create the trace pointing to P_i, and schedule all other nodes to purge traces.

4.3 Architectural Requirements of Supporting the Knowledge Dimension

KD approaches involves the collection and processing of knowledge, and operating the network and network services. Similar functions can be provided by control-oriented network architectures/frameworks such as Software-Defined Networking (SDN) [14] and Network Function Virtualization (NFV) [16], where controller applications or orchestrators process information about the network to dynamically adjust the network behavior. Such architectures have the potential to bridge the gap between the knowledge-driven vision and actual practice, but the details are yet unclear. Here we sketch out several critical requirements to an architectural design that supports the knowledge-driven vision:

- Abstraction: the architecture should provide a way to abstract underlying network resources and application needs;
- Compatibility: the architecture should support existing network architectures that provide packet delivery service like IP and NDN;
- Reactivity: the overall operational process should form a feedback loop: states and observations of the underlying network generate actions, which in turn impact the network.

5 A Quantitative Cross-architectural Mobility Support Performance Evaluation Framework

The proposed framework takes metric definition, network architecture (architecture for short) description, and mobility support approach (approach for short, including KD approaches) description as input, and outputs the formulae for each metric. The formulae can then be used to calculate metric values from given network topologies and movement tracks.

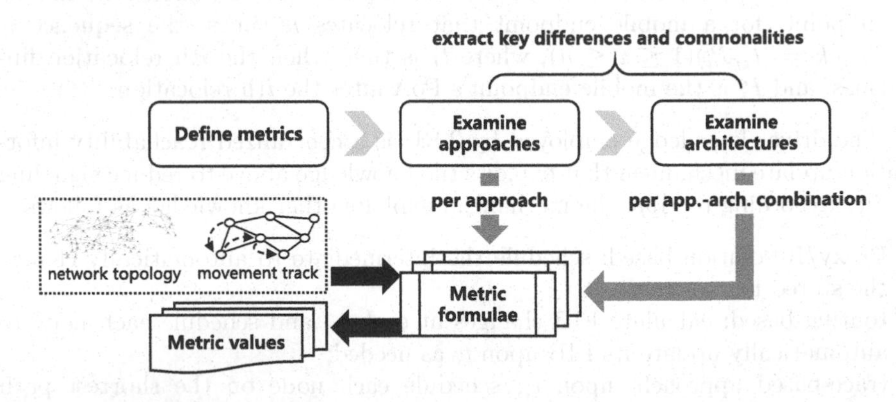

Fig. 3. The thought process of determining metric formulae

The thought process of determining metric formulae is illustrated in Fig. 3. At the core is the definition of metrics, which is independent from architectures and approaches. Based on the definition, how various design choices of each approach and architecture affect metric calculation is carefully examined. Among architectures and approaches, irrelevant differences are ignored, while key differences are reflected in the formulae. Eventually, a list of formulae is given for a metric, respecting each approach, or architecture/approach combination.

5.1 Basic Models

Network Topology: This paper considers an infrastructure network environment consisting of inter-connected nodes. The network topology can be represented by an undirected, weightless and fully-connected graph $T = (V, T)$, where V is the set of vertices in the graph, corresponding to each node in the network, and E is the edges, corresponding to each link between connected nodes.

Network endpoints such as user devices and servers gain access to the network by connecting to nodes. The node that an endpoint connects to is the PoA of the endpoint. Only nodes are considered in shortest path calculation. Thus, the shortest path between endpoints is the shortest path between their PoAs.

Movement Track: A movement track of a mobile endpoint is a temporally ordered sequence of PoAs, that the mobile endpoint attaches to after each relocation. In a topology $T = (V, E)$, the movement track of a mobile endpoint M covering n relocations is identified by $MT(T, M) = (M_0, M_1, M_2, \ldots, M_n)$, where $M_i \in V(0 \leq i \leq n)$, and $M_i \neq M_{i+1}(0 \leq i \leq n-1)$.

5.2 Metric Definition and Calculation

A total of 5 metrics are defined under two mobile communication scenarios to reflect the two critical aspects of mobility support performance: user experience and network overhead. Due to page limit, the full deduction process of each formula is not given. The complete list of formula and important comments are given instead, which should suffice to demonstrate how the formulae differentiate architectures and approaches in terms of mobility support performance.

This section inherits the definitions of M, C, and S in Sect. 3 to represent three types of endpoints. In the formulae, endpoints stand for the starting, ending, or breaking point of shortest paths. $l()$ produces the length of a path, e.g., $l(MS)$ produces the length of MS. $j()$ stands for the junctional node of two paths, e.g., $j(MS, CS)$ stands for the junctional node of MS and CS.

Basic Scenario. In this scenario, C communicates with M by continuously sending packets to M, and those packets lost due to M's movement will eventually reach M after being retransmitted. Four metrics are defined in the basic scenario: path stretch, handover delay, signaling traffic, and maintained states.

Path Stretch (Stretch): Stretch is defined as the ratio of the actual forwarding path length to the shortest path length between C and M. This metric reflects how well an approach optimizes the packet forwarding path: the smaller the value, the smaller the average communication delay experienced by users, indicating better performance. The formulae for each approach is given in Table 1. Proxy and Trace suffer from triangular path, the actual forwarding path may be longer than the optimal path. But because Trace updates FIBs, packets may bypass S if trace is met before reaching S. Neither KD approaches affect the formulae when combined with any approach. TIP only changes the PoA of S, the forwarding path should be calculated in the same way as when TIP is no enabled. TRU does not affect the packet forwarding process at all.

Table 1. Metric calculation formulae - Stretch

Approach/Arch.	IP	NDN
Proxy	$l(CSM)/l(CM)$	
Resolution	1	
Routing		
Trace	$l(Cj(CS,MS)M)/l(CM)$	

Handover Delay (Delay): Delay is defined as the time it takes for the first packet to reach M after relocation. Where the first packet is always the retransmission of the last packet sent to M, which is lost due to relocation. This metric reflects the "seamless" communication capability of an approach under an architecture: the smaller the value, the less disruptions experienced by users, meaning better communication quality. The formulae for each approach and architecture combination is given in Table 2. M' stands for the last PoA of M. The formulae produce the smallest possible value in an ideal situation, so that retransmission takes place as early as possible. Note that we assume that relocation happens instantaneously, so the earliest transmission time is the exact moment M connects to a PoA. For Routing and Trace, the difference between IP and NDN comes from NDN's in-network retransmission capability: a node may retransmit Interest itself when a FIB update points to a new forwarding direction. Considering KD approaches, TIP still does not change the formulae. TRU, however, removes the time required to propagate signaling packets, thus changes the formulae when combined with each combination.

Signaling Traffic (Traffic): Traffic is defined as the total amount of traffic produced by signaling packets. This metric reflects the extra traffic generated by

Table 2. Metric calculation formulae - Delay (X = j(MS, CS), Y = j(M'S, MS))

Approach/Arch.	IP	NDN
Proxy	$max(l(MS),l(CS))+l(MS)$	
Proxy-TRU	$l(MS)$	
Resolution	$max(l(MS),l(CS))+l(CS)+l(CM)$	
Resolution-TRU	$l(CS)+l(CM)$	
Routing	$l(CM)\cdot 2$	$l(j(CM,CM')M)\cdot 2$
Routing-TRU	$l(CM)$	$l(j(CM,CM')M)$
Trace	$max(l(XC),l(XM))+$ $l(XM)$	$min(l(YM)\cdot$ $2, max(l(XC),l(XM)))+l(XM)$
Trace-TRU	$l(YM)$	

an approach: the smaller the value, the smaller the extra traffic, the higher the scalability. The formulae for each approach is given in Table 3. For Routing, the formulae consider a star topology and maximize aggregation ratio, and $\#N$ represents the total number of nodes, i.e., $|V|$. Considering KD approaches, TIP still does not affect the formulae, while TRU completely eliminates signaling packet traffic (Traffic is zero) except for Resolution, which still generates signaling packets in the query process.

Table 3. Metric calculation formulae - Traffic

Approach/Arch.	IP	NDN
Proxy	$l(MS)$	
Proxy-TRU	0	
Resolution	$l(MS) + l(CS) \cdot 2$	
Resolution-TRU	$l(CS) \cdot 2$	
Routing	$\#N - 1$	
Routing-TRU	0	
Trace	$l(MS)$	
Trace-TRU	0	

Maintained States (State): State is defined as the total number of states that need to be maintained to keep the reachability of M up-to-date. This metric reflects the storage cost of each approach, like traffic, smaller value reflects higher scalability. The formulae for each approach is given in Table 4. For Routing, each node needs to maintain one entry. Neither KD approaches affect the formulae, because they do not affect what the reachability information is or how such information is stored.

Table 4. Metric calculation formulae - State

Approach/Arch.	IP	NDN
Proxy	1	
Resolution		
Routing	$\#N$	
Trace	$l(MS)$	

5.3 Compound Scenario - Upload

In this scenario, M uploads a certain amount of data to C. A compound metric upload time (Time) is defined for this scenario. Time is defined as the total amount of time needed to finish the uploading job. The calculation of Time reuses Stretch and Delay defined in the basic scenario.

The following parameters are defined for this scenario ($i \geq 0$):

- UT: total amount of data to upload, in unit size;
- r: upload rate ratio, reflects network bandwidth;
- T_i: stay period, i.e., the period M stays before ith relocation, $T_0 = 0$;
- D_i: relocation period, i.e., the time taken to finish the ith relocation, $D_0 = 0$;

The formula is deducted as follows: the actual upload time during each stay is $T_i - Delay_i$; let U_i be the actual upload amount at the ith stay, then:

$$U_i = \frac{(T_i - Delay_i)r}{L(CM_i)Stretch_i} \text{ (if } T_i < Delay_i, \text{ then } U_i = 0) \tag{1}$$

Assume that after n relocations, $\sum_{i=1}^{n} U_i > UT$, then upload is finished during the last stay period, elapsed time:

$$GTime = \sum_{i=1}^{n-1}(D_i + T_i) \tag{2}$$

The time used to upload data during the last stay is thus:

$$UTime = T_n \frac{UT - \sum_{i=1}^{n} U_i}{U_n} \tag{3}$$

And further:

$$Time = GTime + UTime$$
$$= \sum_{i=1}^{n-1}(D_i + T_i) + T_n \frac{UT - \sum_{i=1}^{n}(T_i - Delay_i)r/(L(CM_i)Stretch_i)}{(T_n - Delay_n)r/(L(CM_n)Stretch_n)} \tag{4}$$

6 Evaluation

6.1 Network Topology Settings

A total of three topologies of three types are used, including:

- Rocketfuel topology (RT): topologies in the Rocketfuel [21] dataset, which are generated from the measurement results on the real Internet; the topology for AS 1775 is used;
- balanced tree topology (BTT): topologies with the balanced tree shape, that can be generated by setting the branching factor and height; the used topology is generated with branching factor of 2, and height of 6;
- random graph topology (RGT): topologies with a specific number of nodes and randomly generated links; the used topology is generated with 256 nodes, and a link growth probability of 0.6.

6.2 Movement Pattern Settings

A movement pattern describes the probability distribution of an endpoint's next PoA, based on its movement track so far. A movement track can be randomly generated according to a movement pattern. The movement patterns used in experiments include:

- Completely random movement (CRM): regardless of the history movement track, the next PoA is chosen uniformly from all the nodes;
- local movement (LM): there is a very high probability that the next PoA is a neighbor of the current PoA, this probability is set to 0.7;
- powerlaw movement (PM): the mobile endpoint will often attach to a few "frequent" nodes, mimicking the real life scenario where a person regularly visits a set of locations; the α parameter is set to 2.

6.3 Results

Experiment results are given for each metric, regarding each evaluated approach and architecture combination. A combination is named by concatenating the corresponding host dimension approach, any combined KD approach or KD approaches, and finally architecture with a "-", e.g., "Trace-TIP-NDN". If the result is the same for both IP and NDN, architecture is ignored. If multiple host dimension approaches share the same value, the approaches are concatenated with a "/", e.g., "Proxy/Resolution-IP".

Table 5. Results for Stretch

Combination	Mo. pattern	RT	BTT	RGT	Avg.
Proxy	CRM	2.738	2.481	2.219	2.513
	LM	2.458	3.18	2.22	
	PM	2.38	2.725	2.216	
Proxy-TIP	CRM	1.156	1.07	1.809	1.288
	LM	1.032	1.032	1.79	
	PM	1.088	1.053	1.562	
Resolution/Routing	CRM	1			1
	LM				
	PM				
Trace	CRM	1.612	1	2.13	1.563
	LM	1.541	1	2.121	
	PM	1.53	1	2.131	
Trace-TIP	CRM	1.101	1	1.805	1.258
	LM	1.023	1	1.79	
	PM	1.05	1	1.55	

Stretch: Results for Stretch are shown in Table 5. Proxy has the highest value across all topologies and mobility patterns, reaching the overall peak value of 3.18 under balanced tree topology (BTT) and local movement (LM). The reason is that Proxy suffers the most from triangular path, especially under tree-like topology. Also suffering from triangular path, the average value of Trace is 62% of Proxy. Because Trace updates FIBs, traffic between C and M does not necessarily traverse S, generating shorter forwarding path than Proxy when C and M are near each other. Regarding KD approaches, TIP results in 48% and 19% reduction respectively when combined with Proxy and Trace, almost leveling the performance of the two. The reason is that TIP migrates S near most of the shortest paths between each pair of C and M's network locations, thus relieves the triangular path issue of Proxy, and weakens the advantage of Trace.

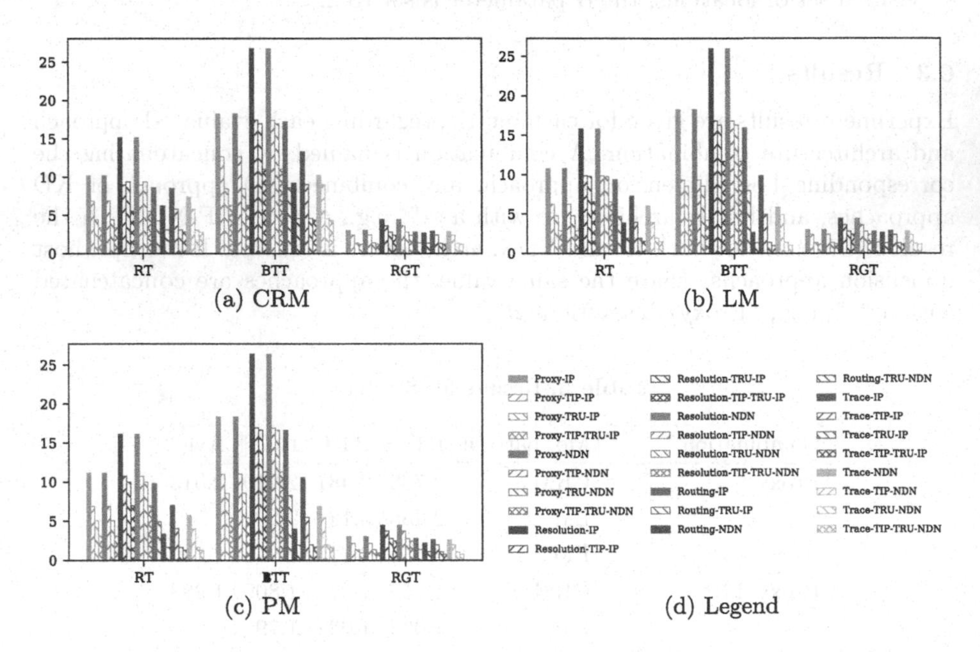

Fig. 4. Results for Delay (with legend). The X axis is topology, the Y axis is the value of Delay, bars represent approach and architecture combinations.

Delay: Results for Delay are shown in Fig. 4. In both IP and NDN, Resolution has the highest delay, averaging 15.517 time units, which is 3.8 times of the lowest average (Routing in NDN). The reason is that Resolution requires a packet exchange between C and S prior to sending a packet to M. Routing and Trace performs considerably better in NDN than in IP, the reason is that NDN's in-network Interest retransmission feature is effective when FIBs are updated.

For KD approaches, TIP reduces the length of the path taken by signaling packets, while TRU completely eliminates the delay caused by the propagation of signaling packets. With both TIP and TRU enabled, the value further drops.

Table 6. Results for Traffic

Combination	Mo. pattern	RT	BTT	RGT	Avg.
Proxy/Trace	CRM	8.484	1.395	4.469	4.903
	LM	8.575	1.38	4.794	
	PM	8.59	1.395	5.048	
Proxy/Routing/Trace-TIP	CRM	6.311	1.225	3.417	3.202
	LM	4.245	1.196	2.922	
	PM	5.425	0.964	3.115	
Routing	CRM	126	255	171	184
	LM				
	PM				
Proxy/Trace-TRU	CRM	0			0
	LM				
	PM				
Resolution	CRM	25.144	4.195	14.649	14.783
	LM	25.235	4.18	14.974	
	PM	25.25	4.195	15.228	
Resolution-TIP	CRM	11.151	3.225	5.837	7.066
	LM	11.765	3.196	6.622	
	PM	11.425	3.144	7.235	
Resolution-TRU	CRM	16.66	2.8	10.18	9.88
	LM	16.66	2.8	10.18	
	PM	16.66	2.8	10.18	
Resolution-TIP-TRU	CRM	4.84	2	2.42	3.864
	LM	7.52	2	3.7	
	PM	6	2.18	4.12	

Traffic: In IP and NDN, Routing produces the most traffic, because the signaling packet needs to be flooded across the network. For KD approaches, TIP reduces the length of the propagation path, reducing traffic for Proxy, Resolution and Trace; and TRU completely eliminates propagation of signaling packets, reducing traffic to 0 except for Resolution (Table 6).

State: Routing has the highest value thus performs the worst, because all nodes need to maintain a FIB entry, while for other approaches, only the intermediate or certain nodes need to maintain a piece of reachability information. When TIP is combined with Trace, the average distance between the mobile entity and the intermediate is shortened, and the value is slightly reduced.

Time: For the Time metric, Upload scenario parameters are set as follows: total upload size is set to 1024; upload rate factor is set to 2; stay periods are chosen uniformly between 16 and 48; and relocation periods are set to the length of the shortest path between the current and previous PoA.

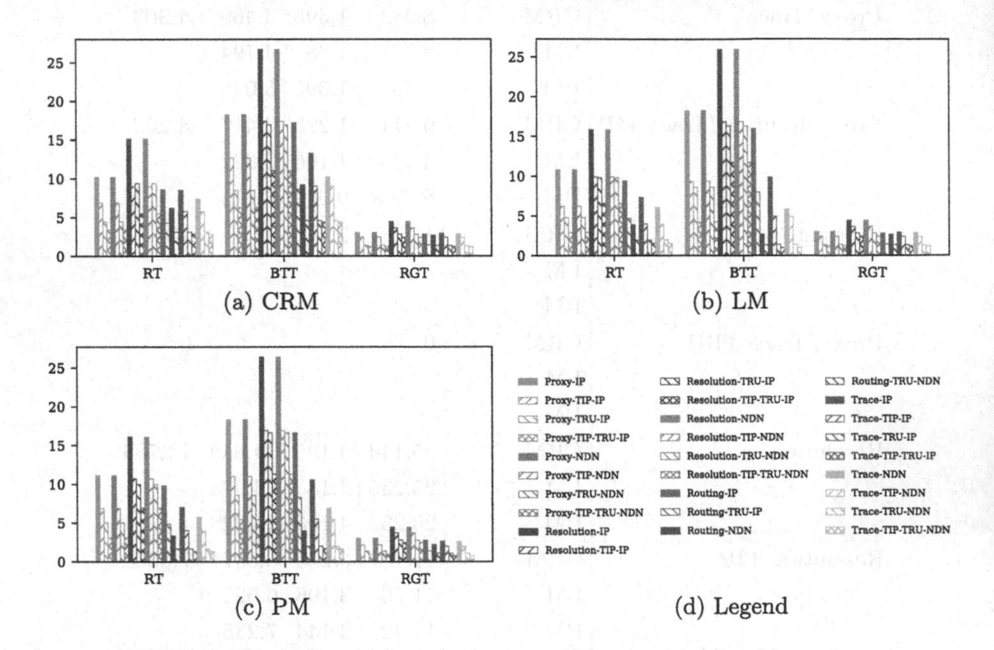

Fig. 5. Results for Time (with legend). The X axis is topology, the Y axis is the value of Delay, bars represent approach and architecture combinations.

Results for Time are shown in Fig. 5. Mobility pattern has a greater impact on this metric, and the value is generally higher (worse performance) for CRM. The main reason is that the mobility pattern determines the average distance covered in each relocation, which will greatly affect the ratio of the stay period to the total time. The larger the average travel distance, the longer the relocation period, thus the stay period is smaller, yielding higher total upload time. Across combinations, the Routing in NDN has an advantage in each topology and mobility pattern. The reason is that NDN produces smaller Delay due to the Interest retransmission mechanism, and Routing generates now path stretch, thus making the effective upload time longer in each stay.

6.4 The Effectiveness of KD Approaches

The way KD approaches affect mobility support performance is analyzed for each approach and architecture combination. For each metric, an Optimization Ratio (OR) is by comparing the value before and after KD approach(es) is enabled on top of each host dimension approach and network architecture combination. A higher positive value of OR indicates a more significant improvement in performance when combining with the corresponding KD approach(es), while a

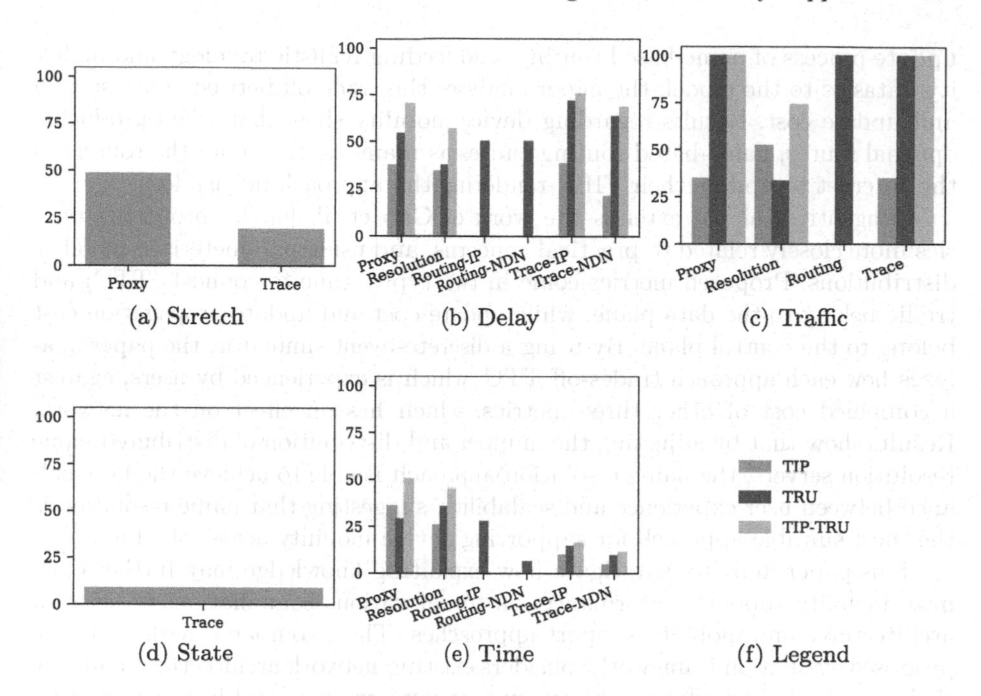

Fig. 6. Results for Optimization Ratio (with legend). The X axis is approach and architecture combinations, the Y axis is the value of OR as percentage, bars represent KD approaches.

negative value indicates worse performance. Results (see Fig. 6) show that KD approaches improve all 5 metrics. The OR is generally over or near 50%, and some even reach 100%. Such a result preliminarily demonstrates the effectiveness of the knowledge-driven paradigm, and is also expected because KD approaches operate in a new dimension to take advantage of a new set of design choices. However, this also suggests the existence of knowledge-related costs not evaluated in this paper. Evaluating such costs is crucial to this line of research and will be covered in future work. Also note that when both KD approaches are enabled, the performance boost is always more significant. This shows that the two KD approaches do not conflict with each other, and should always be enabled together if possible.

7 Related Work

Gao et al. [6] performed empirical evaluation on the mobility support performance of three "puristic" mobility support approaches, aiming at determining the most suitable architectural approach to realizing location-independent communication, i.e., communicating without caring about changing network locations. The paper presents a quantitative methodology that supports three quantifiable metrics: update cost, path stretch, and FIB size. By modeling the route update process of name-based routing, and feeding realistic topology and mobility datasets to the model, the paper analyses the trade-off between path stretch and update cost. Results regarding device mobility show that when producing optimal routes, name-based routing causes as many as 14% of all the routers in the Internet to update their FIBs, rendering the approach infeasible.

Chaganti et al. [5] extends the work of Gao et al. [6], by proposing metrics more closely related to practical concerns, and using parameterized mobility distributions. Proposed metrics come in two types: time-to-connect (TTC) and traffic belong to the data plane, while update cost and update propagation cost belong to the control plane. By using a discrete-event simulator, the paper analyzes how each approach trades-off TTC, which is experienced by users, against a combined cost of other three metrics, which has an effect on the network. Results show that by adjusting the number and distribution of distributed name resolution servers, the name resolution approach is able to achieve the best balance between user experience and scalability, suggesting that name resolution is the most suitable approach for supporting device mobility across the Internet.

This paper aims to investigate how exploiting knowledge may further optimize mobility support performance, on top of various combinations of network architectures and mobility support approaches. Thus, compared with [5,6], the proposed evaluation framework considers existing network architectures and how their design choices affect mobility support performance, and how communication scenarios shape the way mobility support performance should be evaluated.

In terms of quantitative evaluation method, both [6] and our work relies on modeling and numerical simulation. However, [6] only models how FIBs are updated in the routing process, while the proposed one is built upon the careful examination of major design choices of network architectures and mobility support approaches, thus has richer expressive power covering more types of metrics. Chaganti et al. [5] uses discrete-event simulator which are rather straight-forward to develop and has the desired versatility for collecting a multitude of metrics, they are prone to hard-to-detect implementation faults, and the development and adaptation of such an implementation is generally more time consuming.

8 Concluding Remarks

A future Internet must support the rapidly increasing traffic from mobile endpoints in a scalable and low-latency manner. This paper proposes a knowledge-driven dimension for designing mobility support solutions and two novel knowledge-driven approaches. Experiment results show that knowledge driven approaches improves all 5 proposed metrics, demonstrating the advantage of exploiting knowledge. This research makes a preliminary attempt to apply knowledge to the mobility support problem, future research will focus on three directions: expand the framework to support quantitative analysis of knowledge-related costs; propose an architectural design to support the knowledge dimension; and apply knowledge to new research topics, such as in-network mobile computing.

References

1. Afanasyev, A., Yi, C., Wang, L., Zhang, B., Zhang, L.: SNAMP: secure namespace mapping to scale NDN forwarding. In: 2015 IEEE Conference on Computer Communications Workshops (INFOCOM WKSHPS), pp. 281–286. https://doi.org/10.1109/INFCOMW.2015.7179398
2. Atkinson, R., Bhatti, S., Hailes, S.: A proposal for unifying mobility with multi-homing, NAT, & security. In: Proceedings of the 5th ACM International Workshop on Mobility Management and Wireless Access, MobiWac 2007, pp. 74–83. Association for Computing Machinery. https://doi.org/10.1145/1298091.1298105
3. Augé, J., Carofiglio, G., Grassi, G., Muscariello, L., Pau, G., Zeng, X.: MAP-me: managing anchor-less producer mobility in content-centric networks **15**(2), 596–610. https://doi.org/10.1109/TNSM.2018.2796720
4. Azgin, A., Ravindran, R., Wang, G.: A scalable mobility-centric architecture for named data networking
5. Chaganti, V., Kurose, J., Venkataramani, A.: A cross-architectural quantitative evaluation of mobility approaches. In: IEEE INFOCOM 2018 - IEEE Conference on Computer Communications, pp. 639–647. https://doi.org/10.1109/INFOCOM.2018.8485893
6. Gao, Z., Venkataramani, A., Kurose, J.F., Heimlicher, S.: Towards a quantitative comparison of location-independent network architectures. In: Proceedings of the 2014 ACM Conference on SIGCOMM, SIGCOMM 2014, pp. 259–270. Association for Computing Machinery. https://doi.org/10.1145/2619239.2626333
7. Grilo, A., Estrela, P., Nunes, M.: Terminal independent mobility for IP (TIMIP) **39**(12), 34–41. https://doi.org/10.1109/35.968810
8. Han, D., Lee, M., Cho, K., Kwon, T.T., Choi, Y.: Publisher mobility support in content centric networks. In: The International Conference on Information Networking 2014 (ICOIN 2014), pp. 214–219. https://doi.org/10.1109/ICOIN.2014.6799694
9. Henderson, T.R., Jokela, P., Nikander, P., Moskowitz, R.: Host identity protocol. https://tools.ietf.org/html/rfc5201
10. Hermans, F., Ngai, E., Gunningberg, P.: Global source mobility in the content-centric networking architecture. In: Proceedings of the 1st ACM Workshop on Emerging Name-Oriented Mobile Networking Design - Architecture, Algorithms, and Applications, NoM 2012, pp. 13–18. Association for Computing Machinery. https://doi.org/10.1145/2248361.2248366
11. Hu, X., Li, L., Mao, Z.M., Yang, Y.R.: Wide-area IP network mobility. In: IEEE INFOCOM 2008 - The 27th Conference on Computer Communications, pp. 951–959. https://doi.org/10.1109/INFOCOM.2008.148
12. Huang, K., Zhang, Q., Zhou, C., Xiong, N., Qin, Y.: An efficient intrusion detection approach for visual sensor networks based on traffic pattern learning. IEEE Trans. Syst. Man Cybern. Syst. **47**(10), 2704–2713 (2017). https://doi.org/10.1109/TSMC.2017.2698457
13. Kim, D.h., Kim, J.h., Kim, Y.s., Yoon, H.s., Yeom, I.: Mobility support in content centric networks. In: Proceedings of the Second Edition of the ICN Workshop on Information-Centric Networking, ICN 2012, pp. 13–18. Association for Computing Machinery. https://doi.org/10.1145/2342488.2342492
14. Kim, H., Feamster, N.: Improving network management with software defined networking **51**(2), 114–119. https://doi.org/10.1109/MCOM.2013.6461195

15. Li, D., CHuah, M.C.: SCOM: a scalable content centric network architecture with mobility support. In: 2013 IEEE 9th International Conference on Mobile Ad-hoc and Sensor Networks, pp. 25–32. https://doi.org/10.1109/MSN.2013.44
16. Mijumbi, R., Serrat, J., Gorricho, J.L., Bouten, N., De Turck, F., Boutaba, R.: Network function virtualization: state-of-the-art and research challenges **18**(1), 236–262. https://doi.org/10.1109/COMST.2015.2477041
17. Perkins, C., Bhagwat, P.: A mobile networking system based on internet protocol **1**(1), 32–41. https://doi.org/10.1109/98.911984
18. Ramjee, R., Varadhan, K., Salgarelli, L., Thuel, S., Wang, S.Y., La Porta, T.: HAWAII: a domain-based approach for supporting mobility in wide-area wireless networks **10**(3), 396–410. https://doi.org/10.1109/TNET.2002.1012370
19. Rekhter, Y., Li, T.: A border gateway protocol 4 (BGP-4). https://tools.ietf.org/html/rfc1771
20. Rodríguez Natal, A., Jakab, L., Portolés, M., Ermagan, V., Natarajan, P., Maino, F., Meyer, D., Cabellos Aparicio, A.: LISP-MN: mobile networking through LISP **70**(1), 253–266. https://doi.org/10.1007/s11277-012-0692-5
21. Spring, N., Mahajan, R., Wetherall, D.: Measuring ISP topologies with rocketfuel. ACM SIGCOMM Comput. Commun. Rev. **32**(4), 133–145 (2002)
22. Teraoka, F., Uehara, K., Sunahara, H., Murai, J.: VIP: a protocol providing host mobility **37**(8), 67-ff. https://doi.org/10.1145/179606.179657
23. Wang, L., Waltari, O., Kangasharju, J.: MobiCCN: mobility support with greedy routing in content-centric networks. In: 2013 IEEE Global Communications Conference (GLOBECOM), pp. 2069–2075. https://doi.org/10.1109/GLOCOM.2013.6831380
24. Yang, Y., Xiong, N., Chong, N.Y., Défago, X.: A decentralized and adaptive flocking algorithm for autonomous mobile robots. In: 2008 The 3rd International Conference on Grid and Pervasive Computing - Workshops, pp. 262–268 (2008). https://doi.org/10.1109/GPC.WORKSHOPS.2008.18
25. Zhang, L., Wakikawa, R., Zhu, Z.: Support mobility in the global internet. In: Proceedings of the 1st ACM Workshop on Mobile Internet Through Cellular Networks, MICNET 2009, pp. 1–6. Association for Computing Machinery. https://doi.org/10.1145/1614255.1614257
26. Zhang, Q., Zhou, C., Xiong, N., Qin, Y., Li, X., Huang, S.: Multimodel-based incident prediction and risk assessment in dynamic cybersecurity protection for industrial control systems. IEEE Trans. Syst. Man Cybern. Syst. **46**(10), 1429–1444 (2016). https://doi.org/10.1109/TSMC.2015.2503399
27. Zhang, Y., Afanasyev, A., Burke, J., Zhang, L.: A survey of mobility support in named data networking. In: 2016 IEEE Conference on Computer Communications Workshops (INFOCOM WKSHPS), pp. 83–88. https://doi.org/10.1109/INFCOMW.2016.7562050
28. Zhang, Y., Xia, Z., Mastorakis, S., Zhang, L.: KITE: producer mobility support in named data networking. In: Proceedings of the 5th ACM Conference on Information-Centric Networking, ICN 2018, pp. 125–136. Association for Computing Machinery. https://doi.org/10.1145/3267955.3267959
29. Zhu, Z., Zhang, L., Wakikawa, R.: Understanding apple's back to my mac (BTMM) service

An Efficient Energy Efficiency Power Allocation Algorithm for Space-Terrestrial Satellite NOMA Networks

Yanan Wu[1] and Lina Wang[1,2](✉)

[1] School of Computer and Communication Engineering, University of Science and Technology Beijing, Beijing 100083, China
wyn_ustb@163.com, wln_ustb@126.com

[2] Shunde Graduate School, University of Science and Technology Beijing, Foshan, China

Abstract. Due to the shortage of spectrum resources, Non-orthogonal multiple access (NOMA) has been considered as a forward-looking technology to enhance the performance for space-terrestrial satellite networks. In this paper, the power allocation algorithm is proposed on account of Stackelberg game to apply in the space-terrestrial satellite NOMA networks. The terrestrial base stations (BSs) and satellites are leaders and followers, respectively. The alternative direction method of multipliers (ADMM) algorithm is applied in BSs layer and satellites layer to acquire optimal power allocation scheme. The results indicate that the system energy efficiency has great promotion by the proposed algorithm.

Keywords: NOMA · Stacklberg · ADMM · Energy efficiency

1 Introduction

In the wake of the promotion and application of massive mobile terminal devices, it is expected that the whole number of devices linked to the global mobile communication network will reach 100 billion in the future. In order to carry more than 100 billion mobile data in the future, mobile communication technology is faced with enormous challenges. Satellite networks can be used to supplement terrestrial networks access, pool satellite and terrestrial network resources to provide faster broadband services. The terrestrial network and satellite network are integrated to supply services for terminals [1]. NOMA has been proposed as a forward-looking multiple access mechanism for future wireless communication network [2,3].

This work was supported by the Scientific and Technological Innovation Foundation of Shunde Graduate School, USTB under Grant No. BK19BF009 and the National Natural Science Foundation of China under Grant No. 61701020.

S. Shi et al. (Eds.): AICON 2020, LNICST 356, pp. 217–226, 2021.
https://doi.org/10.1007/978-3-030-69066-3_19

In NOMA system, the channel resource can be shared by multi-users to obtain higher spectral efficiency. [2] evaluated the NOMA system performance and pointed out that the total throughput and user capacity are superior to OMA system. The proposed fixed power allocation and user grouping algorithm can enhance system performance. Z. Xiao, etc. [4] studied the sum rate problem of maximized two user NOMA system, and proposed a suboptimal scheme to resolve this problem. The authors find the beamforming vector to two users simultaneously subject to an analog beamforming structure, meanwhile, allocate corresponding power to them. In the future, the satellite will be investivaged to ensure ubiquitous coverage. In [1], the downlink transmission of NOMA on account of terrestrial-satellite networks was studied. Based on the matching of ground users and satellite users, the power allocation schemes of two different types of users are developed by using Lagrange multiplier method. By contrast with the suboptimal allocation algorithm and the average power algorithm, the results indicated that the proposed algorithm can enhance the system capacity. As far as we know, many researchers have assumed a single-tier NOMA network model, and research on terrestrial satellite NOMA network model is also scarce. Inspired by this, we investigate the power allocation algorithm of space-terrestrial satellite NOMA networks.

In this paper, the power allocation algorithm on account of Stackelberg game is studied in the space-terrestrial satellite networks. Section 2 presents the NOMA based space-terrestrial satellite networks model and shows the problem formulation scheme. In Sect. 3, The problem for energy efficiency optimization is proposed, in which Stackelberg game model is applied to resolve power allocation scheme. The Dinkelbach-style algorithm is applied to transform the non-convex optimal function into convex-form to resolve the proposed problem, and ADMM technology is introduced to reduce the time delay and reduce the number of iterations. Performance of the proposed algorithms is evaluated in Sect. 4 by many simulations. In the end, Sect. 5 summarizes the paper.

2 System Model and Problem Formulation Scheme

2.1 System Model

A communication scenario for space-terrestrial satellite networks is proposed, where S satellites and R terrestrial base stations (BSs) provide services together for ground users. Symbol s and r represent sth satellite and rth terrestrial BS, respectively, in which $s \in \{1, 2, \cdots, S\}$ and $r \in \{1, 2, \cdots, R\}$. The satellite is LEO satellite and equipped with M antennas to serve terrestrial users under its coverage and each BS is equipped with N antennas to serve users under coverage. Symbol m and n represent mth sunchannel of the satellite and n subchannel of a BS, respectively, in which $m \in \{1, 2, \cdots, M\}$, $n \in \{1, 2, \cdots, N\}$. In the terrestrial networks, NOMA is applied for multi-users and multi-users can be applied to the same subchannel with SIC technology, and each BS services T users on each subchannel. However, the research of NOMA technology in satellite networks is still in its infancy, and many problems need to be solved and studied urgently. Therefore, NOMA isn't implemented in satellite network, and each satellite services a user on each subchannel. In this paper, we believe that all users are either served by the BSs or served by the satellite (Fig. 1).

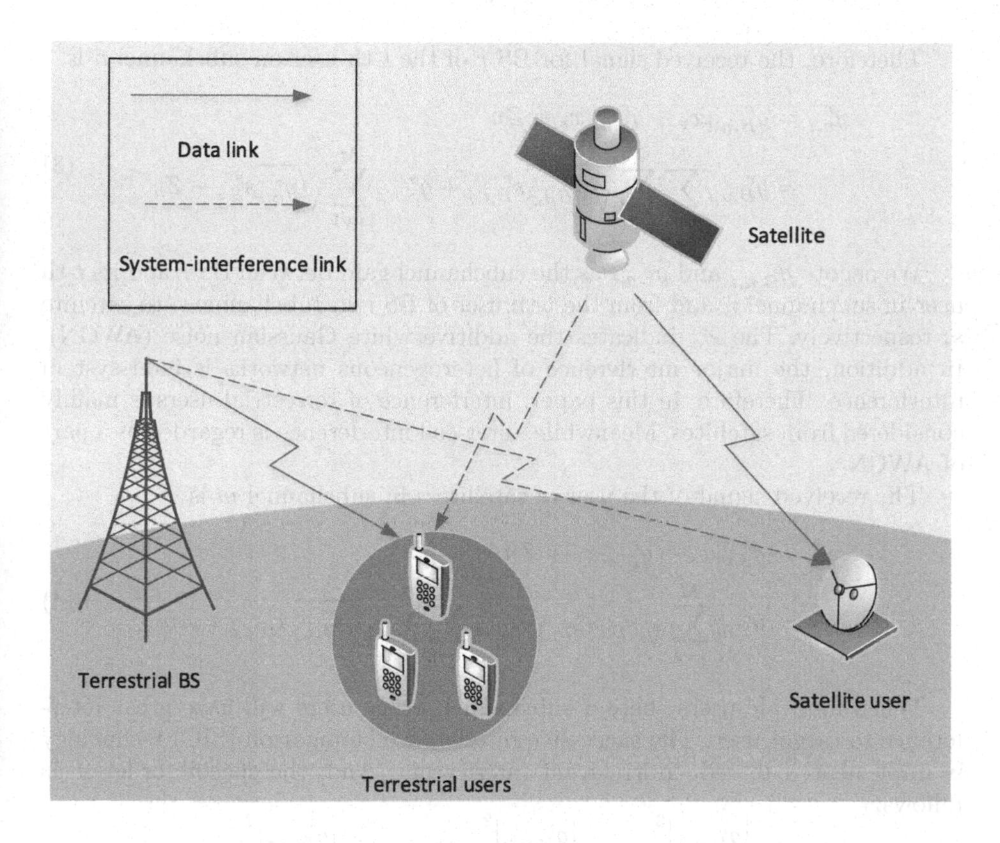

Fig. 1. System model.

2.2 Problem Formulation Scheme

Just as indicated from the above system model, the transmit signal for BS r represents as follows

$$x_r = \sum_{i=1}^{N} \sum_{j=1}^{T} \sqrt{p_{B,i,j}^r} s_{B,i,j}^r \tag{1}$$

in which $p_{B,i,j}^r$ denotes the allocated powers for the j-th user of BS r on subchannel i, and $s_{B,i,j}^r$ represents the messages for the j-th user of BS r over subchannel i.

In the same way, The transmit signal for satellite s represents as follows

$$x_s = \sum_{i=1}^{M} \sqrt{p_{S,i}^{s}} s_{S,i}^{s} \tag{2}$$

Therefore, the received signal for BS r of the t-th user on subchannel n is

$$\begin{aligned} y_{n,t}^{r} &= g_{B,n,t}^{r} x_r + g_{S,n,t}^{s} x_s + Zn \\ &= g_{B,n,t}^{r} \sum_{i=1}^{N} \sum_{j=1}^{T} \sqrt{p_{B,i,j}^{r}} s_{B,i,j}^{r} + g_{S,n,t}^{s} \sum_{i=1}^{M} \sqrt{p_{S,i}^{s}} s_{S,i}^{s} + Z_n \end{aligned} \tag{3}$$

We denote $g_{B,n,t}^{r}$ and $g_{S,n,t}$ as the subchannel gain between BS r and its t-th user in subchannel n and from the t-th user of BS r in subchannel n to satellite s, respectively. The Z_n indicates the additive white Gaussian noise (AWGN). In addition, the major interference of heterogeneous networks is inter-system interference. Therefore, in this paper, interference of terrestrial users is mainly considered from satellites. Meanwhile, inter-cell interference is regarded as a part of AWGN.

The received signal of the user of satellite s in subchannel m is

$$\begin{aligned} y_{m}^{s} &= g_{S,m}^{s} x_s + g_{B,m}^{r} x_r + Zn \\ &= g_{S,m}^{s} \sum_{i=1}^{M} \sqrt{p_{S,i}^{s}} s_{S,i}^{s} + g_{B,m}^{r} \sum_{i=1}^{N} \sum_{j=1}^{T} \sqrt{p_{B,i,j}^{r}} s_{B,i,j}^{r} + Zn \end{aligned} \tag{4}$$

When multiple users share a subchannel, other users will have great interference to target user. The successive interference elimination (SIC) technology is implemented to reduce irrelevant interference. And, the specific order is as follows:

$$\begin{aligned} &\frac{\left|g_{B,n,1}^{r}\right|^2}{I_{B,n,1}^{CR} + Z_n} \leq \frac{\left|g_{B,n,2}^{r}\right|^2}{I_{B,n,2}^{CR} + Z_n} \leq \cdots \leq \frac{\left|g_{B,n,t}^{r}\right|^2}{I_{B,n,t}^{CR} + Z_n} \\ &\leq \frac{\left|g_{B,n,t+1}^{r}\right|^2}{I_{B,n,t+1}^{CR} + Z_n} \leq \cdots \leq \frac{\left|g_{B,n,T}^{r}\right|^2}{I_{B,n,T}^{CR} + Z_n} \end{aligned} \tag{5}$$

in which $I_{B,n,t}^{CR} = \left|g_{S,n,t}^{s}\right|^2 \sum_{i=1}^{M} p_{S,i}^{s}$ is the inter-system interference. The t-th user can decode and eliminate the data for user k $(k < t)$ according to the SIC, and the signal for user k $(k > t)$ is considered as interference. Hence, $I_{B,n,t}^{N} = \left|g_{B,n,t}^{r}\right|^2 \sum_{i=t+1}^{T} p_{B,n,i}^{r}$ is called superposition interference.

The SINR of the t-th user of BS r in subchannel n is given by

$$SINR_{n,t}^{r} = \frac{\left|g_{B,n,t}^{r}\right|^2 p_{B,n,t}^{r}}{I_{B,n,t}^{N} + I_{B,n,t}^{CR} + Z_n} \tag{6}$$

Correspondingly, the SINR of the user of satellite s in subchannel m is given by

$$SINR_m^s = \frac{\left|g_{S,m}^s\right|^2 p_{S,m}^s}{I_{S,m}^{CR} + Z_n} \tag{7}$$

in which $I_{S,m}^{CR} = \left|g_{B,m}^r\right|^2 \sum_i^N \sum_j^T p_{B,i,j}^r$ indicates the inter-system interference, and denotes inter-system interference from the terrestrial BSs for the satellite users.

On the subchannel n, the capacity for the t-th user served by the BS r is indicated by

$$C_{n,t}^r = \frac{B}{N}\log_2(1 + SINR_{n,t}^r) \tag{8}$$

similarly,

$$C_m^s = \frac{B}{M}\log_2(1 + SINR_m^s) \tag{9}$$

The energy efficiency of the BS r is indicated as the ratio of the total capacity and the power consumed by the BSs, and is denoted by

$$\eta_{EE}^r = \frac{C_B^r}{P_B^r} \tag{10}$$

in which $C_B^r = \sum_{n=1}^N \sum_{t=1}^T C_{B,n,t}^r$, and $P_B^r = \sum_{n=1}^N \sum_{t=1}^T p_{B,n,t}^r + p_{B,static}$. $p_{B,static}$ is the circuit consumption at every BS. Similarly,

$$\eta_{EE}^s = \frac{C_S^s}{P_S^s} \tag{11}$$

in which $C_S^s = \sum_{m=1}^M C_{S,m}^s$, and $P_S^s = \sum_{m=1}^M P_{S,m}^s + p_{S,static}$. The $p_{S,static}$ is similar to $p_{B,static}$, which is a constant value. Optimizing the energy efficiency based NOMA of the space-terrestrial satellite networks is our target. The energy efficiency maximization of BS r can be formulated as follows:

$$\begin{aligned} &\max_{p_{B,n,t}^r > 0} \eta_{EE}^r \\ &s.t. C1 : \sum_{n=1}^N \sum_{t=1}^T p_{B,n,t}^r \le P_{MAX}^R, \forall r \\ &C2 : C_{B,n,t}^r \ge C_{\min}^r, \forall r, n, t \end{aligned} \tag{12}$$

where the capacity of each user has a lowest limit $C_{\min}^r$. P_{MAX}^R is the power upper limit of BSs. Similarly, the energy efficiency maximization problem of the s-th satellite is equivalent to

$$\begin{aligned} &\max_{p_{S,m}^s > 0} \eta_{EE}^s \\ &s.t. C1 : \sum_{m=1}^M P_{S,m}^s \le P_{MAX}^S, \forall s \\ &C2 : C_{S,m}^s \ge C_{\min}^s, \forall s, m \end{aligned} \tag{13}$$

where P_{MAX}^{S} is the satellite s's power upper limit, and the $C_{\min}^{s}$ is lowest limit capacity of every satellite user. However, in space-terrestrial satellite networks, power allocations for BSs and satellites is interactive with each other to obtain maximum system energy efficiency.

3 Stacklberg Game Formulation and Energy Efficiency Optimization

3.1 Stacklberg Game

In this paper, the Stackelberg game aims to achieve a balance between BSs and satellite. Stacklberg was originally proposed in the field of economics [5]. Game is made up of two parts, namely the leader and the follower. Followers will follow the leader's decision and change their own decision, and the decision between them is influenced by each other. In this paper, the leaders and followers of the system for competition is regarded as the maximization of the system's energy efficiency [10], in which BSs are followers, and satellites are leaders. In the Stacklberg game, each base station tries to maximize its own energy efficiency. After the base stations make the decisions, the satellite makes a strategy to maximize each satellite's energy efficiency. The iterative calculation between the base stations and the satellite will attain the equilibrium point of Stackelberg game.

3.2 Problem Optimization

We first try to solve the optimal solution of the follower subgame, and the optimal power allocation scheme of BSs should be obtained. The process of solving (12) or (13) is complicated. In order to make the solution more efficient, the original fractional non-convex function is transformed into a convex-form function to get the optimal solution for the proposed problem by Dinkelbach-style algorithm [6]. Where we could convert the optimization problem of (12) in the non-convex form as

$$\begin{aligned} F(\lambda) = & \max_{p_{S,m}^{s}>0, p_{B,n,t}^{r}>0} C_{B}^{r}(P^{R}, P^{S}) - \lambda P_{B}^{r} \\ & s.t. C1, C2, C3, C4 \end{aligned} \tag{14}$$

When the solution of the equation F(λ) = 0 is obtained, the optimal solution of the function (14) is also obtained, where the parameter λ is set as an adjective variable. Therefore, the power allocation scheme will be formulated on the basis of F(λ). Where we could convert the optimization problem of (14) in the non-convex form as

$$\min_{p_{S,m}^{s}>0, p_{B,n,t}^{r}>0} \lambda P_{B}^{r} - C_{B}^{r}(P^{R}, P^{S}) \tag{15}$$

For the sake of solving the optimal power allocation scheme, we introduce the ADMM algorithm. A more detailed introduction to the ADMM method is available in [7]. And we introduce the adjective vectors of U_{BS} and V_{BS}, while the elements for BS constitute U_{BS} and V_{BS} are global adjective vectors where each one corresponds to each in X_{BS}. The Γ is defined to satisfy constraint $C2'$. Then, the target function is imported by $g(V_{BS}) = 0$ if $V_{BS} \in \Gamma$, otherwise $g(V_{BS}) = +\infty$. On the account of the above description, the power optimization function (15) is rewritten to

$$\min_{U_{BS}, V_{BS}} \lambda P_B^r - C_B^r(P^R, P^S) + g(V_{BS}) \tag{16}$$

The transformation is as follows

$$F_\rho^{BS} = f(U_{BS}) + g(V_{BS}) - \frac{\rho_{BS}}{2} \|\phi_{BS}\|_2^2 + \frac{\rho_{BS}}{2} \|U_{BS} - V_{BS}^t + \phi_{BS}\|_2^2, \tag{17}$$

in which ρ_{BS} represents the constant penalty parameter, and μ_{BS} is the dual variable. The optimization problem should be solved by following steps

$$\begin{aligned} U_{BS}^{t+1} &:= \arg\min_{U_{BS}} \left\{ f(U_{BS}) + \tfrac{\rho}{2} \|U_{BS} - V_{BS}^t + \phi_{BS}\|_2^2 \right\} \\ V_{BS}^{t+1} &:= \arg\min_{V_{BS}} \left\{ \left\|U_{BS}^{t+1} - V_{BS} + \phi_{BS}^t\right\|_2^2 \right\} \\ \phi_{BS}^{t+1} &:= \phi_{BS}^t + (U_{BS}^{t+1} - V_{BS}^{t+1}) \end{aligned} \tag{18}$$

The power allocation algorithm based ADMM for BS is concluded in the following Algorithms 1, and the satellite users power allocation algorithm has same scheme:

Algorithm 1. Power Allocation algorithm based ADMM for BS users

1: Initialize U_{BS}^0, $V_{BS}^0 \in C2$ and $\phi_{BS}^0 > 0$, iteration index $t = 0$, penalty parameter $\rho_{BS} = 0$, stop criteria $\xi > 0$;
2: **while** $f(U_{BS}) > \xi$ **do**
3: Updates U_{BS}^{t+1}
4: Updates V_{BS}^{t+1}
5: Updates ϕ_{BS}^{t+1}
6: $t := t + 1$;
7: **end while**

The optimal energy efficiency of the BSs obtained by controlling the power allocation is in Algorithm 1, that is, the followers make their own decision. According to the characteristics of stacklberg game process, the leader will make decisions accordingly. Algorithm 2 shows the game between the leader and the follower. On the wake of the game progresses and both will attain an equilibrium point, then, the power distribution scheme will be acquired.

4 Simulation Results and Analysis

We assume the space-terrestrial satellite NOMA networks including $S = 2$ satellites with $NS = 8$ orthogonal subchannels and $R = 2$ BSs with $NR = 4$ subchannels. And each BS subchannel is shared by $T = 4$ users. The bandwidth $B = 30$ MHz and the carrier frequency indicates 2 GHz, and the noise power is -174 dBm/Hz. It is assumed that the coverage radius of the satellite is 500 km, and the remaining parameters about the satellite are defined on account of [8]. The maximum transmit power is indicated as $P^R_{MAX} = 30$ W, and $P^S_{MAX} = 90$ W. Considering the great difference between satellite channel and BS channel, satellite channel is simulated as Rician channel [8], BS channel is simulated as Rayleigh channel [9].

Algorithm 2. Obtaining the Equilibrium Point Algorithm of a Stackelberg Game.

1: Initialization: $k = 0$, given the maximum iterative number $K_{\max}$ of the Stacklberg game
2: Initialize the value of BSs $P^R(k)$, satellites $P^S(k)$ at iteration $k = 0$.
3: **repeat**
4: Obtaining the power allocation scheme of BSs $P^R(k+1), P^S(k)$ by Algorithm 1.
5: Given the above $P^R(k+1), P^S(k)$, calculate $P^R(k+1), P^S(k+1)$ of satellite users power allocations scheme from Algorithm 1.
6: **until** $k = K_{\max}$.

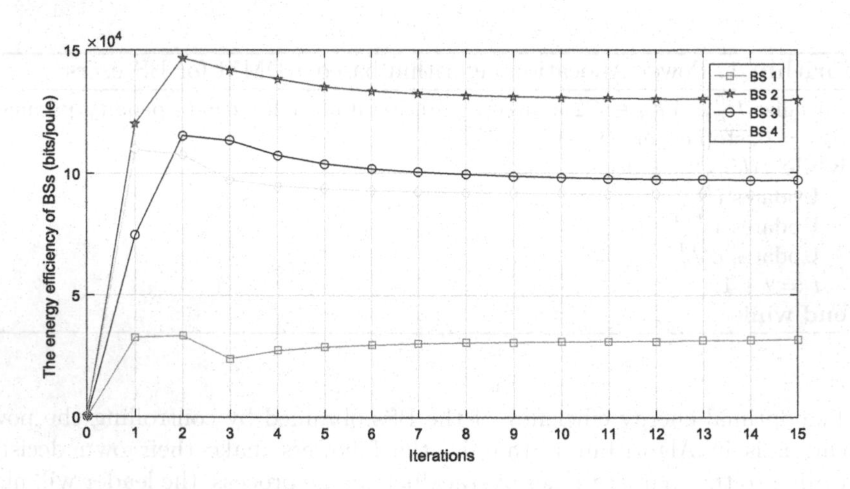

Fig. 2. The energy efficiency of different followers at each iteration.

Figure 2 indicates the energy efficiency of different followers under each of iteration. The energy efficiency of each follower increases with the iterations, and the curve converges when the number of iterations is about 5. It can be seen that iterative Algorithm 1 can remarkably improve the energy efficiency for followers in the sub-game. Similarly, this algorithm is also applicable to the leader subgame.

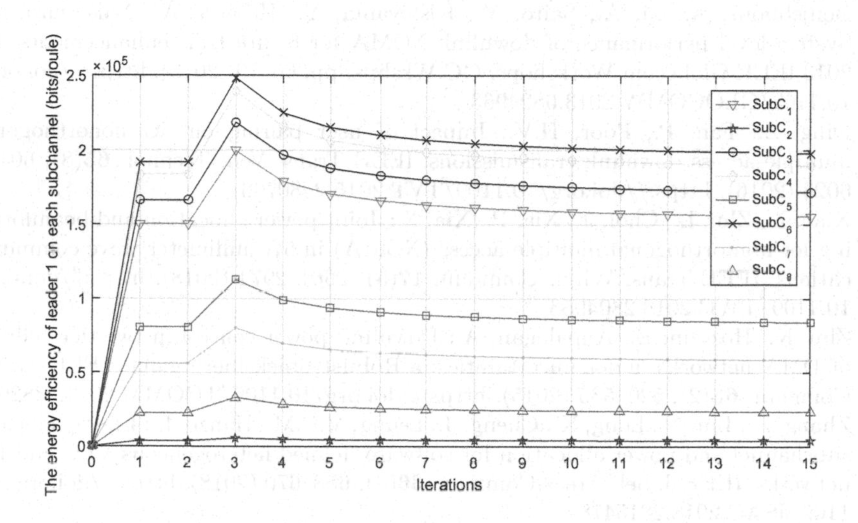

Fig. 3. The energy efficiency for satellite 1 on each subchannel at each iteration.

Figure 3 shows the energy efficiency in each subchannel for 1st satellite (leader 1) under each of iteration. The energy efficiency of the satellites (leaders) is increasing and reaching the equilibrium point. The figure indicates the proposed algorithm can improve the energy efficiency of leaders. Because of the different allocation of each user on each subchannelthe, the energy efficiency of each subchannel is different. The energy efficiency in the Stackelberg game is obviously perfect.

5 Conclusion

In this paper, the space-terrestrial satellite NOMA network was investigated to maximize system energy efficiency by introducing Stackelberg game. The leaders (BSs) and followers (satellites) make corresponding strategys according to the other party's strategys to obtain the maximum system energy efficiency, respectively. The ADMM algorithm is applied to obtain power allocation scheme in BSs layer and satellites layer, respectively. At the same time, the balance between BSs and satellite is achieved by Stackelberg game. Significantly improve the energy efficiency of the proposed network.

References

1. Zhu, X., Jiang, C., Kuang, L., Ge, N., Lu, J.: Energy efficient resource allocation in cloud based integrated terrestrial- satellite networks. In: IEEE ICC, pp. 1–6 (2018). https://doi.org/10.1109/ICC.2018.8422663
2. Benjebbour, A., Li, A., Saito, Y., Kishiyama, Y., Harada, A., Nakamura, T.: System-level performance of downlink NOMA for future LTE enhancements. In: 2013 IEEE Globecom Workshops, GC Wkshps, pp. 66–70 (2013). https://doi.org/10.1109/GLOCOMW.2013.6824963
3. Ding, Z., Fan, P., Poor, H.V.: Impact of user pairing on 5G nonorthogonal multiple-access downlink transmissions. IEEE Trans. Veh. Technol. **65**(8), 6010–6023 (2016). https://doi.org/10.1109/TVT.2015.2480766
4. Xiao, Z., Zhu, L., Choi, J., Xia, P., Xia, X.: Joint power allocation and beamforming for non-orthogonal multiple access (NOMA) in 5G millimeter wave communications. IEEE Trans. Wirel. Commun. **17**(5), 2961–2974 (2018). https://doi.org/10.1109/TWC.2018.2804953
5. Zhu, K., Hossain, E., Anpalagan, A.: Downlink power control in two-tier cellular of DMA networks under uncertainties: a Robust Stackelberg game. IEEE Trans. Commun. **63**(2), 520–535 (2015). https://doi.org/10.1109/TCOMM.2014.2382095
6. Zhang, H., Liu, N., Long, K., Cheng, J., Leung, V.C.M., Hanzo, L.: Energy efficient subchannel and power allocation for software-defined heterogeneous VLC and RF networks. IEEE J. Sel. Areas Commun. **36**(3), 658–670 (2018). https://doi.org/10.1109/JSAC.2018.2815478
7. Boyd, S., Parikh, N., Chu, E., Peleato, B., Eckstein, J.: Distributed optimization and statistical learning via the alternating direction method of multipliers. Found. Trends Mach. Learn. **3**(1), 1–122 (2011). https://doi.org/10.1561/2200000016
8. Lutz, E., Cygan, D., Dippold, M., Dolainsky, F., Papke, W.: The land mobile satellite communication channel-recording, statistics, and channel model. IEEE Trans. Veh. Technol. **40**(2), 375–386 (1991). https://doi.org/10.1109/25.289418
9. Further Advancements for E-UTRA Physical Layer Aspects, 3rd Generation Partnership Project (3GPP), December 2016
10. Gao, D., Liang, Z., Zhang, H., Dobre, O.A., Karagiannidis, G.K.: Stackelberg game-based energy efficient power allocation for heterogeneous NOMA networks. In: IEEE Global Communications Conference (GLOBECOM), Abu Dhabi, United Arab Emirates 2018, pp. 1–5 (2018). https://doi.org/10.1109/GLOCOM.2018.8647786

Research and Equilibrium Optimization of AODV Routing Protocol in Ad Hoc Network

Zhongyue Liu[1(✉)], Shuo Shi[1,2], and Xuemai Gu[1,3]

[1] School of Electronic and Information Engineering, Harbin Institute of Technology, Harbin 150001, Heilongjiang, China
liuzhongyue_hit@163.com, {crcss,guxuemai}@hit.edu.cn
[2] Network Communication Research Centre, Peng Cheng Laboratory, Shenzhen 518052, Guangdong, China
[3] International Innovation Institute of HIT in Huizhou, Huizhou 516000, Guangdong, China

Abstract. In the Ad Hoc network of AODV protocol, local node failure will occur when a node dies due to high cost or the network energy consumption is too large and the problem is serious. In this paper, based on the routing cost design, an optimized design scheme of AODV protocol (E-AODV) is proposed through the establishment and maintenance method. Simulation results show that this method improves the performance of AODV network, reduces the death process of nodes and reduces the energy consumption.

Keywords: AODV · Energy consumption · Route maintenance · E-AODV

1 Introduction

Ad Hoc network is a multi-hop self-organizing and self-managing network composed of a group of mobile terminals with wireless transceiver devices. Ad Hoc network does not require a fixed base station, and has the characteristics of strong survivability, and extremely convenient creation and movement. It makes up for the shortcomings of cellular systems and wired networks. It can be easily and flexibly networked and used in public services, emergency search and rescue, and intelligent Transportation and other fields have broad application prospects.

The nodes of the Ad Hoc network are both communication terminals and routers, which can move freely. In the Ad Hoc network, the network topology will often change due to the irregularity of the wireless channel changes, the movement of nodes, joining, and exiting, and due to the limitations of the wireless coverage of the nodes, two nodes that cannot be directly connected It needs to rely on the message forwarding of other nodes to communicate [1–4]. The role of the routing protocol is to monitor changes in the network topology in this network environment, exchange routing information, locate the location of the destination node, generate, maintain and select routes, and forward data according to the selected route. Ensuring network [5] connectivity is the basis for mobile nodes to communicate with each other, and is an important and core issue in Ad Hoc networks.

S. Shi et al. (Eds.): AICON 2020, LNICST 356, pp. 227–235, 2021.
https://doi.org/10.1007/978-3-030-69066-3_20

At present, dozens of Ad Hoc network single-path routing protocols have been proposed to solve the routing problem in Ad Hoc networks. According to different ways of establishing routes, routing protocols [6] can be divided into a priori routing protocols and reactive routing protocols. DSDV is a more typical a priori routing protocol, and DSR, AODV and TORA are more typical reactive routing protocols. Research shows that in the case of node movement, the reactive routing protocol has lower routing overhead, and its performance is better than the a priori routing protocol; among the reactive routing protocols, the AODV protocol has [7] moderate routing overhead and fast convergence. The advantages are obvious. It is one of the promising routing protocols in the Ad Hoc network and has become the basis for the expansion of multipath protocols.

Compared with other routing protocols, the AODV routing protocol starts routing search only when needed, thereby improving the utilization of network resources and reducing delay. The AODV routing protocol can quickly respond to network topology changes and will not form a loop. Due to the continuous improvement of people's requirements for network performance, many scholars have optimized the AODV routing protocol in different aspects in recent years.

Lu Wei et al. [8] proposed a new improvement mechanism for energy loss and constructed a mathematical model, which included the signal attributes of the nodes. Choose the path according to the link quality and energy consumption. This protocol can not only ensure the minimum energy consumption but also select the optimal path in communication.

Literature [9] proposed a power estimation strategy, which uses the composite expected number of transmissions and the remaining energy of the node as the routing metric. The improved algorithm effectively reduces energy loss and prolongs the network lifetime. This new mechanism is superior to traditional AODV in terms of routing performance.

Due to the limited energy availability of each wireless node in the Ad Hoc network, the energy of the node will continue to decrease until it is exhausted during the communication process, so it is very necessary to consider energy in the data forwarding process. To ensure stable transmission of the link, congestion must be considered. The consequences of congestion can lead to data loss and even network breakdown. Therefore, the choice of routing metric is critical to communication quality.

Based on the working process of AODV, an improved AODV routing protocol (E-AODV) was proposed to solve routing cost, low node survival rate and unstable link connection problems. Establish multiple routing nodes to destination nodes based on cost path. Use this method to improve energy utilization rate and link quality to extend the life of the network. Figure 1 shows the classification block diagram of Ad Hoc network routing protocols:

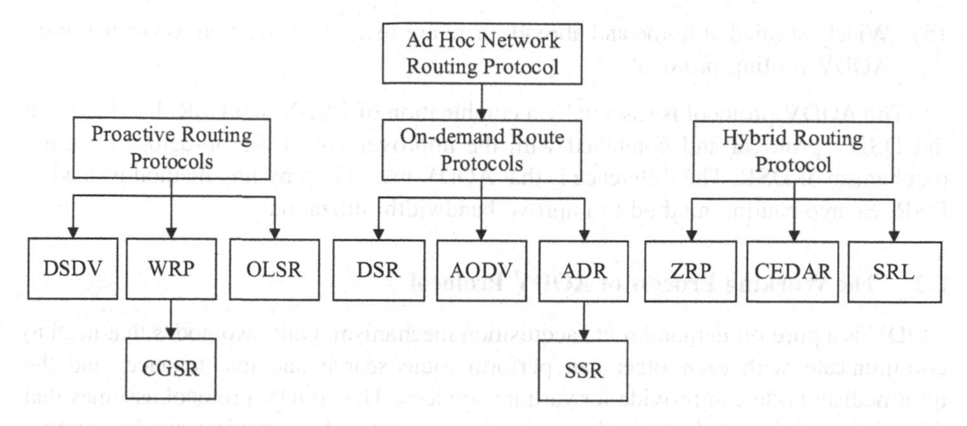

Fig. 1. Classification block diagram of Ad Hoc network routing protocol

2 AODV Protocol and Working Principle

2.1 The Advantages and Inadequate of AODV Protocol

The AODV routing protocol is designed for nodes in the Ad Hoc network to perform mutual data transmission. It is an on-demand routing protocol. On-demand means that the node does not store the routing information of all nodes in the network. When the destination node transmits data, it will check the routing table. If there is no route, it will broadcast the routing request to the network. Route request (RREQ), route reply (RREP), route error (RERR) and active route detection (HELLO) are the four types of information defined by the AODV routing protocol. The RREQ, RREP, and RERR information messages all contain a sequence number field. The use of the serial number allows the node to distinguish between the old and new information messages, so that the node can update the routing information generated by the old information messages in the routing table with the new information messages. HELLO information is a special case of RREP information. By broadcasting HELLO information, the connection between a node and its directly connected node can be detected. This information is transmitted by UDP, so the IP protocol address can be used.

The AODV routing protocol supports operation in a small-scale network, with the number of nodes ranging from tens to thousands, and complete trust between nodes that require mutual communication, because the data may need other intermediate nodes in the process of transmitting to the destination node Analyze data and forward. In general, the AODV routing protocol has the following advantages:

(1) There is no need to maintain the routing table in real time, and only seek routing and update the routing table when needed.
(2) The intermediate node can replace the destination node to reply, reducing the delay of the route discovery process and improving the convergence speed.
(3) All nodes and information messages have sequence numbers, which avoids the problem of routing loops and counting to infinity.
(4) Support multicast, good scalability.

(5) Widely studied at home and abroad, there are many improved protocols based on AODV routing protocol.

The AODV protocol is essentially a combination of DSDV and DSR. It is based on the DSDV protocol and combined with the improvement of the on-demand routing mechanism of DSR. The difference is that AODV uses a hop-by-hop method instead of DSR. Source routing method to improve bandwidth utilization.

2.2 The Working Process of AODV Protocol

AODV is a pure on-demand route acquisition mechanism. Only two nodes that need to communicate with each other can perform route search and maintenance, and the intermediate node can provide forwarding services. The AODV protocol assumes that the wireless link is bidirectional, and its routing protocol mechanism can be summarized as two processes, route discovery and route maintenance.

Generally speaking, when the source node needs to establish a communication link with the destination node, it must first broadcast RREQ (routing request) within the communication range. The intermediate node will decide whether to forward according to whether the RREQ is received for the first time, and the destination node receives After arriving at RREQ, it needs to reply RREP (routing reply) to the source node. The message RREP will pass through the intermediate node and return to the source node. At this time, the routing path is established. When the link is interrupted during the communication process, AODV will initiate local route repair for maintenance. The following describes each process in detail.

(1) Route discovery. When the source node has data to send and needs to communicate with the destination node but the route to the destination node is not found, the source node broadcasts a route request RREQ message in order to find a route to the destination node, and the neighboring node receives the RREQ message Firstly, judge whether the same RREQ packet has been received before sending, if yes, discard it, if not, use the information in the RREQ packet to establish a reverse route. If the intermediate node contains a route to the destination node, it sends a route response RREP message to the source node, otherwise it broadcasts the RREQ message to surrounding nodes in turn. When the destination node of the RREQ receives the RREQ packet, it also establishes a reverse route, and then sends the RREP packet in unicast form to the source node of the RREQ packet. Figure 2 shows the establishment process of the reverse route:

(2) Route maintenance. The node periodically broadcasts the hello message through the MAC layer to determine the link status. If the node does not receive the hello response message for three consecutive times, the link is considered to be disconnected, the routing information containing the link is deleted, and routing is initiated The wrong RRER message informs the neighboring node and the corresponding upstream node to delete routing information that causes the destination node to be unreachable due to link disconnection.

Figure 3 shows the establishment process of the forward route:

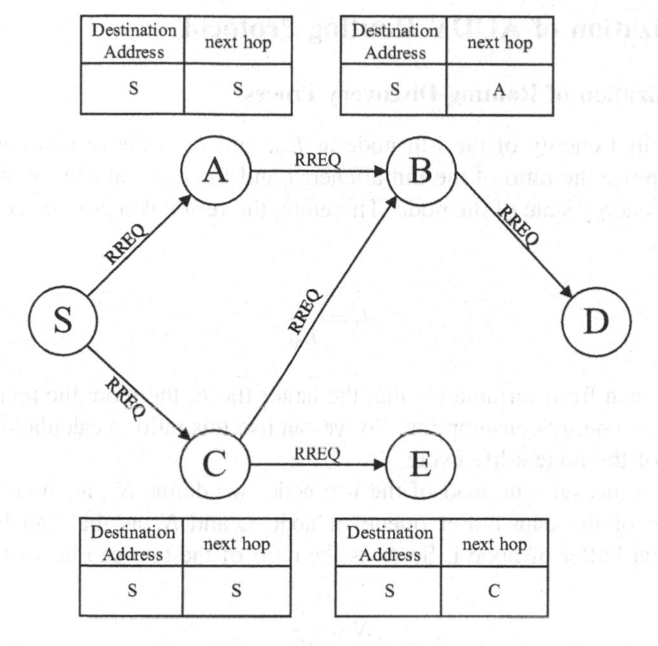

Fig. 2. Reverse route establishment

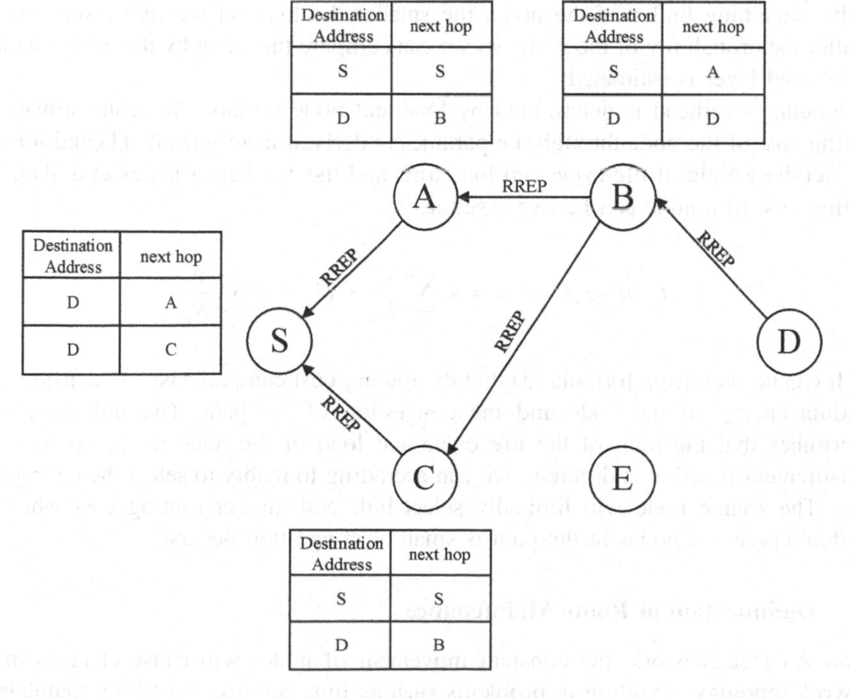

Fig. 3. Forward routing establishment

3 Optimization of AODV Routing Protocol

3.1 Optimization of Routing Discovery Process

Define the initial energy of the i-th node as E_{i0}, and E_i is the residual energy of the node. We can use the ratio of the initial energy and the residual energy of the node to represent the energy state of the node. Therefore, the residual energy percentage can be expressed as:

$$E = \frac{E_i}{E_{i0}} \tag{1}$$

It can be seen from formula (1) that the larger the E, the more the relative residual energy, the less energy consumption, So we can use this ratio to calculation the energy expenditure of the node's life cycle.

In order to measure the load of the i-th node, we define N_{i0} as the original maximum volume of the data buffer queue of node i, and N_i as the data length in the remaining data buffer of node i. So N as the ratio of the two can be written as:

$$N = \frac{N_i}{N_{i0}} \tag{2}$$

It can be seen from formula (2) that the N become larger, the bigger of data length in the remaining buffer of the node, the smaller the load on the link layer, and the smaller the probability of blocking, so we can compare the chain by this ratio The load of the road layer is estimated.

Routing overhead is determined by load and node lifetime. We can estimate the routing cost of the node through the parameters derived from formula (1) and formula (2). Set the weight of life cycle and load to α, and the number of nodes to d, then The routing cost of a node can be expressed as:

$$Routing\_Cost = \alpha \times \sum_{i=1}^{d} \frac{E_i}{E_{i0}} + (1 - \alpha) \sum_{i=1}^{d} \frac{N_i}{N_{i0}} \tag{3}$$

It can be seen from formula (3) that the routing cost comprehensively considers the residual energy of the node and the congestion of the path. The difference of α determines that the ratio of the life cycle and load of the node in the routing cost measurement function is different. we can accroding to reality to select the appropriate path. The source node will limitedly select link with higher routing cost when the residual energy of nodes in the path is small or congestion occurs.

3.2 Optimization of Route Maintenance

In an Ad Hoc network, the constant movement of nodes will cause changes in the network topology, resulting in problems such as link failures. Therefore, maintaining the stability of routing lines is a necessary function of routing protocols. The maintenance of AODV routing table includes three parts: route repair, HELLO message, route buffer timer.

Due to the movement of nodes, the routing table of each node is constantly updated. If the update is stopped for a period of time, the route will fail and the line will be cleared. The AODV protocol detects the failure of the link by broadcasting the HELLO data packet message regularly. If it fails, the routing information will be cleared immediately.

Once the notification of the broken link is received, the source node restarts the route discovery process and establishes a new route. However, the establishment of a new route will also cause the increase of network energy consumption, which will interrupt the link. In order to avoid this phenomenon If it happens, we optimize the AODV protocol so that it can track and delete nodes that may be interrupted to establish a more stable route.

If the remaining energy of a node in the network is small (less than a certain percentage of the remaining energy of the maximum node) or the load is too large, this node will broadcast the RRER, and the upstream node will immediately delete the selected node as the RRER after receiving its broadcast RRER The routing link of the next hop. At this time, the data can be transmitted through the alternate route. If there is no alternate route (the node is unique), the RREQ is repaired from the last hop node to obtain new routing information. If a new route cannot be found, the route will eventually be rediscovered from the source node. This method can greatly reduce network delay and routing overhead.

4 Simulation Analysis

NS2 is a network simulation software widely used in academia at present. This experiment uses the NS2 simulation platform to simulate and analyze the AODV protocol of the Ad Hoc network. The configuration options include: routing protocol, protocol stack, channel, topology, transmission model, and whether to open the wired routing, whether to open the trace of each layer, etc.

The simulation scenario used in this article is to configure 50 nodes in a space of 800 m $\times$ 800 m, the node's moving speed is between 15 and 20 m/s, the effective wireless transmission range is 250 m, and the packet transmission rate is 1 packet/sec. The value of α is 0.5, the critical value of routing cost is 10%, and the simulation time is 600 s. These parameters are used to compare the performance of traditional AODV and E-AODV routing protocols.

We evaluate the impact of E-AODV on reliability and effectiveness by comparing the changes in energy consumption between AODV and E-AODV protocols as the nodes increase, and evaluate the improvement of E-AODV by simulating the average node life.

Figure 4 shows that the energy consumption of both protocols increases as the number of nodes increases, but the total energy consumption of AODV is always higher than that of E-AODV because E-AODV uses the multi-path routing mechanism optimizes the original route, reduces the flooding of the original route, and also reduces the route cost and total energy consumption.

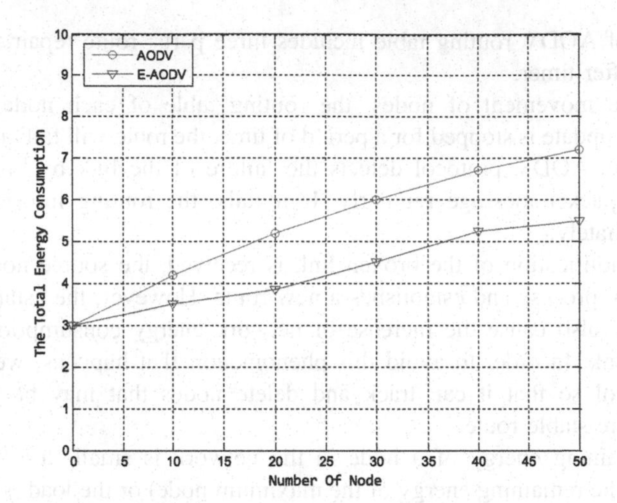

Fig. 4. Total energy consumption

It can be seen from Fig. 5 that the number of dead nodes increases as the simulation time increases. The number of dead nodes under the AODV protocol is significantly higher than that of the E-AODV protocol, and the gap between the number of dead nodes under the E-AODV and AODV protocols at different times is getting bigger and bigger. This is because the E-AODV protocol balances network overhead through node energy and load.

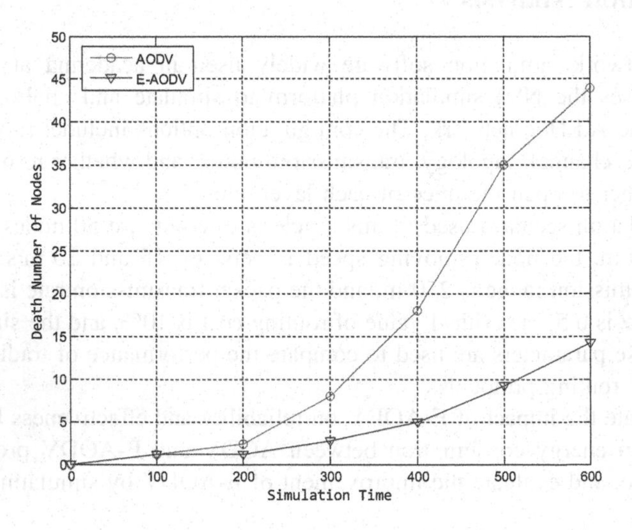

Fig. 5. Number of dead nodes

5 Conclusions

Due to the defects of the AODV protocol in Ad Hoc network, we finally proposed a new AODV optimization protocol E-AODV. We comprehensively consider the remaining energy of the node and the load of the node in the process of establishing the link. Maintain and optimize multi-protocol networks by deleting nodes with lower routing costs. The simulation results show that compared with the original AODV, the E-AODV protocol can reduce the total energy consumption, balance the network overhead. Moreover, the scale factor can be freely selected according to the situation to adapt to more network environment.

References

1. Mingqing, F., Gangyi, D., Yanling, Z.: Research on effective routing for AODV. Modern Electron. Technol. **42**(06), 47–50 (2019)
2. Kaur, K., Kad, S.: Enhanced clustering based AODV-R protocol using ant colony optimization in VANETS. In: IEEE International Conference on Power Electronics Intelligent Control and Energy Systems (2016)
3. Aggarwal, N., et al.: Relative analysis of AODV & DSDV routing protocols for MANET based on NS2. In: International Conference on Electrical Electronics and Optimization Techniques, pp. 3500–3503 (2016)
4. Sureshbhai, T.H., Mahajan, M., Rai, M.K.: An investigational analysis of DSDV, AODV and DSR routing protocols in mobile ad hoc networks. In: 2018 International Conference on Intelligent Circuits and Systems (ICICS). IEEE Computer Society (2018)
5. Mianlu, H.E., Wei, C.H.U., Huizhou, L.I.U.: Research and improvement of AODV routing protocol. Comput. Eng. **41**(01), 110–114 + 120 (2015)
6. Jianw, L.: Research and optimization of AODV routing protocols in mobile ad hoc network. J. Chongqing Univ. (2015)
7. Jinghe, H., Ang, G., Guohui, Z.: Effectiveness evaluation of two reactive routing protocols in MANET. Inf. Commun. **05**, 225–226 (2016)
8. Wei, L., Yuwang, Y.: Energy-saving routing protocol in mobile ad hoc network. Comput. Eng. Des. **39**(10), 3013–3017 (2018)
9. Wangji, H., Yuyuan, Ma., Xin, L., Weisheng, T.: Energy-balanced low-power lossy network routing protocol. Comput. Appl. **38**(04), 1095–1101 (2018)
10. Bhattacharya, A., Sinha, K.: An efficient protocol for load-balanced multipath routing in mobile ad hoc networks. Ad Hoc Networks, **63**, 104–114 (2017)
11. Saeed, T., et al.: Formal modeling of traffic based flooding procedure of AODV for mobile ad hoc networks. In: International Conference on Emerging Technologies, pp. 1–6 (2016)
12. Kuang, J., et al.: An improved AODV protocol based on extension lifetime of the Ad Hoc networks. In: International Conference on Cloud Computing, pp. 1013–1016 (2012)
13. Singh, M., Sharma, J.: Performance analysis of secure & efficient AODV (SE-AODV) with AODV routing protocol using NS2. In: International Conference on Computer Communications, pp. 1–6 (2014)
14. Paranavithana, P., Jayakody, A.: Compromising AODV for better performance: improve energy efficiency in AODV, 201–204 (2017)

AI in UAV-Assisted Wireless Communications

Optimization of OLSR Protocol in UAV Network

Kunqi Hong[1], Shuo Shi[1,2](✉), Xuemai Gu[1,3], and Ziheng Li[1]

[1] School of Electronic and Information Engineering, Harbin Institute of Technology, Harbin 150001, Heilongjiang, China
thiebluesky@163.com, crcss@hit.edu.cn

[2] Network Communication Research Centre, Peng Cheng Laboratory, Shenzhen 518052, Guangdong, China

[3] International Innovation Institute of HIT in Huizhou, Huizhou 516000, Guangdong, China

Abstract. In order to make the OLSR routing protocol more suitable for the self-organizing network of UAVs, this article optimizes the transmission method of HELLO messages and TC packets based on the change of the MPR selection method of the OLSR protocol. In order to adapt to the high dynamics and low density of UAV self-organizing network, the relative moving speed and link transmission quality are taken as the selection criteria of MPRs, so that the stability of MPRs can be improved. At the same time, in order to alleviate the routing overhead and energy problems caused by the increase of HELLO data packets, this article monitors the changes in the network and changes the sending frequency of HELLO messages and TC packets to reduce routing without affecting network updates. The problem of overhead and energy consumption. The network simulation is performed under the PPRZM motion model, and the protocol optimization effect is judged by the comparison of the packet delivery rate, the average end-to-end delay and the routing control overhead.

Keywords: OLSR · HELLO · UAV network · MPRS · End-to-end delay · Routing control overhead

1 Introduction

With the continuous development of hardware processors and wireless communication technology, the performance of UAVs has been significantly improved in all aspects, the price is continuously reduced, and the application fields have become more and more extensive. From the previous full-time military field to the civilian field, the company has excellent performance in fire supervision, emergency response, pesticide spraying, information collection, and target tracking. We can foresee that with the improvement of drone communication level and hardware capabilities, drones will gradually develop towards clustering, miniaturization, and intelligence. Compared with a single UAV, the UAV cluster network has a high degree of flexibility, a great

S. Shi et al. (Eds.): AICON 2020, LNICST 356, pp. 239–248, 2021.
https://doi.org/10.1007/978-3-030-69066-3_21

information coverage, and a very high capacity limit that the UAV cannot compare. It can complete tasks that a single drone cannot.

Compared with the ordinary cellular mobile wireless network, the UAV ad hoc network does not require a base station, and has an incomparable high degree of flexibility and adaptability. It plays an incomparable role in earthquake relief, field exploration, battlefield information confrontation, etc. UAV self-organizing network has the characteristics of ordinary mobile self-organizing network without center, self-organizing network, multiple communications, etc., but also the high mobility of nodes, the sparseness of node density, and the energy autonomy of nodes. These characteristics bring about the extremely high dynamics of the topology of the UAV self-organizing network. Through these characteristics of the UAV self-organizing network, we choose the OLSR active routing protocol as the basis for our optimization. The OLSR protocol uses MPR to forward information and update routing in real time, which is in line with highly dynamic and large-scale UAV networks.

In order to make the OLSR protocol more suitable for the self-organizing network of drones, the OLSR protocol is optimized as follows. First, the selection criteria of MPRs are changed, and the MPR is selected based on mobile similarity and link transmission quality, so that MPRs have stronger stability Performance, reducing the number of updates. At the same time, in order to reduce the routing control overhead caused by the increase of HELLO data packets, we change the transmission frequency of control information by tracking the update status of the network, so as to reduce the routing control overhead while ensuring the timely update of network information.

The rest of paper is organized as follows: Sect. 2 introduces changes to the selection criteria for MPRs, Sect. 3 introduces the tracking of network status and changes in the frequency of control information is given in Sect. 3, Sect. 4 compares the optimized agreement with the original agreement through packet delivery rate, routing control overhead and end-to-end delay. The paper is briefly summarized in Sect. 5.

2 MPRs Selection Algorithm

The original OLSR routing protocol used the connection degree of one-hop nodes as the standard to select MPRs. The purpose is to achieve full coverage of two-hop neighbor nodes by selecting the fewest MPR nodes. However, due to the high dynamics of FANETs, the network topology changes frequently. MPRs need to be updated frequently, which increases the routing control overhead and network response time. Therefore, in order to meet the characteristics of the UAV self-organizing network, We use mobile similarity and transmission stability as the criteria for selecting MPRs to keep the stability of the network link as far as possible, reduce the update frequency of MPRs, and improve link quality. The selection process will be described in detail below.

In FANETs, we can characterize the existence time and stability of the link between the two drones through the relative speed between the two drones, and can obtain the corresponding data through the drone's GPS system, so that each node can know Real-time position and speed information of oneself, in order to calculate the movement similarity between nodes. The mobility similarity of a node refers to the degree of similarity between a UAV node and another adjacent UAV node in the moving speed. Generally speaking, the greater the degree of similarity in the movement behavior between nodes, the link between the two nodes The longer the existence time, the two nodes are considered to be able to maintain a better connection status, otherwise the link between the two nodes is considered to be easily broken. Because in practical applications, the UAV does not change significantly in altitude and is not frequent, so our mobile scene only considers the two-dimensional situation. The speeds of the two drones are v_j and v_i. Then the movement similarity characterizing the relative movement of the two drones is:

$$\theta_{ij} = 1 - \frac{|v_i - v_j|}{|v_i| + |v_j|} \tag{1}$$

For the link transmission quality mentioned above, we characterize the link's evaluation index Expected Transmission Count (ETX) in reference [1], ETX calculates the expected number of retransmissions that are required for a packet to travel to and from a destination. The link quality, LQ, is the fraction of successful packets that were received by us from a neighbor within a window period. The neighbor link quality, NLQ, is the fraction of successful packets that were received by a neighbor node from us within a window period. So the ETX is calculated as follow:

$$ETX = \frac{1}{LQ \times NLQ} \tag{2}$$

In order to calculate and collect data more conveniently and quickly, we divide *ETX* into *MLQ* and *MLQ_L*.*MLQ* is the ratio of the HELLO message sent by the node from the neighbor node to the HELLO message sent by the node to the neighbor node. *MLQ_L* is the ratio of the HELLO message sent by the neighbor node to the HELLO message sent to the node by the neighbor node. And because the value range of the mobile similarity we defined is [0,1], in order to facilitate the calculation, we define a new value *R_ETX* that characterizes the link quality. The calculation formula is as follows:

$$R\_ETX = MLQ\_L \times MLQ \tag{3}$$

Therefore, we consider the selection criteria of MPRs based on the above two characteristics, and because the MPR node is a relay node that transmits information to a two-hop node, we also need to consider the transitivity of the selection indicator and the amplification during the transmission process. Based on the above considerations, we define a comprehensive link evaluation index L to replace the original link connectivity as the selection criterion for MPR nodes. The formula for calculating L is as follows:

$$L(y_i) = \alpha[\theta_{Ay_i}\ \text{average}\,(\theta_{y_is_j})] + \beta[R\_ETX_{Ay_i}\ \text{average}\,(R\_ETX_{y_is_j})] \tag{4}$$

Among them, A is the node performing MPR set calculation, yi is the 1-hop neighbor node of node A, and sj is the strictly symmetric 2-hop node reachable via yi in the 2-hop neighbor set of node A; represents node A and node yi performing calculation The link stability measure between represents the link transmission quality measure between node A and node yi; average() represents the link index between node yi and all strictly symmetrical 2-hop neighbor nodes sj reachable through this node The arithmetic mean value of; α and β are weight coefficients, and satisfy α + β = 1, and the weight can be adjusted for different network focus directions.

Because of the change in the selection criteria of MPRs, we need to change the format of the HELLO packet. Fill in the moving speed of the node in the X and Y directions, the R_ETX between the selected node and the MLQ of the critical point of the selected node. At the same time, we also need to modify the format of the node's local link information database, the one-hop neighbor table and the two-hop neighbor table. The results of the changes are as follows (Figs. 1, 2, 3, 4):

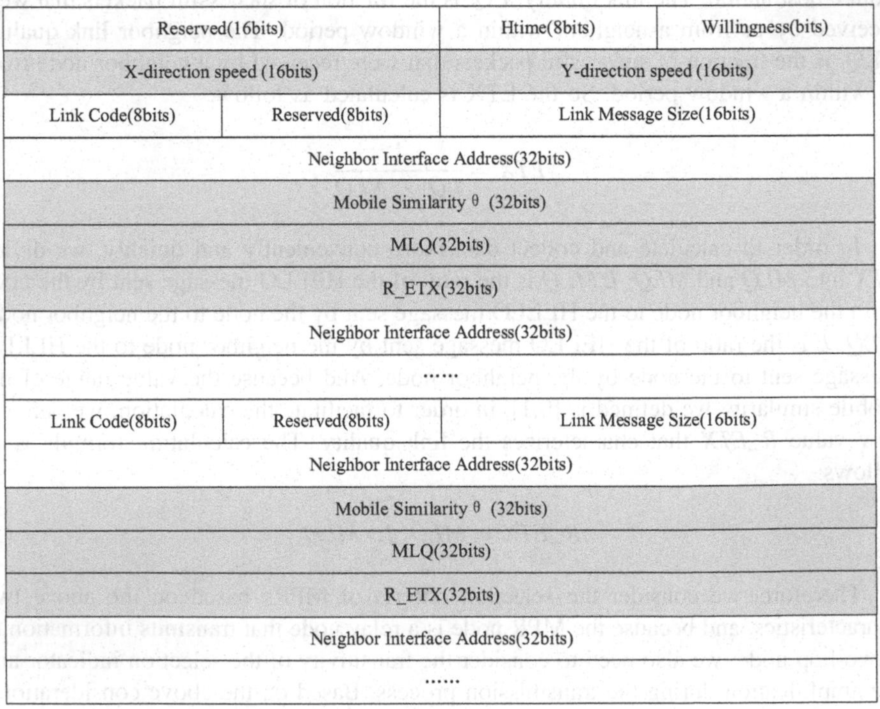

Reserved(16bits)		Htime(8bits)	Willingness(bits)
X-direction speed (16bits)		Y-direction speed (16bits)	
Link Code(8bits)	Reserved(8bits)	Link Message Size(16bits)	
Neighbor Interface Address(32bits)			
Mobile Similarity θ (32bits)			
MLQ(32bits)			
R_ETX(32bits)			
Neighbor Interface Address(32bits)			
......			
Link Code(8bits)	Reserved(8bits)	Link Message Size(16bits)	
Neighbor Interface Address(32bits)			
Mobile Similarity θ (32bits)			
MLQ(32bits)			
R_ETX(32bits)			
Neighbor Interface Address(32bits)			
......			

Fig. 1. Modified HELLO packet format

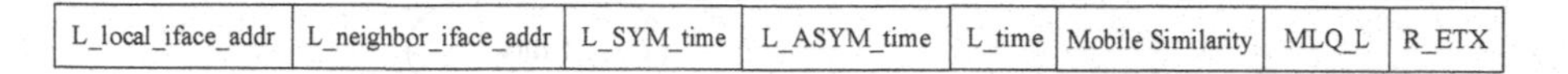

L_local_iface_addr	L_neighbor_iface_addr	L_SYM_time	L_ASYM_time	L_time	Mobile Similarity	MLQ_L	R_ETX

Fig. 2. Modified local link information database format

N_neighbor_main_addr	N_status	N_willingness	Mobile Similarity	R_ETX

Fig. 3. Modified one-hop neighbor table format

N_neighbor_main_addr	N_2hop_addr	N_time	Mobile Similarity	R_ETX

Fig. 4. Modified two-hop neighbor table format

The entire selection process of MPRs is shown in the Fig. 5.

3 Optimization of HELLO Message and TC Packet Transmission

After changing the selection criteria of MPRs in Sect. 2, we have increased the stability of MPRs and reduced the update frequency. But because of the changes to the HELLO packet, the routing overhead for sending HELLO messages has increased. In order to solve the problem of increased routing overhead and energy consumption, we changed the sending interval of HELLO and TC packets, and changed their sending frequency by tracking the real-time status of the network. The fixed transmission interval of the OLSR protocol leads to the following two situations: when the network structure changes rapidly, the OLSR protocol cannot update the network status in time, which greatly reduces the network performance; when the network topology is relatively static, routing control packets It is still being sent relatively frequently, causing the entire network to be flooded with a large number of redundant messages, resulting in a great waste of network resources. However, the number of nodes in the UAV network is usually large and has high mobility. It is very possible that the local topology of the network changes too quickly or the local topology remains stable. Therefore, we learn the changes of the network topology by monitoring the changes of the node link and the MPR selection set, and adjust the sending frequency of HELLO and TC packets in real time and adaptively. In this way, we can reduce routing control overhead while ensuring communication performance.

3.1 Optimization of HELLO Message Sending Interval

The HELLO message is the most basic and most important message structure of the OLSR protocol. It contains the neighbor type, primary address, and link status information of all neighbor nodes of the sending node. It is mainly used in the link awareness and neighbor discovery phases. Local link information base, 1-hop and 2-hop neighbor set establishment and node MPR set calculation. So we can adjust the sending interval of the HELLO message according to the change of the link status.

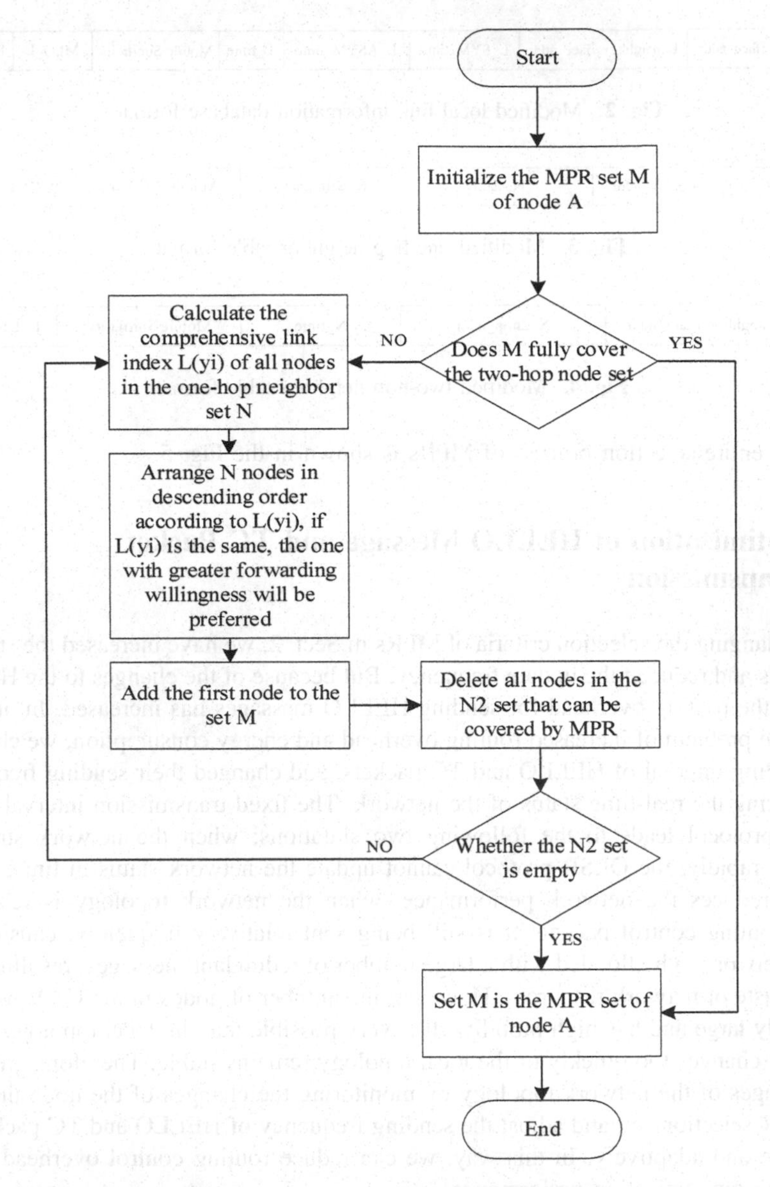

Fig. 5. MPRs selection algorithm process

We set the flag to indicate the three states of the link. SYM_LINK represents that the link is a two-way symmetric link, and both nodes can send and receive data packets to each other; ASYM_LINK represents that the link is an asymmetric link and can only receive each other The message sent by the node; LOST_LINK represents that this link has failed, and both nodes cannot send data packets. After that, we set a link change counter (LCC) inside the node to monitor the link status of the node. The LCC calculation formula is as follows, where New represents the number of new elements in the local link information database, ASYM is the number of times the link identifier changes to ASYM_LINK, SYM This is the number of times the link ID has changed to SYM_LINK.

$$LCC = 3 * New + 2 * ASYM + 1 * SYM \tag{5}$$

The HELLO message sending interval HELLO_INTERVAL set by the OLSR protocol is 2 s. This paper assumes that the lower limit of the HELLO message sending interval is HI_MIN = HELLO_INTERVAL − ΔHI, and the upper limit value is HI_MAX = HELLO_INTERVAL + ΔHI, and ALCC is the average value of LCC in a sending interval of a node. Then, the calculation formula of the adaptive HELLO message sending interval HI is as follows:

$$HI = \begin{cases} HELLO\_INTERVAL + \Delta HI,\ 0 \leq x < 0.5 \\ -2 * \Delta HI * x + 2 * \Delta HI + HELLO\_INTERVAL,\ 0.5 \leq x < 1.5 \\ HELLO\_INTERVAL \text{ - } \Delta HI,\ 1.5 \leq x \end{cases} \tag{6}$$

x = LCC = ALCC.

3.2 Adaptive Optimization of TC Grouping

The TC packet (topology control packet) is used in the topology discovery phase of the OLSR protocol, broadcast to the entire network through the MPR mechanism, and provides topology connection information for subsequent node routing table calculations. In the network, only the nodes that are added to the MPR set can generate and send TC packets; and in addition to the preset TC packet update cycle, if the MPR selection set of the MPR node changes, the MPR node will send TC group for routing maintenance process. Based on the above two reasons, the adaptive transmission frequency of TC packets can be measured by the change of the node's MPR selection set.

An MPR selection change counter MSCC is set in the node, which is activated when the node is selected as the MPR node. It is used to monitor the change status of the node MPR selection set within a sending interval TC_INTERVAL to complete the adaptive adjustment of the TC packet sending interval. The counting rule of this counter is that when the MPR selection set of the node changes, the counter MSCC will increase by 1.

The default TC packet sending interval of the OLSR protocol is TC_INTERVAL, and its value is 5 s. Now suppose that the maximum sending interval TCI_MAX of TC packets is 8 s. Then the calculation method of the transmission interval TCI of the TC packet is as follows:

$$TCI = \begin{cases} TC\_INTERVAL,\ MSCC \neq 0 \\ TCI_{last} + 1,\ MSCC = 0 \end{cases} \tag{7}$$

TCI_{last} is the last TC packet transmission interval, and $TCI \in$[TC_INTERVAL, TCI_MAX].

4 Simulation and Conclusion

This paper mainly focuses on the simulation of three parameters. The end-to-end delay represents the time consumed by the data packet sent by the source node in the network from the moment it was sent to the destination node. The average end-to-end delay is the average of the time consumed by all successfully received data packets. The packet delivery rate represents the percentage of the total number of data packets successfully received by the destination node to the number of data packets sent by the source node during the network data transmission process. The routing control overhead index refers to the ratio of the number of bits of the data packet received by the destination node to the number of bits contained in the routing control packet, that is, the number of bits of the routing control packet required to successfully transmit 1 bit of data. OLSR is the original OLSR protocol, OLSR-M is the OLSR protocol optimized for the MPR selection standard, and OLSR-MI is the OLSR protocol optimized for the transmission frequency of OLSR-M.

From the above simulation results, we can see that by changing the selection criteria of MPRs, we have improved the packet delivery rate of the OLSR protocol and reduced the end-to-end delay. But because we added speed and other information in the HELLO data packet, the control information overhead becomes larger. However, by optimizing the sending interval of the TC group and the HELLO message, the control overhead has been significantly reduced, while maintaining an excellent end-to-end delay and packet delivery rate. Through the above analysis, we can see that our optimization of OLSR is successful (Figs. 6, 7, 8).

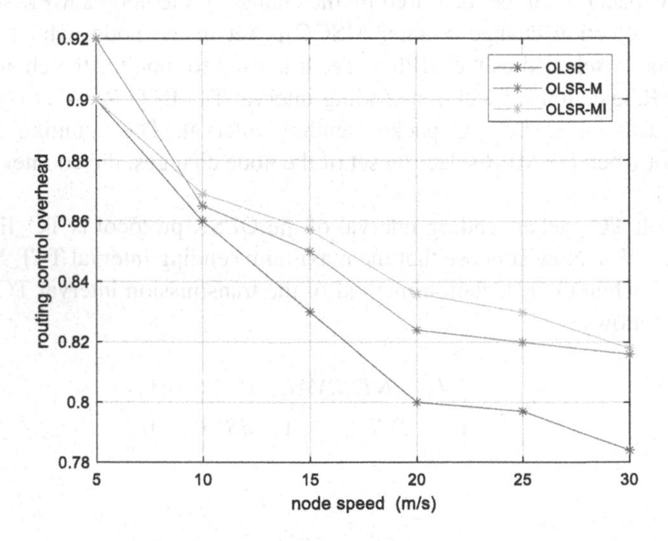

Fig. 6. Packet delivery rate

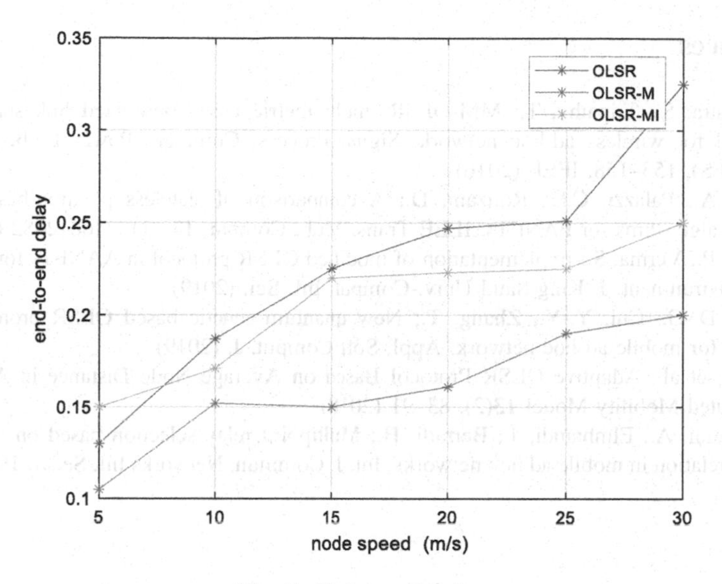

Fig. 7. End-to-end delay

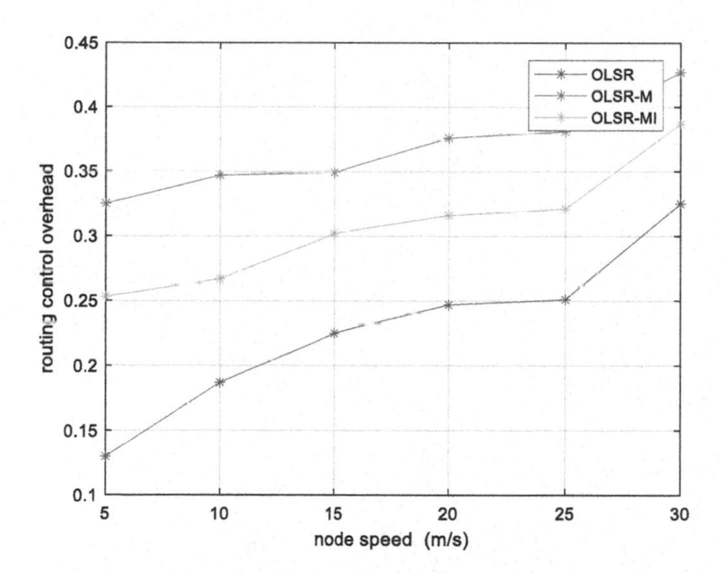

Fig. 8. Routing control overhead

References

1. Mohapatra, S., Tripathy, T.: MM-OLSR: multi metric based optimized link state routing protocol for wireless ad-hoc network. Signal Process. Commun. Power Embedded Syst. (SCOPES), 153–158. IEEE (2016)
2. Bujari, A., Palazzi, C.E., Ronzani, D.: A comparison of stateless position-based packet routing algorithms for FANETs. IEEE Trans. Mob. Comput. **17**(11), 2468–2482 (2018)
3. Kumar, P., Verma, S.: Implementation of modified OLSR protocol in AANETs for UDP and TCP environment. J. King Saud Univ.-Comput. Inf. Sci. (2019)
4. Zhang, D.-G., Cui, Y.-Y., Zhang, T.: New quantum-genetic based OLSR protocol (QG-OLSR) for mobile ad hoc network. Appl. Soft Comput. J. (2019)
5. Lee, T., et al.: Adaptive OLSR Protocol Based on Average Node Distance in Airdropped Distributed Mobility Model **13**(2), 83–91 (2018)
6. Abdellaoui, A., Elmhamdi, J., Berradi, H.: Multipoint relay selection based on stability of spatial relation in mobile ad hoc networks. Int. J. Commun. Networks Inf. Secur. **10**(1) (2018)

Delay Minimization in Multi-UAV Assisted Wireless Networks: A Reinforcement Learning Approach

Chenyu Wu[1(✉)], Xuemai Gu[1,3], and Shuo Shi[1,2]

[1] School of Electronic and Information Engineering, Harbin Institute of Technology, Harbin 150001, Heilongjiang, China
{wuchenyu,guxuemai,crcss}@hit.edu.cn

[2] Network Communication Research Centre, Peng Cheng Laboratory, Shenzhen 518052, Guangdong, China

[3] International Innovation Institute of HIT in Huizhou, Huizhou 516000, Guangdong, China

Abstract. Unmanned Aerial Vehicles (UAVs) assisted communications are promising technology for meeting the demand of unprecedented demands for wireless services. In this paper, we propose a novel framework for delay minimization driven deployment of multiple UAVs. The problem of joint non-convex three dimensional (3D) deployment for minimizing average delay is formulated and solved by Deep Q network (DQN), which is a reinforcement learning based algorithm. Firstly, we obtain the cell partition by K-means algorithm. Then, we find the optimal 3D position for each UAV in each cluster to provide low delay service. Finally, when users are roaming, the UAVs are still able to track the real-time users. Numerical results show that the proposed DQN-based delay algorithm shows a fast convergence after a small number of iterations. Additionally, the proposed deployment algorithm outperforms several benchmarks in terms of average delay.

Keywords: Unmanned Aerial Vehicles · Delay minimization · Deployment · Reinforcement learning

1 Introduction

With the pullulating and landing deployment of wireless skills, as well as the birth of killer apps, users' pursuit of service quality is higher, and the existing skills cannot meet the needs of tomorrow communication. People are looking for ever-increasing turnkey solutions, including exploration on higher airways, better encoding and transmission skills, and a large-scale connection that incorporates multiple networks. UAVs are thought to be killers of auxiliary communication [1]. Rather than the orthodox ground wireless-skills, UAV assisted communication has the preponderances of high movability, low expenditure, especially better LOS positioning ability. Therefore, the employment of UAVs to acquire high rate is expected to play a pivotal role.

S. Shi et al. (Eds.): AICON 2020, LNICST 356, pp. 249–259, 2021.
https://doi.org/10.1007/978-3-030-69066-3_22

The deployment of UAVs as locomotive BS to assist surficial infrastructure has been deemed as an prominent technology for handling cellular network discharging and offloading in hot spots, such as prompt renew after infrastructure damage, important recreational gathering, high level meeting and natural disasters. Under the premise of dependency and adjustability, these criticisms can be solved universally by UAVs. In this paper, UAV, as a relay node, not only improves the total throughput of the system, but also provides reliable connection for remote users without perfect direct link [2, 3]. In addition, UAVs can also be used to assist the Internet of things network to ensure large-scale connectivity and low latency [4, 5].

In reference [6], the air ground model is given and the altitude problem of UAV is well solved. In this paper, we can seek out the emblematic parameters of the air-ground model and bring inspiration to the deployment of UAV. Recent strategy is not only about maximizing coverage, but also on algorithms that try to cover the largest number of users. In order to improve the system coverage, the deployment layout of single UAV and multi UAV has been studied [7, 8]. The layout algorithm can be synchronous or asynchronous [9]. However, due to the high computational complicacy, especially in dealing with dynamic circumstances such as roaming users and ever-changing channel conditions, the three-dimensional layout of multiple UAVs is defiant. RL reduces the complexity of convex optimization by means of iteration and interaction, and has great effect in shaping planning and multi-objective and constraint problems. In reference [10], the author proposes a deep reinforcement learning algorithm for UAV control, which considers fairness, energy consumption and connectivity. The object is to seek a tactic to control the movement mode of each UAV. However, the three-dimensional layout of multiple UAVs is ignored.

The indicators of user relationship are various, such as delay, flux, the number of users meeting the threshold, file hit rate and so on. However, they can not be separated from each other. It can be summarized by the quality of user service, which is nothing more than choosing the best service target and service mode according to the user's needs, location, channel information, etc. In this paper, we consider a scenario that multiple UAVs serve ground users for delay minimization. Firstly, we obtain the user association to reduce the impact of user interleaving by K-means algorithm. Then, we find the optimal 3D placement bourn for each UAV to minimize the sum delay of the users. Finally, when users are moving, the UAVs are still able to track the real-time users and provide low latency service.

2 System Model and Problem Formulation

We consider the downlink of UAV assisted ground users in urban as shown in Fig. 1. Multiple UAVs act as BSs in the air to carry files of users' interest and serve the users in the target region. There exists K UAVs serving users set $\mathcal{U}$ with the total number of U. Users are separated in K clusters. We assume that there are U^k users in the k-th cluster and the specific user u_i^k is the i-th user in class k, $i \in \{1, 2, \ldots, U^k\}$.Each user

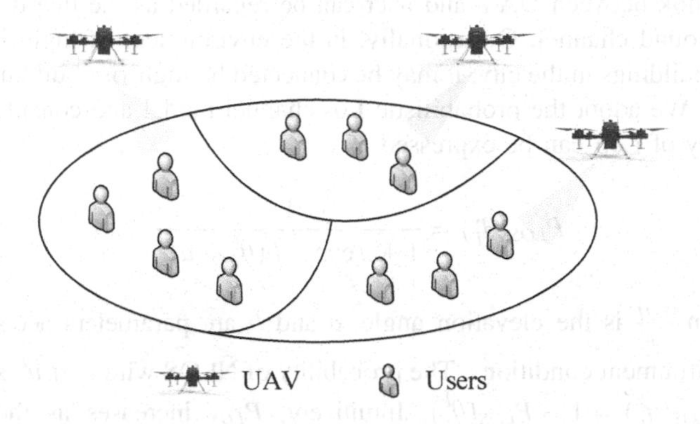

Fig. 1. System model for delay minimization. Each UAV serves one cluster.

belongs to the coverage of only one UAV. The users in the city are very dense, so the delay will be greatly increased if the time division method is used to serve the users in turn. So that we assume at the same time, UAV can serve multiple users and adopt multiple access based on frequency.

2.1 Transmission Model

The user's location is random. In some literatures, the user is modeled as a PPP or a uniform distribution around the center of a circle by using statistical methods. This is not the most important because our algorithm is scalable and can be applied to different user distributions without changing the model. Users can move continuously during the service period of UAVs. Due to the high mobility, It is difficult to study the control of UAV in large time scale. Thus, we use technique called time discretization, which is to divide the time T into equal slots with length δ and index t. The UAV flies at an appropriate fixed altitude H and the maximum speed is $V_{\max}$. The 2D position of a specific user u_i^k at each time slot is $[x_i^k(t), y_i^k(t)]^T$, and the 3D coordinate of the k-th UAV is $[x^k(t), y^k(t), h^k(t)]^T$. Compared with the whole mission cycle, the moving distance of UAV in a short time is relatively small, which can be approximately static in the initial or terminal position. Making use of time discretization, the instantaneous distance between UAV k and the specific user u_i^k can be fixed in a small time:

$$d_i^k(t) = \sqrt{[h^k(t)]^2 + [x^k(t) - x_i^k(t)]^2 + [y^k(t) - y_i^k(t)]^2} \tag{1}$$

The downlink between UAV and user can be regarded as the line of sight dominated air-to-ground channel. Occasionally, in the environment of high-rise buildings and high-rise buildings in the city, it may be connected by high-rise buildings and high-rise buildings. We adopt the probabilistic Los channel model and consider occlusion. The probability of LoS can be expressed as

$$P_{LOS}(\theta_i^k) = \frac{1}{1 + a\exp(-b(\theta_i^k - a))} \tag{2}$$

where $\theta_i^k = \sin^{-1}\frac{H}{d_i^k}$ is the elevation angle, a and b are parameters according to the change of environment conditions. The probability of NLOS with user u_i^k's feedback is given by $P_{NLOS}(\theta_i^k) = 1 - P_{LOS}(\theta_i^k)$. Intuitively, P_{LOS} increases as the UAVs fly directly on the target and approximate one when θ_i^k becomes large enough.

Then, the path loss for user u_i^k is

$$PL_{LOS} = (\frac{4\pi f_c}{c})^{-2}(d)^{-\alpha}10^{\eta_{LOS}} \tag{3}$$

$$PL_{NLOS} = (\frac{4\pi f_c}{c})^{-2}(d)^{-\alpha}10^{\eta_{NLOS}} \tag{4}$$

$$PL{=}P_{LOS} \times PL_{LOS} + P_{NLOS} \times PL_{NLOS} \tag{5}$$

where f_c stands for the carrier frequency.c is speed constant of light. α is the exponent indicating loss, η_{LOS} and η_{NLOS} are the attenuation factors according to the existence of LoS and NLoS.

Many assume that the number of spectrum is variable and can be continuously allocated. This assumption has certain truth, but it is very difficult to practice.

We discuss simple scheme of FDMA and assume the bandwidth B is allocated to users belonging to the same sphere in an equal manner, thus the spectrum for U^k user is $B_i^k = B/U^k$. The maximum power carried is equally distributed similarly with each user u_i^k having $P_i^k = P/U^k$. By estimating from the receiver along with the SNR, the service rate for user u_i^k with bit/s in unit of measurement:

$$r_i^k(t) = B_i^k \log_2(1 + \frac{p_i^k}{PL_{d_i^k(t)}\sigma^2}) \tag{6}$$

where $\sigma^2 = B_i^k N_0$ is the AWGN var, N_0 is power spectral density for general noise.

2.2 Problem Formulation

We consider the UAV hovering over the user with variable altitude when the user is stationary or continuously moving. The bandwidth and transmission power of each UAV are uniformly allocated to each user. Therefore, the optimization problem is simplified as a region segmentation problem, and its formula is as follows

$$\max_{x,y,h} \ \delta_{\text{sum}} = \sum_{k=1}^{K} \sum_{i=1}^{U_k} \sum_{t=1}^{T} s/r_i^k(t) \tag{7}$$

Where s is the standardized file size of content to transfer. It can be seen from Eq. (7) that the altitude and horizontal coordinates of UAV have influence on the delay of users. This is because both the distance and the Los probability are related to the altitude of the UAV. Increasing the flight altitude of UAV will lead to greater path loss, but also will obtain higher Los probability.

Due to the combination of user association and optimal location search, exhaustive search algorithm is a direct method to obtain the optimal result. However, this is computationally complex. Therefore, a low complexity 3D deployment algorithm based on DQN is proposed. In addition, when the optimal position of UAV is fixed, the acquisition of dynamic tracking is also very important due to the nonconvex problem of sum delay.

3 Deployment and Movement of UAVs Using DQN

In the actual scene, the user roams continuously, which leads to the increase of delay. Traditional methods tend to predict with high complexity solutions. Therefore, RL is employed to tail after users.

Reinforcement learning (RL) is a forceful tool to solve decision-making problems. In recent years, reinforcement learning has reached the limit of human cognition in many aspects in the field of game, and can be used as an auxiliary means to solve optimization problems. In this part, we first introduce some basic knowledges of RL, and then we propose an algorithm to minimize average delay based on Deep Q-Network (DQN).

Reinforcement learning contains basic elements including: environment which is preset and can not be changed, agent which is trained, state which stands for the status of robots that are being trained, action that the robots take using their habits, and reward gained after each step. In RL, agents interact with the atmosphere in a way of action and reward. The process is a MDP $\mathcal{M} = <\mathcal{S}, \mathcal{A}, \mathcal{R}, \Pr(s_{t+1}|s_t, a)>$, where $\mathcal{S}$ is state set, $\mathcal{A}$ is action set, $\mathcal{R}$ is the set of reward. When taking action a_t, there is a transition probability of $\Pr(s_{t+1}|s_t, a_t)$ from state s_t to s_{t+1}.

The aim of RL is to conceive a policy that maximizes the total rewards observed during the episodes. Value is a common term in RL which stand for the set of policies that evaluate the long-term reward of the policy. Q-learning is a basic value-based algorithm of RL, which maintains a Q-table to record and minimize the discounted cumulative reward which is

$$\min C = \mathbb{E}^{\pi}(\sum_{t=1}^{\infty} \gamma_d^{t-1} r(s_{t+1}|s_t, a_t)) \tag{8}$$

The integral is from the present moment to the infinite future which is the final state of other restrictions, where $\pi = \arg\max_{a_t \in \mathbb{A}} Q(s_t, a_t)$ is the policy to choose action, γ_d is the discount factor. Allowing agents to choose actions according to the maximum value cannot achieve good results, because it will destroy the balance between exploration and optimization. An excellent tutorial tip to explore the environment is the ε-greedy policy. The Q table which is also known as value function is updated by

$$Q_{t+1}(s_t, a_t) = (1 - \alpha)Q_t(s_t, a_t) + \alpha(r_t + \gamma_d \min_{a'} Q_t(s_{t+1}, a')) \tag{9}$$

where α is learning rate. However, since this algorithm holds a big form for each action-state pair, it is intolerable for large scale problem. For example, when we play chess, the state is the current chess piece, the actions set is to drop a piece randomly in the blank position of the current chessboard. Considering the size of the board, the total action space is equal to the length times the width, which still does not include some actions that can and cannot be done according to the rules of the game. So we can see that maintaining a table consumes huge resources and sometimes can't solve problems. Neural network is a good substitute, because large-scale network can approximate any nonlinear function to meet our needs (Table 1).

Table 1. Simulation Parameters.

Parameter	Description	Value
U	Number of users	80
K	Number of clusters and UAVs	4
P	Total transmit power	0.1 W
δ	Time slot length	1 s
N_0	Noise power spectral density	−174 dBm/Hz
f_c	Carrier frequency	2 GHz
B	Total bandwidth of each UAV	1 MHz
a, b	Environmental parameters	10.39,0.05(urban)
η_{LOS}, η_{NLOS}	Additional path loss for LOS, NLOS	1,20 (dB)

Moreover, the control of UAVs is a continuous control problem. Many works regard the UAV as a static base station, which plays the same role as the small base station and studies the optimal solution in statistical sense. I don't think this assumption is very reasonable because the UAV is a mobile agent, so it is necessary to give full play to its mobility advantages to carry out path planning. DQN take example by neural networks to reckon the value. The NN target is minimizing the loss:

$$L(\theta^Q) = \mathbb{E}\left[r_t + \gamma_d Q'(s_{t+1}, \pi(s_{t+1})|\theta^{Q'}) - Q(s_t, a_t|\theta^Q)\right]^2 \tag{10}$$

where the first part $y_t = r_t + \gamma_d Q'(s_{t+1}, \pi(s_{t+1})|\theta^{Q'})$ is the target value to reach, θ^Q is the weight of NN. The network back propagates and updates θ^Q using gradient decent with derivative $\nabla L(\theta^Q)$.

In addition, DQN adopts two kinds of technologies: experience playback and target network to reduce the influence of data correlation. The correlation between data can not make neural network learn useful knowledge well. With the introduction of stochastic gradient, this problem is solved well. Experience playback is selecting batch size B_s experience from buffer in a random manner. In addition, DQN tries identical target network Q' as the NN of the original one. The weight of the original NN is to update the parameters in a delay manner of the target network.

We explain the important elements:

1) Agent: Agent is one of the core of RL. At present, the mainstream research direction has been extended to multi-agent learning. It considers the multi-objective cooperation or competition game, which itself is a difficult problem to see the optimal solution, because there are still many challenges. In contrast, single agent has been proved to be a good solution to some simple decision-making problems, and the distributed single-agent solution is also a choice. Because the interference is not considered, there is no cooperation and competition between UAVs. The training agent: each UAV
2) State: During each training step t (also time index for epochs of the whole progress), $s_t = [x^k(t), y^k(t), h^k(t), x_1^k(t), x_2^k(t), \ldots, x_{U_k}^k(t)]$. The state is the 3D site of UAV and 2D coordinates for ground customers.
3) Action: In order to provide continuous control of the UAVs, we denote the operating direction as the action. Also, the agent can suspend in a still manner. There are 6 directions available: left, forward, up, backward, right, as well as down.
4) Reward: Reward is a common term, which is suitable for our goal related. In the actual scene, users can't give us immediate feedback because the user's experience is delayed, but in the simulation and training, we can choose experience data according to the parameters. Data generation is one of the benefits of RL, which does not rely on training data sets, but through experience. However, it also brings about the problem of data utilization. The reward of epoch t is defined as:

$$r(t) = \sum\nolimits_{i=1}^{U_k} s/r_i^k(t) \tag{11}$$

which is the current sum delay.

Using DQN, the UAVs can quickly and efficiently find the location and moving direction to obtain the minimum delay. The progress of the whole algorithm is shown in Algorithm 1.

Algorithm 1 Deep Reinforcement Learning for Delay Minimization

1: Initialize value Q with random weights $\boldsymbol{\theta}$
2: Initialize target value Q' with same parameters $\boldsymbol{\theta}^{-}=\boldsymbol{\theta}$
3: Initialize N capacity memory $\mathcal{D}$, and buffer size is set as B_s.
4: **for** episode $m = 1, 2, ..., M$ **do**:
5: Initialize initial state s_1 and prepare training environment
6: **if** random$< \varepsilon$:
7: choose action $a_t = \arg\min_a Q(s_t, a; \boldsymbol{\theta})$
8: **else**:
9: randomly choose an action
10: execute a_t and observe s_{t+1}, r_t
11: store transition (s_t, a_t, r_t, s_{t+1}) in $\mathcal{D}$
12: sample random mini-batch (s_j, a_j, r_j, s_{j+1}) with size B_s from $\mathcal{D}$
13: Set value for target: $y_j = r_j + \gamma_d \min Q'(s_{j+1}, a \mid \boldsymbol{\theta}^{-})$
14: Loss function $L(\boldsymbol{\theta}^{Q}) = \sum_{j=1}^{B_s} [y_i - Q(s_j, a_j \mid \boldsymbol{\theta})]^2$
15: update $\boldsymbol{\theta}$ using $\nabla L(\boldsymbol{\theta}^{Q})$ using GD
16: Set $\boldsymbol{\theta}^{-}=\boldsymbol{\theta}$ in a repetitive manner for every B_{up} steps
17: **end for**

4 Results and Analysis

First of all, we introduce the simulation platform and the specific super parameters in machine learning. We conduct our experiments in Tensorflow with version 1.0. It is a time-consuming and laborious process to find the suitable hyperparameters. In order to simplify, we only give the best hyperparameters which represent the best performance of the system, but we don't talk about testing and selecting the parameters

The main hyperparameters are as follows: rate for learning α is 0.001, memory size $\mathcal{D}$ is 5000, factor of discount as 0.9, repetitive update B_{up} = 300 steps. The neural network adopts two-layer fully connected architecture, because in lots of experiments, the single-layer network can not fit the model well, and the three-layer network also has the problem of over fitting and slow training speed. Our algorithm is also compared with the traditional exhaustive-based algorithm and random deployment algorithm in terms of convergence and system performance.

Figure 2 depicts the instantaneous delay for each ground user. We draw the three-dimensional equipotential diagram of all users' delay. Intuitively, it is a concave surface. The cluster has 20 users and is served by one UAV. It can be observed that with the increase of the distance between UAV and ground user, the delay of the ground user also increases.

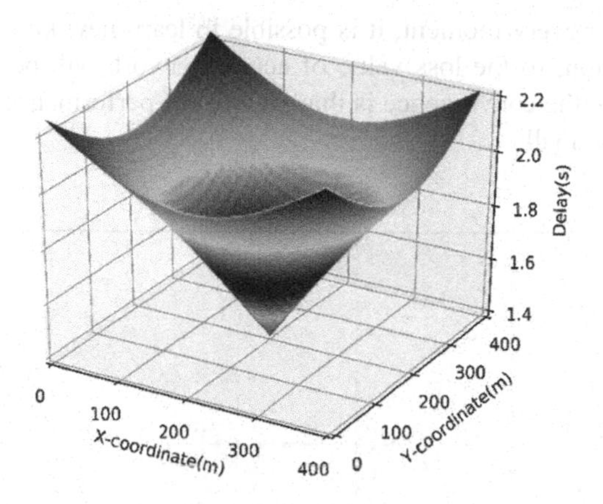

Fig. 2. Minimum delay versus user location

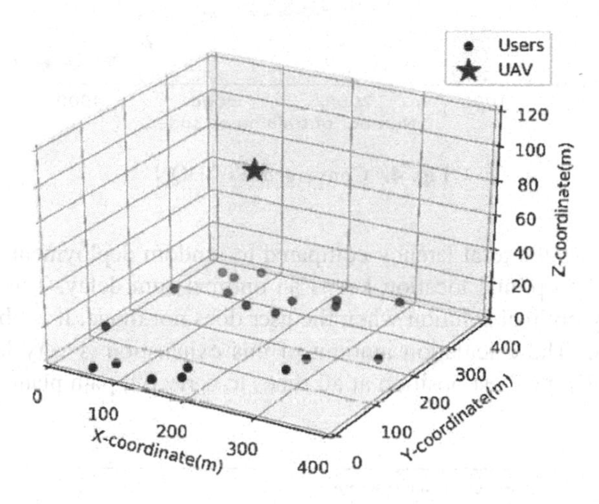

Fig. 3. Optimal UAV location versus user distribution

Figure 3 draws the optimal 3D map of UAV from the position of the first fleet and ground user. The blue star represents the best location for the UAV. The horizontal coordinates and height of UAVs are determined by the user's position, because they affect the line of sight probability and path loss.

Figure 4 depicts the relationship between total delay and training times. It can be seen that the UAV can perform its actions in an iterative manner and learn from the mistakes, thus improving the and delay. It can be seen that the algorithm converges after a certain number of iterations. Despite the initial position of the UAV, it was integrated after about 5000 sets. The process of convergence is not a straight line or has been declining, but a fluctuating decline, which is one of the basic common sense of RL, because RL constantly carries out trial and error and iteration to complete learning from experience. At every moment, it is possible to learn new knowledge to optimize the objective function, so the loss value of neural network will be increased. Finally, the method to judge the convergence is that the overall performance tends to be stable, and the variance is small.

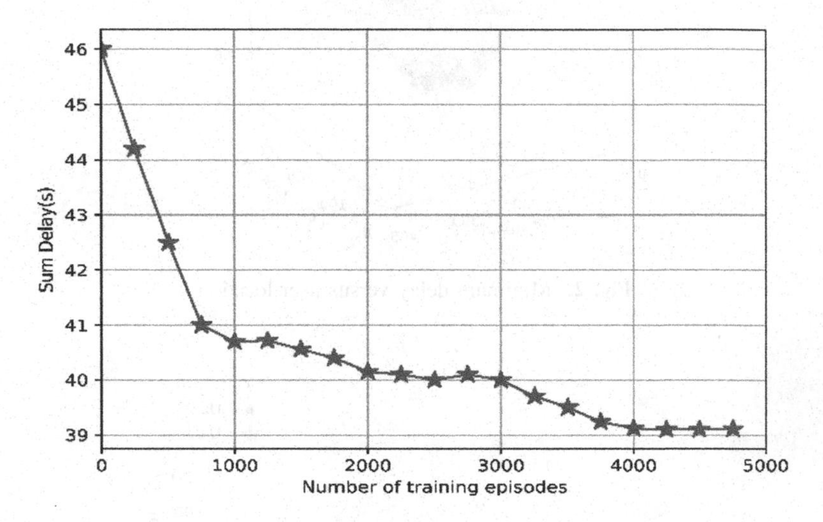

Fig. 4. Convergence of DQN

Figure 5 shows the total latency compared to random deployment. When the user remains static, the optimal location keeps an optimal sum delay. The green line represents the delay optimal solution when the user does not move. It is obtained by brute force exhaustion. The calculation amount of this exhaustion is very large. When the user moves, we list the best position at all times to carry out path planning, and a more

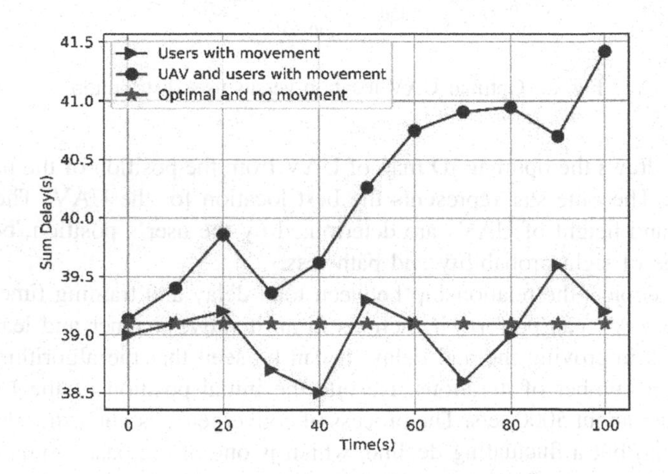

Fig. 5. Sum delay when users are moving and static

intelligent algorithm is needed to give the optimal decision in real time. The blue line represents the real-time total flux under the path planning given by our algorithm when the user moves. It can be seen that the value fluctuates around the initial value. This is because the user's movement is random and sometimes tends to gather near the UAV, so the total flux must be relatively large. When the ground user moves according to the random walk model, the UAV should move along the user's direction. Otherwise, as the user leaves the initial point, and the delay increases. As can be seen from the figure, our algorithm is suitable for dynamic environment.

References

1. Zeng, Y., Zhang, R., Lim, T.J.: Wireless communications with unmanned aerial vehicles: opportunities and challenges. IEEE Commun. Mag. **54**(5), 36–42 (2016)
2. Zeng, Y., et al.: Throughput maximization for UAV-enabled mobile relaying systems. IEEE Trans. Commun. **64**(12), 4983–4996 (2016)
3. Zhang, S., Zhang, H., He, Q., Bian, K., Song, L.: Joint trajectory and power optimization for UAV relay networks. IEEE Commun. Lett. **22**(1), 161–164 (2018)
4. Qin, Z., Fan, J., Liu, Y., Gao, Y., Li, G.Y.: Sparse representation for wireless communications: a compressive sensing approach. IEEE Signal Process. Mag. **35**(3), 40–58 (2018)
5. Qin, Z., Li, F.Y., Li, G.Y., McCann, J.A., Ni, Q.: Low-power wide-area networks for sustainable IoT. IEEE Wireless Commun. **26**(3), 140–145 (2019)
6. Al-Hourani, A., Kandeepan, S., Lardner, S.: Optimal LAP altitude for maximum coverage. IEEE Wireless Commun. Lett. **3**(6), 569–572 (2014)
7. Lyu, J., Zeng, Y., Zhang, R., Lim, T.J.: Placement optimization of UAV-mounted mobile base stations. IEEE Commun. Lett. **21**(3), 604–607 (2017)
8. Mozaffari, M., Saad, W., Bennis, M., Debbah, M.: Efficient deployment of multiple unmanned aerial vehicles for optimal wireless coverage. IEEE Commun. Lett. **20**(8), 1647–1650 (2016)
9. Sun, J., Masouros, C.: Deployment strategies of multiple aerial BSs for user coverage and power efficiency maximization. IEEE Commun. Lett. **67**(4), 2981–2994 (2019)
10. Liu, C.H., Chen, Z., Tang, J., Xu, J., Piao, C.: Energy-efficient UAV control for effective and fair communication coverage: a deep reinforcement learning approach. IEEE J. Sel. Areas Commun. **36**(9), 2059–2070 (2018)

Trajectory Planning Based on K-Means in UAV-Assisted Networks with Underlaid D2D Communications

Shuo Zhang[1], Xuemai Gu[1,3], and Shuo Shi[1,2](✉)

[1] School of Electronic and Information Engineering, Harbin Institute of Technology, Harbin, Heilongjiang 150001, China
crcss@hit.edu.cn

[2] Network Communication Research Centre, Peng Cheng Laboratory, Shenzhen, Guangdong 518052, China

[3] International Innovation Institute of HIT in Huizhou, Huizhou, Guangdong 516000, China

Abstract. Unmanned aerial vehicles (UAV) has become a popular auxiliary method in the communication field due to its mobility and mobility. The air base station (BS) is one of the important roles of UAV. It can serve the ground terminals (GTs) without being restricted by time and space. When GTs are scattered, trajectory optimization becomes an indispensable part of the UAV communication. In this paper, we consider a UAV-assisted network with underlaid D2D users (DUs), where the UAV aims to achieve full coverage of DUs. Trajectory planning is transformed into the deployment and connection of UAV stop points (SPs), and a K-means-based trajectory planning algorithm is proposed. By clustering DUs, the initial SPs is determined. Then add new SPs according to the coverage, and construct the trajectory. The simulation analyzes the validity of the algorithm from the distribution of DUs and the number of initial cluster centers. The results show that the proposed algorithm is compared favorably against well-known benchmark scheme in terms of the length of the trajectory.

Keywords: Unmanned aerial vehicles · Trajectory planning · K-means algorithm

1 Introduction

With the rapid development of aviation and electronic technology, unmanned aerial vehicles (UAVs) will play an important role in the field of wireless communication due to its mobility and portability [1, 2]. UAVs can be used as aerial BS to provide reliable wireless communication in scenes of battlefield or disaster area. Different from traditional ground BSs, UAVs can be deployed flexibly and move along a given trajectory which is determined by their aviation characteristics to cover ground terminals (GTs) [3]. Therefore, trajectory planning has become a basic prerequisite to ensure that the UAV successfully completes its mission [4].

S. Shi et al. (Eds.): AICON 2020, LNICST 356, pp. 260–271, 2021.
https://doi.org/10.1007/978-3-030-69066-3_23

Trajectory planning has always been an important research hotspot in the UAV field. However, previous research work is mainly about UAV navigation applications under various environmental constraints, such as obstacle collision avoidance [5]. When faced with communication missions with a wide distribution and a large area, serving as many users as possible or achieving full coverage is one of the goals of UAVs. The authors in [6] study the 3-dimensional (3D) deployment problem of a single aerial BS under the probabilistic line-of-sight (LoS) channel model to realize the offloading of as many GTs as possible from the terrestrial BS. Zeng, Y. et al. [7] design the UAV's trajectory to minimize mission completion time on the basis of ensuring that each GT restores files with a high probability of success.

Due to the limited coverage of a single UAV, multiple UAVs are tried to achieve full coverage. A polynomial-time algorithm with successive vehicle-mounted mobile base stations placement is proposed in [3]. In addition, clustering algorithm is also one of the commonly used methods to place multiple UAVs. In [8], the authors adopt k-means algorithm to classify terrestrial users to be served by multiple UAVs, and GTs that are not covered by the UAVs are supported by the fixed ground BSs. As an effective means to improve spectrum efficiency, device-to-device (D2D) applications are becoming more frequent. There are three mode in D2D communications, i.e., cellular mode, underlay mode and overlay mode. To improve spectrum efficiency, underlay mode is the one with the highest usage. Mohammad Mozaffari et al. [9] investigate the deployment of a mobile UAV in UAV and D2D coexistence networks. In order to completely cover the area, the disk covering problem is computed to obtain the minimum number of stop points (SPs) that the UAV needs. Furthermore, it is often used to relieve cache pressure and expand communication range. Based on [3], authors [10] realize the optimization of the UAV trajectory in the cache network in which the UAV and D2D coexist. This provides a new way to study the trajectory planning, that is, firstly determine the SPs of the UAV, and then construct the trajectory.

In this letter, we assume that ground terminals exist in the form of D2D and a UAV serving as aerial BS. The UAV's task is to cover all D2D users (DUs). We propose a trajectory planning algorithm by utilizing K-means and geometric relationship. In the proposed algorithm, the initial cluster centers obtained by K-means is used as the initial SPs of UAV and then update the number of SPs by judging the positional relationship between DUs and SPs to complete the trajectory planning. Finally, simulation results prove the validity of the algorithm.

2 System Model and Problem Formulation

We study a UAV-assisted network with underlaid D2D communications, where a UAV acts as an aerial BS to provide wireless connections to terrestrial DUs as shown in Fig. 1. The DUs exist in the form of a homogeneous Poisson Point Process (HPPP) φ_D with density λ_D, which are denoted by the set $\mathcal{N} = \{1, 2, \cdots, N\}$ and at known locations given by $\{\mathbf{w}_n\}_{n\in\mathcal{N}}$, where $\mathbf{w}_n \in \mathbb{R}^{2\times 1}$ denotes the two-dimensional (2D) coordinates of the n-th DU. We consider that the DUs are motionless and known for the UAV trajectory planning [7].

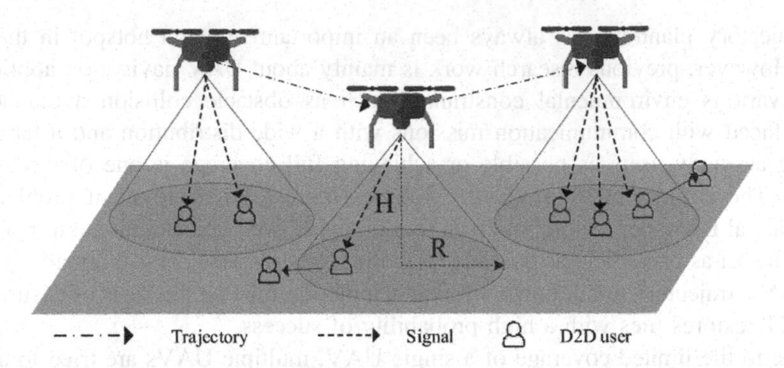

Fig. 1. A UAV based wireless communication system, where a UAV acts as a flying BS and GTs appear as D2D pairs.

We aim to design the trajectory of the UAV to achieve the purpose of covering all ground DUs. The trajectory can be discrete into several line segments. The two endpoints of each track are the stop points of the UAV. The idea of this paper is to construct the UAV trajectory by finding the right SPs.

Denoting by $\mathcal{M} = \{1, 2, \cdots, M\}$ the set of SPs, UAV can cover multiple DUs at each SP. Therefore, the trajectory planning can be transformed into an optimization problem of the locations of the SPs.

Assume that the UAV starts from the origin and flight height is H in meter. $\mathbf{w}_U[m] \in \mathbb{R}^{2\times 1}$ denotes the horizontal coordinates of SPs. There is no need to go back to the origin after completing the coverage. Therefore, the distance between m-th SP and n-th DU is written as

$$d_n[m] = \sqrt{H^2 + \|\mathbf{w}_U[m] - \mathbf{w}_n\|^2} \tag{1}$$

Due to the high altitude characteristics of UAVs, we consider that the UAV-DU communication channels are dominated by LoS links. Under the LoS model, UAV-GT link distance is the dominant factor for the air-to-ground (A2G) channel power gain. The average channel power gain from the UAV to n-th DU at m-th SP can be modeled as

$$\beta_n[m] = \beta_0 d_n^{-\alpha}[m] = \frac{\beta_0}{\left(H^2 + \|\mathbf{w}_U[m] - \mathbf{w}_n\|^2\right)^{\alpha/2}} \tag{2}$$

Where the meaning of β_0 is the channel power gain at the reference distance which is 1 m, and α is the path loss exponent.

The transmitting power of the UAV is denoted by P_U. The received signal-to-noise ratio (SNR) by DU n is given by

$$\gamma_n[m] = \frac{P_U\beta_n[m]}{\sigma^2} = \frac{\gamma_0^U}{(H^2 + \|\mathbf{w}_U[m] - \mathbf{w}_n\|^2)^{\alpha/2}} \tag{3}$$

where σ^2 denotes the additive white Gaussian noise (AWGN) power and $\gamma_0^U \triangleq \frac{P_U\beta_0}{\sigma^2}$ is the SNR at the reference distance.

Define the threshold of SNR at the DU is γ_{th}, the maximum transmitting power of the UAV is $P_U^{\max}$. Then, we have the maximum coverage radius of the UAV.

$$R_C^* = \sqrt{\left(\gamma_0^{U\max}/\gamma_{\text{th}}\right)^{2/\alpha} - H^2} \tag{4}$$

We aim to minimize the number of UAV SPs while each DU is covered by UAV at least once within its communication radius R_C^*. This does not rule out the possibility that some DUs will be covered by UAV multiple times. The problem can be formulated as follows

$$(\text{P1}): \begin{cases} \min_{\{\mathbf{w}_U\}_{m\in\mathcal{M}}} & |\mathcal{M}| \\ \text{s.t.} & \min_{m\in\mathcal{M}}\|\mathbf{w}_U[m] - \mathbf{w}_n\| \le R_C^*, \quad \forall n \in \mathcal{N} \end{cases} \tag{5}$$

where $|\mathcal{M}| = M$ denotes the cardinality of the set $\mathcal{M}$ and the Euclidean norm $\|\mathbf{w}_U[m] - \mathbf{w}_n\|$ is the distance between n-th DU and m-th SP projected on the ground plane.

Since the coverage of the UAV can be regarded as a disk, so the problem (P1) is also called the geometric disk coverage problem [11]. The goal is to minimize the total number of disks on top of ensuring that each user is covered by at least one disk. Usually, the above problem is an NP problem [3].

3 Proposed Solution

To facilitate solving this problem, we propose an efficient heuristic algorithm based on K-means algorithm. The main idea is to take the locations of cluster centers obtained through K-means as the initial UAV SPs, and then determine the coverage of UAV. If the coverage is completed, the UAV trajectory is established according to the minimum path selection principle; otherwise, add new SPs to further coverage and update the UAV trajectory.

3.1 K-Means Algorithm

K-means [12] is a cluster-based clustering algorithm. Assuming that there is no label data set as (6), the task of the algorithm is to cluster the data set into K clusters $C = C_1, C_2, \ldots, C_K$.

$$X = \begin{bmatrix} x^{(1)} \\ x^{(2)} \\ \vdots \\ x^{(m)} \end{bmatrix} \tag{6}$$

In the K-means algorithm, Euclidean distance is used to measure the similarity between data. In other words, the smaller the distance between data, the higher the data similarity. At the same time, the denser the data distribution, the greater the possibility of forming clusters. Therefore, the sum of squared errors (SSE) is used as the objective function to measure the clustering quality, as shown below [4],

$$E = \sum_{i=1}^{K} \sum_{x \in C_i} \|x - \mu_i\|^2 \tag{7}$$

where μ_i is the center of C_i, as shown in formula (8).

$$\mu_i = \frac{1}{|C_i|} \sum_{x \in C_i} x \tag{8}$$

In this paper, $\{\mathbf{w}_n\}_{n \in \mathcal{N}}$ is the no label data set and initial SPs is equivalent to the clusters.

3.2 Trajectory Planning Base on K-Means Algorithm

Select the Initial Cluster Centers. The two most critical parameters in K-means algorithm are K-value and initial cluster centers. For the purpose of better observation of SP selection process, the value of K is increased from 3 until the initial SPs can achieve full coverage. In order to acquire more dispersed initial cluster centers and avoid local optimization of the algorithm, we use formula (9) to confine the distance between the initial cluster centers [13],

$$d_\gamma = \sqrt{\frac{S}{K}} \tag{9}$$

where the area of the research region is denoted by S. In other words, the distance between the initial cluster centers should satisfy $d_{ij} \geq d_\gamma$.

Stop Point Selection Strategy. When the initial SPs obtained through K-means cannot achieve full coverage, that is, when there are DUs who cannot effectively communicate with the UAV, it is necessary to determine whether new SPs need to be added according to the users' locations, and if necessary, calculate the locations of SPs.

Using D2D communication without adding new SP. As shown in Fig. 2, DU_2 is uncovered and DU_1 is covered. In this case, DU_2 can get the content from DU_1 without adding new SP, thereby shortening the trajectory.

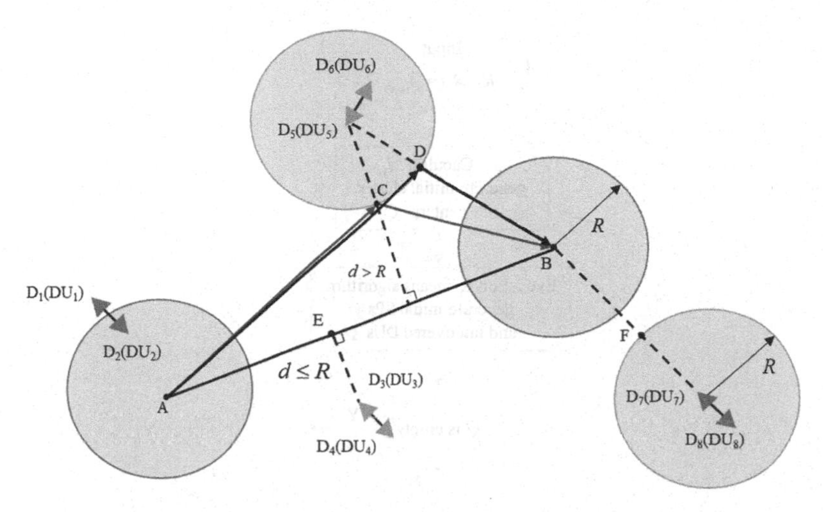

Fig. 2. Illustration of stop point selection strategy. Both A and B are stop points and triangles represent D2D users.

Adding new SP, Trajectory Unchanged. Both DU_3 and DU_4 are uncovered in Fig. 2, a new SP is required to add. It can be seen from Fig. 2 that DU3 is closer to AB than DU_4, and the distance is less than R. In this case, we only need to find the E, which $D_3E \perp AB$. Namely, E is the newly added SP.

Adding new SP, Trajectory Changed. The distance between the 3rd pair and the 4th pair DUs to AB is greater than R. If uncovered users located like the 3rd pair DUs, firstly connect D_5 to B, and perpendicular to AB from D_5. Then make a circle with D_5 as the center and R as the radius, and find the intersection of the above two lines and the circle, i.e., C and D in Fig. 2. Finally, compare the lengths of AC + CB and AD + DB, and the position (C or D) corresponding to the trajectory with the minimum distance is adopted as the new SP. While B is the last SP, there is no need to compare the lengths of the polyline segment. Instead, choose F as the new SP.

Minimum Path Selection Strategy. After obtaining all SPs, how to construct the trajectory is also an important issue. In order to make the trajectory length as small as possible, we adopt the minimum path selection strategy, that is, each SP chooses the nearest SP as the next SP.

Specific Planning Steps. The procedures of the algorithm are shown in Fig. 3 and specific planning steps are in follows:

Step 1. According to K and d_γ, determine the initial cluster centers.

Step 2. K-means algorithm is executed through the initial cluster centers obtained in step 1, and then the initial SPs is obtained.

Step 3. Place the UAV at the positions of the initial SPs to determine the coverage. If the full coverage has been completed, turn to step 5, otherwise proceed to step 4.

Step 4. Determine the relationship between uncovered users and the initial SPs, and then calculate the locations of the newly added SPs according to Stop Point Selection Strategy.

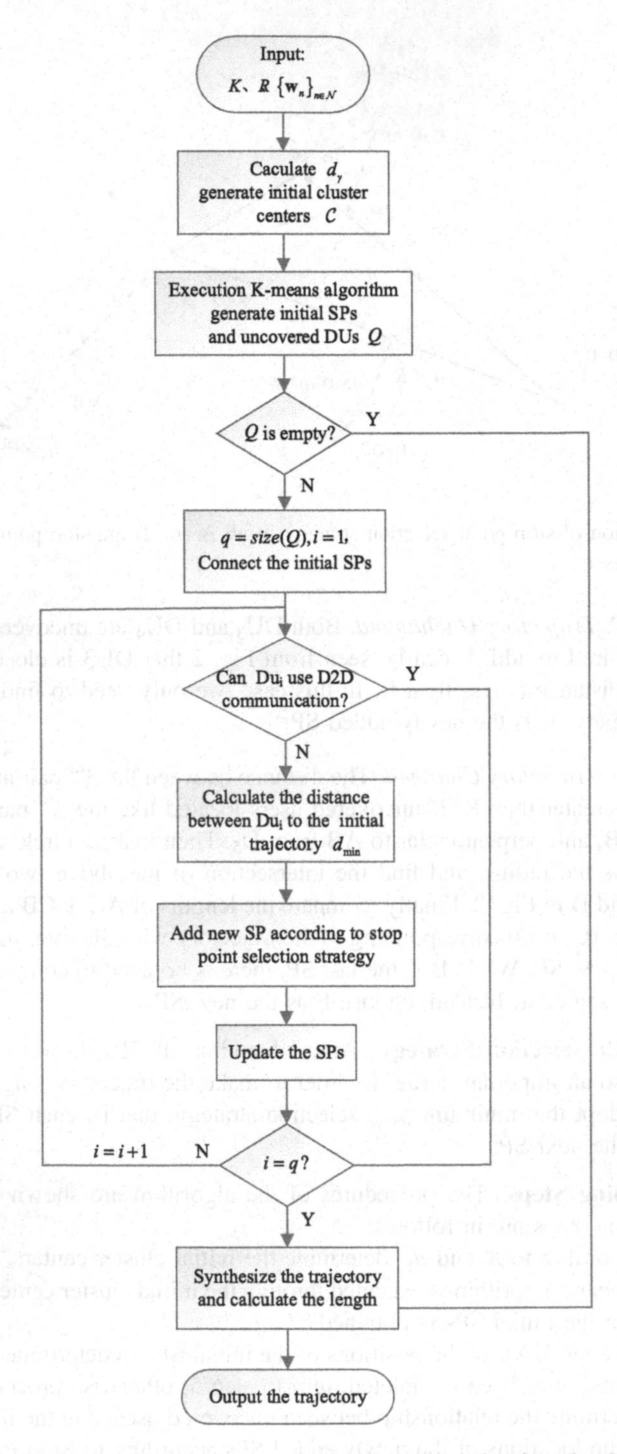

Fig. 3. The flow chart of the proposed trajectory planning algorithm

Step 5. Synthesize the UAV trajectory according to the minimum path selection strategy, calculate and compare the changes of the trajectory.

4 Simulation and Analysis

In this section, simulation results are illustrated to evaluate the performance of the proposed trajectory design algorithm. The D2D pairs are assumed to distribute in a square region of area 4 km^2 and the UAV flight altitude is $H=100$ m, which generally complies with the rules set by the FAA [14]. According to (4), the maximum coverage radius R_c^* is 430 m. To facilitate calculation, we adopt 400 m as the UAV coverage radius in the simulation. To verify the validity of the strategy,we analysis the influence of distribution of DUs and the number of initial cluster centers on trajectory planning.

4.1 Trajectory Planning in Different Distribution Scenarios

The UAV trajectory planning results for four different D2D pair distributions are shown in Fig. 4. Obviously, the initial trajectory obtained through K-means may not be able to achieve full coverage. If there are users who have not been covered, it is necessary to determine whether it is necessary to add a new SP or change the trajectory.

Case 1: Full Coverage after K-Means. The first trajectory planning scene is shown in Fig. 4a), only one D2D user has not been covered by UAV in the initial trajectory. Owe to D2D communication, the uncovered user can obtain contents from corresponding D2D user. If the distribution of DUs is shown like Fig. 4b), the initial trajectory can achieve full coverage when $K=4$.

Case 2: Adding Stop Points, Trajectory Remains Unchanged. As shown in Fig. 4b), four DUs have not be covered by UAV after K-means, of which two users can apply the D2D link, while the other pair of DUs need to add SPs to be covered. Since the distance between the pair of DUs and trajectory of the first stage is less than R, the coverage can be achieved by adding a SP to the original trajectory.

Case 3: Adding Stop Points, Trajectory Changes. The case shown in Fig. 4(d) is slightly more complicated than the above. The distances between DUs who have not been covered and UAV's initial trajectory are greater than R, which means that the UAV cannot add SPs on the initial trajectory to achieve coverage. Therefore, it is necessary to find locations near the initial trajectory where the distance from the uncovered users is R. Due to the addition of new SPs, the UAV's trajectory has changed and the total length of the trajectory has also increased.

4.2 Trajectory Planning in the Same Distributed Scenario

In this subsection, we analyze the influence of the number of initial cluster centers on UAV trajectory planning for the same DUs. In scenarios where the density of DUs is relatively high, the selection of K is vital for UAV trajectory planning.

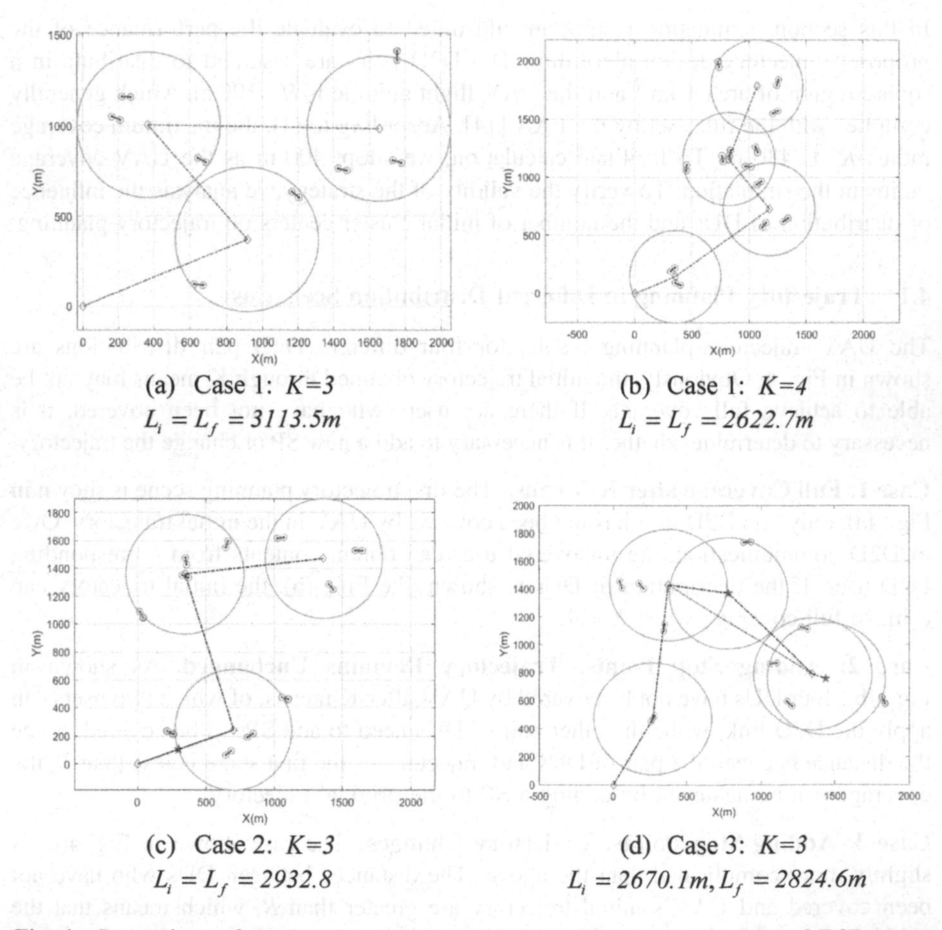

Fig. 4. Comparison of the UAV trajectories with four different distributions of D2D users. Small hollow circles represent D2D pairs, in which pink denotes they have been covered by the initial trajectory, and black denotes conversely. The red dotted line represents the initial trajectory, and the black dotted line represents the final trajectory. Red diamonds represent SPs, and green circles represent the UAV coverage boundary. Stars represent the added SPs. (Color figure online)

We consider a benchmark planning, called "strip-based waypoints", where the UAV's trajectory is devised to realize full coverage [7]. First, obtain the smallest rectangle containing all DUs, and then divide the area into a plurality of rectangular strips with width R. Finally, determine the location of each SP, and the UAV moves along the rectangular strip, as shown in Fig. 5.

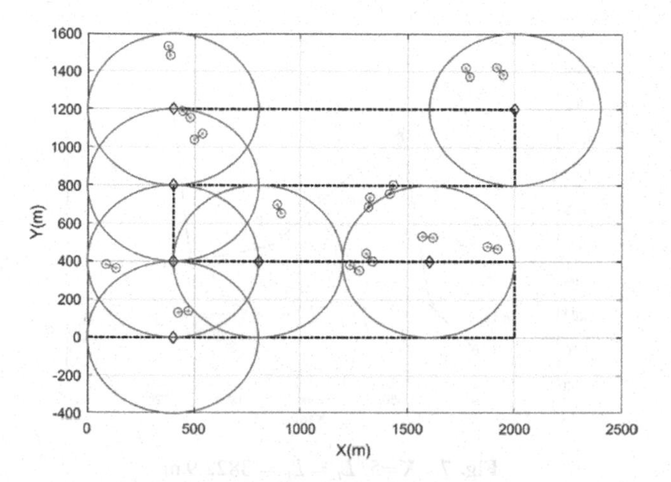

Fig. 5. Strip-based waypoints, $L = 8000\,\mathrm{m}$

From Fig. 6 and Fig. 7, we can clearly observe the influence of K on the UAV trajectory. As K increases, the total length of the initial trajectory and the final trajectory of the UAV increases. Compared with the benchmark scheme, the K-means-based UAV trajectory planning algorithm we proposed has obvious advantages. Regardless of the number of SPs or the length of the trajectory, our proposed algorithm is smaller than the benchmark scheme.

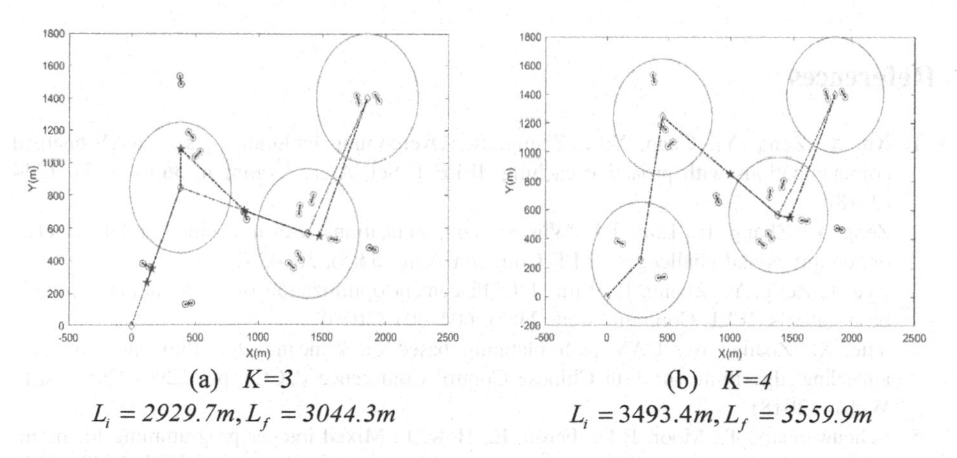

(a) K=3 $L_i = 2929.7m, L_f = 3044.3m$

(b) K=4 $L_i = 3493.4m, L_f = 3559.9m$

Fig. 6. Comparison of the UAV trajectories with K=3, 4. The meaning of all shapes is consistent with Fig. 3.

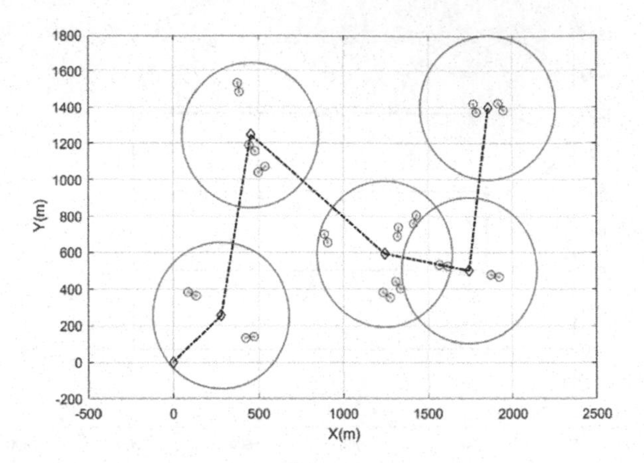

Fig. 7. K=5, $L_i = L_f = 3822.9\,\text{m}$

5 Conclusion

This letter proposed a trajectory planning algorithm based on K-means for a UAV-assisted network with underlaid D2D communications. In our algorithm, the initial stop points are obtained through the K-means algorithm and then determine the coverage. If the coverage is not complete, then D2D communication is selected or the trajectory is changed by adding the new stop points. Compared with the known benchmark planning, the proposed design algorithm has good advantages in terms of trajectory length.

References

1. Xu, X., Zeng, Y., Guan, Y.L., Zhang, R.: Overcoming endurance issue: UAV-enabled communications with proactive caching. IEEE J. Sel. Areas Commun. **36**(6), 1231–1244 (2018)
2. Zeng, Y., Zhang, R., Lim, T.J.: Wireless communications with unmanned aerial vehicles: opportunities and challenges. IEEE Commun. Mag. **54**(5), 36–42 (2016)
3. Lyu, J., Zeng, Y., Zhang, R., Lim, T.J.: Placement optimization of UAV-mounted mobile base stations. IEEE Commun. Lett. **21**(3), 604–607 (2016)
4. Yue, X., Zhang, W.: UAV path planning based on k-means algorithm and simulated annealing algorithm. In: 37th Chinese Control Conference (CCC), pp. 2290–2295. IEEE, Wuhan (2018)
5. Schouwenaars, T., Moor, B.D., Feron, E., How, J.: Mixed integer programming for multi-vehicle path planning. In: 2001 European control conference (ECC), pp. 2603–2608. IEEE, Karlsruhe (2001)
6. Bor-Yaliniz, R.I., El-Keyi, A., Yanikomeroglu, H.: Efficient 3-D placement of an aerial base station in next generation cellular networks. In: 2016 IEEE International Conference on Communications (ICC), pp. 1–5. IEEE, Kuala Lumpur (2016)
7. Zeng, Y., Xu, X., Zhang, R.: Trajectory design for completion time minimization in UAV-enabled multicasting. IEEE Trans. Wireless Commun. **17**(4), 2233–2246 (2018)

8. Galkin, B., Kibilda, J., DaSilva, L.A.: Deployment of UAV-mounted access points according to spatial user locations in two-tier cellular networks. In: 2016 Wireless Days (WD), pp. 1–6. IEEE, Toulouse (2016)
9. Mozaffari, M., Saad, W., Bennis, M., Debbah, M.: Unmanned aerial vehicle with underlaid device-to-device communications: performance and tradeoffs. IEEE Trans. Wireless Commun. **15**(6), 3949–3963 (2016)
10. Ji, J., Zhu, K., Niyato, D., Wang, R.: Probabilistic cache placement in UAV-assisted networks with D2D connections: performance analysis and trajectory optimization. IEEE Trans. Commun. (2020). https://doi.org/10.1109/TCOMM.2020.3006908
11. Srinivas, A., Zussman, G., Modiano, E.H.: Construction and maintenance of wireless mobile backbone networks. IEEE/ACM Trans. Netw. **17**(1), 239–252 (2009)
12. Yang, Y., Wang, Z.: Robotic obstacle avoidance control based on improved artificial potential field method and its realization by MATLAB. J. Univ. Shanghai Sci. Technol. **35** (05), 496–500 (2013)
13. Zhou, Y., et al.: Joint distribution center location problem for restaurant industry based on improved K-Means algorithm with penalty. IEEE Access **8**, 37746–37755 (2020)
14. Federal Aviation Administration (FAA).: Summary of small unmanned aircraft rule (part 107) (2016)

A Multi-source Fused Location Estimation Method for UAV Based on Machine Vision and Strapdown Inertial Navigation

Jiapeng Li[1], Shuo Shi[1,2](✉), and Xuemai Gu[1,3]

[1] School of Electronic and Information Engineering, Harbin Institute of Technology, Harbin, Heilongjiang 150001, China
lijiapeng_atlantis@163.com, crcss@hit.edu.cn
[2] Network Communication Research Centre, Peng Cheng Laboratory, Shenzhen, Guangdong 518052, China
[3] International Innovation Institute of HIT in Huizhou, Huizhou, Guangdong 516000, China

Abstract. In recent years, unmanned aerial vehicle (UAV) technology has been widely used in industry, agriculture, military and other fields, and its positioning problem has been a research hotspot in this field. To solve the problem of invalidation of integrated navigation of global positioning system (GPS) and strapdown inertial navigation system (SINS) in indoor and other areas, this paper presents a multi-source information fusion location algorithm based on machine vision positioning and SINS. Based on image coordinate system (ICS), body coordinate system (BCS) and navigation coordinate system (NCS), combined with AprilTags recognition and positioning technology, this paper builds NCS with AprilTags array to get the position observation of UAV. Based on the idea of multi-source information fusion, this paper applied third-order fused complementary filter algorithm, which combines with the SINS to obtain accurate three-axis speed and position estimation. Finally, the reliability is verified by the test of the UAV experimental platform.

Keywords: Unmanned aerial vehicle · Strapdown inertial navigation system · Multi-Source information fusion

1 Introduction

As a new member of small Unmanned Aerial Vehicles (UAVs), quad-rotor UAVs have many advantages, such as small size, flexible flight, vertical takeoff and landing, hovering at fixed points, and portability. It has been widely used in military surveillance, disaster prediction, agricultural mapping and civil life, and has gradually become a hot topic for researchers and scholars.

A stable control and execution system is a prerequisite for the normal operation of an unmanned aerial vehicle. Accurate, low-latency, low-noise estimation for attitude, speed and position are necessary for the normal operation of the controller. So far, no sensor has been able to measure the flight attitude, speed and position of an UAV at anytime and anywhere with precision and no delay in the navigation coordinate system.

S. Shi et al. (Eds.): AICON 2020, LNICST 356, pp. 272–282, 2021.
https://doi.org/10.1007/978-3-030-69066-3_24

There are many sensors that can be used to estimate the flight status of an UAV, but they have different working principles, measuring objects, working conditions and data delays. Therefore, it is difficult to estimate the flight status of an UAV accurately, with low delay and low noise through a single sensor or through multiple sensors without any processing.

In recent years, the concept of multi-source information fusion has been proposed, and the location method for UAV based on this concept has gradually become a research hotspot in this filed. Based on an inertial measurement unit (IMU) consisting of a three-axis gyroscope and accelerometer [1], as well as an array of magnetic angular rate and gravity (MARG) sensors including a three-axis magnetometer, the direction of gyroscope measurement error is calculated as a quaternion derivative, and accelerometer and magnetometer data are allowed to be used to analyze. Therefore, reliable estimation of UAV flight attitude can be achieved. The strapdown inertial navigation system has periodic oscillation error in pure inertial navigation mode, which seriously affects the navigation accuracy. In [2], according to the principle of equivalence, the external horizontal damping network of SINS is designed, and the periodic oscillation is suppressed by the difference between the velocity of the system itself and the velocity of the electromagnetic log, which improves the accuracy of the system.

It is difficult to achieve accurate and reliable state estimation by IMU alone. In [3], the accuracy of the airborne GPS in a static environment is evaluated and its availability in low-cost projects is demonstrated. The combination of GPS and strapdown inertial navigation is a feasible method for state estimation. However, the update frequency of GPS is much lower than that of SINS. The two streams are out of sync, which affects navigation accuracy. In [4], a digital high-pass filter is used to pre-filter the measured signal and to filter the Schuler period of the difference between SINS and GPS discrete velocity, which greatly improves the navigation accuracy. At present, the common method of SINS/GPS integrated navigation system is based on ground speed, which has some limitations and is interfered by abnormal measurements. In [5], a dynamic coarse alignment method for SINS/GPS integrated system based on location track is presented, which is proved to be more robust than the current popular methods through simulation and measurement. Aiming at the integration of SINS with GPS and the possible violation of Gaussian assumption of process noise [6], a new process uncertainty robust Student's t-based Kalman filter for process uncertainty is presented, and its robustness in suppressing process uncertainty is proved.

However, in many cases, GPS does not work properly, and SINS requires other location observation sensors to participate in multisource information fusion. By calculating the time difference of arrival (TDOA) of the transmitted signal [7], a new positioning method based on multipoint positioning is proposed to replace the GPS positioning method. In [8], a colored noise model is proposed using the received signal strength (RSS) by the onboard communication module and applied to the extended Kalman filter (EKF) for distance estimation. In [9], a particle swarm optimization algorithm is proposed for wireless self-organizing sensor networks. The UAV location search model is described as a constrained optimization problem of a multi-objective utility function to dynamically obtain the optimal location of multi-UAVs.

Machine vision-based indoor positioning of robots, including UAV positioning, is also a hot research direction. The abundant parallel lines and corner points on the ceiling can be used as visual positioning features for indoor mobile robots. In [10], based on the natural characteristics of the ceiling, a new visual positioning method is proposed, and its validity is verified by error analysis and experiments. Redundant navigation systems are essential for the safe operation of UAVs in high-risk environments. In [11], a visual-based path tracking system is proposed for the autonomous and safe return of UAVs under major navigation failures such as GPS interference.

Based on the above observation, this paper aims to designing an indoor navigation algorithm for UAV based on multi-source information fusion of machine vision positioning and strapdown inertial navigation. Specifically, the focus of this study includes:

i. Based on the ground positioning tag array and the flight attitude of the UAV, a real-time positioning method based on the onboard visual unit is constructed, and the estimation of yaw is further corrected.
ii. Visual localization has some problems, such as slow update rate, high delay, easy mutation and failure. Combining strapdown inertial navigation with advantages of low latency, high update rate and long-time stability, an optimal estimation of the UAV's attitude, speed and position can be obtained by means of multi-source information fusion. The effectiveness of the method is proved by the actual measurement.

2 Inertial Navigation

Strapdown inertial navigation requires a mathematical platform coordinate system constructed from the flight attitude angle of the UAV, and then the platform inertial navigation estimation is performed. SINS is built on the body coordinate system (BCS) and navigation coordinate system (NCS). The rotation matrix R_b^n describes the process that the coordinate of a point changing from BCS to NCS. Defining pitch (θ), roll (γ) and yaw (ψ) as the angles of rotation about X, Y, Z axes, we can obtain R_b^n and R_n^b as follows:

$$R_b^n = \begin{bmatrix} \cos\gamma\cos\psi & \sin\theta\sin\gamma\cos\psi - \cos\theta\sin\psi & \cos\theta\sin\gamma\cos\psi + \sin\theta\sin\psi \\ \cos\gamma\sin\psi & \sin\theta\sin\gamma\sin\psi + \cos\theta\cos\psi & \cos\theta\sin\gamma\sin\psi - \sin\theta\cos\psi \\ -\sin\gamma & \sin\theta\cos\gamma & \cos\theta\cos\gamma \end{bmatrix} \tag{1}$$

$$R_n^b = \left(R_b^n\right)^T = \left(R_b^n\right)^{-1} \tag{2}$$

The acceleration vector measured by the onboard IMU is rotated to NCS by a fictitious mathematical coordinate system. The acceleration of the body motion is expressed on NCS by platform inertial navigation, and then the speed and position estimation of the UAV is obtained by integrating as follows:

$$a_{vn} = \left(\lambda_R + R_b^n\right)(a_{mb} + \lambda_m) - \lambda_g g_n \tag{3}$$

$$\ddot{s}_n = \dot{v}_n = a_{vn} \tag{4}$$

Where a_{vn} denotes motion acceleration vector in NCS, a_{mb} denotes measuring acceleration vector in BCS, g_n denotes gravity acceleration vector in NCS, λ_R denotes the error component relative to real rotation matrix, λ_m denotes the measurement error of IMU, λ_g denotes rate of variation of gravity acceleration with height, s_n and v_n respectively denote speed and position in NCS.

It can be seen from (3) that although SINS can provide three-axis speed and position estimation during the flight of an UAV, it is difficult to conform to the real results due to various errors and interference.

3 Visual Orientation

AprilTags is a visual reference library, which can quickly realize 3D positioning and inclination measurement (see Fig. 1), and is widely used in robot positioning technology. Considering the power consumption and load-carrying capacity of small UAVs, this paper creates an AprilTags visual localization algorithm based on monocular vision and AprilTags array.

Fig. 1. AprilTags location and recognition

3.1 AprilTags Location Algorithm

AprilTags identification and positioning system [12] is mainly composed of tag detector and coding system. The tag detector is used to locate the tag and detect the tilt angle. The encoding system is used to extract the information contained in the tag.

AprilTags recognition and positioning is mainly done in three steps [13]. Firstly, calculate the gradient of each pixel and further detect the segments in the image. Secondly, a depth-first search method based on depth 4 is used to find quadrilateral in the image. Compare the standard libraries to determine if tags exist in the quadrilateral and the ID of the tag. Thirdly, the isotropic matrix is calculated by direct linear transformation [14], and the relative position and rotation angle between the label and the camera are calculated.

3.2 AprilTags Array Location Algorithm

In this paper, several AprilTags are arranged into a two-dimensional array at regular equal distances (see Fig. 2). According to the tag position and ID number in the field of view, as well as the UAV attitude angle, the relative position relationship is constructed (see Fig. 3).

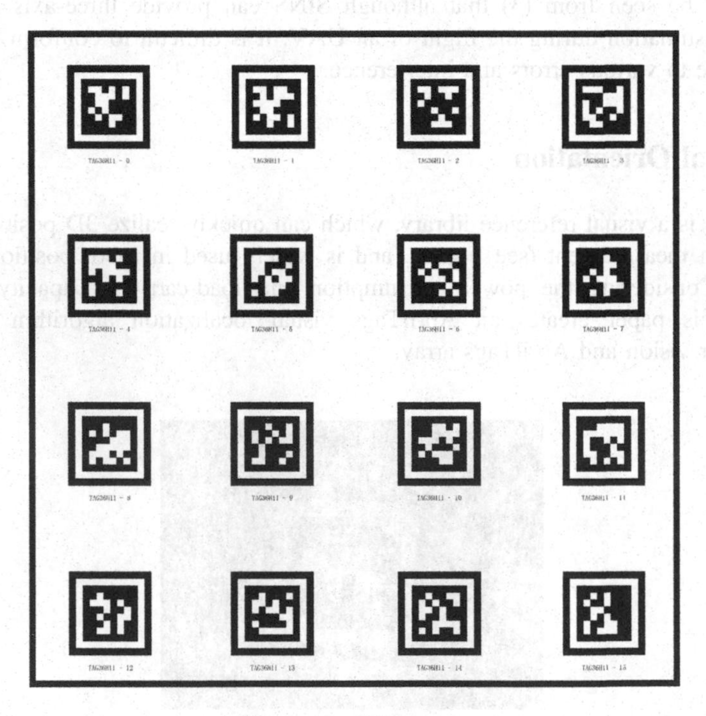

Fig. 2. AprilTags array

In Fig. 3, $OXYZ_s$ denotes horizontal body coordinate system (HBCS), $OXYZ_b$ denotes BCS, α denotes the AprilTags array plane, β denotes the image plane with Z axis of BCS as normal, T denotes the center of the identified AprilTag, A denotes the projection of camera normal on β, B denotes the projection of UAV on α. The position deviation between tag center and image center can be represented by orthogonal T_iC and CA.

Assuming that the angle between vector $\vec{O}\,C$ and vector $\vec{O}\,P$ is θ_T, the angle between vector $\vec{O}\,C$ and vector $\vec{O}\,T$ is γ_T, and the positive direction of rotation is defined by the right-handed helix rule, the relationship can be obtained as below:

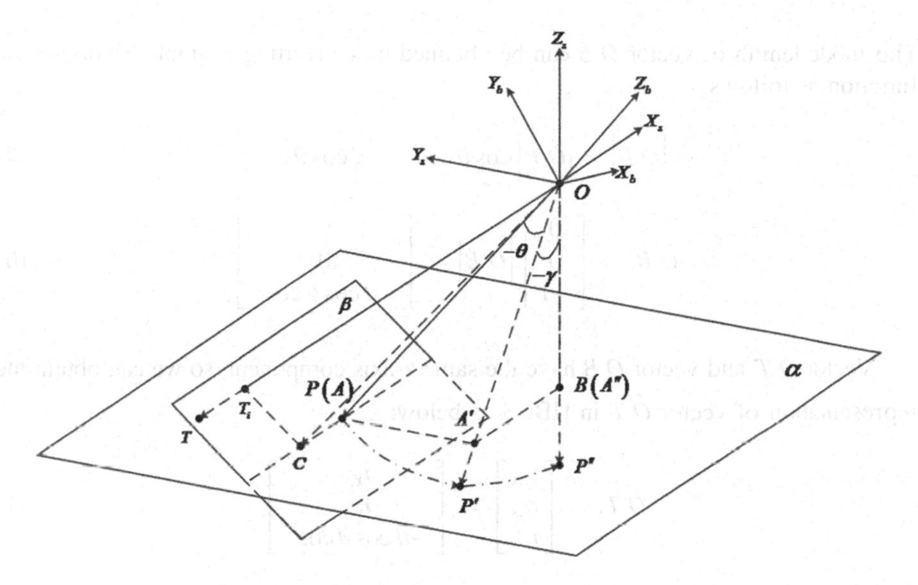

Fig. 3. Image coordinate conversion algorithm

$$\frac{\vec{O}\,T}{\left|\vec{O}\,T\right|_b} = R_{x,b}\left(\delta\left|\vec{C}\,P\right|\right)R_{y,b}\left(\delta\left|\vec{T_i}\,C\right|\right)\frac{\vec{O}\,P}{\left|\vec{O}\,P\right|_b} = R_{x,b}(\theta_T)R_{y,b}(\gamma_T)\frac{\vec{O}\,P}{\left|\vec{O}\,P\right|_b} \tag{5}$$

where δ is a constant greater than zero, which is determined by the camera resolution and viewing angle. Vector $\vec{O}\,T$ can be expressed as below:

$$\frac{O\,T}{\left|\vec{O}\,T\right|_b} - R_{x,b}(\theta_T)R_{y,b}(\gamma_T)\begin{bmatrix}0\\0\\-1\end{bmatrix} = \begin{bmatrix}-\sin\gamma_T\\ \sin\theta_T\cos\gamma_T\\ -\cos\theta_T\cos\gamma_T\end{bmatrix} \tag{6}$$

Vector $\vec{O}\,T$ and $\vec{O}\,B$ can be expressed respectively as follows:

$$\frac{\vec{O}\,T}{\left|\vec{O}\,T\right|_s} = R_{z,s}(\psi)R_{y,s}(\gamma)R_{x,s}(\theta)\frac{\vec{O}\,T}{\left|\vec{O}\,T\right|_b} \tag{7}$$

$$\frac{\vec{O}\,B}{\left|\vec{O}\,B\right|_s} = \begin{bmatrix}0\\0\\-1\end{bmatrix} \tag{8}$$

Vector $\vec{O}\,B$ points to the negative direction of z-axis of HBCS. Its mode length is measured directly by the laser ranging module of the UAV, which is denoted as d.

The mode length of vector $\vec{O}B$ can be obtained by converting a simple trigonometric function as follows:

$$\left|\vec{O}B\right| = \left|\vec{O}P\right| \cos\theta \cos\gamma = d\cos\theta\cos\gamma \tag{9}$$

$$\vec{O}B_s = \begin{bmatrix} 0 \\ 0 \\ -1 \end{bmatrix} \left|\vec{O}B\right| = \begin{bmatrix} 0 \\ 0 \\ -d\cos\theta\cos\gamma \end{bmatrix} \tag{10}$$

Vector $\vec{O}T$ and vector $\vec{O}B$ have the same z-axis component, so we can obtain the representation of vector $\vec{O}T$ in HBCS as below:

$$\vec{O}T_s = \begin{bmatrix} t_{x,s} \\ t_{y,s} \\ t_{z,s} \end{bmatrix} = \begin{bmatrix} t_{x,s} \\ t_{y,s} \\ -d\cos\theta\cos\gamma \end{bmatrix} \tag{11}$$

Assuming that the actual coordinate of point T in the array is $T_\alpha(t_{x,n}, t_{y,n}, 0)$, then the coordinate of point B in the array can be expressed as $B_\alpha(t_{x,n} - t_{x,s}, t_{y,n} - t_{y,s}, 0)$, and the position of the UAV can be obtained as below:

$$\vec{p}_n = \begin{bmatrix} t_{x,n} - t_{x,s} \\ t_{y,n} - t_{y,s} \\ d\cos\theta\cos\gamma \end{bmatrix} \tag{12}$$

4 Multi-source Information Fusion

4.1 Heading Direction Estimation

To ensure that the NCS defined by SINS is consistent with that defined by the AprilTags array, the heading angle of the UAV needs to be corrected by the rotation angle of AprilTags. In this paper, the error of two angles is used as compensation, the heading angle is corrected, and the estimation is obtained, which ensures the complete synchronization of the two NCS. The fusion estimation method is shown below:

$$\hat{\psi}(k) = \hat{\psi}(k-1) + \mu e(k) \tag{13}$$

$$e(k) = \begin{cases} e_1(k) \\ e_2(k) \\ e_3(k) \end{cases} = \begin{cases} \psi_m(k) - \psi(k-1) & , min\{abs[e(k)]\} = abs[e_1(k)] \\ \psi_m(k) - \psi(k-1) + 360 & , min\{abs[e(k)]\} = abs[e_2(k)] \\ \psi_m(k) - \psi(k-1) - 360 & , min\{abs[e(k)]\} = abs[e_3(k)] \end{cases} \tag{14}$$

where μ denotes a constant between 0 and 1 as correction rate, $e(k)$ denotes the error of two heading angles.

4.2 Velocity and Position Estimation

Velocity and position estimation of UAV obtained from SINS has the advantage of low delay, but it is difficult to directly apply to control system because of measurement error and interference. The location estimation of UAV obtained by the machine vision location algorithm proposed in Sect. 3 has the advantages of high delay and poor stability, and it is also difficult to apply directly to the control system. To solve this problem, third-order fused complementary filter algorithm is applied (see Fig. 4), which can get reliable speed and location estimates by complementing each other.

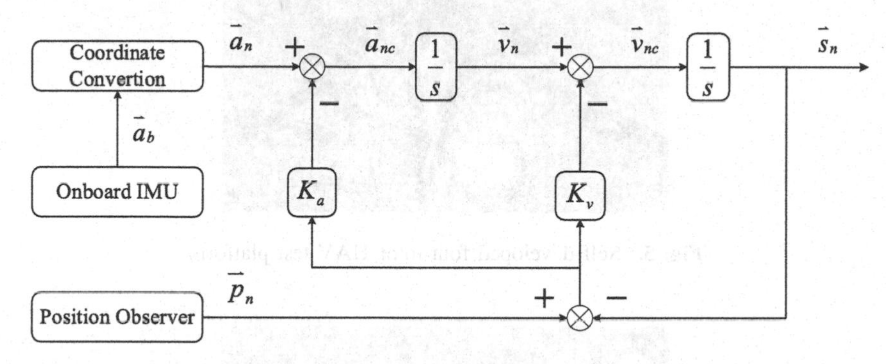

Fig. 4. Third-order fused complementary filter algorithm

In Fig. 4, $\vec{a}_n$ denotes the motion acceleration in NCS from SINS, $\vec{p}_n$ denotes the position estimation from location algorithm proposed in Sect. 3. The discrete iteration of this algorithm can be represented as follows:

$$a_n(k) = a_{vn}(k) + K_a(p_n(k) - s_n(k)) \tag{15}$$

$$v_n(k) = v_n(k-1) + a_n(k-1)T + K_v(p_n(k) - s_n(k)) \tag{16}$$

$$s_n(k) = s_n(k-1) + v_n(k-1)T + \frac{1}{2}a_n(k-1)T^2 \tag{17}$$

where T denotes the iteration period of the algorithm, $a_{vn}(k)$ denotes the motion acceleration in NCS at time k, $a_n(k)$, $v_n(k)$ and $s_n(k)$ respectively denote the estimation of acceleration, velocity and position of multi-source information fusion algorithm, K_a and K_v are both compensation factors (Fig. 5).

5 Experimental Testing

In this chapter, a planar array consisting of 16 AprilTags is designed according to the rules of Chapter 3. The positioning algorithm proposed in this paper is tested by a self-developed four-rotor UAV experimental platform, and the test results are actually sampled (see Fig. 7).

Since all three axes are visually located to obtain position estimates, this result can represent the X and Z axes. We can see in Fig. 7, Fig. 8 and Fig. 9 that location estimation after multi-source information fusion has faster response speed and almost the same positioning accuracy than machine vision location estimation. This is sufficient to demonstrate the reliability of the algorithm presented in this paper (Fig. 6).

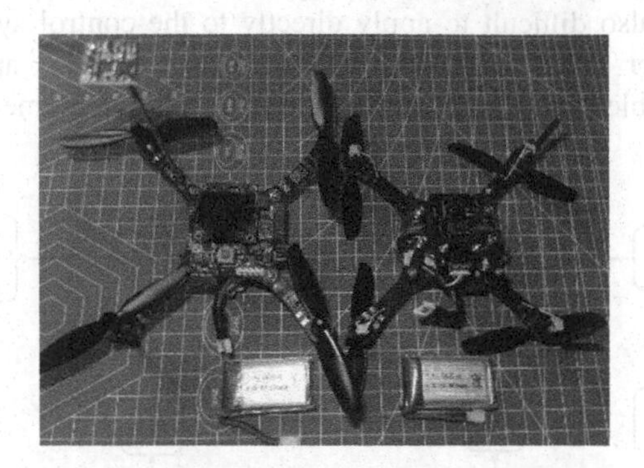

Fig. 5. Self-developed four-rotor UAV test platform

Fig. 6. UAV hovering in the AprilTags array area

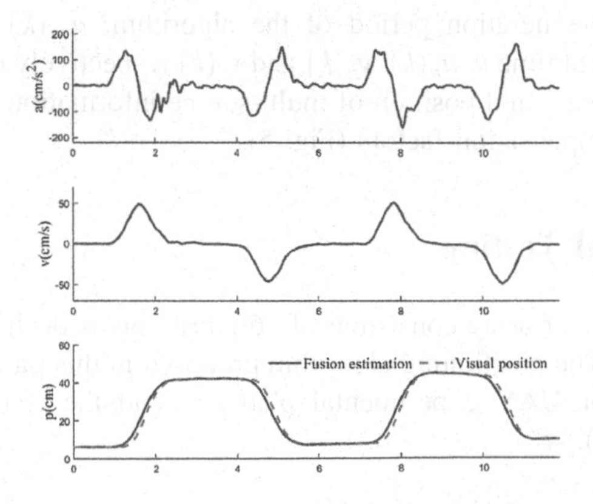

Fig. 7. Estimation of motion acceleration, velocity and position of x-axis

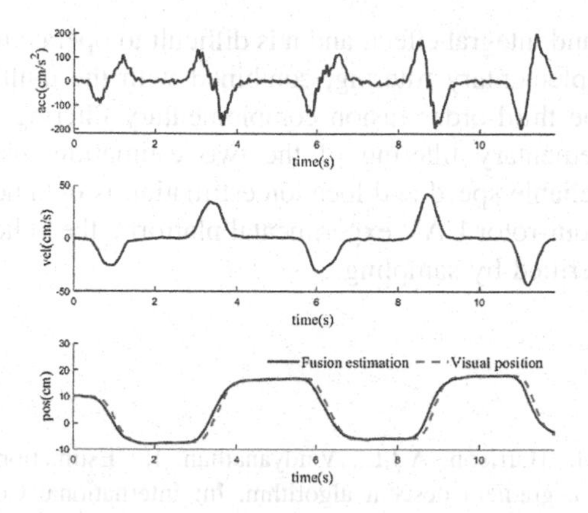

Fig. 8. Estimation of motion acceleration, velocity and position of y-axis

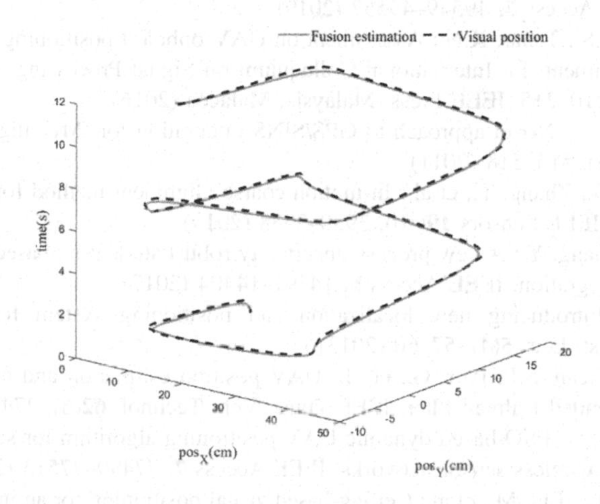

Fig. 9. Estimation of position of x-axis and y-axis over time

6 Conclusions

In this paper, a multi-source fusion location estimation method based on machine vision and strapdown inertial navigation is presented for indoor positioning of small UAVs with GPS invalid or unable to load. Based on AprilTags positioning and recognition technology, this paper defines NCS with AprilTags array, identifies tags with onboard camera, and obtains the spatial location of the UAV through the algorithm proposed in Sect. 3. Due to processor performance and power constraints, the location estimation obtained by the onboard image processing unit running the visual localization algorithm is more stable and not affected by the integral effect, but with high latency and unpredictable mutations, it is difficult to directly apply to the cascade controller. The position estimation of strap-down inertial navigation has fast response speed and strong anti-mutation ability, but it is susceptible to serous interference from measurement error and integral effect, and it is difficult to operate independently. Based on the idea of complementary filtering, combined with the multi-source information fusion algorithm, the third-order fusion complementary filtering algorithm is used to achieve the complementary filtering of the two estimation results, and the corresponding fast and reliable speed and location estimation is obtained. Finally, based on the self-developed four-rotor UAV experimental platform, the validity and reliability of the algorithm are verified by sampling.

References

1. Madgwick, S.O.H., Harrison, A.J.L., Vaidyanathan, R.: Estimation of IMU and MARG orientation using a gradient descent algorithm. In: International Conference on Rehabilitation Robotics, pp. 1–7. IEEE Press, USA, Piscataway (2011)
2. Deng, Z., Sun, J., Ding, F., et al.: A novel damping method for strapdown inertial navigation system. IEEE Access **7**, 49549–49557 (2019)
3. Kamarudin, S.S., Tahar, K.N.: Assessment on UAV onboard positioning in ground control point establishment. In: International Colloquium on Signal Processing & Its Applications (CSPA), pp. 210–215. IEEE Press, Malaysia, Malacca (2016)
4. Sun, W., Sun, F.: Novel approach to GPS/SINS integration for IMU alignment. Syst. Eng. Electron. **22**(3), 513–518 (2011)
5. Xu, X., Xu, D., Zhang, T., et al.: In-motion coarse alignment method for SINS/GPS using position loci. IEEE Sensors **19**(10), 3930–3938 (2019)
6. Huang, Y., Zhang, Y.: A new process uncertainty robust student's t based Kalman filter for SINS/GPS integration. IEEE Access **5**, 14391–14404 (2017)
7. Mazidi, E.: Introducing new localization and positioning system for aerial vehicles. Embedded Syst. Lett. **5**(4), 57–60 (2013)
8. Luo, C., McClean, S.I., Parr, G., et al.: UAV position estimation and collision avoidance using the extended Kalman filter. IEEE Trans. Veh. Technol. **62**(6), 2749–2762 (2013)
9. Na, H.J., Yoo, S.: PSO-based dynamic UAV positioning algorithm for sensing information acquisition in wireless sensor networks. IEEE Access **7**, 77499–77513 (2019)
10. Xu, D., Han, L., Tan, M., et al.: Ceiling-based visual positioning for an indoor mobile robot with monocular vision. IEEE Trans. Ind. Electron. **56**(5), 1617–1628 (2009)
11. Warren, M., Greeff, M., Patel, B., et al.: There's no place like home: visual teach and repeat for emergency return of multirotor UAVs during GPS failure. IEEE Robot. Autom. Lett. **4** (1), 161–168 (2019)
12. Li, Z., Chen, Y., Lu, H.: UAV autonomous landing technology based on apriltags vision positioning algorithm. In: Chinese Control Conference, pp. 8148–8153. IEEE Press, China, Guangzhou (2019)
13. Rivest, R.L., Leiserson, C.E.: Introduction to Algorithms. McGraw-Hill Inc., New York, USA (1990)
14. Olson, E.: AprilTag: a robust and flexible visual fiducial system. In: IEEE International Conference on Robotics and Automation, pp. 3400–3407. IEEE Press, China, Shanghai (2011)

A Summary of UAV Positioning Technology in GPS Denial Environment

Junsong Pu[1], Shuo Shi[1,2](✉), and Xuemai Gu[1,3]

[1] School of Electronic and Information Engineering,
Harbin Institute of Technology, Harbin 150001, Heilongjiang, China
crcss@hit.edu.cn

[2] Network Communication Research Centre, Peng Cheng Laboratory,
Shenzhen 518052, Guangdong, China

[3] International Innovation Institute of HIT in Huizhou,
Huizhou 516000, Guangdong, China

Abstract. In recent years, the capabilities of UAV systems have continued to improve, and they have emerged in military and civilian fields such as urban counter-terrorism reconnaissance, disaster monitoring, logistics distribution, and traffic diversion, and their application prospects are particularly broad. UAV positioning is a necessary link for UAVs to perform tasks and an important manifestation of UAV's autonomous capabilities. How to meet the positioning requirements of UAVs in environments with weak or no GPS signals such as urban buildings/forests/indoors has become a research hotspot in the UAV field. This paper introduces several UAV positioning methods that can work in GPS denial environment, analyzes their advantages and disadvantages, the current challenges in UAV positioning and finally looks forward to the future development.

Keywords: UAV · Positioning · GPS denial

1 Introduction

Unmanned aerial vehicles came out in the United Kingdom in 1917. They have significant characteristics such as no risk of casualties, low cost, light weight, good maneuverability, and strong concealment capabilities. Therefore, its development is highly valued by many countries. UAVs can be used in aerial photography, transportation, intelligent irrigation, etc., and have broad civilian prospects. At the same time, UAVs have played an important role in several high-tech local wars in recent years, and their military application value is equally significant.

Today, positioning, route planning, control, and environmental perception are still major problems in the field of drone technology. Among these problems, positioning is the primary link in autonomous flight and an important prerequisite for the successful completion of various tasks by drones. GPS positioning is currently the most commonly used method for outdoor drone positioning. However, in some mission scenarios, such as cities, forests, or environments with complex electromagnetic fields, GPS signals are blocked by tall obstacles or electromagnetic interference, and cannot

S. Shi et al. (Eds.): AICON 2020, LNICST 356, pp. 283–294, 2021.
https://doi.org/10.1007/978-3-030-69066-3_25

be used for drones. Provide continuous and effective positioning information. In addition, GPS itself also has accuracy problems. This poses a huge challenge for the UAV to achieve full and reliable positioning during the mission. Therefore, the UAV positioning technology that does not rely on GPS has important research value and has received widespread attention in recent years.

This paper introduces several UAV positioning methods that do not rely on GPS in popular research fields, analyzes their advantages and disadvantages, the current challenges in UAV positioning and finally looks forward to the future development.

2 Ground-Based Binocular Drone Positioning

According to different sensor configurations, the UAV positioning system can be divided into two working modes: ground-based and airborne. The ground-based positioning method is usually used in the UAV guided take-off and landing stage. The position and attitude data of the target UAV is obtained through a combination of ground-based sensors, and the image coordinates of the UAV target are extracted from it, and the spatial location of the UAV is calculated in real time according to the principle of vision measurement. At this time, as long as the ground servo turntable is driven according to the calculated space position of the drone, the camera can track the aerial drone target in real time. Ground-based visual positioning mode has been widely used in some specific scenarios, such as UAV take-off and landing, with its sufficient computing resources and stable operating performance.

The UCARS [1] system developed by Sierra Nevada of the United States in 2006 uses a millimeter-wave radar that is installed on land or ships to automatically locate and track the transponder of the drone. The DT-ATLS [2] system developed by the company in 2008 combines the advantages of various sensors such as differential GPS, millimeter-wave radar, and infrared imager. It not only achieves improved guidance for drones landing on the ground and aircraft carriers, but also It can be applied to various types of drones. The Deck Finder [3] system developed by the Austrian Siebel Electronic Equipment Company in 2013 uses another method-radio frequency ranging, which can achieve precise spatial positioning of the UAV through six radio frequency transmitters installed on the ship.

After studying these systems, we can learn that a simple implementation plan is to set up two 2-degree-of-freedom turntables (PTU) on the ground, and one camera is fixed on each of them, as shown in Fig. 1.

In order to ensure that the UAV remains in the field of view of the two cameras during the flight, it is controlled by the ground station computer. The rotation of the turntable causes the camera to rotate in conjunction to track the flight of the drone. The ground station computer collects the image data of the two cameras and the attitude data of the turntable in real time, and uses the fusion positioning algorithm to calculate the space position of the drone, and sends it to the drone in real time through the data transmission link as the positioning data for the autonomous flight of the drone.

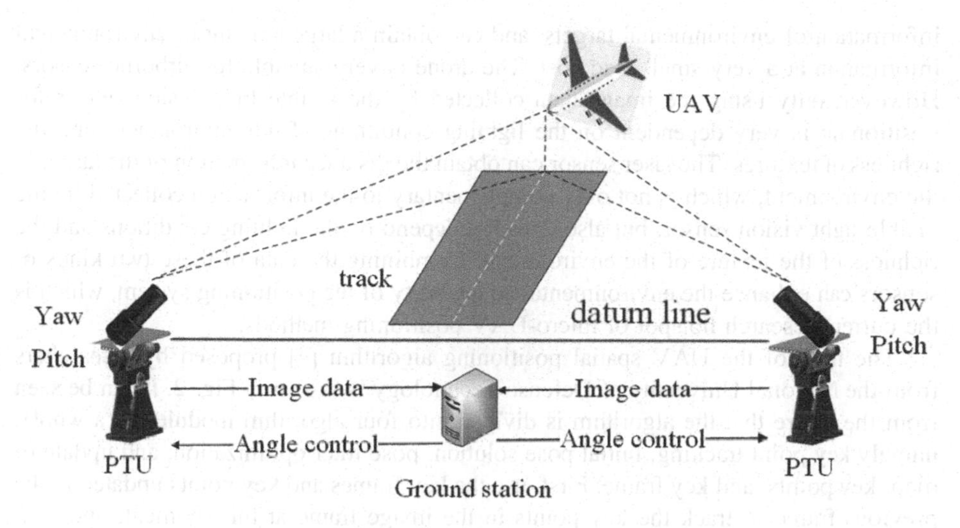

Fig. 1. Ground-based binocular drone positioning

The UAV spatial position calculation algorithm based on the ground-based binocular vision system usually requires input of the following parameters: image data, camera internal and external parameters, attitude data of the two turntables, estimated initial contour center and radius, estimated ROI (region of interest) Center and side length, the state of the drone at the previous moment. Through image preprocessing, region segmentation and filtering, and introducing an extended Kalman filter model for parameter estimation, the state of the UAV at time k can be iteratively obtained.

3 Airborne Positioning

Aerial positioning is often used in the flight phase of drones. It uses a set of sensors (including inertial sensors, visible light vision sensors, laser sensors, etc.) on the drone to obtain the drone's position and movement status information. At the same time, a certain data fusion model is used to correct and predict the collected data samples to ensure small measurement errors. This positioning method can be performed in a small indoor area, and some methods can also obtain indoor environment information at the same time, which has good application value.

3.1 Airborne Visual/Laser Fusion Positioning

Due to the limitations of the UAV's flight environment and load capacity, higher requirements are placed on the sensors on board. Especially small and micro UAVs have weak load capacity, which further limits the types of sensors that can be mounted. In addition to traditional micro drone airborne sensors such as IMU and barometer, visible light vision sensors can collect environmental images, obtain the orientation information of environmental targets, and can obtain a large amount of environmental information at a very small load cost. The drone is very suitable for airborne sensors. However, only using the image data collected by the visible light vision sensor for positioning is very dependent on the lighting conditions of the environment and the richness of textures. The laser sensor can obtain the distance information of the target in the environment, which is not only complementary to the information collected by the visible light vision sensor, but also does not depend on the lighting conditions and the richness of the texture of the environment. Combining the data of these two kinds of sensors can enhance the environmental adaptability of the positioning system, which is the current research hotspot of micro-UAV positioning methods.

The flow of the UAV spatial positioning algorithm [4] proposed by researchers from the National University of Defense Technology is shown in Fig. 2. It can be seen from the figure that the algorithm is divided into four algorithm modules as a whole, namely key point tracking, initial pose solution, pose filter optimization, and update of map, key points, and key frame. First, use the key frames and key points updated in the previous frame to track the key points in the image frame at this moment, and then calculate the initial pose according to the key points. Then, the pose filter optimization module uses the barometer and IMU data to optimize the calculated initial pose and output the UAV's spatial positioning results. Finally, add optical flow information to the laser scanning points according to the calibration parameters, add it to the incremental 3D map, and update the keyframe and key point.

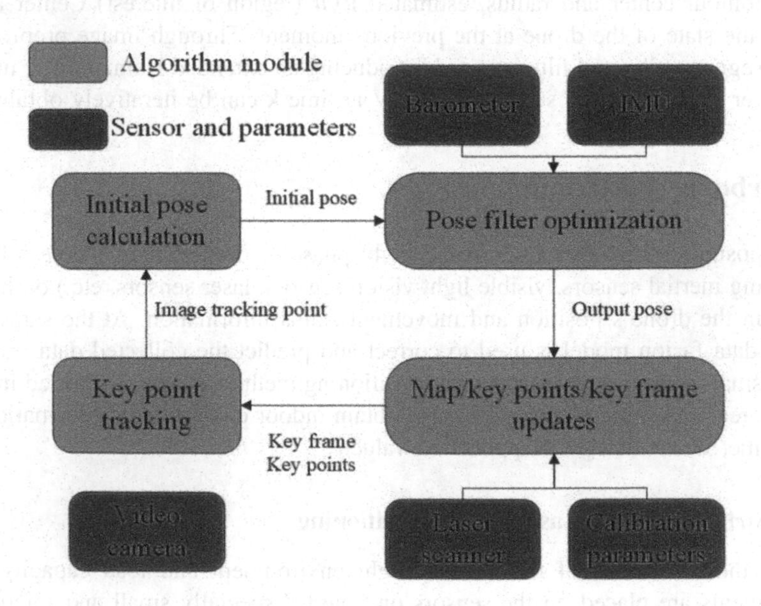

Fig. 2. Algorithm flow of airborne UAV spatial positioning

3.2 Fusion Positioning of Visual SLAM and IMU

Visual SLAM

The problem of visual SLAM (simultaneous localization and mapping) originates from the field of robotics. It can perceive and model the surrounding environment in an unknown environment to obtain the robot's own pose information. The classic visual SLAM framework mainly includes four aspects: visual odometry [5], back-end optimization, loop detection and mapping [6]. The flowchart is shown in Fig. 3.

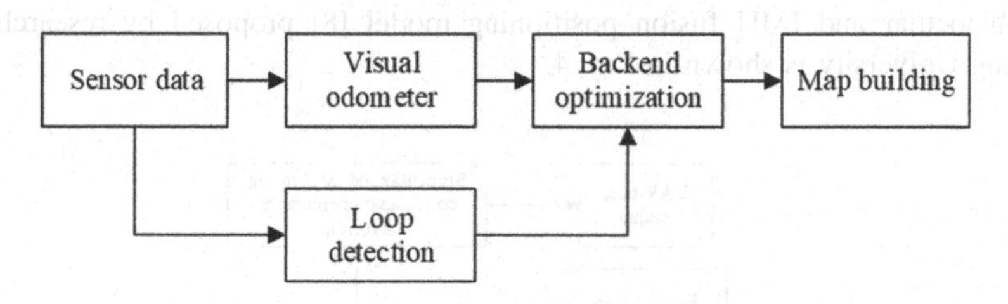

Fig. 3. Visual SLAM flowchart

1. Sensor data collection: mainly image data in visual SLAM.
2. Visual odometer: The function of the visual odometer is to estimate the motion of the camera corresponding to two adjacent frames of images. The visual odometer is also called the visual SLAM front end.
3. Back-end optimization: The back-end receives odometer information and loop information, optimizes the camera's motion trajectory, and obtains a globally consistent trajectory and map.
4. Loop detection: It is judged whether the robot has moved to the vicinity of the previous position. If a loop is detected, the loop information is provided to the back end for optimization processing.
5. Map building: According to the estimated trajectory, create a map corresponding to the task requirements. According to different processing methods, visual SLAM is usually divided into two methods: filter-based and optimization-based. Filter-based methods are generally based on Markov assumptions, using two steps of prediction and update to achieve real-time positioning and environment modeling. Based on the optimization method, the system is generally described as a nonlinear least squares problem, and the objective function is minimized by continuously optimizing the state quantity.

SLAM Integrating Vision and IMU

If only binocular vision is used for UAV positioning, because the visual information is not robust enough under harsh conditions such as moving objects, obstructing the line of sight, and drastic changes in illumination, there will usually be large positioning errors or even positioning failures. Therefore, we usually need to use the drone itself or external auxiliary sensors to achieve positioning in a GPS rejection environment [7]. Inertial measurement units (IMU), cameras, laser scanners, ultra-sonics, etc. are some commonly used drone positioning sensors. These sensors calculate their own position and attitude by sensing themselves or outside information and using some geometric constraints and other relationships. The IMU is mainly used to measure the acceleration and angular velocity information of the carrier. There is basically no requirement for the environment. The IMU can provide high-frequency (200–1000 Hz) pose estimation, while the vision can only provide low-frequency (10–50 Hz) pose estimation. But if only relying on the IMU for positioning, there will be movement drift in a short time (2–5 s). Image visual information and IMU information have good complementarity, and fusion positioning can achieve accurate, reliable, high-frequency UAV pose estimation.

A binocular and IMU fusion positioning model [8] proposed by researchers at Zhejiang University is shown in Fig. 4.

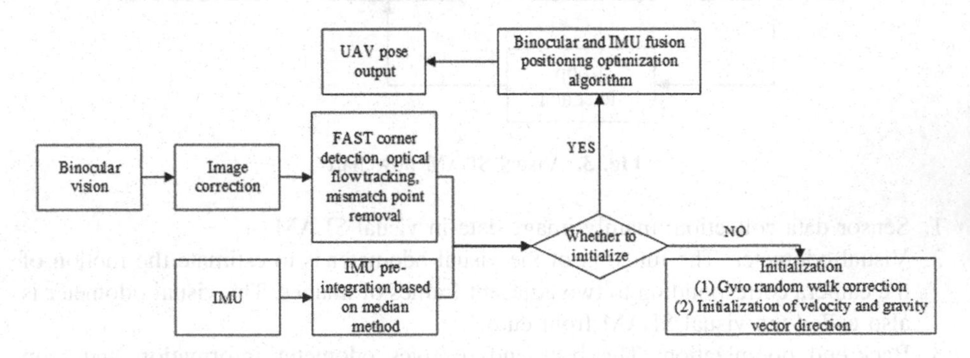

Fig. 4. Binocular vision and IMU fusion positioning system framework

The model is composed of a motion equation and an observation equation, and converts the fusion positioning problem into restoring state quantities from noisy data. For all motions and arbitrary observation data, define the error between the measured data and the predicted data, and use the LM algorithm to iteratively solve to obtain the state to be solved. In order to improve the robustness of the results, the binocular and IMU data are fused at the end.

3.3 Fusion Algorithm of UWB Positioning and Inertial Navigation

UWB (Ultra-Wide Band) technology is essentially a wireless communication technology, and the positioning technology suitable for wireless signals mainly includes methods based on ranging and direction finding. According to the parameters measured in the positioning process, the ultra-wideband positioning method is mainly divided into the Received Signal Strength Indication (RSSI) method, the Angle of Arrival method (AOA) and the signal receiving time method. The method of receiving time can be divided into time-of-arrival method (TOA) and time-difference-of-arrival method (TDOA).

The positioning algorithm based on ranging can be divided into two steps, ranging and position calculation. Having used the TDOA algorithm to calculate the arrival time difference between the node to be located and multiple positioning base stations, the next step is how to use these TDOA data solves the position coordinates. If the Chan algorithm [9] or Taylor algorithm [10] is directly used to analyze the TDOA measurement data, it will be found that the error of the TDOA measurement value in some areas is large, which will cause the algorithm to fail to converge in the area with large errors or the solution result is insufficiently consistent. Considering that the UAV has an IMU unit, if the multi-sensor fusion technology is adopted, when the measurement error of individual sensors is large, the fusion algorithm can still more accurately reflect the system state. Therefore, the ultra-wideband positioning system and IMU inertial measurement unit are essential to provide stable and reliable positioning results for the UAV formation in the indoor environment through the EKF fusion algorithm.

Extended Kalman Filter (EKF) is essentially a nonlinear version of Kalman Filter [11]. The original intention of the Kalman filter algorithm is to use the state equation of the linear system to obtain the optimal estimation of the system state from a series of noise-containing measurements. Therefore, the Kalman filter is aimed at the linear system, and the noise of the system is required to be Gaussian white noise. However, in practical engineering applications, many systems are nonlinear, and the Kalman filter algorithm is not effective in dealing with these nonlinear systems. In order to make Kalman filter applicable to nonlinear systems, extended Kalman filter is derived. At present, when the state equation is determined, EKF has become the de facto standard in the industry for state estimation of nonlinear systems. The extended Kalman filter first uses Taylor's first-order expansion to linearize the state equation and measurement equation of the nonlinear system, and then uses the partial derivative matrix (Jacobian matrix) of the equation to update the covariance matrix.

Researchers at Xiamen University draw on the idea of time division and design a TDOA algorithm based on time division [12], as shown in Fig. 5. Each UWB positioning base station is assigned a time slot. Each base station can only broadcast TODA data packets when it is its time slot, otherwise it will continue listening. When the base station finishes sending data, it will automatically switch to the receiving state, and then it will update the time slot when it receives broadcast data packets from other base stations or when the reception times out (a base station does not send packets in its time slot). By allocating time slots and making reservations for receiving and sending, we have achieved time synchronization between different base stations to ensure the reliability of information transmission. After the drone receives the data packet broadcast by the base station, it first calculates the TDOA value, and then combines the information from the IMU sensor to use EKF for data fusion to calculate the positioning coordinates. Each time EKF is executed, the average value of the accelerometer and gyroscope values of the IMU sensor will be obtained (this is because the update period of the IMU is greater than the update period of the EKF), and then use the state equation to predict the movement state of the drone based on these data, calculate covariance, and then introduce process noise to the covariance matrix. Since TDOA is geometrically the difference between the two sides of a triangle, the wrong data can be filtered out by using the relationship between the sides of the triangle. Then use the measurement equation of TDOA to calculate the Jacobian matrix, calculate the extended Kalman gain, and update the state information, and finally update the error covariance, enter the next calculation, and so on.

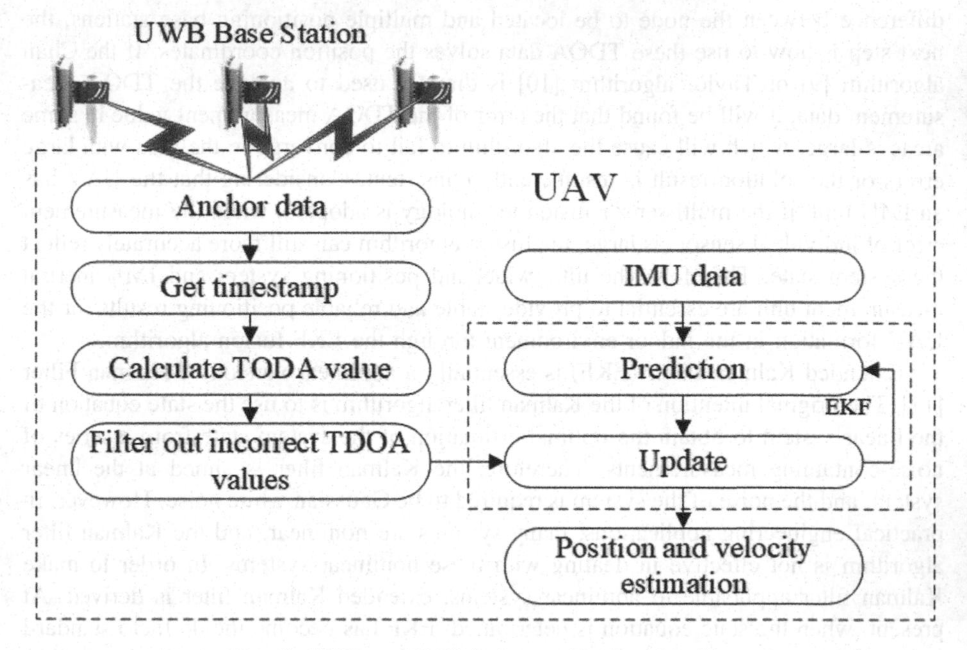

Fig. 5. UWB and IMU fusion positioning algorithm

3.4 Fusion Positioning of Lidar and Inertial Navigation

Lidar navigation is usually achieved through synchronous positioning and composition technology. After years of development, the Lidar SLAM method [13] has been widely used in the field of robotics. Lidar is divided into two-dimensional lidar and three-dimensional lidar. Limited by factors such as volume, load, and cost, micro-small UAVs usually use two-dimensional lidar for SLAM.

At present, LIDAR SLAM often uses scan matching method to estimate the carrier's pose. The so-called scan matching is to register the current environmental laser scan points with reference scan data. As far as ground robots are concerned, the methods for estimating pose through two-dimensional lidar scan matching are quite mature, but for UAVs with six degrees of freedom, the use of these scan matching methods has greater limitations. On the one hand, there is a change in the attitude of the aircraft during the flight, which causes the environmental information detected by the lidar to be out of the same plane at certain moments, which may cause errors in SLAM positioning. On the other hand, because the two-dimensional radar can only scan a plane of the environment, and the aircraft is moving in the height direction, the two-dimensional environmental structure scanned by the two-dimensional lidar will undergo abrupt changes, and the reference scan data in the matching will be compared with the current. The scanning points are not in the same plane, which results in large errors in matching. The SLAM with the aid of inertial information proposed by researchers from Nanjing University of Aeronautics and Astronautics can perform

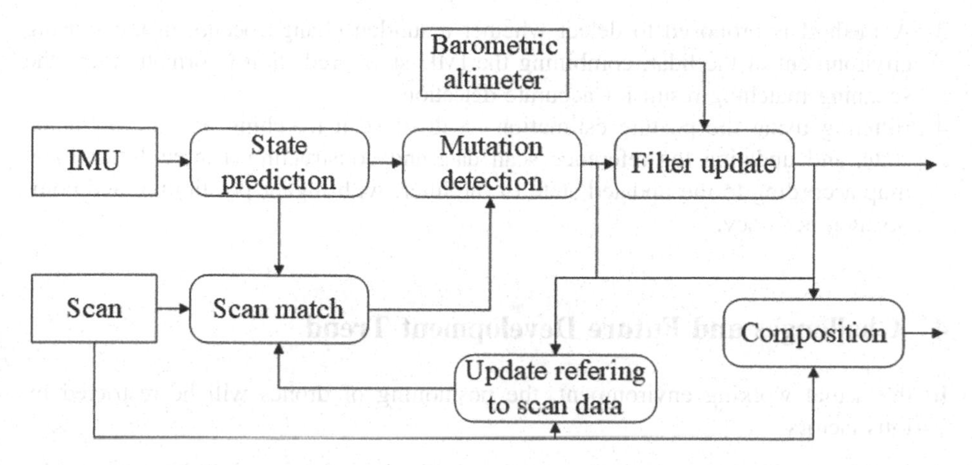

Fig. 6. Inertial navigation and lidar fusion positioning method

reliable and accurate navigation and positioning in a three-dimensional environment [14], as shown in Fig. 6.

First, use IMU to predict the attitude quaternion, velocity, position, accelerometer zero offset and gyroscope zero offset of the UAV at the current time according to the state of the UAV at the previous moment. Then the pose estimation is performed according to the projection points of the lidar scanning points on the same horizontal plane, so as to compensate the pose estimation errors caused by the scanning points not on the same horizontal plane caused by the attitude movement of the aircraft. And use IMU information to predict the initial value of the UAV's pose. At the same time, in order to avoid changes in altitude during the flight of the aircraft, the two-dimensional environmental structure sensed by the two-dimensional lidar mounted on the aircraft will undergo abrupt changes. This mutation will cause a large error in the pose estimated by the scan matching, and it needs to be detected and eliminated by combining the information of the IMU and the scan matching. Subsequently, the overlap between the current scan data and the reference scan data is used for registration, the pose is estimated, and the UAV state is updated according to the EKF model. Finally, make a three-position composition.

Compared with the traditional method, the inertial assisted lidar mainly adds three modules: state prediction, mutation detection and filter update. Its main advantages are:

1. Use IMU information to compensate for the influence of carrier attitude changes on lidar scanning, and perform scan matching on the basis of IMU prediction of carrier attitude, which speeds up the matching iteration process and can avoid the matching process from falling into the local optimum to a certain extent.
2. The mutation detection link is introduced to detect whether the Lidar sensing environment has a sudden change, and the update strategy of the filter and reference scan data is determined according to the detection result, which can overcome the problem of large or even impossible pose estimation errors when the two-dimensional environment structure changes suddenly. Therefore, we can realize precise and robust navigation.

3. A method is proposed to detect whether a sudden change occurs in the sensing environment of the lidar, combining the IMU state prediction information and the scanning matching result for accurate detection.
4. Filtering using the posture estimation result of scan matching and the predicted state, and updating the reference scan data and constructing a three-dimensional map according to the updated state of the filter, with higher positioning and composition accuracy.

4 Challenges and Future Development Trend

In the actual working environment, the positioning of drones will be restricted by various factors.

1. In the actual working environment, the positioning of drones will be restricted by various factors. The limited load of the UAV has greatly restricted the configuration of the sensor and the computing power of the onboard processor, which limits the performance of all aspects of the positioning system. For micro-rotor drones, their load generally does not exceed 5 kg, and there are not many sensors that can be mounted on them. These low-mass sensors have certain limitations in range, range, and accuracy, which greatly limits the working airspace and working environment of the positioning system.
2. Each positioning method has its own advantages and disadvantages.

- Using the image data collected by the visible light vision sensor for positioning can obtain a large amount of environmental information at a very small load cost, but it is very dependent on the lighting conditions of the environment and the richness of textures.
- The IMU is mainly used to measure carrier acceleration and angular velocity information, but if only the IMU is used for positioning, there will be movement drift after a period of time.
- UWB positioning uses the TDOA method and does not rely on synchronization between base stations, but requires a large number of base stations and is sensitive to occlusion in the channel.
- Although lidar does not rely on illumination and is accurate in two-dimensional positioning, for a six-degree-of-freedom UAV, there is a large positioning error.

3. The fusion model itself has certain limitations [15]. At present, the most commonly used model in the data fusion of UAV positioning is the EKF model, but when the prediction function and update function of this model are nonlinear, (especially in SLAM problems), EKF cannot guarantee the global optimum. If the estimated value at the time when the processing of the previous frame is completed is not yet accurate, the prior information transmitted to the next frame will contain errors. Since the state of the previous frame does not change anymore, the error in the prior information cannot be eliminated, and the error is continuously transmitted backwards causing error accumulation.

In order to solve these problems in the positioning of drones in the GPS denial environment, the future development direction may have the following aspects:

1. The introduction of new positioning methods. It includes the invention of new positioning technology and the miniaturization of existing positioning technology, making it easy to carry on drones.
2. Reduction of errors in traditional positioning methods. For example, reduce the degree of dependence of visual positioning on ambient light, or reduce the time accumulation error of IMU.
3. The update of the fusion model of multiple positioning technologies. For example, Mourikis improved EKF in 2007 and proposed MSCKF [16] to better solve the SLAM problem.
4. In addition, deep learning, which has developed rapidly in recent years, can be used to further improve the accuracy of positioning. Or use the idea of a priori map to further eliminate errors.

5 Conclusions

With the development of science and technology and the diversification of demand, UAVs have been applied in more and more occasions. In many cases, only relying on GPS for positioning can no longer meet the positioning needs of different scenarios. Focusing on the positioning technology of drones in GPS denial environments, this paper mainly discusses ground-based binocular vision positioning, fusion positioning of airborne vision/lidar/inertial navigation and UWB base station, analyzes the current challenges in UAV positioning and future development trends. Since the existing positioning method is limited by its own robustness and the loading conditions of the UAV itself, fusion positioning is required. The introduction of new positioning methods, the optimization of traditional positioning methods, and the updating of fusion models will always be valuable research directions in this field.

References

1. UCARS-V2: UAV Common Automatic Recovery System-Version 2 For Shipboard Operations [DB/OL]. https://www.sncorp.com/pdfs/cns_atm/UCARSV2Product%20Sheet.pdf. 11 Jul 2013
2. DT-ATLS: Dual-Thread Automatic Takeoff and Landing System [DB/OL]. https://www.ion.org/publications/abstract.cfm?articleID=7005,2011
3. Zhang, Y.: European positioning system: flight test anti-GPS jamming ability [DB/OL]. Beijing, https://www.dsti.net/. 09 Jul 2013
4. Tang, D.Q.: GPS Free Localization and Implementation in Robot Operating System for Unmanned Aerial Vehicles. National University of Defense Technology, 2015: 28.
5. Nister, D., Naroditsky, O., Bergen, J.: Visual odometry. In: IEEE Computer Society Conference on Computer Vision & Pattern Recognition (2004)
6. Ho, K.L., Newman, P.: Detecting loop closure with scene sequences. Int. J. Comput. Vision **74**(3), 261–286 (2007)

7. Qin, T., Li, P., Shen, S.: VINS-mono: a robust and versatile monocular visual-inertial state estimator. IEEE Trans. Rob. (99), 1–17 (2017)
8. Yang, T.: Research and system design of UAV localization system based on the fusion of Binocular vision and IMU. Zhejiang University (2019)
9. Chan, Y.T., Hang, H.Y.C., Ching, P.: Exact and approximate maximum likelihood localization algorithms. IEEE Trans. Veh. Technol. **55**(1), 10–16 (2006)
10. Foy, W.H.: Position-location solutions by Taylor-series estimation. IEEE Trans. Aerosp. Electron. Syst. **2**, 187–194 (1976)
11. Li, M., Mourikis, A.I.: High-precision, consistent EKF-based visual-inertial odometry. Int. J. Rob. Res. **32**(6), 690–711 (2013)
12. Li, H.B.: Design and Implementation of Indoor Nano-quadcopter Formation Flight System Based on UWB Positioning Technology. Xiamen University (2019)
13. Hess, W., Kohler, D., Rapp, H., et al.: Real-time loop closure in 2D LIDAR SLAM In: IEEE International Conference on Robotics and Automation, pp. 127–1278. IEEE (2016)
14. Shi, P.: UAV autonomous navigation technology based on Fusion of LIDAR and IMU in double-blind environment. Nanjing University of Aeronautics and Astronautics, (2019)
15. Liu, H.M., Zhang, G.F., Bao, H.J.: A survey of monocular simultaneous localization and mapping. J. Comput. Aided Des. Graphics **28**(06), 855–868 (2016)
16. Mourikis, A., Roumeliotis, S.: A multi-state constraint Kalman filter for vision-aided inertial navigation. In: Proceedings of IEEE International Conference on Robotics and Automation, Los Alamitos, pp. 3565–3572. IEEE Computer Society Press (2007)

Energy-Efficient Multi-UAV-Enabled Computation Offloading for Industrial Internet of Things via Deep Reinforcement Learning

Shuo Shi[1,2](✉), Meng Wang[1], and Xuemai Gu[1,3]

[1] School of Electronic and Information Engineering, Harbin Institute of Technology, Harbin 150001, Heilongjiang, China
crcss@hit.edu.cn

[2] Network Communication Research Centre, Peng Cheng Laboratory, Shenzhen 518052, Guangdong, China

[3] International Innovation Institute of HIT in Huizhou, Huizhou 516000, Guangdong, China

Abstract. Industrial Internet of things (IIoT) has been envisioned as a key technology for Industry 4.0. However, the battery capcity and processing ability of IIoT devices are limited which imposes great challenges when handling tasks with high quality of service (QoS) requirements. Toward this end, in this paper we first use multiple unmanned aerial vehicles (UAVs) equipped with computation resources to offer computation offloading opportunities for IIoT devices due to their high flexibility. Then we formulate the multi-UAV-enabled computation offloading problem as a mixed integer non-linear programming (MINLP) problem and prove its NP-hardness. Furthermore, to obtain the energy-efficient solutions for IIoT devices, we propose an intelligent algorithm called multi-agent deep Q-learning with stochastic prioritized replay (MDSPR). Simulation results show that the proposed MDSPR converges fast and outperforms the normal deep Q-learning (DQN) method and other benchmark algorithms in terms of energy-efficiency and tasks' successful rate.

Keywords: UAV-enabled IIoT · Deep reinforcement learning · Computation offloading

1 Introduction

Industrial Internet of things (IIoT), as an emergying communication paradigm, is expected to evolutionize manufacturing and also drive growth in productivity across various types of applications such as smart factories and smart grids [1]. It is important to provide the required quality of service (QoS) in terms of reliability and real-time to IIoT applications [2]. However, the battery capacity and processing ability of IIoT devices are limited, which imposes great challenges when handling tasks with characteristics of computation-sensitive and latency-sensitive.

S. Shi et al. (Eds.): AICON 2020, LNICST 356, pp. 295–305, 2021.
https://doi.org/10.1007/978-3-030-69066-3_26

Recently, with the rapid developement of mobile edge computing in IIoT, local devices can choose to offload the latency-sensitive tasks to edge servers through wireless access which can effectively alleviate the computation stress in local [3]. Due to the high flexibility and controllbility, unmanned aerial vehicles (UAVs) equipped with computation capabilities can act the role of edge servers to enable edge computing even in the absence of wireless infrastructures [4]. Furthermore, the communication links between devices and UAVs are largely line of sight (LoS) links [5], which will improve the reliability in both uplink and downlink transmission. The studies of energy-efficient computation offloading in IIoT with UAVs assisted are still remain as an open issue. In [4], a single UAV servers as a moving cloudlet for mobile users, aiming to minimize the total energy consumptions by jointly optimizing the bits allocation for uplink and downlink communications. [5] considers a stochastic task arrival model and obtains an energy-efficient offloading solution through Lyapunov optimization [6]. Considers a multi-UAV scenario and solving the non-convex computation offloading problem with iterative optimization algorithm. [7] studies the computation rate maximization problem under both partial and binary computation offloading modes subject to the energy constraints.

However, the solutions to the complex computation offloading problems in the aforementioned works [5–7] are based on one-shot optimization and fail to maximize the long-term performance of computation offloading. Moreover, these solutions are obtained under a specific model and require a priori knowledge of the network model, if the channel fading is fast, they cannot meet the requirement of real-time decision. Furthermore, the computation capabilities in devices are limited in IIoT, algorithms with high time complexity cannot run efficiently in the processor of devices. To obtain long-term rewards and real-time offloading decisions, deep reinforcement learning (DRL), as an emerging artificial intelligence technique, has been introduced to complex computation offloading problems with low complexity [8,9] proposes a deep dynamic scheduling algorithm to solve the offloading decisions problem with minimum energy costs using deep Q-learning (DQN). [10] considers a computation offloading problem with stochastic vehicle traffic, dynamic computation requests and time-varying communication conditions, and uses Q-learning as offline solution, DQN as online solution.

Motivated by the aforementioned issues, in this paper, we focus on the energy-efficient binary computadtion offloading problem in IIoT with UAV assited. IIoT devices can choose to execute the tasks locally or offload the tasks to UAVs through LoS links or offload the tasks to the remote cloud center through base stations. Compared with [4,5,7] which focus only on a single UAV, we use multiple moving UAVs as edge servers to provide computation offload opportunities, which is more practical due to the limited battery capacity of UAVs. We first propose the multi-UAV-enabled network model and formulate the computation offloading problem as a mixed integer non-linear programming (MINLP) problem. Branch and bound (BB) method can be used to solve the problem [8] but with prohibitively high computational complexity. [9] proposes an iterative optimization method to solve the MINLP problem, however, the complexity grows rapidly as the size of the network increases. To address this, we propose our multi-agent deep Q-learning with stochastic prioritized replay (MDSPR) to solve the problem distributedly and efficiently. Compared with [9,10] using normal DQN and DDQN, our method choose training experiences with high priorities instead of choosing randomly, which can accelerate the training process and improve convergence stability.

2 System Model and Problem Formulation

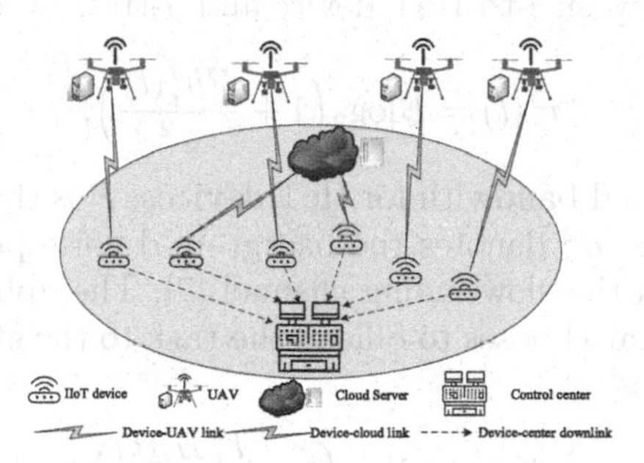

Fig. 1. Multi UAV-Enabled IIoT devices computation offloading network model.

We consider a three-layer UAV-Enabled IIoT computation offloading network model where IIoT devices process the tasks assigned from IIoT control centers, each IIoT device can choose to execute the tasks locally or offload the tasks to the UAVs or offload the tasks to the remote cloud center through terristrial base stations, and send the execution results back to control centers, as shown in Fig. 1. UAVs with communication and computation abilities act the role of MEC servers. $\mathcal{M} = \{1, 2, ..., M\}$, $\mathcal{N} = \{1, 2, ..., N\}$ denote the devices' set and UAVs' set, respectively. IIoT devices are distributed in an area with a radius of K and UAVs hover above the devices with a height of H. The locations of devices and UAVs during a single time interval $t \in \{1, 2, ..., T\}$ can be expressed as $\mathbf{q}_i(t) = (x_i(t), y_i(t)), i \in \mathcal{M}$ and $\mathbf{p}_j(t) = (X_j(t), Y_j(t), H), j \in \mathcal{N}$, respectively. Moreover, tasks can be modelled as $W_i(t) = (s_i(t), c_i(t), d_i(t))$ according to [3], where $s_i(t)$ denotes the total bits of the task, $c_i(t)$ denotes the total CPU circles required by the task and $d_i(t)$ denotes the tolerant delay of the task. In this paper, we assume UAVs flying with a random speed v in the fixed area, $\mathbf{q}_i(t)$, $\mathbf{p}_j(t)$ remain unchanged in a single time slot and there is a coming task for each IIoT device in a single time interval.

2.1 Communication and Computation Model

According to [8] the line of sight (LoS) channel gain between i-th IIoT device and j-th UAV can be modelled as

$$h_i^j(t) = \beta_0 d^{-2}(t) = \frac{\beta_0}{H^2 + ||\mathbf{q}_i(t) - \mathbf{p}_j(t)||^2}, \tag{1}$$

where β_0 denotes the channel gain at the reference distance $d_0 = 1$ m. Tasks can be offloaded to UAVs or the cloud center through wireless channle. Each IIoT device is associated with UAVs using orthogonal frequency-division multiplexing (OFDM). Tasks can be successfully offloaded to UAVs if the distance d between UAVs and devices is less than R, where R denotes the communication range of UAVs.

Since the size of the processed tasks can be negligible compared with the original tasks, in this paper, we only consider the uplink transmission. The uplink transmission rate between i-th IIoT device and j-th UAV can be expressed as

$$r_i^j(t) = B\log_2\Big(1 + \frac{Ph_i^j(t)}{\sigma^2}\Big), \tag{2}$$

where B is the allocated bandwith for IIoT devices, P is the uplink transmission power of IIoT devices, σ^2 denotes the background noise power, and we assume that it is constant on the slow fading channel [7]. The uplink transmission rate of an IIoT device i that chooses to offload the task to the cloud through wireless link can be denoted as

$$r_i^c(t) = \omega\log_2\Big(1 + \frac{P_c H_i(t)}{\sigma^2}\Big), \tag{3}$$

where w, P_c, $H_i(t)$ denote the bandwith, transmission power , the channel gain between i-th IIoT device and cloud, respectively. For UAV computation offloading, the task transmission time can be expressed as

$$L_i^j(t) = \frac{s_i(t)}{r_i^j(t)}, \tag{4}$$

for cloud computation offloading, according to [7] the transmission time can be denoted as

$$L_i^c(t) = \frac{s_i(t)}{r_i^c(t)} + \tau, \tag{5}$$

where τ denotes the uplink propogation delay factor due to congestion of the bachhaul link. To simplify the analysis, all the UAVs are assumed to have the same computation capacity, denoted as C_u. The computatioin capacity of IIoT devices and the remote cloud center are denoted as C_l and C_c, respectively. Thus the local task computation time can be expressed as

$$L_l(t) = \frac{c_i(t)}{C_l}, \tag{6}$$

according to [7], the energy consumption of local computation can be given as

$$E_i^l(t) = \theta c_i(t)^2, \tag{7}$$

where θ is the factor denoting the consuming energy per CPU cycle. The task computation time in UAV can be denoted as

$$D_u(t) = \frac{c_i(t)}{C_u}. \tag{8}$$

Note that,the cloud processing time can be negligible since the computation capacity is enormous in cloud center. Thus, for UAV offloading cases, the total communication energy can be expressed as

$$E_i^j(t) = P(L_i^j(t) + D_u(t)), \tag{9}$$

for cloud offloading cases, the total energy consumptions can be modelled as

$$E_i^c(t) = P_c(L_i^c(t) + \tau), \tag{10}$$

2.2 Problem Formulation

In each time interval IIoT devices should make offloading decisions, we define $\mathbf{I} = (I_1(t), I_2(t), ..., I_M(t))$ as the offloading indicated vector in a single time interval, where $I_i(t) \in \{I_i^{-1}, I_i^0, I_i^1, ..., I_i^M\}, \forall t \in T, \forall i \in \mathcal{M}$ is an indicator which can be used to describe an offloading decision, $I_i^k \in \{0,1\}, k \in \{-1,0,1,2,...,N\}$. Let $E(t) \in \left\{E_i^c(t), E_i^j(t), E_i^l(t)\right\}$, $L(t) \in \left\{L_i^c(t), L_i^j(t), L_i^l(t)\right\}$. The energy consumption of a singe task can then be derived as

$$E_i(t) = I_i(t)E(t) = \begin{cases} E_i^c(t), & k = -1 \\ E_i^l(t), & k = 0 \\ E_i^j(t), & k \in \mathcal{M} \end{cases} \tag{11}$$

Our objective is to minimize the total energy cosumptions of all IIoT devices in a time period T through optimizing the offloading indicated vector while satisfying the basic constraints. Thus, we can formulate the energy-efficient multi-agent computation offloading problem as follows

$$\begin{aligned} \min_{I,p,q} \quad & \sum_{t=1}^{T}\sum_{i=1}^{M} I_i(t)E_i(t). \\ \text{s.t.} \quad & C1: I_i^k \in \{0,1\}, \forall i \in \mathcal{M}, k \in \{-1,0,1,2,...,N\} \\ & C2: \sum_{k=1}^{N} I_i^k \leq 1, \forall i \in \mathcal{M}, k \in \{-1,0,1,2,...,N\} \\ & C3: \sum_{i=1}^{M} I_i(t) \leq W, \quad I_i(t) \in \{I_i^1, ..., I_i^M\} \\ & C4: I_i(t)L(t) \leq d_i(t), \forall i \in \mathcal{M}, \forall t \in \mathcal{T} \\ & C5: H^2 + ||\mathbf{q}_i(t) - \mathbf{p}_j(t)||^2 \leq R^2, \forall i \in \mathcal{M}, \forall t \in \mathcal{T} \end{aligned} \tag{12}$$

C1 indicates that I_i^k is a binary variable. C2 shows that if a task is choosen to offload to UAVs, it can only be offloaded to a single UAV. C3 gives the constraints that tasks need to be executed in UAVs cannot exceed the maximum limit. C4 guarantees that the task can be finished within the tolerant delay. C5 means if i-th IIoT devices offload the task to j-th UAV, the distance between the device and UAV should less than the communication range of the UAV.

3 Proposed Intelligent Computation Offloading Solution : MDSPR

In this section, we present our solutions using multi-agent deep Q-learning with stochastic prioritized replay (MDSPR) for the energy-efficent multi-agent computation offloading problem.

3.1 Markov Decision Process Modeling

We first model the problem as a markov decision process (MDP). A MDP involves a agent that repeatedly observes the current state s_t during a time period. Agents interact with the environment through making actions a among all the available actions in that state, and the environment gives a feedback reward r_t to agents. Then, the agent will transfer to a new state s_{t+1} with a transition probability $P(s_{t+1}|s_t, a)$. A typical MDP tuple can be modeled as $< \mathcal{S}, \mathcal{A}, \mathcal{F}, \mathcal{R}, \mathcal{S}' >$.

State Space. A state vector should reflect the agents' perception of the environment. According to our network model, we define $s_t = \{\mathcal{S}_1, \mathcal{S}_2, \mathcal{S}_3, \mathcal{S}_4\}$ as the state vector. $\mathcal{S}_1 = \{s_i(t), c_i(t), d_i(t)\}$ reflects the task properties in each time interval. $\mathcal{S}_2 = \{x_i(t), y_i(t)\}$, $\mathcal{S}_3 = \{X_1(t), Y_1(t), X_2(t), Y_2(t), ..., X_N(t), Y_N(t)\}$ denotes the current locations of the IIoT device and UAVs.

$\mathcal{S}_4 = \{H_i(t), h_i^1(t), h_i^2(t), ..., h_i^N(t)\}$ gives the channel conditions between i-agent and different offloading objectives. The dimension of a state vector is $3N+6$.

Action Space. $\mathcal{A} = \{-1, 0, 1, 2, 3, ..., N\}$ represents the action space. Here $a = -1$ means to offload the task to cloud center, $a = 0$ means to execute the task locally, $a = \{1, 2, 3, ..., N\}$ denotes that the agent chooses to offload the task to UAVs.

Transition Probability. $\mathcal{F}$ denotes the probability distribution $P(s_{t+1}|s_t, \{a\}_{a \in \mathcal{A}}) \in [0, 1]$ when state transition happens. Note that, in our network model, the agent cannot predict the state and reward of next state before making actions, so the MDP problem needs to be solved by model-free reinforcement learning method, in which the transition probabiliy distribution remains unknowm.

Reward Function. The reward function is a immediate feedback for agents' actions from the environment. In our optimization problem, the objective is to minimize the total energy consumptions of agents, so the reward function should be inverse proportional to energy consumptions, which can be defined as

$$r(t) = \begin{cases} \dfrac{1}{I_i(t)E(t)}, & I_i(t)E(t) \leq d_i(t) \\ \dfrac{1}{I_i(t)E(t)} - m, & I_i(t)E(t) > d_i(t) \end{cases} \tag{13}$$

where m is a punishment factor when the delay constraint is not satisfied.

3.2 Deep Q-Learning Structure

In this paper, we use deep Q-learning structure to solve the MDP. Deep Q-learing (DQN) is a promising reinforcement learning technique which combines deep learning with Q-learing aiming to solve complex problems in communication and networking [9]. Compared with Q-learning, DQN uses deep neural networks to replace the traditional Q-table, so that DQN can overcome the curse of dimensionality in Q-learning. The key idea of DQN is to use neural network as a function approximator. Given input state s_t, the output of neural network is $Q(s, a_i; \mathbf{w})$, where $\mathbf{w}$ is the weights of DQN.

Experience replay is an important technique to improve training efficiency in DQN. Since there is no Q-table in DQN, in a series of actions, the training samples may have strong correlations which will make learning process fall into local optimum. To solve this, DQN introduces a replay memory $\mathcal{D}$, once the agent obtain a new experience $< s_t, a_t, r_t, s_{t+1} >$, the experience is stored in $\mathcal{D}$. Each training step a mini-batch with size L is randomly sampled from D instead of a one step experience.

In the training stage, despite the current network $Q(s, a_i; \mathbf{w})$, DQN introduces another target network $Q(s, a_i; \mathbf{w'})$ to produce target value. The weights of $Q(s, a_i; \mathbf{w'})$ are updated every Z training steps. The target value is defined as

$$y_t = r_t + \gamma \max_{a' \in \mathcal{A}} Q(s', a'; \mathbf{w'}), \tag{14}$$

where γ denotes the reward decay in DQN. However, recent research has proved that the greedy iteration method may cause overestimation. Double DQN (DDQN) is proposed to eliminate overestimation which can be expressed as

$$y_t = r_t + \gamma Q(s', \operatorname*{argmax}_a Q(s, a_i; \mathbf{w}); \mathbf{w'}), \tag{15}$$

DDQN eliminates the problem of overestimation by decoupling the selection of target Q action and the calculation of target Q into two neural networks. Temporal difference error (TD-error) is defined as the difference between target value and current value which can be denoted as

$$\delta_t^j = |y_t - Q(s_t, a_1, a_2, ..., a_N)|, \forall t \in \mathcal{T}, \forall j \in \mathcal{N} \tag{16}$$

The loss function is defined as the least squared loss of TD-error denoted as

$$\mathbb{L}(\mathbf{w}) = \mathbb{E}[(y_{target} - Q(s, a; \mathbf{w}))^2]. \tag{17}$$

$\mathbb{L}(\mathbf{w})$ can be minimized through back propogation methods in deep learning, in this paper we use the classic gradient descent method

$$\mathbf{w} = \mathbf{w} - \alpha \bigtriangledown_{\mathbf{w}} \mathbb{L}(\mathbf{w}), \tag{18}$$

where α is the learning rate.

3.3 Stochastic Prioritized Replay

Traditional experienced replay method selects L experiences randomly from D, however, the importance of experiences in replay memory are different. Experiences with larger TD-errors may have better training effect. Furthermore, simply choosing L experiences with the largest TD-errors (greedy prioritized) in each training step will cause high time complexity since the size of $\mathcal{D}$ is large. Therefore, according to [2], in this paper we use deep Q-learning with stochastic prioritised replay. We adopt the binary tree structure to store and sample experiences with priorities to accelerate the training process with relative low time complexity. The top node stores the sum of all priorities in $\mathcal{D}$, in sampling stage, the sum is divided equally into L intervals. In sampling stage, each interval generates a random count in its range, and then starts to search downwards with the top node as the parent node. If the input value is less than the left child node, the left child node is regarded as the parent node to continue the downward search, otherwise, the input value is subtracted by the value of left child node, then, the right child node is denoted as the parent node, and the difference is used as an input to continue searching downwards until the last layer, and the experience corresponding to the parent node is the output sample. The chosen probability distribution of s_k can be denoted as

$$P(k) = \frac{p_k}{\sum_{d=1}^{O} p_d}. \tag{19}$$

where O denotes the capacity of $\mathcal{D}$. (19) shows that the higher the priority, the greater the probability of being selected. Moreover, the iterations of stochastic prioritized replay during each sampling process is $L * log_2(O)$, compared with greedy prioritized $L * O$.

4 Simulation Results

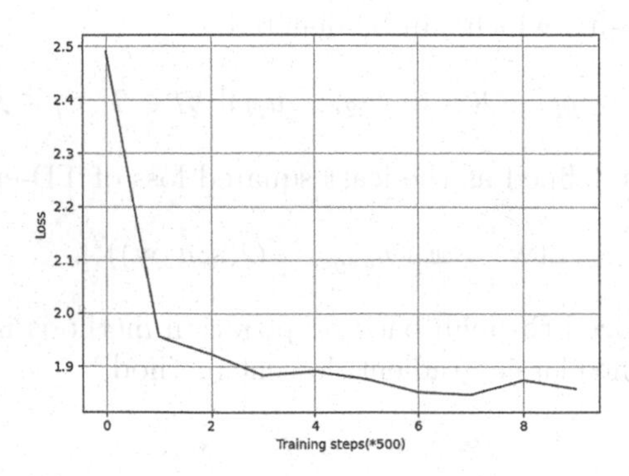

Fig. 2. The convergence performance of the proposed MDSPR.

First, we show the convergence performance of the MDSPR. Figure 2 shows the changes of loss over training steps, it is observed that the loss reduce rapidly at the beginning. This is because at the beginning of the training, agents take actions randomly and the approximation is not precise. After 600 steps the loss become stable, which indicates the convergence of MDSPR. We then compare the performance of our MDSPR with three benchmark algorithms, namely random algorithm, DQN algorithm and DDQN algorithm. Figure 3 shows the average energy consumptions of a mobile user when dealing with a single task. Since the tasks are generated with random s_i, c_i, d_i, an inappropriate offloading decision may cause huge additional energy comsumptions. For example, if a task with high computation requirement and small size is executed in local, the cost will increase significantly compared with that the task offloaded to the remote cloud center. In the first n training steps, the agent makes offloading decisions according to ϵ-greedy policy, after the learning stage, the average energy consumptions become stable due to the convergence of neural networks. Figure 4 shows the total energy consumptions under different number of tasks after the learning stage. From Fig. 4, it can be seen that the proposed MDSPR outperforms the three benchmark algorithms. The reason is that in each training step, MDSPR chooses more valuable experiences to learn, thus maximizing the overall long-time reward. Furthermore, it is observed that with the increase of tasks, the energy consumptions increase approximately linearly. This is because agents make offloading decisions upon each task arrival according to the maximum Q-value. After the convergence of the neural network, the action with highest Q-value has a high probability to obtain minimize energy consumptions.

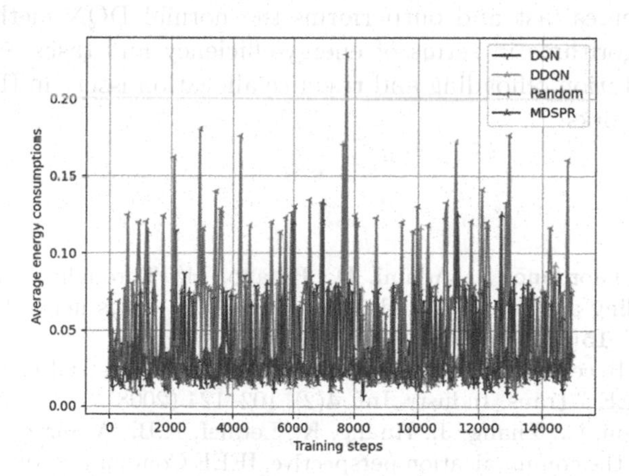

Fig. 3. The average energy consumptions of a mobile user when dealing with a single task.

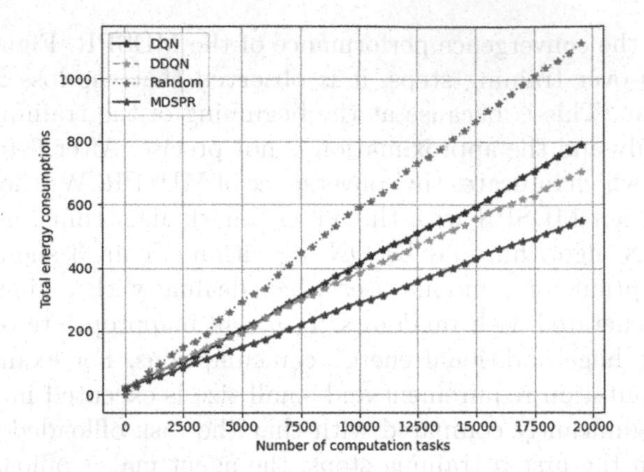

Fig. 4. The average energy consumptions of a mobile user when dealing with a single task.

5 Conclusion

In this paper, we study the energy-efficient computation offloading problem in UAV-enbaled IIoT network. To release the computation burden of IIoT devices, we use UAVs as edge servers to provide computation offloading opportunities and formulate the multi-UAV-enabled computation offloading problem as an MINLP problem and prove its NP-hardness. To obtain energy-efficent decisions, we propose our MDSPR solution. Simulation results show that the proposed MDSPR converges fast and outperforms the normal DQN method and other benchmark algorithms in terms of energy-efficiency and tasks' successful rate. Partial computation offloading and resource allocation issues in IIoT will be left as our future works.

References

1. Ojo, M.O., Giordano, S., Adami, D., Pagano, M.: Throughput maximizing and fair scheduling algorithms in industrial internet of things networks. IEEE Trans. Industr. Inf. **15**(6), 3400–3410 (2019)
2. Willig, A.: Recent and emerging topics in wireless industrial communications: a selection. IEEE Trans. Industr. Inf. **4**(2), 102–124 (2008)
3. Mao, Y., You, C., Zhang, J., Huang, K., Letaief, K.B.: A Survey on mobile edge computing: the communication perspective. IEEE Commun. Surv. Tutorials **19**(4), 2322–2358 (2017)
4. Jeong, S., Simeone, O., Kang, J.: Mobile edge computing via a UAV-mounted cloudlet: optimization of bit allocation and path planning. IEEE Trans. Veh. Technol. **67**(3), 2049–2063 (2018)
5. Zhang, J., et al.: Stochastic computation offloading and trajectory scheduling for UAV-assisted mobile edge computing. IEEE Internet Things J. **6**(2), 3688–3699 (2019)

6. Zhang, J., et al.: Computation-efficient offloading and trajectory scheduling for Multi-UAV assisted mobile edge computing. IEEE Trans. Veh. Technol. **69**(2), 2114–2125 (2020)
7. Zhou, F., Wu, Y., Hu, R.Q., Qian, Y.: Computation rate maximization in UAV-Enabled wireless-powered mobile-edge computing systems. IEEE J. Sel. Areas Commun. **36**(9), 1927–1941 (2018)
8. Mnih, V., et al.: Human-level control through deep reinforcement learning. Nature **518**(7540), 529–533 (2015)
9. Wang, Y., Wang, K., Huang, H., Miyazaki, T., Guo, S.: Traffic and computation Co-offloading with reinforcement learning in fog computing for industrial applications. IEEE Trans. Industr. Inf. **15**(2), 976–986 (2019)
10. Liu, Y., Yu, H., Xie, S., Zhang, Y.: Deep reinforcement learning for offloading and resource allocation in vehicle edge computing and networks. IEEE Trans. Veh. Technol. **68**(11), 11158–11168 (2019)

Smart Education: Educational Change in the Age of Artificial Intelligence

Campus Bullying Detecting Algorithm Based on Surveillance Video

Liang Ye[1,2,3](✉), Susu Yan[1,3,4], Tian Han[2,4], Tapio Seppänen[2], and Esko Alasaarela[2]

[1] Harbin Institute of Technology, Harbin 150080, China
yeliang@hit.edu.cn
[2] University of Oulu, 90014 Oulu, Finland
[3] Science and Technology on Communication Networks Laboratory, Shijiazhuang, China
[4] Harbin University of Science and Technology, Harbin 150080, China

Abstract. In recent years, more and more violent events are taking placc in campus life. Campus bullying prevention is already the focus of current education. This paper proposes a campus bullying detecting algorithm based on surveillance video. It can actively monitor whether students are being bullied on campus. The authors use Openpose to extract bone information from video. According to the coordinate information of bone points, they extract static and dynamic features. Support vector machine (SVM) is used to classify different actions. The recognition accuracy of the classification model is 88.57%. In this way, the campus surveillance camera is able to realize real-time monitoring of bullying behavior. It is conducive to the construction of a harmonious campus environment.

Keywords: Campus bullying · bone points · Openpose · Support vector machine

1 Introduction

Campus bullying has become a common social phenomenon [1]. It has caused great harm to the society and education in many countries. In many school bullying cases, the bullied hide the fact that they were bullied because of fear. They do not inform their parents or teachers in time, which leads to more and more bullying. As a result, bullied students have a serious psychological trauma. The research of deep learning and computer vision are more and more mature. However, most of the application scenarios of computer vision are limited to industry, transportation and commerce.

Through a literature survey [2, 3] in the field of computer vision, it is known that there are relatively few researchers who are researching on the application of campus scenes. The Kinect device [4] developed by Microsoft can capture the dynamic posture of two-dimensional human body. It leads to a lot of research work on gesture recognition techniques based on Kinect. However, the system is highly dependent on the Kinect device. It increases the hardware cost of the system. This research designs a bullying detecting method based on Openpose. It uses ordinary cameras to collect the

S. Shi et al. (Eds.): AICON 2020, LNICST 356, pp. 309–315, 2021.
https://doi.org/10.1007/978-3-030-69066-3_27

campus surveillance video of students, and uses Openpose to obtain the bone points of the human bodies from the video. It can remove redundant information in the image. Only the bone information of human bodies is retained for later action recognition. Thus, the information of an image is significantly reduced, and it is convenient for data transmission. The collected video information is transmitted to the background for processing. Distances and angles between bone points are used to extract motion features. Support vector machine is used to classify actions. The following sections will describe the algorithm in details.

2 Campus Bullying Detecting Algorithm

2.1 Bone Points of Human

In this study, bone points are used to identify human actions. It overcomes the influence of external environmental factors such as light changes and clothing changes. Human bone points represent the positions of human bodies (heads, limbs and trunks) in two-dimensional coordinates. It not only shows the local shapes of human bodies, but also describes the topological information. Therefore, it is of great significance to use the bone information of human bodies to recognize human actions.

In fact, human action is mainly the relative movement of human bones around their own joints. The human body is made up of 206 bones, which can be divided into three parts, namely skull, trunk and limb bones. There are 29 skulls, 51 trunk bones and 126 limb bones. It is unnecessary to use all the bones of the human body for action reconstruction. Instead, the authors choose the simple bone points model. In this study, the coco data set [5] is used as the annotation model of bone points. It selects 18 bone points. Table1 gives the specific bone points and their corresponding labels.

Table 1. Bone points and labels of the coco model.

Label	0	1	2	3	4	5	6	7	8
Bone	Nose	Neck	Right shoulder	Right elbow	Right wrist	Left shoulder	Left elbow	Left wrist	Right hip
Label	9	10	11	12	13	14	15	16	17
Bone	Right knee	Right ankle	Left hip	Left knee	Left ankle	Left eye	Right eye	Left ear	Right ear

2.2 Bone Points Detection Based on Openpose

Openpose multi-person pose estimation model was proposed by researchers from Carnegie Mellon University [6]. The model uses a deep neural network to extract the original feature xtraction is an important step map of the image. The model input is divided into two branches. In one branch, a convolution neural network (CNN) is used to predict the heat map of human joint points. In the other branch, another CNN is used to obtain the partial affinity domain of all the connected joint points. The whole network diagram is shown in Fig. 1. The authors use the coco human skeleton model to extract bone information from key frames in the video. Set the key points numbered 0, 1, 2, 3, 4, 5, 6, 7, 8, 15, 16, and 17 as the activation points in practical application (Fig. 2), and obtain the coordinates of bone points as raw data for post-processing.

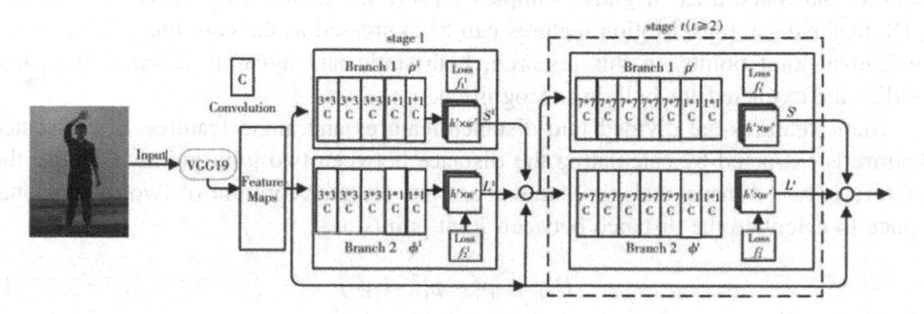

Fig. 1. Structure diagram of the whole network. The overall network architecture is divided into six stages. The upper branch is responsible for predicting the positions of bone points. The lower branch is responsible for predicting the affinity region between bone points. It can improve the accuracy of bone point prediction after multi-stage operation.

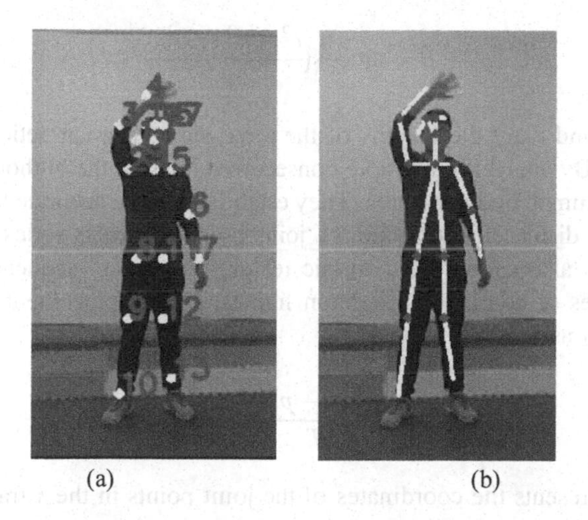

(a) (b)

Fig. 2. Output map of human bone points. The main function of Openpose is divided into two parts. (a) The first part is to identify the joints of the human body in the input image. (b) The second part is to connect the corresponding joint points belonging to each person.

2.3 Feature Extraction and Classification

Feature extraction is an important step in human action recognition. Firstly, the authors catalog the collected video data samples. Violent behaviors are regarded as bullying and marked as "positive samples". Daily-life behaviors are regarded as non-bullying actions and marked as "negative samples". There are totally 83 positive samples and 118 negative samples. Motion features can be expressed as the coordinate information of human joint points. In this research, both static and dynamic features of human bodies are extracted for bullying recognition.

Static features are divided into distance features and angle features. The distance feature is extracted by calculating the distance between two joint points. Because the picture is two-dimensional, the authors use the distance formula of two-dimensional space to calculate the distance between joint points as,

$$D_{i,j}^{s} = |p_i^s - p_j^s| \quad i \neq j \tag{1}$$

where p_i^s and p_j^s represent the coordinates of different joint points in the same action sequence frame, respectively, and $D_{i,j}^{s}$ is the Euclidean distance between joint points.

The authors use the coordinates of three points to calculate the angle. Calculate the angle by the cosine theorem,

$$\theta = \arccos\left(\frac{c^2 + a^2 - b^2}{2ac}\right) \tag{2}$$

where a, b, and c are the lengths of the three sides. Human actions differ in both time and space. By analyzing multiple consecutive images, the authors summarize the change rules of human body postures. They establish a more accurate topology for each joint point. The displacement vector of joint points on time series is an important feature, which is also called the dynamic feature. An action sequence can be represented by a series of continuous skeleton frames. The displacement vector sequence can be expressed as follows:

$$\omega_i^s = \left|\frac{p_i^{s+1} - p_i^s}{\Delta T}\right| \quad 1 < s < \tau \tag{3}$$

where p_i^s represents the coordinates of the joint points in the s frame of the action sequence, and ΔT is the time interval between the two frames.

Table 2 shows the extracted features. Campus bullying is commonly attended by multiple persons, and the interaction of these persons is relatively strong. Therefore, the authors use circumscribed rectangular frames to separate different persons in one image. Figure 3 shows the circumscribed rectangular frame target separation and bone analysis.

Table 2. Distance features, angle features, and dynamic features.

Features	Specific features	Quantity
Distance features	Hand - hand	1
	Foot - foot	1
	Hand - waist	2
	Hand - shoulder	2
	Hand - knee	2
	Knee - knee	1
	Elbow - knee	2
	Hand - foot	2
Angle features	Hand - elbow- shoulder	2
	Foot - knee- waist	2
	Elbow - shoulder-neck	2
	Knee - waist - neck	2
Dynamic features	Hands	2
	Feet	2
	Waists	2

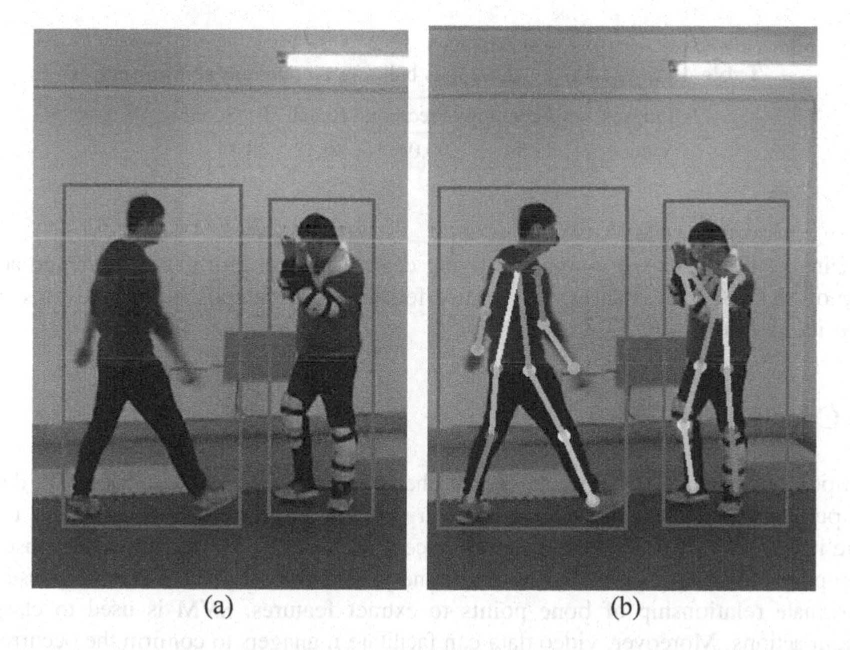

Fig. 3. Target separation and bone extraction. (a) Target separation with circumscribed rectangular frames. (b) Bone extraction.

As mentioned above, label the feature sequences extracted from campus bullying fragments as positive, and those from daily-life actions as negative. Thus, this study is a 2-class classification. The authors choose SVM [7] for classification. Based on the minimization loss function, it looks for a hyperplane to distinguish samples of different classes. The authors used five-fold cross validation to estimate the classification performance, and Table 3 shows the confusion matrix.

Table 3. Confusion matrix of campus bullying detection (%).

	Bullying (predicted)	Non-bullying (predicted)
Bullying (real)	**86.72**	13.28
Non-bullying (real)	9.59	**90.41**

According to Table 3, 86.72% of bullying actions were accurately identified as violent actions, and 90.41% of daily actions were recognized as non-violent actions.

Then the authors calculated the four indexes of accuracy, precision, recall, and F1-score. Table 4 shows the results.

Table 4. Four indexes of campus bullying recognition performance.

Indexes	Accuracy	Precision	Recall	F1-Score
Value (%)	88.57	90.04	86.72	88.35

Finally, the proposed campus bullying detection algorithm gets an average accuracy of 88.57%, which shows a promise for detecting campus bullying events with bone information.

3 Conclusions

Campus bullying is a common social phenomenon in many countries. To detect campus violence, this research proposes a campus bullying detecting method using bone information extracted from surveillance video images. Firstly, the authors use the Openpose model to extract the human bone points in the video. Then, they use the coordinate relationship of bone points to extract features. SVM is used to classify violent actions. Moreover, video data can facilitate managers to confirm the occurrence of violence. Finally, the proposed method gets an average accuracy of 88.57%, which shows a promise for detecting campus bullying events with surveillance cameras.

Acknowledgements. This paper was funded by the National Natural Science Foundation of China under grant number 41861134010, the Key Laboratory of Information Transmission and

Distribution Technology of Communication Network (HHX20641X002), National Key R&D Program of China (No. 2018YFC0807101).

References

1. Sung, Y.-H., Chen, L.-M.: Double trouble: the developmental process of school bully-victims. Child. Youth Serv. Rev. **91**(01), 279–288 (2018)
2. Hammami, S.M., Alhammami, M.: Vision-based system model for detecting violence against children. MethodsX **2**(4), 7–8 (2020)
3. Hao, M., Cao, W.H., Liu, Z.T.: Visual-audio emotion recognition based on multi-task and ensemble learning with multiple features. Neurocomputing **12**(07), 390–391 (2020)
4. Li, G., Li, C.: Learning skeleton information for human action analysis using Kinect. In: Signal Processing: Image Communication, vol. 115814 (2020)
5. Newell, A., Huang, Z., Deng, J.: Associative embedding: end-to-end learning for joint detection and grouping. In: Advances in Neural Information Processing Systems, pp. 2274–2284 (2017)
6. Cao, Z., Simon, T., Wei, S E.: Realtime multi-person 2D pose estimation using part affinity fields. In: IEEE Conference on Computer Vision & Pattern Recognition. IEEE Computer Society (2017)
7. Accattoli, S., Sernani, P., Falcionelli, N.: Violence detection in videos by combining 3D convolutional neural networks and support vector machines. Appl. Artif. Intell. **34**(4), 202–203 (2020)

The Applications and Drawbacks of Emerging AI Framework in Online Education Field

Ming Jiang(✉), Zhenyu Xu, and Zhanhong Shen

Hui Zhou Engineering Vocational College, Huizhou, China
mjiang@hit.edu.cn

Abstract. With the rise of the fourth revolution of science and technology, artificial intelligence is pushed to the forefront of the world; Under the attention of the world again and again to upgrade innovation, gradually permeability; Under the attention of the world again and again to upgrade innovation, gradually into the people of various fields, including the education industry. Education is the foundation of training talents, its purpose is to improve the person's intelligence and a kind of activity, and artificial intelligence are pretty much the same.

Keywords: Artificial intelligence · Education · Mixed reality education

1 Introduction

"Artificial Intelligence" (Artificial Intelligence, AI) jargon in 1956 at the earliest, At the time, artificial intelligence is defined as to be their own simulation can be accurately describe the learning ability of features or intelligent robot [1]. In the past 60 years AI made great progress. Kai fu lee, Wang Yong-gang defined in "Artificial Intelligence" is: "deep learning + big data = artificial intelligence, artificial intelligence rely on artificial intelligence algorithm put knowledge and meaning, education and human nature into '0' and '1' machine code, and provide learning solutions for people." [2]. Artificial intelligence is the study of human intelligence activity rule, the structure has a certain intelligent artificial system, is the study of how to apply the computer hardware and software to simulate the human. The emergence of artificial intelligence and its rapid development made great contribution to education reform. If the object of education is mutual learning and communication, So the education object of artificial intelligence can do P2P, between people and machines, M2M learn from each other. In recent years, the prestige of the artificial intelligence is more and more big, the industry is also more and more. Education as a first step in training talents, nature is to follow in the footsteps of time development, the introduction of artificial intelligence is also keep up with the trend of The Times development. This article will revolve around under the new situation, The application of AI education way, AI education existing problems and solutions to these problems in-depth discussion.

S. Shi et al. (Eds.): AICON 2020, LNICST 356, pp. 316–325, 2021.
https://doi.org/10.1007/978-3-030-69066-3_28

2 The Background

Under external pressure of competition of the international environment, a new round of international competition is mainly around the core technology, top talent, and innovative talents competition; under the new situation has led us to develop internationally competitive and innovative, creative, combining the artificial intelligence of intelligent high-tech talent. To drive the progress of science and technology of personnel training, promoting economic development and industrial transformation, achieve the goal of improve the international competitiveness of our country. Once upon a time, in the artificial intelligence in our country's talent gap is too large, the slow development of artificial intelligence, and with the world development orbit, But after the reform and opening-up policy, a series of reform for our country and the world smaller and smaller, today's China is occupies a majority say countries in the world. In the past, in May 2019 the People's Daily interview, said: "the advantage of education talent intelligence, boost the development of intelligence technology industry" [4]. Emphasizes the AI in the prominent influence to the education, only to learn the core technology, shortened to make technological innovation process, display the person's intelligence to the maximum. Along with the people of a new generation of artificial intelligence research and development, in health care, pension, education, culture consists of people of various fields such as are for use with artificial intelligence, and education as the pioneer of the popularization of artificial intelligence, natural is toward intelligent direction.

3 The Application of AI in Education Field

Is by the end of the 2019 market research report,T he application of artificial intelligence in education industry form mainly embodied in the following five aspects: Intelligent evaluation, Intelligent tutor, Education Simulation, Education robot and Mixed reality education [5]. Here we selected the now use the highest frequency and the latest education situation is analyzed. Construct and optimize the content model, and establish a knowledge graph, so that users can more easily and more accurately find content that suits them. A typical foreign application in this regard is a graded reading platform, recommending appropriate reading materials to users, and linking reading with teaching, with a quiz at the end of the text, and generating reports of relevant reading data, so that teachers can keep track of students' reading conditions.

3.1 Education Robot

Education robot. It in addition to the robot body itself, and the corresponding control software and teaching textbooks, etc. Because education robot to adapt to the new curriculum, the cultivation of student's scientific literacy and play the positive role to improve education quality, In many primary and secondary schools to promote, and to the features of "play school" is popular with teenagers,

Robot into schools and computers on campus, has become a necessary trend, Robot education has become the primary and secondary school education in the field of the new curriculum, the education robot will become a trend in the future. Robot education prevailed earlier in the United States, Japan, the United Kingdom, Germany and other countries, and some countries have included robot education in the curriculum of primary and secondary schools. For example, the United States already has relatively complete robotics education courses in primary and secondary schools, including robotics courses in the classroom, robotics extracurricular practical activities, robotics-themed summer camps, and related courses that use robots as tools to cultivate students' abilities. There is also a project "Robotics in K12 Education" supported by the National Natural Science Foundation of the United States, which aims to help K12 educators develop or improve robots to support STEM and develop robotics courses. And this practice-based education model is obviously recognized by many parents and children. According to a joint survey conducted by BSM and Digital Kids Media, 65% of parents said that compared to traditional toys, they are more willing to spend money on education that can learn STEAM education. On the robot. Former US President Barack Obama has personally participated in programming to encourage young people to learn programming, and Melinda Gates, Chairman of the Gates Foundation, has also expressed his admiration for educational robots.

3.2 Online Autonomous Learning Platform

Online autonomous learning platform. Autonomous learning platform is different from previous forms of classroom teaching, students able to log in to the learning anytime and anywhere through the network platform for a course. Login process is based on students from the network learning platform to learn course knowledge to finish the homework, tests and comprehensive tests - end of course evaluation. Autonomous learning platform can be used as auxiliary teaching, cannot completely replace the classroom teaching. It also has the content management, resource management, case management, work management, and management and learning management function module, able to complete online lessons, online teaching and learning control and so on a series of teaching activities. The online education platform uses all tools to conduct educational activities based on the premise of improving efficiency. It is the essence of online education research to use advanced Internet technology to change the way teachers and students communicate in class, to further improve the efficiency of students' knowledge acquisition, and to further cultivate their abilities. Pre-class teachers collect relevant materials based on educational goals and educational content, and make them into web pages; teachers in class explain classroom tasks, allowing students to explore, seek, and exchange answers to questions based on their own understanding of tasks and their weak links Or the solution; then the teacher gives the answer to the problem or the method to solve the problem. This way of teaching fully makes the students change from passive receivers of knowledge to explorers of knowledge; students focus on solving the knowledge that they have not yet mastered according to their own situation, thus avoiding that students who have already mastered the knowledge in the previous class have nothing to do Or students who have not mastered the previous part of the knowledge are in a passive situation at a loss. Online courses are best at solving the problems of hierarchical education and individualized education in the current class teaching system. Online education allows some students to take the lead, making it possible for students to learn in accordance with their aptitude, and the efficiency of education will be greatly improved. This is where the strongest vitality of online education lies under the premise of the class teaching system.

3.3 MR Technology

MR (Mixed reality) technology. MR techniques include augmented reality and enhanced virtual, the technology through the introduction of reality in the virtual environment, Set up between virtual and reality, and the user can real-time interactive information loop, thus increasing reality [8]. This is the further development of virtual reality technology, the technology by introducing real scene information in virtual environment, In the virtual world, build an interaction between real world and user information feedback loop, it using augmented reality and enhanced virtual reality and virtual world together and produce a new visual environment. Information and virtual reality simple superposition was the result of AR technology, and virtual information exist alone is the embodiment of the VR technology. MR technology is the key to the new visual environment coexistence of real and virtual digital objects, and real-time interaction. The key point of the core technology of MR technology is how to create a high-definition, virtual object that can be accurately focused on the front and back objects like the human eye. This is also the biggest difference and advantage compared to ordinary AR/VR equipment. At present, the most typical civilian MR equipment on the market is Samsung Xuanlong MR. According to many users who have personally experienced Xuanlong MR, the device will find that the distance between people and visual objects seems to disappear, giving a more realistic holographic feeling. And this kind of visual technology, which is similar to the rapid construction and transformation of virtual models, and can provide content richness, can indeed be used in many fields.

4 AI Education are Faced with the Problem

Despite the exciting future development potential of AI in education field, there are still many problems remain unsolved. For example, the technology of context-awareness in AI is not yet mature, there remains tens of technical issues to solved in the AI system. At the application field level, the integration of artificial intelligence technology and education is not close enough. At present, most products only focus on a single narrow field of adaptive learning, and pay less attention to student growth, comprehensive ability development, and physical and mental health.

4.1 Shortcomings of Existing AI Technology

The use of artificial intelligence in a special primary intelligence phase. This phase refers to the development of artificial intelligence technology in our country is just a simulation on a person's behavior, The machine through a lot of learning and memory to simulate the function of the human behavior, Performance for the intelligent products on the market data is missing, the teaching model of a single problem. Teachers will never be replaced. Because teachers are to promote the growth of people and have two functions, one is the function of teaching, the other is the function of educating people. In the future, if it is only knowledge-based teaching, knowledge-based teaching will be increasingly improved by artificial intelligence, but it is impossible to completely replace it. Because people need communication between people, face-to-face communication, this emotional communication is still different from our communication on the screen. People will never replace. But the efficiency of many of our lectures will be greatly improved. In addition, in addition to teaching, teachers also educate people, as well as solve various problems in the growth of students. The solution of this problem requires artificial intelligence to enhance. Teachers are very important in education and teaching. I think the relationship between teachers and artificial intelligence is a relationship of mutual empowerment and mutual enhancement.

4.2 The Contradiction Between Technological Development and Real Need

The talent cultivation of artificial intelligence is too broad. Artificial intelligence is a branch of computer science, its purpose is to understand the essence of the computer intelligence, and use this function to create closer to human intelligence machine. And its composition is very broad, and contains the psychology branch, machine vision, mathematics, control theory, and so on, but most of the colleges and universities just opened but to open the course system of artificial intelligence far couldn't keep up with the cultivation of the talent post standard, led to the theory and practice, then there is the artificial intelligence development is rapid, the corresponding disconnected for teaching content teaching material and teacher's teaching experience of imperfect is artificial intelligence and education. Data is the "nourishment" of artificial intelligence. The key bottleneck of artificial intelligence in education is data. Data between different education systems and platforms is not open and shared. The phenomenon of information islands is serious. It is difficult to collect data throughout the entire process of student learning. Without data, there is no intelligence. At the level of intelligent decision-making, a single intelligent algorithm cannot adapt to complex and changeable education scenarios, and there is a risk of partial generalization. It requires multiple intelligent systems to make joint decisions, strengthen manual intervention, and realize human-machine joint decision-making; the value of artificial intelligence There are polarizations in understanding. There are not only the overall affirmation of artificial intelligence and the idea of "only artificial intelligence" in the process of educational application, but also the overall denial of it and the idea of turning a blind eye to the benefits of artificial intelligence. In terms of artificial intelligence education applications, Human-machine coordination should be emphasized, not overestimated or underestimated.

4.3 The Combination is not Close Enough

In the field of application level, The combination of artificial intelligence and education is not close enough. People increase the attention was focused on the education quality of teaching and digging their own learning ability from two aspects, To the students' physical and mental health, in the heart not enough comprehensive quality and ability development. In addition, artificial intelligence makes cognition occur not only in the mind, but also in the interaction between people and intelligent tools. In terms of educational relations, artificial intelligence has broken the balance of knowledge dissemination in education and strengthened the "student-centered" relationship. And virtual tutors, virtual learning partners, virtual teams, virtual coaches, virtual classmates, etc., are the extension, reinforcement and supplement of human brain intelligence, changing the interaction between the learning subjects, learning subjects and the environment in the past. Learning ecology. "But no matter what changes, the general trend of education development is to let students learn to learn and create.

4.4 Lack of Applicability

In the aspect of judgment too careless. For the diversification of teaching scenario, single teaching schemes are biased, machine linkage access to information is not complete in a series of make a wrong judgment for students' learning situation decision-making. The knowledge learning of artificial intelligence has a gradual process. The core of artificial intelligence is the automation of intelligence. Just like machinery is the extension of our physical strength, artificial intelligence is the extension of our brain power. It allows us to handle complex things that we could not handle before. The above is to improve the efficiency of our teachers and learn these knowledge appropriately. Like in our lives, such as holding a mobile phone to record voice every day, is that complicated? Not complicated. However, the technology behind it is very complicated.

4.5 Ethical Issues

For the presence of the artificial intelligence produced ideological polarization. Both to negate its emergence and ignore the benefits of artificial intelligence and the into artificial intelligence will replace human negative thoughts. A study by Ruhr University in Bochum, Germany showed that the brain changes its structure within 3 h after learning new things. Artificial intelligence-supported learning methods such as personality learning, collaborative learning, experiential learning, and inquiry learning will change the brain structure more significantly. In particular, the in-depth experience and inquiry learning supported by artificial intelligence will deeply activate different brain nerve areas in many ways, which means that the human brain is constantly being reshaped by intelligent technology.

5 The Development of AI in Education

The current application technologies of artificial intelligence in the education field mainly include image recognition, speech recognition, and computer interaction. For example, through image recognition technology, artificial intelligence can free the teacher from the heavy batch work and scoring work; speech recognition and semantic analysis technology can assist teachers in English oral test assessment, and can also correct and improve students' English pronunciation; and human-computer Interactive technology can help teachers answer questions for students online. Last year, the media reported that robotic assistants at Georgia Institute of Technology replaced human assistants to communicate with students online, but no students found out, indicating the application potential of artificial intelligence in this area. In addition, the educational applications of artificial intelligence such as personalized learning, intelligent learning feedback, and robot remote teaching support are also promising. Although the application of artificial intelligence technology in education is still in its infancy, with the advancement of artificial intelligence technology, its application in education may deepen in the future, and the application space may be even greater.

5.1 Promote Education System Innovation

Artificial intelligence will drive the future innovation of education system, and artificial intelligence can't depart from the support of big data. Education is no longer one-sided in the future, But under the circumstances of data flow, in the artificial intelligence technology as the support of the education system in artificial intelligence, Only combining the artificial intelligence and big data, to maintain the normal operation of the education system. In the education industry, artificial intelligence is not only used to save teachers' manpower and improve teaching efficiency, but also to drive changes in teaching methods. Take artificial intelligence-driven personalized education as an example, collect student homework, classroom behavior, exams and other data, make personalized diagnosis of different students' academic conditions, and further develop targeted counseling and exercises for each student, so as to teach students in accordance with their aptitude This has become a direction for educational artificial intelligence to explore personalized education. However, a key point for realizing artificial intelligence to lead personalized teaching is data collection and analysis. Regarding the relationship between data and artificial intelligence, there is a saying that data is a kind of "nutrient" for artificial intelligence. In the same way, we can also say that educational data is the "nutrient" of educational artificial intelligence.

Educational data is generated from various educational activities and the entire teaching process. If artificial intelligence is to be better applied to education, the first thing to face is the problem of data collection. At present, there are two sources of educational data. One is from the digital teaching environment, where teaching and learning data are naturally produced in this digital environment, and the other is to collect educational information from traditional teaching behaviors and transform it into data. The advantage of the former is that it collects data in real time, is efficient and saves manpower. Now that the breadth and depth of Internet + education needs to be further promoted, a large part of the source of education data depends on the latter. In the future, education data may become A major factor in the development of artificial intelligence in education.

5.2 Make Things Easier

Determine the role of artificial intelligence. The rapid development of artificial intelligence technology, on the one hand, study, life and work of for mankind causes all sorts of convenient, on the other hand will also let the human of artificial intelligence. So, artificial intelligence should exist in a kind of what kind of role? "Afar" dog won lee se-dol century war, is the artificial intelligence can be intellectually through deep learning and imitating human brain thinking, But a big part of artificial intelligence called artificial intelligence is derived from the machine for rely on, Artificial intelligence is not everything. There's mastercard it is the person will lose the meaning of existence, Artificial intelligence is the essence of imitation, he isn't it like humans have the ability to perceive things and independent thinking ability to solve problems, from this perspective, the machine can't replace human innate judgment and originality is the same with artificial intelligence. Human made robots are not used to replace human, but to help humans, to extend the ability of human beings. Robots are artificial, need people to maintain, and the robot has a lot of ability is not as good as. As some dangerous environment, people can't go to the robot can. But the robot in many unknown and complex environment risk environment such as earthquake, unable to make the right decisions, Then will need to have rich experience and knowledge of human cooperation with it, to complete the task together, therefore the relationship between human and robot is cooperation.

5.3 Change Roles Between Teachers and Students

Positive change the role of teachers and students. The emergence of artificial intelligence in a certain extent impact the position of teachers' professional, Spur teachers to some extent, this also should be from the traditional "book centered" teaching idea of a single into a comprehensive role, Learn to use the information collected by the education robot. To make system analysis, use according to their aptitude, Targeted and personalized education mode. And on the other hand, students have to do is reasonable use of resources, improve their own quality and competitiveness, Strive to become a "will use, to use, with good"

[6] intelligent science and technology innovation talents. In this new learning culture, teachers have also changed. In the future, teachers cannot be replaced by artificial intelligence. But it is believed that the role of teachers must be changed from knowledge disseminators to learning facilitators; teachers' ability structure must also be changed. Teachers who do not understand technology will be replaced by teachers who understand technology; there must be a reasonable division of labor between humans and machines.

5.4 Establishing and Perfecting the Education Teaching System

Establishing and perfecting the education teaching system of artificial intelligence. In one hundred, education for this. Schools as a talent training base, to keep pace with The Times, bring into play the function of education to move. First, the school will accept artificial intelligence into the campus, to facilitate the construction of wisdom campus, fundamentally enable students to experience the convenience brought by artificial intelligence, "To learn, use high school". Second, the school take responsibility to cultivate innovative talents, to meet the needs of The Times development for talents. Third, the state to strengthen the support of vocational colleges as well as the appropriate courses in STEAM at different ages, pay attention to develop the students' ability of innovation, creative thinking and opened the AI of colleges and universities, from experience, perfect the relevant teaching system. In traditional artificial intelligence learning systems, we find that more is to meet the learning needs of a specific field, the purpose is to promote learners to obtain specific knowledge and skills, and these systems are often used as a supplement to school education, fail to go deep Affect students' daily study and life. With the development of artificial intelligence technology, the education field puts forward higher requirements for artificial intelligence technology, and it is expected that artificial intelligence technology will have a revolutionary impact on education. In general, artificial intelligence can still bring many opportunities to education, can provide personalized learning, promote the improvement of learners' literacy and ability in all aspects, accelerate the popularization of global classrooms, and reduce teachers' repetitive tasks.

6 Conclusion

Now, with internal and overseas for artificial intelligence research paper literature gradually reduce the proportion of the artificial intelligence, in the field of artificial intelligence research in China mature fine and our country has been in 5G communication technology achievements is good to promote the development of artificial intelligence in the education industry. Artificial intelligence has entered the machine period of the term, our country also vigorously support the development of artificial intelligence, in the new era, the development of artificial intelligence is the core of talent. Artificial intelligence industry still faces challenges in education, the cluster of the new technology breakthrough, the study way of change, and human vision is to drive education for a better education.

References

1. Mccarthy, J., Minsky, M.L., Rochester, N., et al.: A proposal for the Dartmouth summer research project on artificial intelligence. J. Mol. Biol. **1**, 279–289 (2006)
2. Lee, K., Gang, W.: Artificial intelligence. Culture Press, Beijing (2017)
3. Lin his life. The philosophical thinking of intelligent machine, vol. 21. Jilin University, Changchun (2017)
4. To promote artificial intelligence education three big problems of the People's Daily
5. Artificial intelligence application in the field of education, Finance trade vocational college, gansu province, gansu province, lanzhou, PeiLiLi, 730000
6. Learning technology black technology: artificial intelligence will bring the disruptive innovation education aixia Li, xiaoqing Gu (east China normal university education information technology department, Shanghai 200062)
7. Net of paper of hot point of artificial intelligence technology and development of the original 2019
8. Introduction to the application of mixed reality technology in university library, ocean university of China library, zhen Sun, xiaochen Ji, 266100

AI Applications in Education

Zhengyu Xu(✉), Yingjia Wei, and Jinming Zhang

Hui Zhou Engineering Vocational College, Guangdong, China
379830002@qq.com

Abstract. In recent years, led by the wave of artificial intelligence, "artificial intelligence + education" has become a very hot topic. More and more traditional educational institutions have begun to organize and layout the field of ARTIFICIAL intelligence education. Training artificial intelligence talents will become an important mission of education. Meanwhile, educational methods will change with the development of artificial intelligence, and the deep integration of artificial intelligence and education will become the development trend of the future education world. The future has come, when education and artificial intelligence meet, what kind of spark will be produced? This paper mainly discusses the application, research status and development trend of artificial intelligence in the field of education, as well as the deep integration of artificial intelligence and education.

Keywords: Artificial intelligence (AI) · Education · The depth of the fusion · Future educational world

1 Introduction

In July 2017, the State Council issued the Development Plan for The Next Generation of Artificial Intelligence, proposing the development of the artificial intelligence education system and the use of artificial intelligence technology to speed up education reform and new talent training programs. In May 2019, General Secretary Xi Jinping announced the official launch of the 'Artificial Intelligence + Education' strategy at the International Conference on artificial intelligence and Education. The integrated innovation of artificial intelligence and education has become a new form of future education, with a promising future [1].

First, what is artificial intelligence? The definition of artificial intelligence was first proposed in a Dartmouth seminar in 1956. At that time, it was defined as: artificial intelligence is a robot with the ability to simulate learning characteristics or intelligent characteristics that can be accurately described [2]. In recent years, the pace of artificial intelligence development is faster and faster, and it involves more and more industries, and gradually appears in People's Daily life. Thus, the concept of "Artificial Intelligence + X" entered people's lives. They take advantage of artificial advantages and put forward such forms as "Artificial Intelligence + Industry", "Artificial Intelligence + Medical care", "Artificial Intelligence + Agriculture", "Artificial Intelligence + Transportation", "Artificial Intelligence + Education" and so on. People hope that the deep learning, data mining, expert system and other functions of artificial intelligence

S. Shi et al. (Eds.): AICON 2020, LNICST 356, pp. 326–339, 2021.
https://doi.org/10.1007/978-3-030-69066-3_29

Fig. 1. Conceptual diagram of the application of artificial intelligence in education.

can make up, improve and optimize the current work. The artificial intelligence education or educational artificial intelligence developed from "Artificial Intelligence + Education" is one of the active explorations [3].

Second, what is education? In a broad sense, education means to increase people's knowledge and skills, while in a narrow sense, education means that schools have a purpose, a plan and an organization to carry out systematic learning for students.

Third, what is the difference between artificial intelligence education and traditional education? If the object of traditional artificial education is to learn from each other and exchange the culture or experience left by predecessors, then the object of artificial intelligence education can learn from each other, from person to person, from machine to machine and from machine to machine. Artificial intelligence education is a new form of education formed on the basis of artificial intelligence technology, which is integrated in concept, organization, mode, resources and other aspects.

Fourth, what is the relationship between education and artificial intelligence? Secretary Xi Jinping once said, "Education is the foundation of the country". The prosperity of a country's development is closely related to that of the younger generation, a country's education is closely related to its development. Only when the young generation has ideals and responsibilities can it promote the harmonious and stable development of the society and the inheritance of culture [4]. So education has always been the focus of attention of countries all over the world. So with the advent of artificial intelligence, it opens up a new form of education. With its own computing power, storage capacity and big data processing, combined with the Internet platform, the teaching quality and efficiency are improved and the education is more in line with the needs of social development. The relationship between the two is mutual promotion, not mutual substitution. For teachers, the emergence of artificial intelligence has reduced a lot of tedious and repetitive administrative work and reduced their burden. For students, the emergence of artificial intelligence promotes their independent learning ability, thinking ability and innovation ability, and expands their horizon and thinking mode. For education, the difficulties faced by artificial intelligence entering the education industry will accelerate the pace of technological upgrading. The emergence of artificial intelligence brings infinite possibilities to the forms of education, such as virtual and real interaction, online education, intelligence assessment and so on.

2 Significance

2.1 Artificial Intelligence Will Make the Traditional Classroom More Intelligent

The application of artificial intelligence in education can free teachers from the repetitive mechanical work, so that they can have more time and energy to focus on the cultivation of students' high-level thinking and innovation ability, making "teachers" become real "masters" and "coaches". Teachers are no longer merely impeller of knowledge, but teaching service providers to meet students' personalized needs and growth consultants to design and implement customized learning programs.

2.2 Open Education Environment, Education Scene is no Longer Limited to the Classroom

Artificial intelligence can bring unusual experience to teaching. The application of artificial intelligence to education can make education no longer confined to the classroom. At the same time, the school, society and enterprises will be organically integrated, so that students can get more open and diversified educational experience.

2.3 Artificial Intelligence Improves the Quality of Teaching

Artificial intelligence can observe and analyze students' learning patterns and individual differences. And take the student as the center, put my student knowledge advantage, the evaluation analysis student learning ability and carries on the real-time instruction. Can break through the previous teaching content is identical, with the same teaching material and teaching mode to respond to different students teaching method. A lot of data suggests that under this mode, students' academic performance is improved, and they have outstanding performance compared with students under traditional education mode in terms of course completion time, course passing rate and examination results. At the same time, Artificial intelligence education also makes universal quality-oriented education possible. It not only improves the efficiency of exam-oriented education, but also infuses critical thinking training, innovation ability training and other elements to comprehensively cultivate students' abilities.

3 The Main Form of Artificial Intelligence Education Applications

The most direct result of the application of artificial intelligence in education is the birth of an intelligent teaching system. The intelligent teaching system emerged on the basis of computer-assisted teaching. It is an open human-computer interaction system formed by student-centered, computer-based, and computer-simulated thinking processes of teaching experts. At present, the intelligent teaching system has become the main form of artificial intelligence application in education. Intelligent teaching systems mainly apply artificial intelligence principles in knowledge representation, reasoning methods and natural language understanding. Because it integrates the activities of knowledge experts, teachers and students, correspondingly, the intelligent teaching system is generally divided into three basic modules: knowledge base, teaching strategy and student model, plus a natural language intelligent interface. Specifically, the functions of the intelligent teaching system are as follows: to understand the learning ability, cognitive characteristics and current knowledge level of each student; to select appropriate teaching content and teaching methods according to the different characteristics of the students, and to provide students Targeted individual guidance; allowing students to use natural language to conduct man-machine dialogue with the "computer tutor". The design of intelligent teaching system requires not only knowledge of computer science, but also theoretical guidance of educational science.

Application of artificial intelligence in the field of education. One of the biggest challenges in education is that everyone learns differently. It is difficult for teachers to accurately grasp each student's real learning situation, leading to the teaching design and teaching process, difficult to focus on each student's real learning needs, resulting in a waste of energy, time and teaching resources. But the artificial intelligence system can provide each learner with a personalized learning style, so that each student can learn in the most suitable way, accurately record the learning status of each student, assist teachers to achieve hierarchical teaching and precise teaching, and effectively solve the core problems of teaching and learning. At present, the application of artificial intelligence in the field of our education mainly includes image recognition, speech recognition, human-computer interaction and so on [6]. Applications mainly focus on tutoring, online learning, classroom teaching and other aspects. Artificial intelligence application in the field of teaching is mainly manifested in the application of Intelligent Tutoring System.

Intelligent teaching system is set intelligent classroom, intelligent marking, intelligent diagnosis and intelligent treatment, intelligent preview, intelligent operation, intelligent sentiment analysis for the integration of intelligent teaching system is designed to create a good learning environment for students, so that the students can convenient call all kinds of resources, to accept a full range of learning services, to achieve the success of learning. By establishing the subject of teachers, students and teaching management, the corresponding teaching strategies can be formulated and implemented according to the characteristics of different students and personalized teaching services can be provided for students. Distributed intelligent teaching system based on network is the latest development direction of intelligent teaching system. It can make students who are originally separated in different areas learn together in a virtual environment, make full use of network resources, give play to learners' initiative, and bring better teaching effect.

The application of artificial intelligence in the field of education is still in its infancy, and people effectively combine the high efficiency of machines with human intelligence to influence the development of society. In recent years, artificial intelligence technology has always maintained a rapid development speed, and its application in the field of education plays a huge role in education and teaching, and promotes the development of humanized and individualized teaching, and integrates teaching activities. Closely connected with the development of science and technology, this is a major innovation activity in the field of education. For the application of artificial intelligence in education, there are mainly four specific forms as follows.

3.1 Smart Assessment

Under the traditional education model, teachers' work content focuses on two aspects, one is classroom teaching, and the other is correcting homework. Among them, teachers need to spend more time and energy to correct students' homework. However, driven by big data technology, text recognition technology, and semantic analysis technology, automatic correction of homework has been realized in reality. Intelligent evaluation can simplify the correction process to a large extent. This is also a major change to the traditional evaluation method. It is faster, more efficient, and very accurate. It frees teachers from heavy homework corrections. Make it more energy in classroom teaching, effectively promote the improvement of teaching efficiency.

3.2 Smart Tutor System

The intelligent tutor system is one of the adaptive learning systems. It is precisely because of the emergence of this system that the one-way instillation mode of teachers to students under the traditional teaching mode has been changed to a large extent, and better teaching results can be obtained. The system can make targeted learning plans according to different students' mastery of learning content, and at the same time highlight students' personalized learning methods, and help students master knowledge points more quickly through richer learning resources to realize specific learning goals. Through the intelligent tutor system, it is even possible to analyze the expressions of the students and understand the learning status of the students from it. Through the feedback mechanism, the teacher can be more aware of the students' mastery of the classroom teaching content, and use an emotional perception to predict and adjust it. In fact, the development of the intelligent tutor system is still immature at this stage. Basically, it has more applications in self-study and Q&A, but it has relatively few applications in classroom teaching. If you want to apply it better, you still have to pay attention to the improvement and optimization on the technical level.

3.3 Educational Simulation Game

In modern education concepts, quality education is emphasized. Therefore, the classroom atmosphere should not be lifeless, but should be presented in a more entertaining way. Under the background of the rapid development of artificial intelligence, educational simulation games are not entertainment activities in the traditional sense. They are more targeted. They promote the openness of education and teaching through games, and create some digital games based on the simulation environment. Students can have a higher enthusiasm for learning. Through intelligent simulation games, students can form a new understanding of things, and at the same time, their observation and thinking abilities can also be well exercised, which promotes students to discover and solve problems proactively. Based on the simulation game environment, students can be more involved in learning through playing different roles, and participate in learning activities with great interest to gain new knowledge. The introduction of simulation games in teaching can show some abstract knowledge in concrete forms, so that students can form a more intuitive understanding and feelings, can effectively enhance students' attention, and make students' professional knowledge learning more solid and in-depth.

3.4 Educational Robot

Educational robots involve many disciplines. The application of multi-disciplinary knowledge and technology, the role of educational robots developed in assisting teaching is obvious. It can effectively add interest in the classroom, stimulate students' innovative ability, and rely on information technology to enhance students' knowledge and the ability to obtain information. In specific teaching applications, educational robots are an intelligent teaching tool that can form a powerful supplement for teachers to carry out teaching activities. Students can also actively seek answers to questions through this human-computer interaction and promote self-learning capabilities. Educational robots can perceive changes in students' emotions. Educational robots can perceive changes in students' emotions. If there are more exchanges with students, they can more accurately grasp the learning effects of students, which is conducive to teaching students in accordance with their aptitude, so that students can feel knowledge from the communication with intelligent robots charm.

4 A Case Study on the Application of Artificial Intelligence in the Field of Education——Take the Answering Software "Homework" as an Example

As a technology based on mathematics and computers, artificial intelligence has been applied to the study and life of middle school students and has played an important role. At present, in terms of learning, artificial intelligence mainly realizes its application in learning through various learning software. In the following, the author will take the "homework assistant" as an example to analyze the application of artificial intelligence in the field of education from three aspects: learning enthusiasm, precise burden reduction, and learning style reform.

4.1 The Application of Artificial Intelligence from the Perspective of Learning Enthusiasm

The enthusiasm of learning is one of the factors that affect the effect of teaching. In the learning process, if students are active and autonomous, they can receive and master knowledge well; on the contrary, if they are passively procrastinated, it will affect the final effect of learning. So, how does artificial intelligence improve students' learning enthusiasm? In fact, many students cannot clearly understand their own learning status, which leads to more confusion when making learning plans. However, artificial intelligence solves this problem well. Artificial intelligence helps students to summarize test sites and make learning plans, so that students can have their own learning and review plans, which not only saves time and brings convenience to students, but also improves students' enthusiasm for independent learning. For example, in the learning software "Homework Help", the "Homework Help" uses artificial intelligence to analyze the students' problems, and then develop a review plan suitable for the students. Therefore, students do not have to spend too much time on how to make a study plan. In addition, artificial intelligence learning methods such as photo search and voice input also save students a lot of trouble and play an important role in improving students' learning enthusiasm.

4.2 The Application of Artificial Intelligence from the Perspective of Precise Burden Reduction

In recent years, the Ministry of Education has been advocating to reduce the burden on students in the compulsory education stage, that is, simplifying the content of classroom knowledge and reducing homework. However, the consequence of this learning model is that the basic knowledge of students is not solid and the review after class is more difficult which may eventually result in students not being able to firmly grasp the knowledge. So, how to accurately reduce the burden and strengthen the weak points to consolidate knowledge for middle school students when the burden is generally high? The answer is to use artificial intelligence. Through the application of artificial intelligence to learning management, students can effectively understand their own strengths and weaknesses, and the training for strengths is appropriately reduced, and the training for weaknesses is appropriately increased. For example, the learning software "Homework Help", students review mathematics on the "Homework Help", and the "Homework Help" analyzes the math problems that students have done through artificial intelligence, and analyzes the students' current mathematics knowledge. The current learning situation of students in mathematics. At this point, the "homework assistant" can use artificial intelligence to formulate a corresponding review plan for the students, and reasonably allocate the review time for the mastery and weak knowledge points. In this way, students do not need to blindly review and reduce the burden of study, and at the same time, they can grasp the weak points to strengthen and consolidate. From this point of view, artificial intelligence plays an important role in reducing and consolidating the burden of study preparation.

4.3 The Application of Artificial Intelligence from the Perspective of Learning Methods

Take after-school review as an example. In traditional after-school review, students usually follow the teacher to review in class, and the other is to review independently after class. However, in these two review methods, students in the former cannot get the teacher's effective help based on their actual situation; the problem with the latter is that without the teacher's professional guidance, students may not be able to confirm whether their review is reasonable. "Help" solves this problem well. The learning software "Homework Helper" has the function of taking photos to search for questions. Therefore, when students review independently after class, they can take photos of doubtful questions and upload them to the "Homework Helper", which can then be based on the image recognition technology in artificial intelligence. To identify the question on the picture, and then export the answer and answering process of the question from the "Homework Help" database. In addition, the "Homework Gang" also has a voice input function. Students can also input questions by voice. The "Homework Gang" relies on artificial intelligence's speech recognition history to recognize students' voice information, and then analyzes the voice and derives the corresponding Answer and analysis. It can be seen that the efficiency of the artificial intelligence learning method of photographing and voice searching is far more convenient and efficient than the learning method of manually inputting the questions or consulting the teacher one by one.

5 The Problems Facing Artificial Intelligence Education

Although current AI education is not perfect, but it has also had a strong impact on all aspects of education. From the field of practice, People are trying to bring education into line with the requirements of the age of artificial intelligence, Our thinking has been changed, the mode has been updated, the behavior F has been changed, the resource development and other aspects have achieved positive thinking and exploration. However, as the current education is still in the early stage, the role of artificial intelligence and the development of artificial intelligence education is mainly reflected in the "technology", lack of education, and some problems existing in the discipline and artificial intelligence education practice.

5.1 Artificial Intelligence Education Technology Upgrade

Artificial intelligence is not equal to artificial intelligence education. The core technology of artificial intelligence is to simulate the thinking activities and behavior patterns of human beings in some aspects according to the amount of data collected, algorithm characteristics and computing speed. But artificial intelligence education is by no means a simple way to collect and analyze big data, And the subjective analysis of students' learning ability, type, style, specialty and a variety of related relations, and then put forward teaching Suggestions or take intervention measures. There are essential differences between man and machine. The simulated intelligence of machines is different from the natural intelligence of human beings. The intelligence of a machine is that the problem is formalized by man and that the computer can do the calculation. Then, Human intelligence is acquired through learning and practice, and has initiative. But the intelligence of the machine does not have the intelligence of the human mode of thinking.

5.2 Limitations of Artificial Intelligence Education Interaction

Although artificial intelligence has been developed for more than 60 years, it has great limitations in applied education at present. Education to some extent, education, to some extent, is a means of learning by which people communicate and inspire each other according to their own knowledge, but intelligent teaching system is far from reaching this level. Secondly, machines cannot communicate with students as humans can. Machines only judge students' input information and master students' learning situation, which leads to people's wrong information receiving due to the data generated by "machine intelligence", ignoring the real situation.

5.3 Learning Mode Solidification

Artificial intelligence education system is made into a teaching module based on the data of knowledge level, cognitive ability and learning style provided by different students. Through the test results of this module, students are judged and their learning process is evaluated. The level of each student is different, if according to the formal teaching module teaching, rather than according to the specific situation of each student flexible teaching, in the long run, it is not conducive to the personalized development of students.

5.4 The Scope of Artificial Intelligence is Limited

Although the introduction of ARTIFICIAL intelligence into education conforms to the development of The Times, it does not mean that all subjects are suitable for artificial intelligence or the current artificial intelligence education is not able to cover the learning of all subjects. There are obvious differences among various disciplines, which are mainly reflected in the differences in research objects, theoretical framework, discipline thoughts, research methods and expression methods, etc. These differences lead to the natural differences in teaching and learning of different disciplines [5].

5.5 Analysis of the Limitations of Artificial Intelligence in Education

In terms of the current level of development of artificial intelligence and the characteristics of artificial intelligence itself, its application in education also has its limitations.

Can Not Communicate Smoothly With Students. Education is essentially an "interactive" activity, and intelligent teaching systems cannot achieve the most full and true interaction. At present, the research results of natural language understanding are very limited, far from meeting the requirements of everyone's communication. In addition, with regard to educational issues such as attitudes, morals, and emotions, machines can only judge the degree of mastery and internalization through the information input by students into the computer, but cannot judge the true situation of students through the communication and observation of natural states like human teachers. Therefore, "machine intelligence" can easily be blinded by "eyes" and cannot achieve the natural and smooth communication between people.

Imperfect Decision-Making and Reasoning Mechanisms. The key intelligence of the intelligent teaching system lies in its decision-making and reasoning mechanism, that is, the "teaching strategy" module makes flexible decisions through reasoning according to the specific situations of different students. This decision is based on the knowledge level and cognitive characteristics of the students provided by the student module and learning styles, and these cannot be fully formalized. At the same time, with the continuous updating of educational concepts and the continuous improvement of teaching models and teaching methods, the ability of the teaching strategy modules used by the system to evaluate and judge the learning process of students is limited.

6 The Research Status

6.1 National Policy Support

At present, China has fully realized the importance of the integrated development of artificial intelligence and education, and carried out the relevant planning layout. In 2017, the Chinese government issued *the Development Plan for the New Generation of ARTIFICIAL intelligence*, proposing to speed up the training of high-end ARTIFICIAL intelligence talents, build artificial intelligence disciplines and develop intelligent education, said Zhong Denghua, vice minister of Education. In 2018, the Ministry of Education released the AI Innovation Action Plan for Institutions of Higher Learning to promote the implementation of AI in the field of higher education. In February 2019, China Education Modernization 20135 was released, proposing to accelerate the educational reform in the information age, build intelligent campuses, coordinate the construction of integrated intelligent teaching, management and service platform, and use modern technology to accelerate the reform of talent training mode. Under the guidance of these policies, some regions and schools have begun to explore the integration of artificial intelligence and education and teaching.

6.2 Research and Development Boom at Home and Abroad

From a global perspective, many developed countries and international organizations are deploying information-based, digitized and intelligent strategies to maintain their status as major countries and maintain their strategic initiative and autonomy. China's prosperity and development in the field of artificial intelligence technology, including the field of intelligent education, has made significant and even more than expected achievements worldwide. In the current overseas best seller “The Fourth Education Revolution – How Artificial Intelligence changes Education”. The grandiose interpretation of China’s ai boom is typical.” The US lead in AI will slowly wane – and China is eager to fill the gap as it strives to establish itself as the world’s leading AI innovation center by 2030,” the book says. It is noteworthy that although the book involves a lot of topics have some new ideas, but it is not new to the field of education in China, and there are even more current practices than its theoretical width. It can be seen that with the prosperity and development of technology base, China’s “artificial intelligence + education” has been eagerly explored in the first place.

6.3 The Market Share of Artificial Intelligence Education Continues to Increase

In recent years, the artificial intelligence education market has been favored by domestic and foreign investors and attracted much attention from capital, but the application penetration rate of artificial intelligence education is not high. In 2018, well-known investment institutions represented by Zhenfund and Hongshan Capital, education industry represented by Good Future, New Oriental and Internet industry represented by Tencent all joined the “AI + education” boom, and the financing amount reached 1 billion yuan. However, from the perspective of current types of ARTIFICIAL intelligence education, most of them stay in language training and photo search, and there are still no new breakthroughs in the technical shortcomings of AI education. The degree of intelligence is low, and the market share of education is not yet saturated, showing an increasing trend.

6.4 The Promotion of Artificial Intelligence Education Enterprises

The emergence of artificial intelligence education enterprises promotes the integration of artificial intelligence and education. For example, in foreign countries, the artificial intelligence education enterprises represented by Knewto of the United States have surpassed other enterprises of the same type in user scale, ranking the first. They use the education form of “AI + education” to develop self-adaptive education, collect students’ learning data, analyze students’ shortcomings, and propose targeted teaching programs to improve teaching quality. In China, Netdragon is the earliest enterprise to enter the “artificial intelligence + education” industry, and has established a big data and artificial intelligence laboratory as well as a research and development team. They also reached strategic cooperation with universities at home and abroad to deepen the multi-dimensional exploration of ARTIFICIAL intelligence plus education and promote the deep integration of artificial intelligence education.

7 Future Development Trend and Development Direction

Artificial intelligence has been developed by leaps and bounds for more than 60 years, and it is widely used in all walks of life, especially in the education industry. Companies are using AI technology and big data to help students learn and explore personalized education. For example, Polaris USES AI to build the artificial intelligence assisted teaching system "Polaris AI", Ape Guidance launches the AI correction product "Little Ape manual calculation", and the homework box releases the full AI course product "small box classroom", etc. In 2018, Teachers in Teachers' Classroom released the "Squirrel AI" intelligent adaptive system, which focuses on online general practice tutoring in primary and secondary schools. With the help of the intelligent adaptive system, teachers scan knowledge gaps based on students' online learning data, so as to help students find and fill up gaps and improve learning efficiency. This year, the technology President of Good Future Group said that the construction of smart education open innovation platform has achieved preliminary results. Under the framework of smart education national new-generation artificial intelligence open innovation platform, good Future Education open Platform will continue to deepen the openness to unite industrial partners and create a better future for the education industry. On August 12, 2020, the Artificial intelligence Education Alliance was established in Qingdao, and Hisense will contribute to the deep integration of "AI + education". Today, AI adaptive education is no longer a new thing. Up to now, the world has more than 100 companies, more than 100 million users. Enough to see the speed of its development. Under the condition of new technology in determining the quality of innovation education, innovation education teaching system must be thoroughly the traditional education concept, applied in the field of education technology is not only cold, auxiliary, without the connotation of machinery and equipment, and should be with temperature, the connotation, character, emotion and thinking, dynamic and creative interaction partners [7]. With the rapid development of big data, cloud computing and 5G support, it can be predicted that artificial intelligence technology will play an immeasurable role in the future education world, and will completely change the status quo of teaching, making individualized and precise teaching become a reality.

In recent years, with the development of computer technology, network technology, artificial intelligence technology, and modern education and teaching theories, the development of artificial intelligence in education has shown the following trends.

7.1 Start to Break Through the Single Individualized Teaching Mode

For a long time, computer-assisted teaching systems and intelligent teaching systems have emphasized individualized teaching models. This model does have many advantages in giving full play to students' learning enthusiasm, initiative, and individualized guidance. However, with the development of cognitive learning theory, people have found that it is not enough to emphasize individualization in computer-assisted teaching systems and intelligent teaching systems. In some situations (such as problem solving), it is often more effective to adopt a collaborative approach. Therefore, in the intelligent teaching system in recent years, the collaborative teaching mode has received more and more attention and research.

7.2 Intelligent Teaching System is Increasingly Combined with Hypermedia Technology

The hypermedia system has a good development environment, a flexible and convenient user interface, and features such as pictures, texts and sounds, and its information organization is consistent with the associative memory habits of human cognition, and has become the most ideal information carrier at present. And the most effective information organization and information management technology has broad application prospects in many fields, especially education. The introduction of hypermedia technology into the intelligent teaching system and the development of an intelligent hypermedia-assisted teaching system can greatly improve the teaching environment of the computer-assisted teaching system, stimulate students' enthusiasm for learning, and significantly improve the teaching effect.

7.3 The Relationship Between Intelligent Teaching System and the Network is Getting Closer

The application and popularization of the Internet provide a good space for distance education and lifelong education. At present, the combination of intelligent teaching and multimedia network has become an unstoppable development trend of artificial intelligence in education.

7.4 Combination of Traditional Artificial Intelligence and Neural Network Fuzzy Decision Mechanism

Traditional artificial intelligence carries out cognitive simulation from a macro perspective, which can partially simulate the logical thinking process of humans, while the fuzzy decision-making mechanism of neural network performs cognitive simulation from a micro perspective, and strives to mimic the fuzzy processing function of the human right brain and the parallel of the entire brain chemical processing function. In the future, a new intelligent processing model will be explored: combining the fuzzy decision-making mechanism of the neural network with the reasoning ability of the symbolic expert system, using multiple knowledge sources and multiple models for compound collaborative processing.

If the above technologies can be used maturely, it will play a decisive role in the development of artificial intelligence and its application in education.

8 Expectation and Conclusion

With the development of artificial intelligence education in various industries, there are a variety of problems, which makes us have a new prospect for the future form of education. If the application of a technology into a certain industry only satisfies a certain aspect of its needs. For example, "AI + education" alleviates the administrative work of teachers to a greater extent, while the actual help applied in teaching is very vague. For teaching effectiveness, we should pay more attention to the transformation of results instead of just looking at the figures. AI education is not the sum total of single technology, the real artificial intelligence education should be combined with advanced science and technology to consolidate and develop the educational and teaching achievements. Therefore, we expect to introduce MR mixed reality technology in "artificial intelligence + education" to improve the gaps in education. If it can be combined with artificial intelligence system in the future, it will be of great help to students' theoretical knowledge learning and practical ability. The emergence of artificial intelligence has great temptation for all industries. The potential of education market is huge, and the development potential of artificial intelligence is also unpredictable. But what is important is that people will rationally use technology to create value and avoid technical risks.

In fact, artificial intelligence education, in essence, is to simulate an excellent special-level teacher and give students one-on-one personalized tutoring. In the future, artificial intelligence will surely help all of us with lifelong learning. Future education is the collaboration between human and artificial intelligence. The new technology with artificial intelligence as the core will be integrated with teaching and become the next core driving force. Emerging technologies not only change people's study, work and life, but also bring new opportunities to the innovative development of future education. In a word, with the continuous progress of ARTIFICIAL intelligence, artificial intelligence will definitely have a broad development prospect and wider application in the future education world, so as to continuously promote the development of education in China.

References

1. Xianmin, Y.: Integration and innovation of artificial intelligence and education. Educationist **6** (1), 1–4 (2019)
2. Mccarthy, J., Minsky, M.L., Rochester, N., et al.: A proposal for the dartmouth summer research project on artificial intelligence. J. Mol. Biol. **1**, 279–289 (2006)
3. Zhiming, Y., Xiaxia, T., Xuan, Q., et al.: The connotation, key technologies and application trends of educational Artificial Intelligence (EAI)-the United States "preparing for the future of artificial intelligence" and "national artificial intelligence research and development strategic planning" report analysis. Dist. Educ. J. **1**, 26–35 (2017)
4. Yuchen, W., Chunqiang, W.: Thinking of Artificial Intelligence + Education. Vitality (2019)
5. Wu, J.: Several questions about artificial intelligence education. Beijing Technology and Business University
6. Deng, L., Ang, C., Ning, S.: Analysis of the current situation and situation of artificial intelligence education development. Inf. Commun. Technol. Policy (2019)
7. Chen, W., Leyang, C., Li, Y., Xindong, Y.: 4D printing technology and its educational application prospects-and the integration of "artificial intelligence + education. J. Dist. Educ. **244**(1), 29–40 (2018)

The Future Development of Education in the Era of Artificial Intelligence

Zhengyu Xu(✉), Xinlu Li, and Jingyi Chen

Hui Zhou Engineering Vocational College, Guangdong, China
hitusa@126.com

Abstract. In recent years, "artificial intelligence + education" has become a very hot topic under the guidance of the wave of artificial intelligence. More and more traditional education institutions begin to organize and lay out the field of artificial intelligence education. The cultivation of artificial intelligence talents will become an important mission of education. At the same time, the mode of education will also change with the development of artificial intelligence Integration will become the development trend of the future education world. The future has come, when education and artificial intelligence meet, what kind of sparks will collide? This paper mainly discusses the application, research status and development trend of artificial intelligence in the field of education, as well as the deep integration of artificial intelligence and education.

Keywords: Artificial intelligence · Artificial intelligence + education · Deep integration · Future education world

1 Introduction

Artificial intelligence, referred to as AI. In 1965, at a two-month seminar held at Dartmouth University, the term "artificial intelligence" was put forward for the first time, which marked the formal birth of the emerging discipline of "artificial intelligence". Artificial intelligence, referred to as AI. In 1965, at a two-month seminar held at Dartmouth University, the term "artificial intelligence" was put forward for the first time, which marked the formal birth of the emerging discipline of "artificial intelligence". The essence of artificial intelligence is to simulate the information process of human thinking. Its precise definition is: a computer system has human knowledge and behavior, and has the ability to learn, infer, judge, solve problems, memorize knowledge and understand human natural language. One of the main goals of its research is to enable machines to be competent for complex tasks that usually require human intelligence. As a branch of computer science, artificial intelligence technology has been widely used in various fields, such as machine vision, fingerprint recognition, face recognition, intelligent search, intelligent control and other applications, which have brought great changes to human life and social development. Since entering the 21st century, with the deepening of artificial intelligence research and the development of education informatization, the application of artificial intelligence in the field of education has been paid more and more attention. What kind of application does artificial intelligence have in the field of education? What role does it play in education? What's the impact?

S. Shi et al. (Eds.): AICON 2020, LNICST 356, pp. 340–349, 2021.
https://doi.org/10.1007/978-3-030-69066-3_30

Since the 18th National Congress of the Communist Party of China, with the construction of "three links and two platforms" as the starting point, China's education informatization work has made breakthrough progress, and the indicators have generally doubled, contributing to the historic progress of the party and the country. "Broadband network school to school" has developed rapidly. The Internet access rate of primary and secondary schools in China has increased from 25% to 90%, and the proportion of multimedia classrooms has increased from less than 40% to 83%. More than 14 million teachers participated in the "one teacher, one excellent class, one teacher" activity, and 13 million excellent class resources were formed. The use of information technology has also solved the problem that more than 4 million students in remote and poverty-stricken areas do not have enough classes due to the shortage of teachers. "Renren Tong" has achieved a great leap forward development. The number of its opening has increased from 600000 to 63 million, and its application scope has expanded from vocational education to all kinds of education at all levels. "Public service platform for educational resources" has begun to take shape, with more than 68 million registered users in the public service system. Digital educational resources have changed from decentralized services to national interconnected service systems. "Public service platform for education management" has been fully applied, basically realizing "one school, one code" and "one person, one number" for teachers and students, and basically forming an information pattern of "two-level construction and five level application". The information literacy of teachers and students has been significantly improved. More than 10 million primary and secondary school teachers, more than 100000 principals of primary and secondary schools, and more than 200000 teachers of vocational colleges have received education informatization training. The majority of students' awareness and ability to use information technology for learning have been continuously enhanced. China's education informatization has an international influence and has been invited to share China's experience on the international platform for many times (Fig. 1).

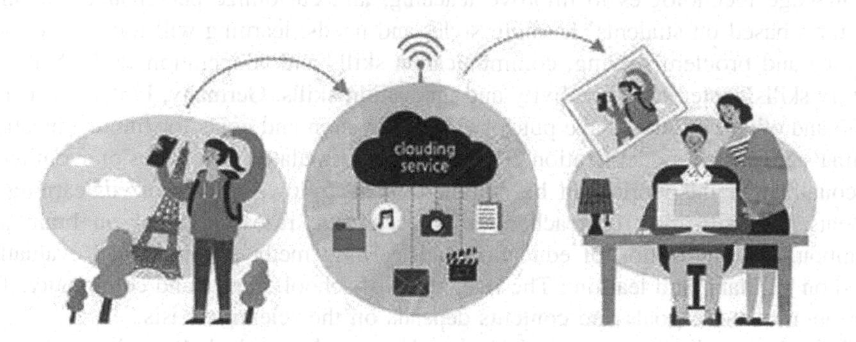

Fig. 1. Future education

2 Research Background of Artificial Intelligence Education

In the traditional teaching environment, most of our traditional education forms still stay in the stage of "exam oriented education", which ignores the high-level ability and knowledge application outside the memory. The high-cost and low-efficiency supplementary training outside the classroom also adds a great economic burden to parents. The consequence is that it increases the burden on Students' learning and even affects the development of multiple abilities And the application of the knowledge learned. It has become an urgent need to integrate artificial intelligence with education.

In 2017, the State Council issued the new generation of artificial intelligence development plan, which proposed to speed up the cultivation of high-end artificial intelligence talents, construct artificial intelligence discipline and develop intelligent education. In 2018, the Ministry of Education issued the "action plan for innovation of artificial intelligence in Colleges and universities" to promote the development of artificial intelligence from the field of higher education. In 2019, "China's education modernization 2035" was released, which proposed to accelerate the education reform in the information age, build intelligent campus, build an integrated intelligent teaching, management and service platform, and accelerate the reform of talent training mode by using modern technology. [1] With the support of relevant policies, artificial intelligence is quietly leading to a "educational reform", which not only brings opportunities and challenges to education, but also brings opportunities and challenges to teachers.

On the basis of the education goals proposed in the 2030 agenda for sustainable development, UNESCO adopted the education 2030 action framework on November 4, 2015, which makes specific plans for the realization of the 2030 education goals, namely "towards inclusive, fair and quality education and lifelong learning for all". Teaching 2030 points out that the teaching ecology will change: the progress of cognitive science and technology enables teachers and students to carry out immersive personalized learning. Teachers will combine the new findings of brain research and cutting-edge technologies to improve teaching, and customize personalized learning programs based on students' learning styles and needs; learning will focus on critical thinking and problem solving, communication skills and cooperation skills And 21st century skills centered on creativity and innovation skills. Germany, Finland, Canada, Japan and other countries have put forward new vision and goals for future education. "China's education modernization 2030" is being formulated. The focus of attention of all countries in the world can be summarized as follows: personalized learning of students, diversification of teachers' roles, scientific research based on brain and cognition, informatization of education and learning methods, diversified evaluation based on big data, and learning The integration of school, family and community. The realization of these goals and contents depends on the scientific basis.

The relevant data shows that in some places, only nearly half of the students in grade two of junior high school are short-sighted, 22% of the students are unwilling to go to school, 15% of the students are seriously tired of learning, and more than half of the students make up classes outside school. These problems directly affect the healthy growth of children and adolescents, affect the happiness of every family, and affect the level of human capital and the long-term development of our country Quantity. There are many problems in education within a certain range, such as "attaching importance to knowledge and neglecting ability", "attaching importance to intellectual education and neglecting moral education (lack of social participation and practice)" "emphasizing results and neglecting process", "emphasizing transmission and light exploration", "emphasizing unity and neglecting individuality". The essence of the contradiction lies in that we still rely on the original experience and our own experience to interpret the new needs of education in the new era, lack of scientific data support, and it is difficult to describe the characteristics of learners in the new era; educators at all levels lack sufficient data model support for the motivation system and educational principle of education in the new era; the paradigm of education system research still stays in the empirical method Can carry out personalized learning in depth on the existing basis. The solution of these problems needs the support of big data and the research of new education theory.

One core idea and two basic principles are the key to promote the modernization of education. One of the core concepts is to promote and serve the reform and development of education as the fundamental purpose of education informatization, and to promote the deep integration of information technology and teaching practice as the core concept. The real essence of educational informatization is integration rather than technology. The key to integration is not from construction, but from application. The two basic principles are to adhere to application driven and mechanism innovation. On the one hand, it is application-oriented, creates application environment through infrastructure construction, expands application channels through teaching and scientific research, promotes application efficiency through training, and improves application level through evaluation. On the other hand, we should pay attention to mobilizing the whole society to promote education informatization, especially to play the role of the industry, not only relying on the power of the government. In recent years, educational informatization has made great achievements, mainly relying on mechanism innovation. Adhering to the "two legs" of the government and the market, "invisible hand" and "visible hand" work together to explore and form a working mechanism of "government policy support, enterprise participation in construction, and schools focusing on application".

Artificial intelligence will accelerate the profound reform of education in the future.

The modernization of education is inseparable from the strong support of modern science and technology. Information technology has a revolutionary impact on the development of education. Driven by the new theories and technologies such as mobile Internet, big data, supercomputing, sensor network and brain science, as well as the needs of economic and social development, the rapid development of artificial intelligence has shown new features such as deep learning, cross-border integration, human-computer collaboration, open group intelligence and autonomous control, which promotes the rapid leap from digitalization, networking to intelligence in all fields of economy and society.

3 Application of Artificial Intelligence in Education

One of the biggest challenges of education is that everyone's learning style is different. It is difficult for teachers to accurately grasp the real learning situation of each student, which leads to the teaching design and teaching process, and it is difficult to focus on the real learning needs of each student, resulting in the waste of energy, time and teaching resources. The artificial intelligence system can provide personalized learning methods for each learner, so that each student can learn in the most suitable way for himself, accurately record the learning status of each student, assist teachers to realize hierarchical teaching and precise teaching, and effectively solve the core problem of teaching and learning. At present, the application of artificial intelligence in our education field mainly includes image recognition, speech recognition, human-computer interaction and so on. The application mainly focuses on tutoring, online learning, classroom teaching and so on.

The application of artificial intelligence in the field of teaching is mainly reflected in the application of intelligent tutoring system. Intelligent teaching system is a smart teaching system which integrates intelligent classroom, intelligent marking, intelligent diagnosis, intelligent treatment, intelligent preview, intelligent homework and intelligent learning situation analysis. It aims to create a good learning environment for students, so that students can easily and quickly transfer various resources and receive all-round learning services, so as to achieve the success of learning. By establishing the subject of teachers, students and teaching management, we can formulate and implement corresponding teaching strategies according to the characteristics of different students, and provide personalized teaching services for students. The distributed intelligent teaching system based on network is the latest development direction of intelligent teaching system. It can make students in different areas learn together in the virtual environment, make full use of network resources, give full play to the initiative of learners, and bring better teaching effect.

As learning becomes more paperless, AI driven learning systems will continue to become more effective and efficient. With the aid of artificial intelligence for intelligent and meaningful dialogue, learning will no longer be boring, but become children's game, truly realize game teaching and improve students' interest in learning.

3.1 Intelligent Tutoring System

Intelligent teaching system is one of the important applications of artificial intelligence technology in education, and it is the further development of computer-aided instruction (CAI) related research. The purpose of intelligent teaching system is to create a good learning environment for students, so that students can easily and quickly transfer all kinds of resources and receive all-round learning services to obtain the success of learning. The current intelligent teaching system mainly relies on the intelligent subject technology to construct. Through the establishment of teachers, students and teaching management subjects, the corresponding teaching strategies can be formulated and implemented according to the characteristics of different students, so as to provide personalized teaching services for students. The distributed intelligent teaching system based on network is the latest development direction of intelligent teaching system.

It can make students in different areas learn together in the virtual environment, make full use of network resources, give full play to the initiative of learners, and bring better teaching effect. The biggest difficulty of traditional teaching is that it is difficult for teachers to accurately grasp the real learning situation of each student, which leads to the teaching design and process, difficulty in focusing on the real learning needs of each student, resulting in the waste of time, energy and teaching resources. The intelligent learning platform can comprehensively and accurately record the learning status and effect of the whole class, quickly and accurately help teachers analyze the gains and losses of each link, so as to timely and effectively adjust teaching strategies, help teachers realize hierarchical teaching and accurate teaching, change from empirical type to scientific type, effectively solve the core problems of both teaching and learning, and truly achieve teaching Long.

3.2 Intelligent Network Examination System

At present, paperless examination has become an important new form of examination. In a broad sense, paperless examination includes the use of computers to establish and manage the question bank, topic selection, examination and marking. It not only innovates the traditional paper examination in form, but also improves the design and evaluation of the examination. The intelligent network test paper system has the advantages of low cost, low efficiency, good confidentiality and high consistency. Even in the case of many restrictions, it can still generate the test paper according to the given strategy. At the same time, the test question bank based on the network can collect the classic exercises written by teachers, centralize and share the success of teachers' labor, and ensure the high quality of test papers. The paper marking system based on artificial intelligence can effectively identify test papers, reduce the possibility of errors, and greatly improve the efficiency of marking process.

3.3 Intelligent Decision Support System

Intelligent decision support system is one of the important applications of artificial intelligence. It combines artificial intelligence with decision support system. The application of expert system enables decision support system to make full use of human knowledge, such as descriptive knowledge about decision-making problems, process knowledge in decision-making process, reasoning knowledge in solving problems, and helps to solve complex problems through logical reasoning Decision making. The intelligent decision support system is mainly composed of database, model base, method base, man-machine interface and intelligent components. At present, the intelligent decision support system has become the main development direction of the decision support system, and shows a strong development potential and bright prospects in the application of network education.

3.4 Intelligent Simulation Technology

In the distance education teaching, experimental teaching is an indispensable teaching link, but at present, the network teaching platform based on teaching and educational administration management rarely involves the experimental teaching content. Intelligent simulation technology is a high integration of artificial intelligence and simulation technology. It strives to overcome the limitations of traditional simulation models and modeling methods, as well as the problems of modeling arduous, monotonous interface and inexplicable results. To some extent, intelligent simulation system can replace simulation experts to complete the steps of modeling, designing experiments, understanding and evaluating simulation results, and has certain learning ability. Using intelligent simulation system to develop experimental teaching courseware can greatly save manpower and material resources, reduce development cost, accelerate development speed and shorten development cycle.

4 The Role of Artificial Intelligence in Teaching

4.1 Artificial Intelligence Makes Traditional Classroom More Intelligent

The application of artificial intelligence to education can make teachers free from the tedious and repetitive mechanical work, so that they can have more time and energy to pay attention to the cultivation of students' high-level thinking and innovation ability, and make "teachers" become real "masters" and "coaches". Teachers are no longer merely imparters of knowledge, but teaching service providers to meet students' personalized needs and growth consultants who design and implement customized learning programs.

When the machine can think, what kind of ability should we cultivate students?

(1) The ability of autonomous learning. The most basic ability of human beings is autonomous learning. In a rapidly changing society with constantly updated knowledge, machines have learned to learn autonomously. If students lack the ability of autonomous learning, they will be difficult to adapt to the challenges of the intelligent age.
(2) Ability to ask questions. It is not difficult for machines to ask simple questions based on experience, but it is difficult to replace human beings with deep-seated problems in the short term, especially driven by curiosity and interest.
(3) Interpersonal skills. Machine to machine communication is mainly realized by machine language, and the core quality of interpersonal communication is mainly based on self emotion control ability and judgment ability of others' emotion.
(4) The ability of innovative thinking. Innovative thinking is reflected in the innovation of methodology to a great extent, while the methods and rules of machine learning are artificially prescribed, so it is difficult to transcend and realize self innovation.
(5) The ability to plan for the future. How to balance the reality and the future is an important embodiment of human wisdom. Artificial intelligence can help us solve many existing problems and provide good services, but it may not be able to make a good judgment on the future.

4.2 Sharing High-Quality Teaching Resources Across Time and Space

International surveys and studies have been conducted to analyze the possibility of more than 360 occupations being replaced by artificial intelligence in the future, ranking all walks of life. The results show that the education profession is very backward, and the possibility of teachers being replaced is only 0.4%. This shows the particularity of education, teachers have the uniqueness that is difficult to be simply replaced by machines. However, if we fully consider the impact of the deep development of artificial intelligence, this situation may change significantly. If intelligence develops further and personalization is realized, it will be a completely different situation. It can be predicted boldly that in a few years, if a highly integrated and personalized robot like intelligent assistant is realized, the teaching team will face another prospect, and many teachers may be replaced. Of course, this replacement is not absolute. The combination of human and computer may be a common form in the intelligent era, and the education of human-computer integration may be the general form of education in the future.

Through information technology to achieve the sharing of high-quality education resources, improve the quality of teaching, which is also the evidence of further promotion of information-based teaching mode in China.

4.3 In the Open Education Environment, the Educational Scene is no Longer Limited to the Classroom

Artificial intelligence can bring unusual experience to teaching. Applying artificial intelligence to education can make education no longer limited to classroom, but also organically connect schools, society and enterprises, so that students can get more open and diversified education.

5 Research Status and Development Trend of Artificial Intelligence Education

Artificial intelligence has been developed for more than 60 years. It is widely used in all walks of life, especially in the education industry. Companies are using AI technology and big data to help students learn and explore personalized education. For example, Polaris uses AI to build an AI assisted teaching system "Polaris AI", ape tutoring has launched AI marking product "little ape mental arithmetic", and homework box has released all AI course products "small box classroom".

In 2018, Yixue Education released the "squirrel AI" intelligent adaptive system. The product will focus on online general practice counseling in primary and secondary schools. With the help of squirrel AI intelligent adaptation system, students can scan knowledge loopholes according to students' online learning data, help students find and fill in the gaps, and improve learning efficiency. This year, the technology president of Tal said that the construction of open and innovative platform for intelligent education has made initial progress Under the framework of the new generation of AI open innovation platform of smart education country, tal education open platform will unite industry partners with continuous and in-depth openness to create a better future for the education industry. On August 12, 2020, the Artificial Intelligence Education Alliance was established in Qingdao. Hisense will contribute to the in-depth integration of “Ai + education”. Nowadays, AI intelligent adaptation education is no longer a new thing. Up to now, there are more than 100 companies and more than 100 million users in the world. Enough to see the rapid development speed.

Under the condition that the quality of innovative education is determined by new technology, the education and teaching system must thoroughly reform the traditional education concept. The technology applied in the field of education should not be just cold, auxiliary and non connotative machines and equipment, but should be interactive partners with temperature, connotation, individuality, emotion, thinking, vitality and creativity. [2] With the rapid development of big data, cloud computing and the blessing of 5g, it can be predicted that artificial intelligence technology will play an immeasurable role in the future education world, and will completely change the teaching situation, so that individualized teaching and precise teaching will become a reality.

6 Conclusion

In fact, the essence of artificial intelligence education is to simulate an excellent super grade teacher and give students one-to-one personalized counseling. In the future, artificial intelligence will certainly help all people to carry out lifelong learning. Future education is the cooperation between human and artificial intelligence. The new technology with artificial intelligence as the core will be integrated with teaching and become the next core driving force. [2] New technology not only changes people's study, work and life, but also brings new opportunities for the innovation and development of education. In a word, with the continuous progress of artificial intelligence, artificial intelligence in the future education world will have broad prospects for development and more extensive application, so as to continuously promote the development of education in China.

References

1. Li, A., Gu, X.: Black technology in learning technology: whether artificial intelligence will bring subversive innovation in Education. Modern Educ. Technol. **29**(5), 13–19 (2019)
2. Chen, W., Chu, L., Yang, L., Ye, X.: 4D printing technology and its educational application prospect – also on the integration with “artificial intelligence + education”. J. Dist. Educ. **36** (2)44(1), 29–40 (2018)
3. Qian, H.: Research on the development of artificial intelligence technology. Modern Telecommun. Technol. **46**, 18–21 (2016)
4. Zhong, Y.: Artificial intelligence: “the doorway” behind the “lively.” Sci. Technol. Herald **34** (7), 14–19 (2016)
5. Zhou, B.: Discussion on the effective design and evaluation of junior middle school mathematics homework. China School Foreign Educ. **26**, 144 (2015)

6. Fei, L.: Emotional education based on artificial intelligence. Tianjin Sci. Technol. **6**, 20–31 (2010)
7. Fei, W.: Analysis of the application status of artificial intelligence in middle school education and teaching. Nat. Med. Educ. Technol. **8**(4), 397–400 (2013)
8. Yang, X.: Integration and innovation of artificial intelligence and education. Educationist **6** (1), 1–4 (2019)

Dormitory Management System Based on Face Recognition

Yu Yang[1,3] and Liang Ye[1,2,3(✉)]

[1] Department of Information and Communication Engineering, Harbin Institute of Technology, No. 2 Yikuang Street, Harbin 150080, China
yeliang@hit.edu.cn
[2] Physiological Signal Analysis Team, University of Oulu, Pentti Kaiteran katu 1, 90014 Oulu, Finland
[3] Science and Technology on Communication Networks Laboratory, Shijiazhuang, China

Abstract. This paper explores the application of face recognition system in dormitory management, and designs an EXE software, which realizes the function of entering and leaving dormitory through face voucher, storing the information of students, and automatically updating relevant data. The face recognition module includes three functions: face image recognition and interception, face alignment, face feature extraction and face verification. The traditional method of HOG is used for recognition and interception, and the method of gray-scale processing and gamma normalization is used for preprocessing to reduce the influence of light. The face alignment uses the 68-point landmarks and similarity transformation to align the face. The face feature extraction uses the pre-trained FaceNet neural network to extract the 128-d feature vector of the face, which is stored in the database or compared with the face in the database to output the Euclidean distance, then compare and find the most possible person according to the distance. The database module includes two functions: storing face information and displaying all information of students MySQL is used to build the database. An automatic dormitory information management system is established through the interconnection interface between MySQL and Python.

Keywords: Face recognition · HOG · FaceNet · MySQL · UI

1 Introduction

At present, with the rapid development of artificial intelligence, the era of AI has quietly arrived, and "showing your face" has gradually become a new trend. With the higher demand of quick and accurate methods of verification, Face recognition is becoming more and more common.

Face recognition is a biometric technology based on human's face feature information. A series of related techniques for face recognition are used to collect images or video streams containing faces with cameras, and to automatically detect and track faces in images, and then to recognize the detected faces. These techniques are also called portrait recognition or facial recognition. 2018 is an important node for the

S. Shi et al. (Eds.): AICON 2020, LNICST 356, pp. 350–361, 2021.
https://doi.org/10.1007/978-3-030-69066-3_31

common application of face recognition technology in China, marking the final arrival of the era of face recognition [1].

Except the field of security and finance, face recognition has been widely used in many other fields, such as transportation, education, medical treatment, police, e-commerce and so on. In order to further grasp the opportunities brought by face recognition technology, China has issued a series of policies to support the research about this new technique.

2 Background

Artificial intelligence (AI) is a advanced and new field. Face recognition is one of the representative directions of AI, which is widely used in our production and life, with high performance of real-time and awareness. Its application is common and its status in access control, payment, authorization and smart-home is increasingly high. At present, the research and development of face recognition is developing rapid and the jurisdiction is getting there and higher, but the use has not been so popular, many occasions that can apply face recognition systems have not installed that for a series of reasons.

Face recognition study became popular in the early 1990s [2] following the introduction of the historical Eigenface method. After this, many holistic approaches are put forward. However, a well-known problem about these holistic approaches is that these methods fail to address the unexpectable facial changes that deviate from their prior assumptions. So local-feature-based face recognition appeared to solve this problem. As the time flies, techniques like learning-based descriptors, deep learning descriptors appeared and the accuracy was dramatically boosted to 99.8% in just several years [3].

The purpose of this paper is to find an quick and accurate method which do not have the demand of systems of high performance to manage students' dormitory in a fast and efficient way. It has three parts as follows:

(1) Face recognition module: FR module includes three functions: face image recognition and interception, face alignment, and face feature extraction and face verification. The traditional method called Histogram of Oriented Gradient is used to detect and intercept a face, and the pre-processing uses the method of gamma normalization to reduce the influence of illumination. Face alignment using a python package called Dlib to align the 68 special point of the face. Face feature extraction adopts pre-trained Facenet neural network to extract 128 dimensional feature vector of face and to store it in database or compare with face in database and calculate Euclidean distance, so as to compare the most similar person according to this distance.
(2) Database module: the database module contains two functions: storing face information and displaying all information of members. Database is established by MySQL database.

(3) UI and software module: the software's user interface is designed by QT designer and encapsulated into a software system that can run by each device. The hardware demand is low and the related software environment is no longer needed.

3 Work Methodology

3.1 FR Module

3.1.1 Face Recognition and Interception

HOG features are called Histogram of Oriented Gradient [4]. The directional gradient histogram feature is a very effective image feature descriptor, which is widely used in various detection and recognition scenes. The algorithm can describe the image features by calculating the gradient of the local image and counting its directional gradient histogram, and can describe the edge texture information of a specific target well. HOG basic idea is that even without a clear understanding of the relative gradient or boundary attitude, the appearance and shape of an object can often be represented by a local strong gradient change or edge direction. For algorithm implementation, this idea is implemented by dividing the image window into small blocks, which are called cell. Units For each pixel in each cell, the one-dimensional gradient histogram is accumulated. In order to have better invariance to different illumination, control normalization is also an important step before using these local data. Normalization can be achieved by accumulating local histogram "energy" in a larger region and using this energy to normalize all cell. This larger region is called block. The algorithm of HOG are as follows:

Algorithm 1:

1. Grayscale processing.
2. Gamma normalization.

   Compute the value of each pixel with the function as follows:

$$I(x, y) = I(x, y)^{gamma} \tag{1}$$

   gamma usually take 0.5

3. Calculating the gradient of each pixel of the image (including size and direction).

   One dimensional gradient of pixel (x, y) in the image can be calculated as follows:

$$\begin{aligned} G_x(x, y) &= H(x+1, y) - H(x-1, y) \\ G_y(x, y) &= H(x, y+1) - H(x, y-1) \end{aligned} \tag{2}$$

$G_x(x, y), G_y(x, y), H(x, y)$ represents the gradient of pixel (x, y) in x orientation, y orientation and the value of the pixel. Then use the one dimensional gradient to calculate the amplitude and phase of each pixel with the following function:

$$
\begin{aligned}
G(x,y) &= \sqrt{G_x(x,y)^2 + G_y(x,y)^2} \\
\alpha(x,y) &= \tan^{-1}\left(\frac{G_y(x,y)}{G_x(x,y)}\right)
\end{aligned}
\tag{3}
$$

4. Segmenting of images into small area called cells (each 16 pixels).
5. Building gradient direction histograms for each cell.

Divide 180 into several bins. Then use these bins as abscissa to do weighted vote for the histograms. The weight are related to the amplitude of each pixel which are calculated by a statistical method called tri-linear interpolation.

Combining cells into large intervals (block), and normalizing gradient histograms within blocks.

First combine several adjacent cells into a larger area called block, then use a function to normalize the histograms within each block. There are four methods to normalize the histograms, this paper use L2-norm normalization as follows:

$$
L2 - norm, v \rightarrow v/\sqrt{\|v\|_2^2 + \varepsilon^2} \tag{4}
$$

v represents the value of each bin.

After the extraction of HOG features, these features can be used to train a SVM [5] classifier to detect a human face.

3.1.2 Face Alignment

Face alignment module is based on the Dlib deep learning library. According to the 68-points face landmarks [6], by finding and multiplying a similarity transformation matrix, the face features can be aligned to the average face. The algorithm can complete the face alignment in a few milliseconds and meet the fairly high accuracy, which can meet the efficiency and accuracy requirements of this program. A practical implementation of the algorithm is provided by the Dlib deep learning library, which can be easily applied to the program.

3.1.3 Face Verification

A deep network structure called Facenet [7] proposed in 2015 is used in this system. The biggest innovation of the network should be to propose a different loss function. What's more, instead of using a classifier, we use the distance of points in the eigenspace to indicate whether the two images are the same class. The network structure is as follows (Fig. 1):

1. Batch are sent to a deep architecture, such as a convolutional neural network. Google Inception ResNet V1 [8] is selected here.
2. 2.After CNN, each feature **x** should be L2-normalized to make, so that the feature of all images is mapped to a hypersphere.
3. Then send the feature into an embedding process, so the image x is mapped to 128-dimensional Euclidean space by function $f(\mathbf{x})$.

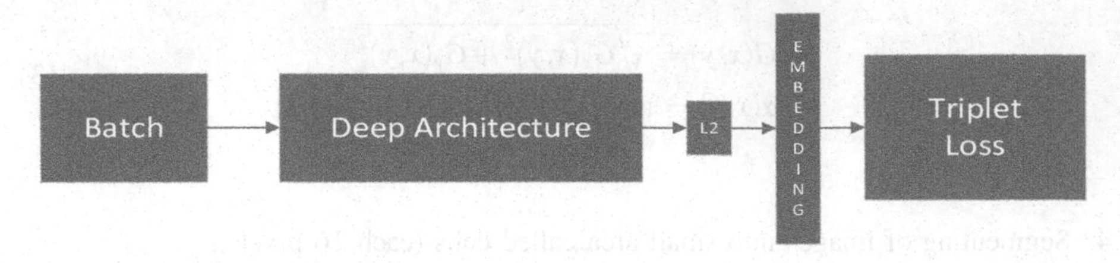

Fig. 1. Net structure

4. After the embedding layer, we optimize the weight of CNN with these features, and here we propose a new loss function, triplet loss function, which is the biggest innovation of the network.

The triple loss function is a loss function with three sample inputs, containing an anchor sample, a positive sample and a negative sample. By training the classification model with LDA method, the intra-class feature distance is getting closer and the inter-class feature distance is getting far away. In order to ensure that the feature distance between the anchor sample and the intra-class image (positive sample) is closer and the feature distance from the inter-class image (negative sample) is farther, we need the triplet loss function to realize it.

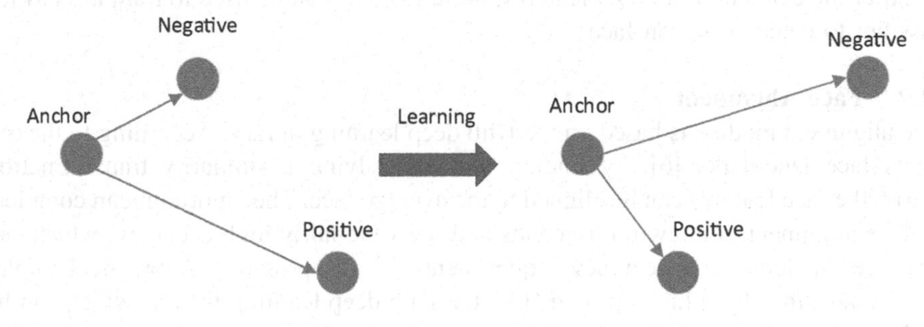

Fig. 2. Training ideas

A constraint can be constructed:

$$\|f(x_i^a) - f(x_i^p)\|_2^2 + \alpha < \|f(x_i^a) - f(x_i^n)\|_2^2, \forall (x_i^a, x_i^p, x_i^n) \in \Gamma \quad (5)$$

The variable α is the distance tolerance between the intra-class pair and the inter-class pair, which can be set artificially. Γ is the set of all triples in the training set.

Then write the formula 5 as a loss (optimization) function, and the model is optimized by reducing the value of the loss function. The loss function can be expressed as:

$$L = \sum_{i}^{N} \left[\|f(x_i^a) - f(x_i^p)\|_2^2 - \|f(x_i^a) - f(x_i^n)\|_2^2 + \alpha \right]_+ \tag{6}$$

By using the transfer learning method, the FaceNet model weights trained by 13000 pictures are transferred to generate 128-dimensional feature vectors of faces. Through comparing the Euclidean distance between the input image and images in the database, we can judge who is the most similar owner of the input face image. The Euclidean distance formula is as follows:

$$\sqrt{\sum_{i=0}^{127} (\vec{x}_i - \vec{y}_i)^2} \tag{7}$$

$(\mathbf{x}, \mathbf{y})$ are the 128 dimensional features of two faces.

3.2 Database Module

Database module is a place to store and display personnel information and face information. To achieve the following functions:

1. Storage and Automatic Updating of Dormitory Personnel Information.
2. Supporting easy to query, modify and import information manually.

MySQL is a relational database management system. Based on the MySQL database, we can establish an quick and high-capacity database module which satisfies the demand above.

4 Results

4.1 Test Platform

Hardware environment: CPU: Intel (R) Core (TM) i5-5257U.

GPU: Intel (R) Iris (TM) Graphics 6100 which does not support CUDA.

Software environment: operating system: Windows10.

Development language: Python.

IDE: Pycharm.

4.2 Variable Configuration and Data Set

The sizes of each element in FR system are as follows:

OpenCV detection window: WinSize = 128 * 64 pixels, the step size of sliding in the image is 16 pixels (both horizontal and vertical).

Block: BlockSize = 64 * 64 pixels, the step size of sliding in the detection window is 16 pixels (both horizontal and vertical).

Cell: Cell size = 16 * 16 pixels.

The training data set of face verification system is selected from CAS-PEAL data set [9] randomly. This test selects 500 images of 50 different people.

4.3 Test Result

After configuring and training, this section carries on a whole analysis to the system, mainly analyzes the face detection accuracy, the recognition speed, the face recognition accuracy rate, the database adds and deletes the check operation speed. LFW [10] (LableFaces in the Wild) data set is used for the test data set.

4.3.1 Face Recognition Performance Test

A trained SVM classifier is used to detect a part of the sampled picture in the LFW data set three times, and the three results are recorded in Table 1:

Table 1. FR test results

Number of tests	Verification accuracy	Recognition time
1	0.971	48.33 s
2	0.972	54.89 s
3	0.971	33.84 s

As a result, the average recognition time of 1454 pictures is about 40 s, the recognition accuracy is about 97.1%, the recognition performance is ideal, and the face detection can be realized quickly and accurately.

4.3.2 Face Verification Performance Test

The LFW data set is first aligned with the face, and the input layer image with FaceNet size should be 160 * 160 pixels. Then the FaceNet is used to verify all the images of the LFW data set, and the total face detection time, face detection accuracy, and the results are as follows:

```
Model directory: C:\Users\apple1\PycharmProjects\evaluation\facenet\src\models\20180402-114759
Metagraph file: model-20180402-114759.meta
Checkpoint file: model-20180402-114759.ckpt-275
Runnning forward pass on LFW images
............
Accuracy: 0.98467+-0.00407
Validation rate: 0.90567+-0.01995 @ FAR=0.00067
Area Under Curve (AUC): 0.998
Equal Error Rate (EER): 0.015
```

Fig. 3. Face verification result

It can be seen that the accuracy rate is 98.46 ± 0.004% and is at a high level. The system can accurately distinguish faces and meet the demand of practical application.

4.3.3 Database Performance Testing

Write 10000 pieces of data into the table, then delete them in batches. The records of the run time of the three programs are as shown in Table 2.

Table 2. Database test results

Number of tests	Runtime
1	129.0 s
2	118.7 s
3	126.5 s

As can be seen, the average time of adding and deleting 10000 pieces of information in the database is about 123 s, and it supports batch deletion, batch import, which also has very fast speed and very convenient operation.

4.3.4 UI Performance

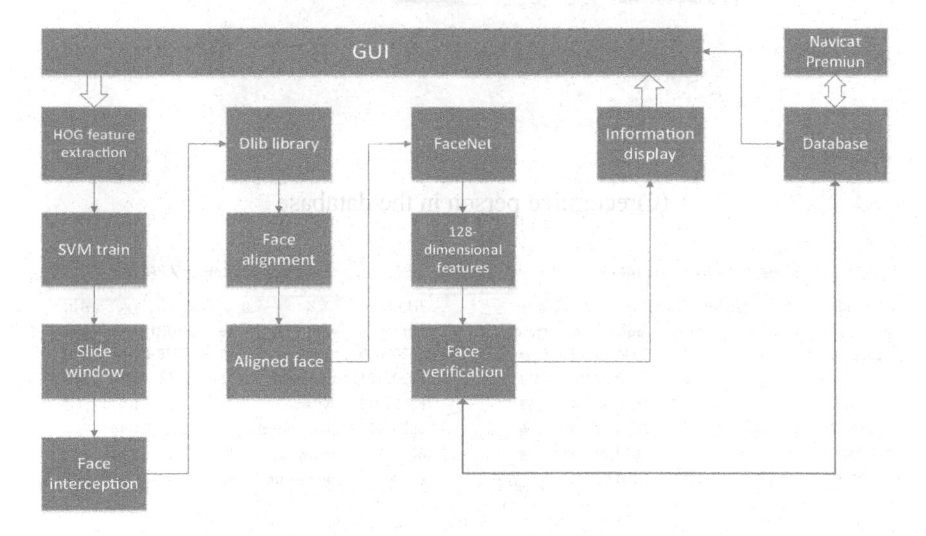

Fig. 4. UI functions

Open the encapsulated software, the overall user interface function is shown in Fig. 4 and design is shown in Fig. 5(a). After turning on the camera, click on the recognition button for real-time recognition. The recognition result of strangers is shown in 5(b), and the recognition result of students in the database and database changes are shown in Fig. 5(c)(d)(e).

The error of the password and the error of the unfamiliar face are all reported normally according to Fig. 6(a)(b). The results of the registration of the stranger's face

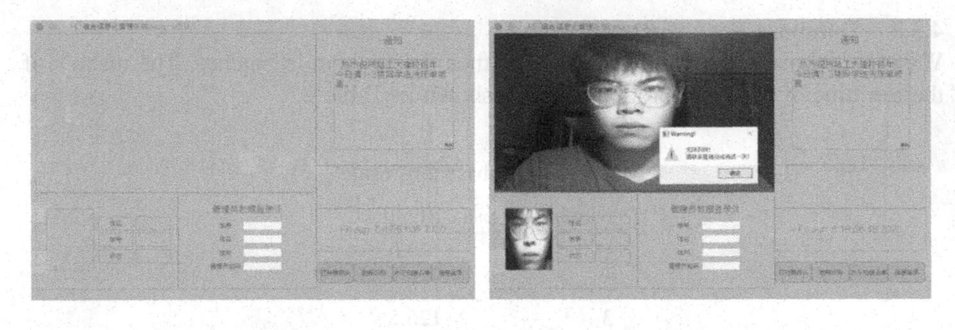

(a)whole UI (b)recognize stranger

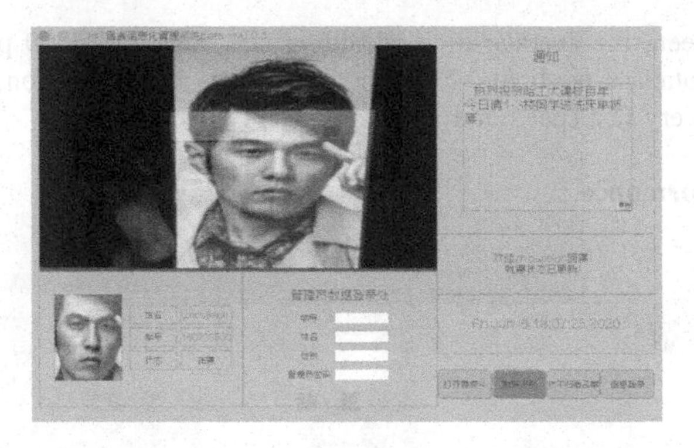

(c)recognize person in the database

对象 students @apartment (APAR...

开始事务 文本 · 筛选 排序 导入 导出

number	name	sex	regist_time	state
160200000	trump	男	2020-04-14	离寝
160200101	zhangziyin	女	2020-04-14	离寝
160201010	liyifan	女	2020-04-14	在寝
160201031	yuanmingzhi	男	2020-04-14	在寝
160200131	zhouyou	男	2020-04-14	在寝
160200500	zhoujielun	男	2020-05-29	离寝

(d) original database

对象 students @apartment (APAR...

开始事务 文本 · 筛选 排序 导入 导出

number	name	sex	regist_time	state
160200000	trump	男	2020-04-14	离寝
160200101	zhangziyin	女	2020-04-14	离寝
160201010	liyifan	女	2020-04-14	在寝
160201031	yuanmingzhi	男	2020-04-14	在寝
160200131	zhouyou	男	2020-04-14	在寝
160200500	zhoujielun	男	2020-05-29	在寝

(e)changed database

Fig. 5. FR results and changes of and database in recognition library

are shown in Fig. 7(a)(b). Click recognition again to identify the newly registered face and it's shown that the recognition is successful by Fig. 7(c).

Because the test time is uncertain, the time can not be used as the trigger signal, so a button is used to simulate the time which is the deadline of backing dormitory, and the click button generates the list of people who are not in bed, as shown in Fig. 8.

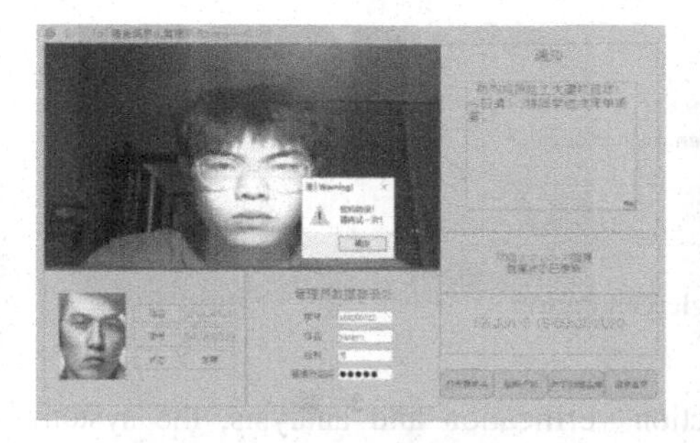

(a)error password

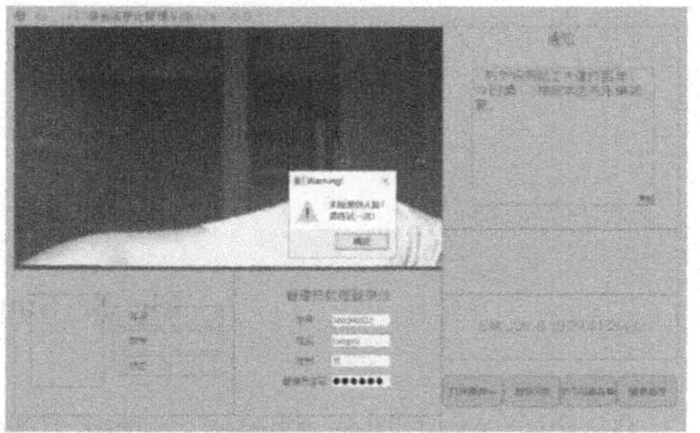

(b)no face

Fig. 6. Error situation

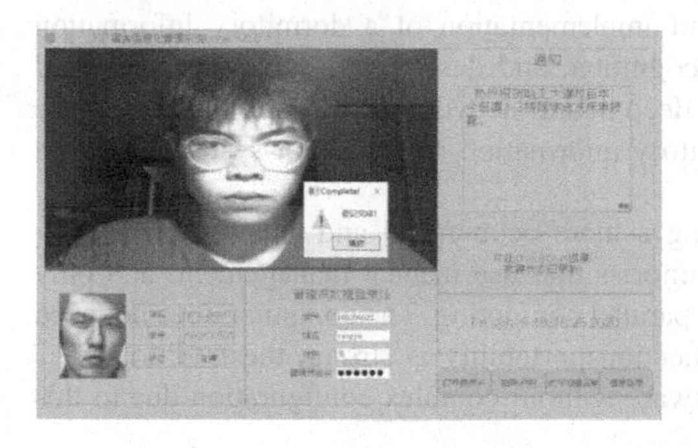

(a)register new face

number	name	sex	regist_time	state
160200000	trump	男	2020-04-14	离寝
160200101	zhangziyin	女	2020-04-14	离寝
160201010	liyifan	女	2020-04-14	在寝
160201031	yuanmingzhi	男	2020-04-14	在寝
160200131	zhouyou	男	2020-04-14	在寝
160200500	zhoujielun	男	2020-05-29	离寝
160200522	yangyu	男	2020-06-05	在寝

(b)changed database

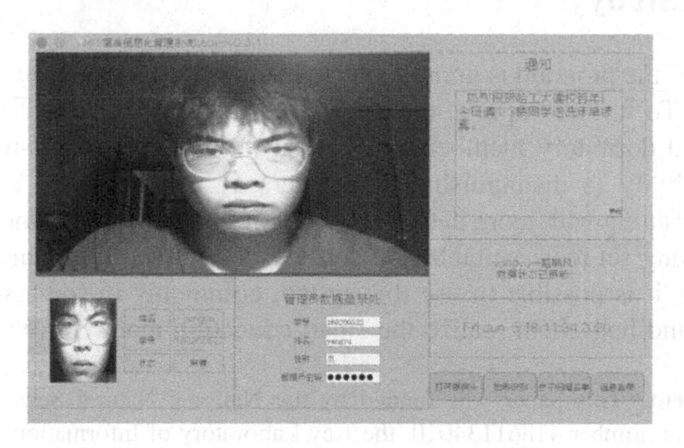

(c)recognize newly registered

Fig. 7. Identification of registered information

Fig. 8. Single generation

At this point, the whole system function verification and analysis, the system successfully completed all the above functions.

5 Conclusion

This paper focuses on the design and implementation of a dormitory information management system based on face recognition, and describes the research of related algorithms with machine learning and deep neural network as the core, then finish the design and implementation of a dormitory information management system based on these theories.

The system is targeted at providing a more economical and effective new management mode for the traditional dormitories that use manual management, and only have computers of low performance in parallel computing. The computer does not need high performance GPU, and the application portability is strong, the traditional PC machine with camera can run this software without complex configuration due to this software's strong compatibility.

6 Further Study

The accuracy of the program is greatly affected by the clarity, distance and illumination of the camera. To further improve the accuracy, it is necessary to collect more images of all registered dormitory members and retrain the FaceNet of pre-training, so as to improve the ability to distinguish the members of the dormitory and the faces of strangers. By training with more data, the CNN can fit the real result more precisely. By collecting training set more suitable and consider the influence of clarity, distance and illumination in, it is possible to use this system commonly in some schools with old infrastructure and help them realize the smart education more conveniently.

Acknowledgement. This paper was funded by the National Natural Science Foundation of China under grant number 41861134010, the Key Laboratory of Information Transmission and Distribution Technology of Communication Network (HHX20641X002), National Key R&D Program of China (No.2018YFC0807101).

References

1. Wei, J., Jian-Qi, Z., Xiang, Z.: Face recognition method based on support vector machine and particle swarm optimization. Expert Syst. Appl. **38**(4), 4390–4393 (2011)
2. Wang, M., Deng, W.: Deep face recognition: a survey. arXiv (2018)
3. Sun, Y., Wang, X., Tang, X.: Deep learning face representation by joint identification-verification. Adv. Neural Inf. Process. Syst. **27**, 1988–1996 (2014)
4. Dalal, N., Triggs, B.: Histograms of oriented gradients for human detection. In: IEEE Computer Society Conference on Computer Vision & Pattern Recognition. IEEE (2005)
5. Saunders, C., Stitson, M.O., Weston, J.: Support vector machine. Comput. Sci. **1**(4), 1–28 (2002)
6. Sun, Y., Wang, X., Tang, X.: Deep convolutional network cascade for facial point detection. In: 2013 IEEE Conference on Computer Vision and Pattern Recognition (CVPR). IEEE (2013)
7. Schroff, F., Kalenichenko, D., Philbin, J.: FaceNet: a unified embedding for face recognition and clustering. In: 2015 IEEE Conference on Computer Vision and Pattern Recognition (CVPR). IEEE (2015)
8. Szegedy, C., Ioffe, S., Vanhoucke, V., et al.: Inception-v4, Inception-ResNet and the Impact of Residual Connections on Learning (2016)
9. Gao, W., Cao, B., Shan, S., et al.: The CAS-PEAL large-scale chinese face database and baseline evaluations. IEEE Trans. Syst. Man Cybern. **38**(1), 149–161 (2008)
10. Huang, G., Mattar, M., Berg, T., Learned-Miller, E.: Labeled faces in the wild: a database forstudying face recognition in unconstrained environments. Technical report (2008)

Factors Affecting Students' Flow Experience of E-Learning System in Higher Vocational Education Using UTAUT and Structural Equation Modeling Approaches

Yunyi Zhang, Ling Zhang(✉), Ying Wu, Liming Feng, Baoliang Liu, Guoxin Han, Jun Du, and Tao Yu

City College of Huizhou, Huizhou 516025, China
68588368@qq.com

Abstract. Higher vocational education has adopted the e-learning system, and scholars have achieved a lot of results in e-learning. Query ID="Q1' Text="This is to inform you that corresponding author has been identified as per the information available in the Copyright form.' However, how to introduce flow ex-perience theory, extract the behavioral intention characteristics of higher vocational students, and how to integrate job requirements and skill certificates into e-learning Design and application need to be discussed in depth. We propose a UTAUT model that combines flow experience, exploring the use of behavior intention as a mediator and flow experience as the target variable. More than 7000 students from City College of Huizhou participated in the questionnaire. The Structural Equation Modeling (SEM) SmartPLS3 software was used to investigate their flow experience to use the e-learning system. The results show that perceived usefulness and facilitating conditions have an important influence on their flow experience and behavioral intentions, both have a partial mediating effect on flow experience through behavioral intention. The e-learning system of higher vocational education should promote the flow experience level of students, and strengthen the elements of employment positions and skills certificates. Suggestion: The e-learning system of higher vocational education should promote the flow experience level of students, and strengthen the elements of employment positions and skills certificates. The model of intention to use e-learning systems for senior students is innovative and effective in practice.

Keywords: UTAUT · Flow experience · Higher vocational students · E-learning · Behavioral intention

1 Introduction

With the widespread use of AI technology and big data technology in e-learning, it has a deepening impact on higher vocational education. Using artificial intelligence technology and big data technology, it becomes a smart education platform (Holmes et al. 2019; Shuguang Liu et al. 2020).

For example, through the e-learning system, Huizhou City College, has a total of 11,329 classes, 700 direct teachers, 20,780 students, and 3,372 courses. The resources deployed in the e-learning system include videos, courses, pictures, documents,

S. Shi et al. (Eds.): AICON 2020, LNICST 356, pp. 362–377, 2021.
https://doi.org/10.1007/978-3-030-69066-3_32

graphics and text pages, and web links. Among them, 36.41% are videos, 14.83% are pictures, 29.48% are documents, and 11.57% are web links, which are the four most important types of resources. Learners can interact with teachers and students through the e-learning system, and can view real-time experience values, access to videos, document resources and web links, submit assignments, and check the status of assignments submitted and graded. e-learning system can give teachers alerts based on students' learning status. However, there are still more than 420 courses with abnormal class attendance, with absenteeism rates ranging from 10% to 47.25%. This semester, 2,416 courses have been deployed in the e-learning platform, but the number of active courses ranges from 1,147 to 1,726.

Students have different abilities, personalities, and motivations. How to implement education according to the different characteristics of students is a key topic of current education. If the student's characteristics are not understood and applied, the technology of wisdom education cannot be used well (Wandasari et al. 2019; Abbas et al. 2019; Huda, 2020). Flow experience comes from work and entertainment activities with people. It is usable and versatile. Therefore, it can better explain the psychological state of people's personal behavior. The process was defined as the state of the best experience, in this state, people would get a sense of pleasure by focusing on a certain task, regardless of external rewards (Csikszentmihalyi, 1990). Flow is very important for active learning. When people are immersed in tasks or activities, at this moment, people will feel happy and enjoy the moment (Chin-Lung Hsu et al. 2003; Shin et al. 2006; Müller and Wulf 2020; Wan et al. 2020). Someone in flow state only has goals and activities, forget all the others (Csikszentmihalyi 1997; Lee 2005; Skadberg et al. 2004).

The UTAUT model has been widely adopted by many countries to study the ac-acceptance of e-learning (Williams 2015; Mohammed 2019; Tarhini 2016; Isaac 2019; Yakubu 2019).

Therefore, this research integrates flow experience and UTAUT model to explore the behavior intentions of higher vocational students using e-learning systems, also try to find the key factors that affect their flow experience and behavior intentions (See Fig. 1). The specific research is as follows:

(i) Explore the UTAUT model and flow experience theory, and propose a flow experience model of vocational students use the e-learning system

(ii) Investigate the key factors influencing the use of e-learning by students in higher vocational schools and the possible intermediary effects.

(iii) In particular, when higher vocational schools link the content of the platform with job requirements and skills certificate requirements, explore the impact of the perceived usefulness on the behavioral intention and higher vocational students' flow experience.

This research plan is to extract the important factors that may exist in the flow experience of vocational students using e-learning. In this study, the UTAUT model will be added to the flow experience, and it will be used as the target variable. In addition, according to the importance of job requirements and skill certificates in higher vocational education, it is planned to add corresponding content to the usefulness index of perceived content. Explore the explanatory power of the newly constructed model for flow experience. At the same time, combined with the research results of previous scholars to explore the influence of behavioral intentions on the target variables of flow

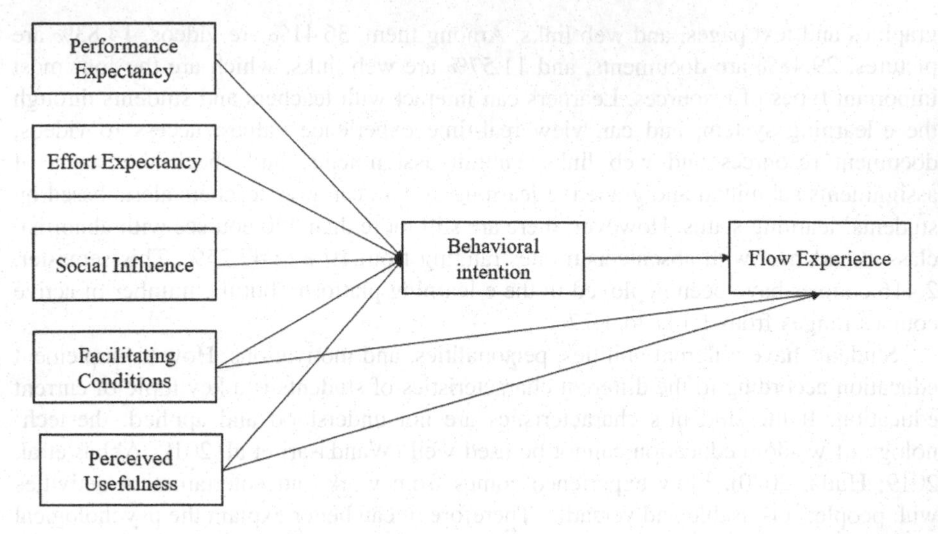

Fig. 1. Research model

experience, and explore the possible mediating effects, methods such as PLS, Bootstrapping, and IPMA will be used, and the key indicators in the model will be used, Taking the students of higher vocational schools as the research objects, explore the possible key indicators of their use of e-learning systems.

2 Literature Review

The UTAUT model has been widely adopted by many countries to study the acceptance of e-learning. It is used to investigate the influencing factors of e-learning websites (Tan 2013; Lin 2020; Alkhuwaylidee 2019). Using the structural equation method, ex-plain the UTAUT model of the e-learning system for students receiving higher education and seek out the influencing factors (Salloum 2018; Gitau 2016). Some scholars had revised the dimensions of UTAUT (Gunasinghe 2019). Under similar circum-stances, some scholars had extended this model and added other dimensions (Khechine 2019). In the early stage, some scholars added the function of this model to teacher characteristics and curriculum characteristics (Almaiah 2019), In Korea, UAE, Those researches have also been effectively carried out (Kang 2015; Salloum 2018), Similarly, Saudi Arabia promotes such research (Alshehri et al. 2019). Of course, the comprehensive measurement level of the model is also very important (Al-Fraihat 2018). Some scholars had proposed the behavioral intention model of students in developing countries in this field (Maldonado 2011; Valencia-Arias 2019). Also in Pakistan's (Kanwal 2017). There are also scholars focusing on investigating the intention of e-learning behavior in mathematics courses (Tarmuji 2019). The UTAUT model is often used to explain the behavioral intentions of the vocational education industry (Nur 2017; Thomas 2013).

The flow experience theory has been adopted by more and more scholars and used flow experience as an important aspect to explain behavioral intentions. (Fraihat 2020; Park, 2020; Ha, 2020; Gao, 2020; Hassan, 2019; Hong, 2019).

Behavioral intention (BE): It is the main goal of the UTAUT model (Venkatesh et al. 2003). It mainly refers to the degree to which someone will perform a certain action in the future. In this study, BE is defined as the degree to which vocational students will adopt e-learning systems in the future.

The PE has a Significant influence impact on BE in the UTAUT model (Venkatesh et al. 2003). PE has been consistently proven to be the most robust and the strongest predictor of BE (Daud et al. 2019). This is supported (Gunasinghe 2019; Alkhuwaylidee et al. 2019; Chao 2019; Garone and Pynoo 2019; Susanto 2019; Handoko 2019).

In this research, EE is defined as the degree to which professional students perceive the e-learning system to help them learn. In past studies, EE has also been shown to be an important predictor of BE (Venkatesh et al. 2003; Yakubu and Dasuki 2019; Rahman and Rosman et al. 2019; Odegbesan and Ayo et al. 2019).

This study SI is defined as the degree to which the vocational students perceive that important people in their lives think they should use one e-learning system. The previous studies have noted the positive influence of SI towards BE (Venkatesh et al. 2003; Olasina 2018; Zhang and Cao 2020; Saragih and Setyowati 2018).

As for PU, it is defined as the degree of improvement in the perceived learning effect of students in vocational schools when using e-learning systems. Previous studies have noted the positive impact of PU on BE (Davis et al. 2003; Al-Fraihat and Joy et al. 2018; Rafiee and Abbasian-Naghneh 2019; Farhan and Razmak et al. 2019).

Facilitating conditions (FC) are perceived enablers or barriers in the environment that influence a student's perception of ease or difficulty of performing a task when using the e-learning system. The previous studies have noted the positive influence of FC towards BE (Davis et al. 2003; Khechine and Raymond 2019; Camilleri and Affiliations 2019; Chopra and Madan 2019).

In this study, flow experience was taken as the target variable, and the BE was regarded as one of the important latent variables, but not as the target variable. This study defined flow experience as the degree to which the vocational student might have flow and focused on the e-learning scene.

Considering this objective, the following hypotheses were framed in this study:

H1: BE significantly FE.
H2: EE significantly BE.
H3: FC significantly BE.
H4: FC significantly FE.
H5: PE significantly BE.
H6: PU significantly BE.
H7: PU significantly FE.
H8: SI significantly BE.
H9: FC has an indirect effect on FE via BE.
H10: PU has an indirect effect on FE via BE.

3 Methodology

3.1 Measuring Instrument

After 5 professional groups of higher vocational schools talked, the questionnaire included 38 questions. This quantitative study used a set of questionnaires containing 38 e-learning questions. According to the 5-point Like force scale, the scores of the questionnaire items increased from completely non-compliant (1) to completely consistent with 5 points. According to the characteristics of higher vocational students, two indicators are added to the latent variable PU, include "the content is closely related to the competence requirements of advertised jobs.", "the content is closely related to the requirements stated in professional certificates."

3.2 Partial Least Square Structural Equation Modeling (PLS-SEM)

Measurement evaluation was analyzed by Least Squares (PLS) with SmartPLS3.0 software. Measurement model indicators are designed to evaluate potential or constructed measurement effects (Ingenhoff and Buhmann 2016). The evaluation of reflectivity measurement models includes convergence validity and discrimination Validity analysis. Convergence validity includes the following: CR, and AVE. The discriminant validity consisted of the following: Cross-loading, Fornell&Larker's criteria. The indicator reliability loadings of Higher than 0.70 means that the structure explains more than 50% of the indicator difference. Internal consistency reliability with a higher value has higher reliability. Convergence effectiveness measures the degree of convergence of the structure on its indicators by explaining the variance of the project through the extracted average variance (AVE). On the other hand, the validity of the discrimination determines the degree of empirical difference between the structure and other structures in the path model in terms of the degree of relevance between the path model and other structures and the difference in the way indicators in a single structure are expressed. (Sarstedt et al. 2014). Fornell&Larcker's criteria were used to determine discriminant validity for the reflective measurement model through cross-loading. This study aimed to analyze the measurement model consisting of seven constructs using SmartPLS 3.0 software.

3.3 Items Validity and Reliability

By selecting 10% of randomly selected students for a small sample test, the purpose is to test the reliability and validity of the questionnaire. It will be launched on July 19, 2020, and the questionnaire will be returned on July 20, 2020. We found that CR > 0.7, Cronbach's Alpha > 0.7, AVE > 0.5 for all indicators, but the observation of the cross-loading table found that the SI indicators were low, and in the differential validity table, the SI variable was not acceptable in Fornell&Larcker, so the judgment was inflated For the question of factors, by observing the VIF value, the three indicators of EE1, EE3, and EE5 with VIF > 5 are deleted from the questionnaire, and the FC7 indicator with VIF < 2 is deleted from the questionnaire. Remove the indicators of SI variables from the questionnaire. Then the items were examined for item redundancy or possible multi-collinearity through item pairs. The results of the pilot test revealed 5 items were discarded and the total of 34 items is accepted with the Cronbach Alpha value is >0.90.

3.4 Population and Sample

The official questionnaire comes from two grade students who have successfully implemented the e-learning system in the year 2019 and 2020. The electronic questionnaire will be launched on July 21, 2020, and the questionnaire will be returned on July 23, 2020. The questionnaire after the test was distributed to 7810 students in the same school. Among the recovered documents, there are 165 unfinished answer sheets. There are 7645 complete answers, so the questionnaire response rate is 97.8%. The sample meets the requirements of the result equation model SmartPLS3. Therefore, it can be used to test hypotheses that we might make.

A totol of 7645 students from City College of Huizhou participated in this research. The respondents were both male (N = 4072, 53.3%) and female (N = 3573, 46.7%). Respondents came from 2 grades, the first year of higher vocational education (N = 3289, 43.0%) and the second year of higher vocational education (N = 4536, 57.0%). Respondents were distributed in 8 different types of majors, including art majors (N = 662, 8.7%), business majors (N = 1752, 22.9%), education majors (N = 541, 7.1%), mechanical and electrical majors (N = 1593, 20.8%), finance majors (N = 836, 10.9%), international docking majors (N = 419, 5.5%), electronic information majors (N = 1287, 16.8%), people's livelihood services majors (N = 555, 7.3%).

4 Findings and Discussion

4.1 Measurement Model Assessment

CR (BE = 0.964, EE = 0.953, FC = 0.959, FE = 0.939, PE = 0.937, PU = 0.955) all meet the recommended index > 0.7, and all of the Cronbach's alpha (BE = 0.959, EE = 0.938, FC = 0.948, FE = 0.902, PE = 0.910, PU = 0.937) all meet the recommended index > 0.7(See Table 1). Table 1 show that construct reliability was accepted (Hair Jr et al. 2016; Gefen et al. 2000; Kannan and Tan 2005).

Table 1. Measurement model result and item

Constructs	Items	Factor loading	AVE	Composite reliability	Cronbach alpha
BE	BE1	0.860	0.751	0.964	0.959
	BE2	0.875			
	BE3	0.869			
	BE4	0.867			
	BE5	0.834			
	BE6	0.873			
	BE7	0.880			
	BE8	0.859			
	BE9	0.883			

(continued)

Table 1. (*continued*)

Constructs	Items	Factor loading	AVE	Composite reliability	Cronbach alpha
EE	EE2	0.889	0.801	0.953	0.938
	EE4	0.903			
	EE6	0.895			
	EE7	0.895			
	EE8	0.893			
FC	FC1	0.849	0.796	0.959	0.948
	FC2	0.917			
	FC3	0.912			
	FC4	0.910			
	FC5	0.898			
	FC6	0.864			
FE	FE1	0.921	0.836	0.939	0.902
	FE2	0.889			
	FE3	0.933			
PE	PE1	0.883	0.788	0.937	0.910
	PE2	0.898			
	PE3	0.906			
	PE4	0.865			
PU	PU1	0.885	0.842	0.955	0.937
	PU2	0.926			
	PU3	0.930			
	PU4	0.929			

Table 2. The results of Fornell&Larcker

	BE	EE	FC	FE	PE	PU
BE	**0.867**					
EE	0.835	**0.895**				
FC	0.867	0.853	**0.892**			
FE	0.806	0.870	0.811	**0.915**		
PE	0.781	0.830	0.808	0.760	**0.888**	
PU	0.867	0.844	0.892	0.813	0.770	**0.917**

Table 3. The results of cross–loadings

	BE	EE	FC	FE	PE	PU
BE1	**0.860**	0.752	0.778	0.726	0.702	0.798
BE2	**0.875**	0.759	0.766	0.736	0.696	0.792
BE3	**0.869**	0.759	0.763	0.737	0.695	0.789
BE4	**0.867**	0.703	0.760	0.668	0.672	0.735
BE5	**0.834**	0.664	0.747	0.628	0.653	0.696
BE6	**0.873**	0.712	0.749	0.690	0.673	0.726
BE7	**0.880**	0.726	0.736	0.717	0.663	0.746
BE8	**0.859**	0.699	0.716	0.674	0.647	0.722
BE9	**0.883**	0.727	0.749	0.704	0.684	0.746
EE2	0.725	**0.889**	0.743	0.734	0.736	0.739
EE4	0.734	**0.903**	0.736	0.794	0.722	0.753
EE6	0.726	**0.895**	0.733	0.772	0.749	0.738
EE7	0.766	**0.895**	0.784	0.793	0.742	0.761
EE8	0.780	**0.893**	0.813	0.797	0.761	0.780
FC1	0.736	0.747	**0.849**	0.737	0.682	0.756
FC2	0.791	0.791	**0.917**	0.745	0.746	0.802
FC3	0.782	0.768	**0.912**	0.724	0.738	0.799
FC4	0.784	0.779	**0.910**	0.739	0.737	0.813
FC5	0.796	0.772	**0.898**	0.731	0.722	0.839
FC6	0.751	0.703	**0.864**	0.660	0.699	0.763
FE1	0.750	0.841	0.753	**0.921**	0.722	0.765
FE2	0.714	0.740	0.726	**0.889**	0.663	0.709
FE3	0.748	0.804	0.746	**0.933**	0.697	0.755
PE1	0.660	0.688	0.686	0.632	**0.883**	0.643
PE2	0.703	0.746	0.731	0.677	**0.898**	0.700
PE3	0.689	0.740	0.703	0.688	**0.906**	0.676
PE4	0.718	0.768	0.747	0.697	**0.865**	0.712
PU1	0.789	0.784	0.869	0.745	0.730	**0.885**
PU2	0.809	0.776	0.812	0.747	0.707	**0.926**
PU3	0.791	0.764	0.783	0.742	0.684	**0.930**
PU4	0.791	0.772	0.809	0.748	0.704	**0.929**

AVE (BE = 0.751, EE = 0.801, FC = 0.796, FE = 0.836, PE = 0.788, PU = 0.842) all meet the recommended index >0.5 (See Table 1). The smallest value of each AVE is 0.751, and the largest is 0.842, which exceeds the recommended value >50. Table 1 show that construct validity was accepted (Hair Jr et al. 2016).

See Table 1, all factor loads in the model exceed 0.7. The discriminative validity is measured using Fornell&Larker and evaluated by dividing the load (Hair Jr et al. 2016). See Table 3, according to the results displayed by the Fornell&Larker method, the square root mean of all AVEs in the table is greater than its correlation with other

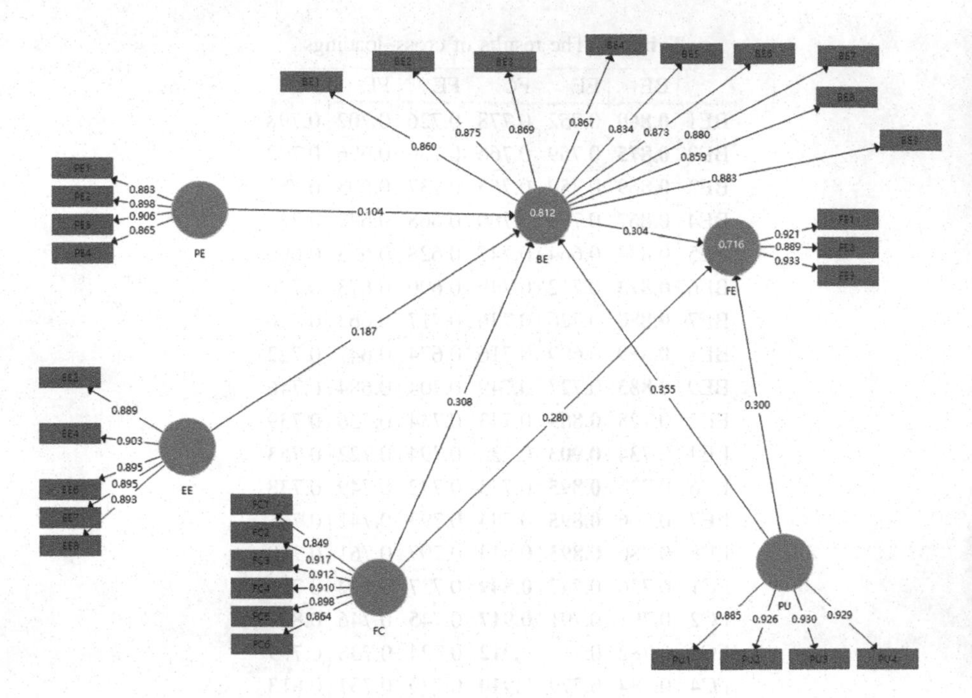

Fig. 2. PLS algorithm results

structures (Fornell&Larcker 1981). See Table 2, the index load on each structure in Table 2 is higher than that of its corresponding structure. Therefore, the discriminative validity of the modified model meets the requirements.

4.2 Structural Model Assessment

In the SmartPLS3.0 software, select the bootstrapping module, select 5000 bootstrap samples, the user can obtain the path coefficient, T value, P value and other indicators of parameters such as R2. Users can directly obtain indicators such as direct effects, indirect effects, and total indirect effects (Hair et al. 2016). See Table 4, Therefore, H1 (BE significantly FE) is received given that ($\beta = 0.304$, $t = 13.529$, $p < 0.001$). H2(EE significantly BE) is received with ($\beta = 0.187$, $t = 11.194$, $p < 0.001$). H3(FC significantly BE) is received with ($\beta = 0.308$, $t = 15.539$, $p < 0.001$). H4(FC significantly FE) is received with ($\beta = 0.280$, $t = 12.891$, $p < 0.001$). H5(PE significantly BE) is received with ($\beta = 0.104$, $t = 7.398$, $p < 0.001$). H6(PU significantly BE) is received with ($\beta = 0.355$, $t = 18.586$, $p < 0.001$). H7(PU significantly FE) is received with ($\beta = 0.300$, $t = 13.873$, $p < 0.001$) (See Table 4, Fig. 3).

In this study, all tests that have an indirect influence on the target variable need to use the method proposed by Preacher and Hayes (2004) and the method of Preacher and Hayes (2008) according to the indirect influence of the guideline.

As shown in Table 5, the indirect influence of FC- > BE- > FE with ($\beta = 0.093$, $t = 11.411$, $p < 0.001$, LL = 0.078, UL = 0.110). It was therefore determined that BE has a meaningful mediating effect between FC and FE, Therefore H9(FC has an indirect effect on FE via BE) is confirmed.

Table 4. Structural assessment result

Hypothesis	Relationship	Std Beta	t-value	p-value	LL	UL	Decision
H1	BE - > FE	0.304	13.529	0.000	0.258	0.347	Supported
H4	FC - > FE	0.280	12.891	0.000	0.237	0.322	Supported
H7	PU - > FE	0.300	13.837	0.000	0.258	0.343	Supported
H2	EE - > BE	0.187	11.194	0.000	0.154	0.220	Supported
H3	FC - > BE	0.308	15.539	0.000	0.267	0.346	Supported
H5	PE - > BE	0.104	7.398	0.000	0.077	0.131	Supported
H6	PU - > BE	0.355	18.586	0.000	0.317	0.392	Supported
H9	FC-BE-FE	0.093	11.411	0.000	0.078	0.110	Supported
H10	PU-BE-FE	0.108	11.777	0.000	0.090	0.125	Supported

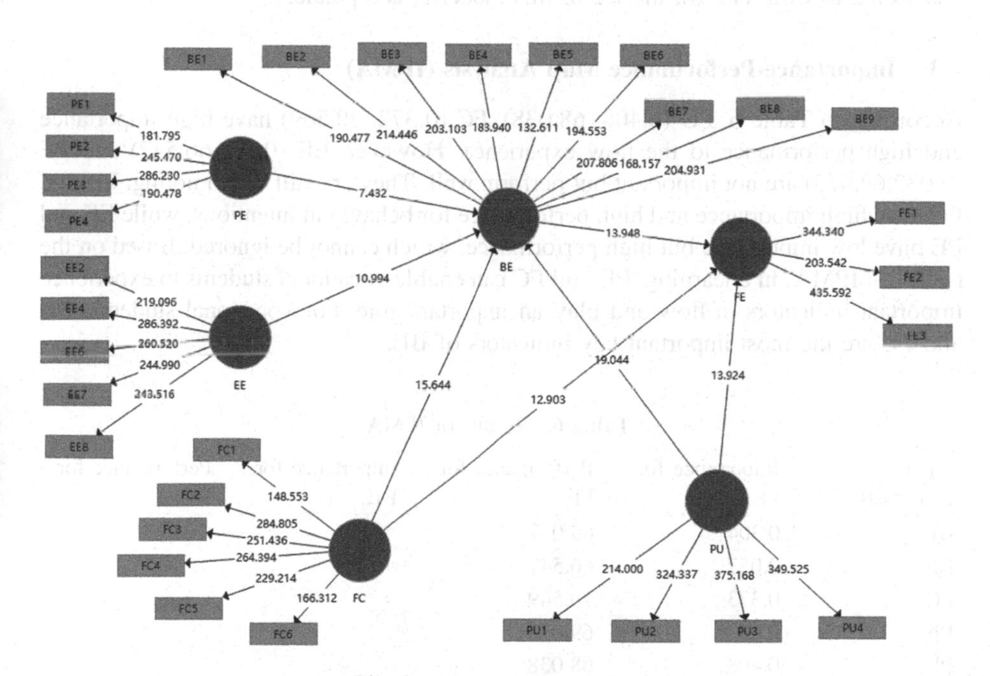

Fig. 3. Bootsrapping results

Table 5. Specific indirect effect

Path coefficient	β	t-value	p-value	2.50%	97.50%
FC - > BE - > FE	0.093	11.411	0	0.078	0.11
PU - > BE - > FE	0.108	11.777	0	0.09	0.125

As shown in Table 5, the indirect influence of PU- > BE- > FE with ($\beta = 0.108$, $t = 11.777$, $p < 0.001$, LL = 0.090, UL = 0.125). It was therefore determined that BE has a meaningful mediating effect between PU and FE, Therefore H10(PU has an indirect effect on FE via BE) is also confirmed.

Williams and MacKinnon (2008) put forward the view of mediation effect. The mediating effect was judged by the VAF score. The intermediate effects of FC and PU are respectively the total effects of FC and PU, and their VAF scores are obtained respectively. When the VAF score is >20% but less than 80%, the latent variable is considered to have a partial mediating effect (Iacobucci, Saldanha, and Deng 2007). From Table 4 and Table 5, The VAF value indicates the result, the VAF of FC- > BE- > FE = 0.249; the VAF of PU- > BE- > FE = 0.264. Therefore, H9(FC has an indirect effect on FE via BE), and therefore H10 (PU has an indirect effect on FE via BE) is also confirmed) have a partial mediating effect.

In Fig. 2, the results revealed that behavioral intention, facilitating conditions, and perceived usefulness had explained 71.6% of the variance in effort expectancy. EE, FC, PE, and PU explained the 81.2% variance of behavioral intention to use e-learning. According to Chin (1998), the R2 of this model is acceptable.

4.3 Importance-Performance Map Analysis (IPMA)

According to Table 6, PU (0.408, 68.038), FC (0.372, 70.569) have high importance and high performance to the flow experience. However, EE (0.057,66.547) and PE (0.032,69.075) are not important but perform well. They are still worth noting. PU and FC have high importance and high performance for behavioral intentions, while EE and PE have low importance but high performance, which cannot be ignored. Based on the results of IPMA, in e-learning, PU and FC can enable vocational students to experience important indicators of flow and play an important role. For vocational students, PU and FC are the most important key indicators of BE.

Table 6. Results of IPMA

Latent constructs	Importance for FE	Performance for FE	Importance for BE	Performance for BE
BE	0.304	69.017		
EE	0.057	66.547		
FC	0.373	70.569		
PE	0.032	69.075		
PU	0.408	68.038		
EE			0.187	66.547
FC			0.308	70.569
PE			0.104	69.075
PU			0.355	68.038

This research mainly extracts the main influencing factors of higher vocational education students' flow experience using e-learning. Based on the UTAUT model, it is proposed that the flow experience theory should be combined, and the behavioral intentions should be aimed at students' ability to produce better flow experience. Therefore, in the basic model, the flow experience structure is added. Also, according to the actual situation of employment competitiveness and skill certificates, which are the focus of higher vocational education, "The content is closely related to the competence requirements of advertised jobs." and "The content is closely related to The requirements stated in professional certificates.", extended the UTAUT model, and proposed a behavioral intention model for higher vocational students using e-learning. The model was analyzed using SmartPLS3.0, The results show the model achieve explain 71.6% of the variance of FE, and 81.2% the variance of BE. Finally, through IPMA analysis, PU has the greatest Performance impact total effect on BE and EE. This data comes from students who have used the e-learning system from City College of Huizhou in the two years. This research result has certain reference value for regional higher vocational education. How to design one e-learning system suitable for higher vocational students? How to promote the flow experience of students? How to best use artificial intelligence technology and big data technology in the smart education platform? How to design and implement courses based on the behavior characteristics of vocational students? This research has practical reference value.

5 Conclusion and Future Works

5.1 Study Contributions and Discussion

The purpose of this study was to extract the factors that influence the intention to use e-learning from the viewpoint of senior vocational students. The results of the data analysis are considered; the relevance of the proposed research model is illustrated and the behavioral intentions of vocational students using e-learning are analyzed under the hypothesis. A structural equation model (PLS-SEM) was used to analyze the research hypothesis. The UTAUT model was modified and the factors of process experience were added. Discriminant validity was analyzed at the prediction phase. It was discovered that the cross-loading of the SI surface and the Fornell&Larker did not satisfy the requirements of the structural equation (PLS-SEM). Therefore, the research model of vocational students' behavioral intentions for e-learning was revised and the association between vocational students' interest in future positions and skill certificates was added in terms of perceived PU. As regards the content usefulness, the eight hypotheses were tested using the PLS technique. It was noted that behavioral intention BE, facilitation FC and perceived content usefulness PU accounted for 71.6% of the variance in flow experience. And it was found that perceived content usefulness PU and facilitation con-dition FC partially mediated the flow experience FE through behavioral intent BE. All eight hypotheses were supported. The study showed that the positive factors influenc-ing the behavioral intentions of vocational students to use e-learning were PU and FC, and the facilitators were EE and PU. For senior students, the positive factors for using e-learning to experience flow experience were BE, PU, and FC. And PU was the most important factor in IPMA analysis.

5.2 Limitations and Future Directions

The study is limited by under-representation as other regions and their findings were not considered. The sample may introduce bias. However, the valid sample covers almost all students in both grades of the school, which could represent the views of students in a particular vocational school. Therefore, the results may not be generalizable to students in other schools, but the views of students in the same type of advanced vocational schools are still informative. The research is aimed at students in regional vocational schools and is exploratory in some nature. The research scope will be expanded in the future.

The results of this study show that practical research on specific factors will continue in the future and provide several opinions as references. First, the future research direc-tion should confirm what are the important factors of vocational school students that have not been detected or verified? Such as behavior habits, hedonic motivation, etc. The second is to study this to further understand whether the samples truly reflect their thoughts during the survey process. If you want to extract the actual behavior of stu-dents in e-learning and include them in the research model? Third, expand the scope of research and understand the generality of this model. Finally, it should be possible to find out that the sample response will be the most efficient at different stages by exper-imental research. For example, different majors, different learning stages, different courses, etc.

References

Holmes, W., Bialik, M., Fadel, C.: Artificial Intelligence in Education. Center for Curriculum Redesign, Bos-ton (2019)

Shuguang, L., Zheng, L., Lin, B.: Impact of artificial intelligence 2.0 on teaching and learning. In: Proceedings of the 2020 9th International Conference on Educational and Information Technology, pp. 128–133 (2020)

Wandasari, Y., Kristiawan, M., Arafat, Y.: Policy evaluation of school's literacy movement on improving discipline of state high school students. Int. J. Sci. Technol. Res. **8**(4), 190–198 (2019)

Abbas, J., Aman, J., Nurunnabi, M., Bano, S.: The impact of social media on learning behavior for sustainable education: evidence of students from selected universities in Pakistan. Sustainability **11**(6), 1683 (2019)

Huda, M.: Empowering application strategy in the technology adoption. J. Sci. Technol. Policy Manage. (2019)

Csikszentmihalyi, M.: Flow and education. NAMTA J. **22**(2), 2–35 (1997a)

Hsu, C.-L., Hsi-Peng, L.: Why do people play online games? an extended tam with social influences and flow experience. Inf. Manage. **41**(7), 853–868 (2004)

Shin, N.: Online learner's 'Flow'experience: an empirical study. Br. J. Educ. Technol. **37**(5), 705–720 (2006)

Müller, F.A., Wulf, T.: Flow experience in blended learning: behind the inconsistent effects of flexibility and interaction. In: Paper presented at the Academy of Management Proceedings (2020)

Wan, Q., Liu, M., Gao, B., Chang, T., Huang, R.: The relationship between self-regulation and flow experience in online learning: a case study of global competition on design for future education. In: Paper presented at the 2020 IEEE 20th International Conference on Advanced Learning Technologies (ICALT), pp. 365–367 (2020)

Csikszentmihalyi, M.: Flow and education. NAMTA J. **22**(2), 2–35 (1997b)

Lee, E.: The relationship of motivation and flow experience to academic procrastination in university students. J. Genet. Psychol. **166**(1), 5–15 (2005)

Skadberg, Y.X., Kimmel, J.R.: Visitors' flow experience while browsing a web site: its measurement, contributing factors and consequences. Comput. Hum. Behav. **20**(3), 403–422 (2004)

Williams, M.D., Rana, N.P., Dwivedi, Y.K.: The unified theory of acceptance and use of technology (Utaut): a literature review. J. Enterp. inf. Manage. (2015)

Almaiah, M.A., Alyoussef, I.Y.: Analysis of the effect of course design, course content support, course assessment and instructor characteristics on the actual use of e-learning system. IEEE Access **7**, 171907–171922 (2019)

Tarhini, A., El-Masri, M., Ali, M., Serrano, A.: Extending the utaut model to understand the customers' acceptance and use of internet banking in Lebanon. Inf. Technol. People (2016)

Isaac, O., Abdullah, Z., Aldholay, A.H., Ameen, A.A.: Antecedents and outcomes of internet usage within organisations in Yemen: an extension of the Unified Theory of Acceptance and Use of Technology (UTAUT) model. Asia Pac. Manage. Rev. **24**(4), 335–354 (2019)

Yakubu, M.N., Dasuki, S.I.: Factors affecting the adoption of e-learning technologies among higher education students in Nigeria: a structural equation modelling approach. Inf. Dev. **35** (3), 492–502 (2019)

Tan, P.J.B.: Applying the UTAUT to understand factors affecting the use of English e-learning websites in Taiwan. Sage Open **3**(4), 2158244013503837 (2013)

Lin, H.-M., Lee, M.-H., Liang, J.-C., Chang, H.-Y., Huang, P., Tsai, C.-C.: A review of using partial least square structural equation modeling in e-learning research. Br. J. Educ. Technol. **51**(4), 1354–1372 (2020)

Alkhuwaylidee, A.R.: Extended Unified Theory Acceptance and Use Technology (UTAUT) for e-learning. J. Comput. Theoret. Nanosci. **16**(3), 845–852 (2019)

Salloum, S.A., Shaalan, K.: Factors affecting students' acceptance of e-learning system in higher education using UTAUT and structural equation modeling approaches. In: Hassanien, A.E., Tolba, M.F., Shaalan, K., Azar, A.T. (eds.) AISI 2018. AISC, vol. 845, pp. 469–480. Springer, Cham (2019). https://doi.org/10.1007/978-3-319-99010-1_43

Gitau, M.W.: Application of the UTAUT model to understand the factors influencing the use of web 2.0 tools in e-learning in Kenyan public universities. University of Nai-robi (2016)

Gunasinghe, A., Abd Hamid, J., Khatibi, A., Azam, S.F.: The Ad-equacy of Utaut-3 in interpreting academician's adoption to e-learning in higher education environments. Interact. Technol. Smart Educ. (2019)

Khechine, H., Augier, M.: Adoption of a social learning platform in higher education: an extended UTAUT model implementation. In: Paper presented at the Proceedings of the 52nd Hawaii International Conference on System Sciences (2019)

Kang, M., Liew, B.Y.T., Lim, H., Jang, J., Lee, S.: Investigating the determinants of mobile learning acceptance in Korea using UTAUT2. In: Chen, G., Kumar, V., Kinshuk, {.}., Huang, R., Kong, S.C. (eds.) Emerging Issues in Smart Learning. LNET, pp. 209–216. Springer, Heidelberg (2015). https://doi.org/10.1007/978-3-662-44188-6_29

Salloum, S.A.S., Shaalan, K.: Investigating students' acceptance of e-learning system in higher educational environments in the Uae: applying the extended technology acceptance model (Tam). The British University in Dubai (2018)

Alshehri, A., Rutter, M.J., Smith, S.: An implementation of the Utaut model for understanding students' perceptions of learning management systems: a study within tertiary institutions in Saudi Arabia. Int. J. Distance Educ. Technol. (IJDET) **17**(3), 1–24 (2019)

Al-Fraihat, D., Joy, M., Sinclair, J.: A comprehensive model for evaluating e-learning systems success. Distance Learn. **15**(3), 57–88 (2018)

Maldonado, U.P.T., Khan, G.F., Moon, J., Rho, J.J.: E-Learning motivation and educational portal acceptance in developing countries. Online Information Review (2011)

Valencia-Arias, A., Chalela-Naffah, S., Bermúdez-Hernández, J.: A proposed model of e-learning tools acceptance among university students in developing countries. Educ. Inf. Technol. **24**(2), 1057–1071 (2019)

Kanwal, F., Rehman, M.: Factors affecting e-learning adoption in developing countries-empirical evidence from Pakistan's higher education sector. IEEE Access **5**, 10968–10978 (2017)

Tarmuji, N.H., Ahmad, S., Abdullah, N.H.M., Nassir, A.A., Idris, A.S.: Perceived resources and technology acceptance model (PRATAM): students' acceptance of e-Learning in mathematics. In: Mohamad Noor, M.Y., Ahmad, B.E., Ismail, M.R., Hashim, H., Abdullah Baharum, M.A. (eds.) Proceedings of the Regional Conference on Science, Technology and Social Sciences (RCSTSS 2016), pp. 135–144. Springer, Singapore (2019). https://doi.org/10.1007/978-981-13-0203-9_13

Nur, M.N.A., Faslih, A., Nur, M.N.A.: Analysis of behaviour of e-learning users by unified theory of acceptance and use of technology (Utaut) model a case study of vocational education in Halu oleo University. Jurnal Vokasi Indonesia **5**(2) (2017)

Thomas, T., Singh, L., Gaffar, K.: The utility of the Utaut model in explaining mobile learning adoption in higher education in Guyana. International Journal of Education and Development using ICT **9**(3) (2013)

Venkatesh, V., Morris, M.G., Davis, G.B., Davis, F.D.: User acceptance of information technology: toward a unified view. MIS quarterly, pp. 425–478 (2003)

Daud, S.M., et al.: A comparison of long-term fouling performance by zirconia ceramic filter and cation exchange in microbial fuel cells. Int. Biodeterior. Biodegradation **136**, 63–70 (2019)

Chao, D., Zhou, W., Ye, C., Zhang, Q., Chen, Y., Lin, G., Davey, K., Qiao, S.-Z.: An electrolytic Zn–Mno2 battery for high-voltage and scalable energy storage. Angew. Chem. Int. Ed. **58** (23), 7823–7828 (2019)

Garone, Anja., et al.: Clustering university teaching staff through Utaut: implications for the acceptance of a new learning management system. Br. J. Educ. Technol. **50**(5), 2466–2483 (2019)

Meiryani, M., Susanto, A., Warganegara, D.L.: The issues influencing of environmental accounting information systems: an empirical investigation of Smes in Indonesia. Int. J. Energy Econ. Policy **9**(1), p. 282 (2019)

Handoko, W., Pahlevani, F., Sahajwalla, V.: Effect of austenitisation temperature on corrosion resistance properties of dual-phase high-carbon steel. J. Mater. Sci. **54**(21), 13775–13786 (2019)

Odegbesan, O.A., Ayo, C., Oni, A.A., Tomilayo, F.A., Gift, O.C., Nnaemeka, E.U.: The prospects of adopting e-learning in the Nigerian education system: a case study of covenant university. In: Paper presented at the Journal of Physics: Conference Series (2019)

Olasina, G.: Factors of best practices of e-learning among undergraduate students. Knowl. Manage. E-Learn. Int. J. **10**(3), 265–289 (2018)

Saragih, A.H., Setyowati, M.S., Hendrawan, A., Lutfi, A.: Student perception of student centered e-learning environment (Scele) as media to support teaching and learning activities at the university of Indonesia. In: Paper presented at the IOP Conference Series: Earth and Environmental Science (2019)

Rafiee, M., Abbasian-Naghneh, S.: E-learning: development of a model to assess the acceptance and readiness of technology among language learners. Comput. Assist. Lang. Learn. pp. 1–21 (2019)

Farhan, W., Razmak, J., Demers, S., Laflamme, S.: E-learning systems versus instructional communication tools: developing and testing a new e-learning user interface from the perspectives of teachers and students. Technol. Soc. **59**, 101192 (2019)

Ingenhoff, D., Buhmann, A.: Advancing Pr measurement and evaluation: demonstrating the properties and assessment of variance-based structural equation models using an example study on corporate reputation. Public Relat. Rev. **42**(3), 418–431 (2016)

Sarstedt, M., Ringle, C.M., Smith, D., Reams, R., Hair Jr., J.F.: Partial least squares structural equation modeling (Pls-Sem): a useful tool for family business researchers. J. Fam. Bus. Strategy **5**(1), 105–115 (2014)

Hair Jr, J.F., Sarstedt, M., Matthews, L.M., Ringle, C.M.: Identifying and treating unobserved heterogeneity with fimix-Pls: part I–method. European Business Review (2016)

Fornell, C., Larcker, D.F.: Structural equation models with unobservable variables and measurement error: algebra and statistics. Sage Publications Sage CA: Los Angeles, CA (1981)

Preacher, K.J., Hayes, A.F.: SPSS and SAS procedures for estimating indirect effects in simple mediation models. Behav. Res. Methods, Instrum. Comput. **36**(4), 717–713 (2004)

Preacher, K.J., Hayes, A.F.: Asymptotic and resampling strategies for assessing and comparing indirect effects in multiple mediator models. Behav. Res. Methods **40**(3), 879–891 (2008)

Williams, J., David, P.M.: Resampling and distribution of the product methods for testing indirect effects in complex models. Struct. Eqn. Model. Multi. J. **15**(1), 23–51 (2008)

Iacobucci, D., Saldanha, N., Deng, X.: A meditation on mediation: evidence that structural equations models perform better than regressions. J. Consum. Psychol. **17**(2), 139–53 (2007)

AI in SAR/ISAR Target Detection

Low Altitude Target Detection Technology Based on 5G Base Station

Yuxin Wu[1], Wenhao Guo[2], Jinlong Liu[1](✉), Bo Yang[3], Lu Ba[1], and Haiyan Jin[1]

[1] Harbin Institute of Technology, Harbin, China
estherwyx@126.com, yq20@hit.edu.cn, baluhit@163.com, jinhyhit@163.com

[2] Shenzhen Foregin Language School, Shenzhen, China
zhaogf2007@126.com

[3] China Institue of Marine Technology and Economy, Beijing, China
b.yang07@foxmail.com

Abstract. With the rising of the civil UAV (Unmanned Aerial Vehicle) industry and the opening of low-altitude airspace, UAVs are frequently used for privacy snooping, terrorist attacks and similar activities, which greatly harm people's safety and social security. However, in the complex urban environments, traditional low-altitude target detection methods are difficult to effectively detect small low-altitude targets which have small size and low speed. With the advent of 5G, intensive networking of 5G base station makes 5G signal is the most abundant resources of electromagnetic in the city, using 5G as external illuminator signal can not only realize the effective detection of low altitude small target, also save the cost, reduce the impact on the urban electromagnetic environment such as the radar, promoted the radar communication integration. Based on the above research background, this paper mainly studied the low altitude target detection scheme based on 5G base stations which is applicable for urban environment. This scheme using 5G base stations as radar transmitters, 5G signal as radiation source, set up the receiver receives the forward scattering signals from the target in order to achieve low altitude target detection and imaging. This paper systematically discusses the feasibility and advantages of 5G signal used as radar radiation signal, studies its radar performance, and the simulation proves its superior speed resolution and range resolution. It provides theoretical basis and support for the application of low-altitude airspace accurate detection and so on, and pushes forward the integration process of radar communication.

Keywords: Low altitude target detection · 5G · Forward scattering · Bistatic radar

1 Introduction

Low-altitude targets mainly refer to "low-altitude, slow-moving and small-sized flying targets" which common feature is that it is difficult to detect when reflected on radar. Therefore, UAVs are often used to spy on intelligence and privacy, causing public

S. Shi et al. (Eds.): AICON 2020, LNICST 356, pp. 381–389, 2021.
https://doi.org/10.1007/978-3-030-69066-3_33

security hazards. For example, some lawless elements use UAVs to peek at some large-scale activities or carry out negative publicity, and even use UAVs to carry out explosions and other terrorist attacks [1].

Recent years, with the rapid rise of UAV high-tech companies such as "DJI" and the further opening of low-altitude airspace, the civil UAV market has further expanded. So how to achieve the effective detection and control of low-altitude small targets has become an urgent security problem.

Low-altitude target detection technology refers to the technology that can effectively detect and track all kinds of low-altitude targets. At present, there are three main technical means: acoustic detection, photoelectric detection and radar detection [2]. The range of acoustic detection is limited to a relatively close range. Optical schemes have high recognition accuracy, but they are easily affected by weather. Traditional radar detection is difficult to separate the echo of low-altitude target from strong ground or object clutter and strong noise, which requires strong signal processing capability. Therefore, it is extremely necessary to find a new system scheme which is suitable for urban low-altitude target detection.

Compared with the traditional monostatic radar, the bistatic radar adopts the design that the receiver and transmitter are separated, and mainly utilizes the forward scattering characteristics of electromagnetic waves [3, 4]. When the target is in the forward scattering region of bistatic radar, its RCS (Radar Cross-Section) can be increased by more than 10db [3], the echo signal will be greatly enhanced. Compared with the traditional monostatic radar, bistatic radar is more suitable for detecting low-altitude small targets with relatively small RCS.

However, it is difficult to implement the traditional bistatic radar in cities for the following reasons:

1. The large-scale setting of high-power radar transmitter in the city will cause serious interference to the original electromagnetic environment of the city.
2. The city is full of high-rise buildings and multipath effect will greatly affect the radar target echo signal and it is difficult to carry out subsequent signal processing.
3. Theoretically, covering the airspace of a standard-sized city requires hundreds of radar receivers and transmitters, but this is only for the purpose of achieving the monitoring of low-altitude small targets or UAVs, and the cost is not proportional to the investment.

But, if the existing communication signal of the city is used as the radar radiation source, electromagnetic pollution can not only be effectively avoided, but also the cost can be save. In the 4G era, it is difficult for the bandwidth and wavelength of communication signals to reach the accuracy and resolution of detecting low-altitude target, but with the development of 5G, 5G signals have wider bandwidth, denser networking and higher carrier frequency, which greatly optimizes the radar detection resolution, detection accuracy and data fusion of bistatic radar based on 5G base stations. It is a good solution for urban low-altitude target detection.

In this paper, a low-altitude target detection system based on 5G base station is proposed which consists of a radar network composed of several 5G base stations and detection radar receivers. Using 5G signals as radar radiation sources, when the target crosses the baseline formed by 5G base stations and receivers, the target will radiate forward scattered waves and be received by the radar receiver. When the target passes the baseline, the intensity of forward scattered waves is theoretically more than 10dB greater than backward reflected waves, which can effectively detect low-altitude small targets (Fig. 1).

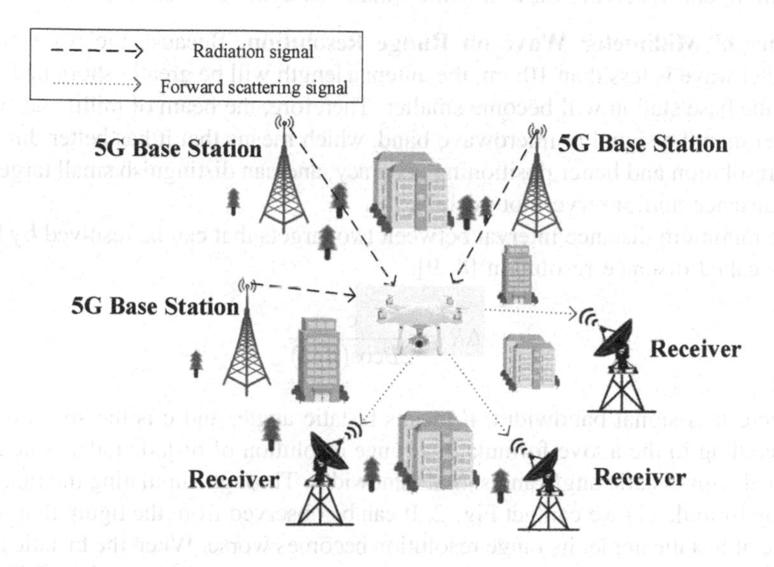

Fig. 1. Low altitude target detection system based on 5G base station

2 Performance of 5G Signal Radar

2.1 Advantages of Communication Radar Integration

The use of 5G signal as the radar radiation source can rely on the rich electromagnetic resources in the urban environment, and make rational use of multiple base stations completes dense networking and multi-angle detection.

In different environments, different communication signals are converted and data fusion is carried out to meet the requirements of width and accuracy for low-altitude target detection, improve the detection performance, and make up for the shortcomings of traditional low-altitude target detection system, such as small detection range and low accuracy. At the same time, the system has the potential of anti-stealth, anti-interference ability and so on.

2.2 Millimeter Wave

The band width of 4G signal is about 20 MHz. According to 3GPP protocol, the two main frequency bands used by 5GNR are FR1 (450 MHz–6 GHz) and FR2 (24.25 GHz–52.6 GHz), and the electromagnetic wave in FR2 is millimeter wave [5]. For FR1 band, the channel width can be up to 100 MHz, while the millimeter band can be up to 400 MHz. According to the theory of digital communication system, the bandwidth is positively correlated with the transmitted signal rate. Combined with other technologies of 5G, the transmission speed of 5G signal will increase by more than 100 times, which greatly speeds up the transmission speed of data, shortens the response time, reduces the data delay, and solves signal synchronization problem of transmitting and receiving base in some traditional bistatic radars [6].

Influence of Millimeter Wave on Range Resolution. Because the wavelength of millimeter wave is less than 10mm, the antenna length will be greatly shortened and the size of the base station will become smaller. Therefore, the beam of millimeter wave is narrower than the signal in microwave band, which means that it has better directivity, higher resolution and better positioning accuracy, and can distinguish small targets with closer distance and observe more details [7].

The minimum distance interval between two targets that can be resolved by bistatic radar is called distance resolution [8, 9]:

$$\Delta R = \frac{c}{2B\cos(\beta/2)} \tag{1}$$

Where B is signal bandwidth, β means bistatic angle, and c is the speed of light.

According to the above formula, the range resolution of bistatic radar is negatively correlated with bistatic angle and signal bandwidth. Through simulating the range resolution by formula (1) we can get Fig. 2. It can be observed from the figure that with the increase of bistatic angle, its range resolution becomes worse. When the bistatic angle is 175 °, that is, when it is about to cross the baseline, its range resolution is about 170 m with the bandwidth is 20 MHz of 4G signal and 8 m with the bandwidth rises to 400 MHz. Under the same bistatic angle, the range resolution is improved by 20 times, which greatly increases the feasibility of distinguishing between two objects that are closer.

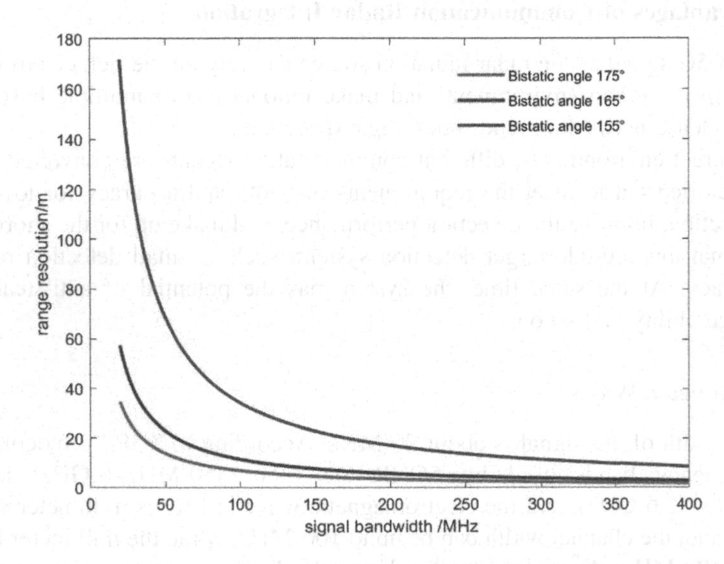

Fig. 2. The relationship between range resolution and signal bandwidth at different bibase angles

When the bistatic angle is a fixed value, with the increase of signal bandwidth, the range resolution is in a sudden drop state from 20 MHz to 100 MHz, and the range resolution is in a low value from 200 MHz to 400 Hz.When the bistatic angle becomes 155 °, the range resolution is as low as 1.7m (when the signal bandwidth is 400MHz). This greatly improves the detection performance of its radar, and makes the 5G signal have more outstanding advantages than other external radiation source signals.

Influence of Millimeter Wave on Velocity Resolution. Velocity resolution refers to the ability of radar to distinguish two moving targets with similar velocities.

Under the condition of fixed bistatic angle, when the target crosses the baseline in the direction perpendicular to the baseline ($\delta = 0^{\circ}$), the velocity resolution of the target is as follows [8, 9]:

$$\Delta V = \frac{\mathrm{c}}{2\mathrm{f}_0 T\cos(\beta/2)} \tag{2}$$

It can be seen from the above formula that the velocity resolution of radar is mainly related to the signal frequency (f_0) and coherent integration time (T). It is known that the frequency of 4G signal is 1.8 GHz–2.6 GHz, while the highest frequency of 5G signal can reach 52.6 GHz (FR2) at present. Figure 4 shows how the velocity resolution of the radar varies with the signal frequency at different coherent accumulation times (Fig. 3).

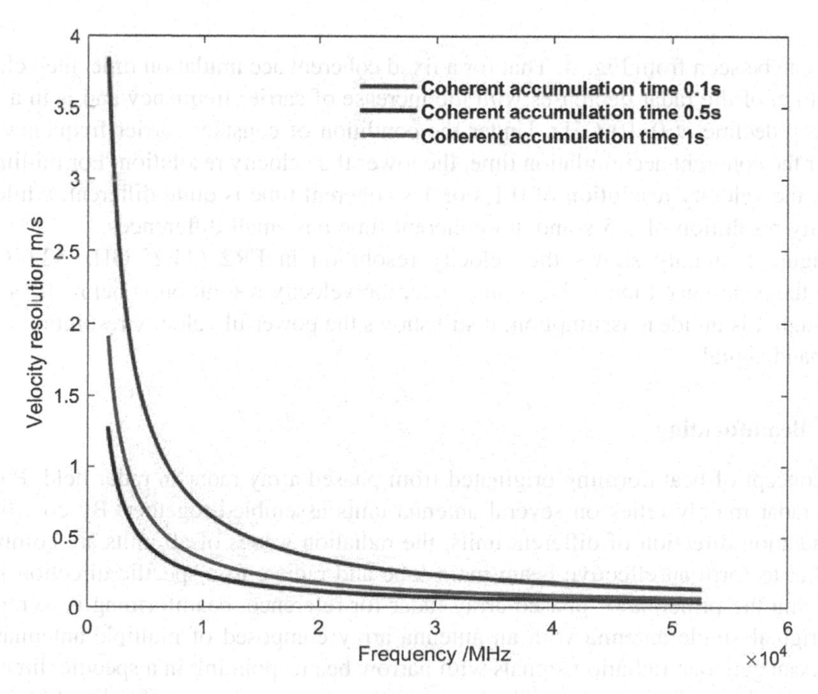

Fig. 3. Relationship between velocity resolution and carrier frequency and accumulation time

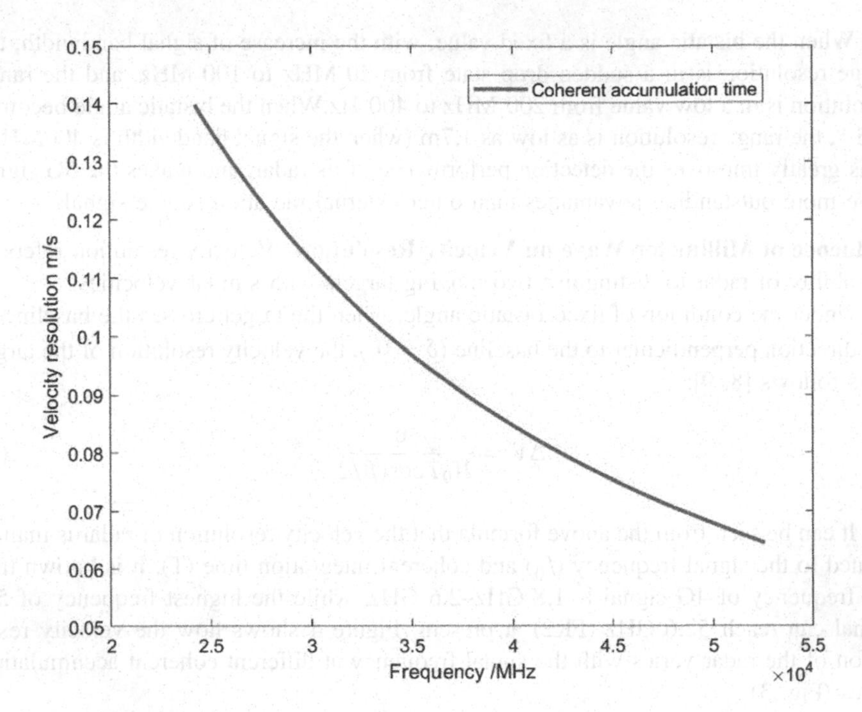

Fig. 4. Velocity resolution when coherent accumulation time is 1s

It can be seen from Fig. 4. That for a fixed coherent accumulation time, the velocity resolution of the radar decreases with the increase of carrier frequency and is in a state of sharp decline at 0–10 GHz. Under the condition of constant carrier frequency, the longer the coherent accumulation time, the lower the velocity resolution. For millimeter wave, the velocity resolution of 0.1 s or 1 s coherent time is quite different, while the velocity resolution of 0.5 s and 1 s coherent time has small differences.

Figure 5 mainly shows the velocity resolution in FR2 (24.25 GHz–52.6 GHz) when the coherence time is 1s. In this state, the velocity resolution is below 0.14 m/s. Although it is an ideal assumption, it still shows the powerful velocity resolution of 5G mm band signal.

2.3 Beamforming

The concept of beamforming originated from phased array radar in radar field. Phased array radar mainly relies on several antenna units assembled together. By controlling the radiation direction of different units, the radiation waves of all units are combined together to form an effective beam main lobe and radiate to a specific direction [6].

Using the principle of phased array radar for reference, beamforming is to replace the original single antenna with an antenna array composed of multiple antennas, so that it can generate radiation signals with narrow beams pointing in a specific direction. The technology allows most of the power to be directed in a specific direction rather than being radiated in a 360-degree direction, wasting most of the energy in free space.

Therefore, beamforming makes the radiation power reaching the designated target higher, and makes the target which with smaller RCS receive greater radiation power than the original one, thus improving the detection ability of 5G signal to small targets. At the same time, because the beam radiates in the specified direction, the multipath effect is weakened, the interference is reduced, and the probability of finding small targets is also improved. This application also reduces the beam width, which brings higher resolution and lower power consumption to radar target detection.

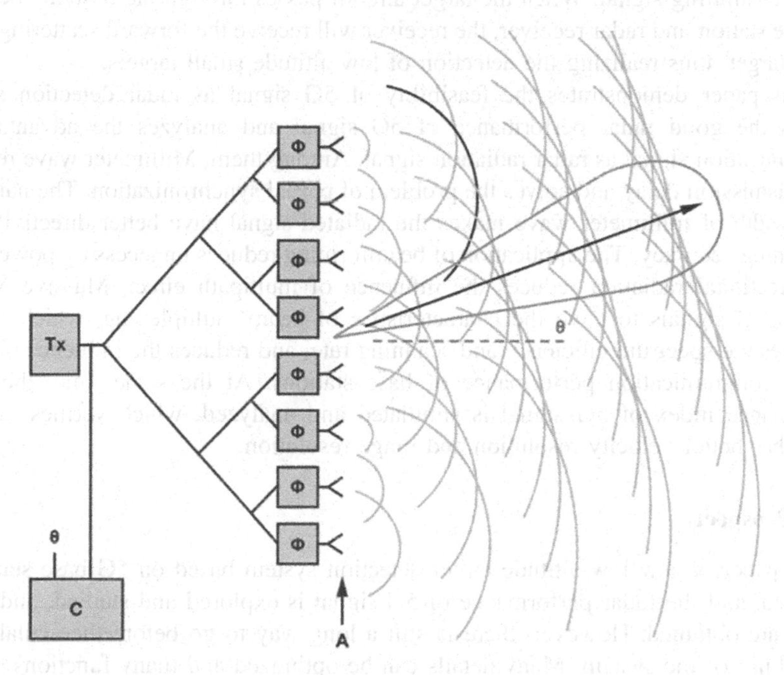

Fig. 5. Schematic diagram of phased array radar

2.4 Massive MIMO

Massive MIMO refers to multi-channel input and output, which can establish contact with multiple users on the same channel, receive and send data at the same time. In the 4G era, the MIMO technology used usually uses 2 or 4 antennas, but in the 5G era, the number of antennas is theoretically unlimited, and the communication capacity is also unlimited. The combination of Massive MIMO and beamforming makes the 5G signal have the characteristics of beamspace multiplexing, which greatly improves the spectral efficiency. The scanning rate of 5G signal as radar signal is also improved, and the target can be found faster without affecting its communication performance.

In 4G era, the antennas are long and few in number, and 5G signals work in millimeter wave band with short wavelength and short antenna length, so more antennas can be integrated in the same base station at the same time to form Massive MIMO. With the increase of the number of antennas, the gain also increases, and the forward scattering power of the target received by the receiver is also larger, which can better detect low-altitude small targets.

3 Conclusion and Prospect

3.1 Conclusion

This paper mainly discusses the low-altitude target detection technology based on 5G base station, and proposes a new urban low-altitude target detection scheme. 5G base station and radar transmitter will be combined into one, and 5G signal will be used as radar transmitting signal. When the target aircraft passes through the baseline between 5G base station and radar receiver, the receiver will receive the forward scattering wave of the target, thus realizing the detection of low-altitude small targets.

This paper demonstrates the feasibility of 5G signal as radar detection signal, verifies the good radar performance of 5G signal and analyzes the advantage of communication signal as radar radiation signal. Among them, Millimeter wave reduces the transmission delay and solves the problem of partial synchronization. The narrower beam width of millimeter wave makes the radiated signal have better directivity and positioning accuracy. The application of beamforming reduces unnecessary power loss, and directional radiation reduces the influence of multipath effect. Massive MIMO enables 5G signals to have the characteristics of beam multiplexing, which greatly improves the spectrum efficiency and scanning rate, and reduces the influence of radar on the communication performance of base stations. At the same time, the radar performance index of 5G signal is simulated and analyzed, which verifies that 5G signal has better velocity resolution and range resolution.

3.2 Prospect

In this paper, a new low-altitude target detection system based on 5G base station is proposed, and the radar performance of 5G signal is explored and studied, and some results are obtained. However, there is still a long way to go before the actual commercial use of the system. Many details can be optimized and many functions can be realized:

1. Although the radar forward scattered signal can well detect low-altitude small targets by increasing the RCS of the target, compared with the backward reflected signal of the target, part of the effective information of target movement is still missed. Therefore, under the premise of urban 5G dense network, the receiver may receive backward reflection and forward scattered signals transmitted to the target and reflected to the receiver by different base stations at the same time. If a data fusion center is established to fuse the two signals received by the receiver, the motion state of the low-altitude target can be obtained more comprehensively, and the future motion direction can be predicted with higher accuracy.
2. The height difference and midline information of the aircraft can be obtained by analyzing the target holographic signal. Different aircraft have different altitude difference and midline information, which can be combined with machine learning to take the altitude difference and midline information as the classification standard, so that the receiver can directly predict the type of aircraft after receiving the forward signal of the target, so as to achieve early warning. Depending on the dense

network of 5G base station, it can explore the radar positioning scheme of multi-external radiation sources and the tracking detection method of the start and maintenance of multi-radar target track. Once the position of the aircraft target is determined, the movement trajectory of the aircraft can be tracked continuously to realize real-time monitoring.

The research and development of low-altitude target detection technology based on 5G base station accelerates the realization of radar communication integration and improves the integration capability of communication network. At the same time, by using the characteristics of low delay and high speed of 5G network, intelligence can be reported quickly, thus make a quick response and create a fast, efficient and low-cost low-altitude target detection solution. It is a realistic and inevitable choice to complete low-altitude target detection by means of a communication network with high coverage and dense networking in cities, with unlimited application potential.

Acknowledgement. This work was supported by the National Natural Science Foundation of China (Grant No. 61601145, 61471142, 61571167, 61871157), Special fund for basic scientific research operating expenses of central Universities (Grant No. HIT.NSRIF.2020008).

References

1. Yang, J.J., Bian, L., Lu, B.B.: New System Radar Technology for detecting low and slow small targets. Electr. Technol. Software Eng. **000**(17):101–101(2018)
2. Zhang, X.K., Liu, S.C., Zhang, C.X., et al.: Bistatic radar detection technology of stealth targets. Syst. Eng. Electr. Technol. **30**(03), 444–446 (2008)
3. Hu, C., Liu, C.J., Zeng, T.: Detection and Imaging of bistatic forward scattering radar. J. Radar **5**(3), 229–243 (2016)
4. Chapurskiy, V.V., Sablin, V.N.: SISAR: shadow Inverse synthetic aperture radiolocation. In: Radar Conference,2000 The Record of the IEEE 2000 International. (2000)
5. 5G Spectrum. https://www.sohu.com/a/330033196_472880 29 July 2019
6. Shi, X.: Application and development of millimeter wave radar. Telecommun. Technol. **000** (1), 3–11 (2006)
7. SKOLNIK M I.: Radar handbook. Radar Antennas (1970)
8. Ding, L.F., Zeng, F.S., Chen, J.C.: Principles of Radar, 4th edn. Electronics Industry Press, Chinese (2009)
9. Zhang, X.K., Zhen, S.C., Chao, G.: Research on low-altitude detection performance of single/double base radar. Syst. Eng. Electr. Technol. **025**(12), 1478–1480 (2003)

Review of Research on Gesture Recognition Based on Radar Technology

Yaoyao Dong[1(✉)] and Wei Qu[2]

[1] Graduate School, Space Engineering University, Beijing 101416, China
jndy940123@163.com
[2] Department of Electronic and Optical Engineering, Space Engineering University, Beijing 101416, China

Abstract. In order to know the development context of radar gesture recognition and predict the possible future development trends, the research and development of gesture recognition based on radar technology in recent years was sorted out. Focusing on key technologies such as dynamic gesture information perception, gesture echo signal preprocessing and feature extraction in radar gesture recognition technology, and classification algorithms for gesture recognition, the relevant literature published at home and abroad is summarized and existing methods are summarized. The performance of the system is analyzed and evaluated; the problems to be solved in the research direction are sorted out and the future research directions are prospected. The results show that radar gesture recognition technology has made great progress in human-computer interaction applications. With the deepening of related research, the gesture recognition system based on radar technology will develop towards intelligence.

Keywords: Human-computer Interaction · Gesture recognition · Machine learning

1 Introduction

In recent years, with the advancement of artificial intelligence and machine learning technology, the relationship between Human-Computer Interaction (HCI) technology and people's daily life has become increasingly close. Traditional contact human-computer interaction methods such as keyboards and mice have been unable to meet the needs of people's lives. Researchers have begun to devote themselves to the development of new non-contact human-computer interaction methods, including gesture recognition, voice recognition, face recognition, and eye iris. Recognition etc. Among them, gesture recognition, as a very important interaction method in the new human-computer interaction, is one of the most powerful and effective methods in human-computer interaction [1]. With the help of this interactive technology, users do not need to operate redundant devices, they can naturally control electronic devices through the movement of fingers and palms [2], and then complete the information exchange between humans and electronic devices such as computers.

S. Shi et al. (Eds.): AICON 2020, LNICST 356, pp. 390–403, 2021.
https://doi.org/10.1007/978-3-030-69066-3_34

Early gesture recognition technology mainly used wearable electronic devices to directly detect and obtain spatial information of human hands and various joints. Among them, more representative devices such as data gloves, using sensors such as accelerometers and gyroscopes [3], can be enriched by operators. In addition, wearable devices based on optical marking method [4] also have good detection performance and robustness. However, the above two gesture recognition technologies are complicated to operate and the equipment is expensive, and they have not been widely used in daily life. Then, gesture recognition technology based on visual images has gradually developed. Compared with wearable gesture recognition systems, visual gesture recognition technology abandons additional wearable systems, allowing users to interact with humans with bare hands [5, 6]. Visual gesture recognition technology is mainly used by computers to detect, track and recognize user gestures through video input devices (cameras, etc.) and computer vision technology to understand user intentions [7]. Although the high-resolution camera enables the recognition rate of visual gesture recognition technology to be as high as above [8, 9], the technology is largely limited by light conditions, and there are also security issues of privacy leakage.

With the rapid development and wide application of radar technology, radar gesture recognition technology has become an important branch of human-computer interaction technology research [10]. Compared with traditional optical sensors, radar sensors can work normally in severe weather conditions such as rain, snow, fog and haze, or under dark conditions, and have the advantages of all-weather and all-weather; secondly, radar sensors can be embedded in the equipment, thereby improving the reliability and design flexibility of the equipment; in addition, the radar signal also has a greater advantage in privacy and security, which can effectively protect user privacy information. In view of the above advantages, the research of gesture recognition based on radar technology has been paid more and more attention. This technology is also a popular direction of computer vision research at present and in the future.

Gesture recognition can be roughly divided into two categories: static gesture recognition and dynamic gesture recognition [11]. This article mainly combs and summarizes the research on radar-based dynamic gesture recognition and related technologies in recent years, and discusses the shortcomings of existing methods. And the next research direction.

2 The Key Technology of Radar Gesture Recognition

Generally, gesture recognition based on radar technology can be divided into 3 major steps: First, the radar sensor is used to collect the user's dynamic gesture information; then, certain preprocessing operations need to be performed on the received radar echo signal to maximize the extraction of the target Gesture features and remove interference signals; finally, according to the gesture feature set, select the appropriate algorithm for gesture recognition. The basic process of a gesture recognition system based on radar technology is shown in Fig. 1. This article takes the radar gesture recognition step as the main pulse to analyze the key technology of gesture recognition.

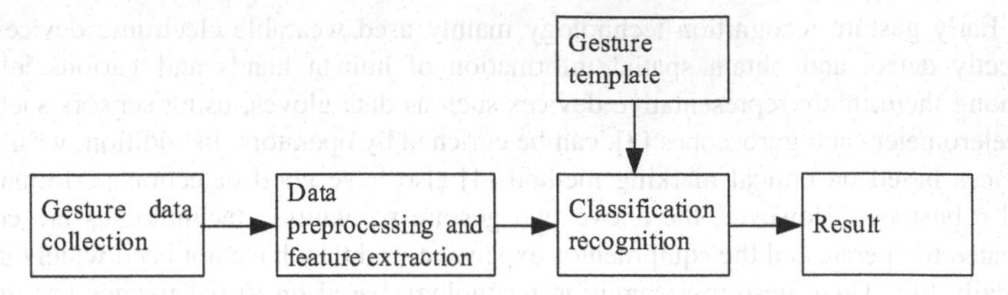

Fig. 1. Block diagram of radar gesture recognition process

2.1 Gesture Data Collection

As the premise foundation of the radar gesture recognition system, the collection of gesture data is directly related to the performance of the entire system. At present, for the radar gesture recognition system, the radar transmission signals used by the researchers can be mainly classified into two categories according to the frequency band: one is concentrated in the K-band and below, and the other mainly uses millimeter waves. Regardless of the frequency band, most of the radar signal waveforms currently used are continuous wave (CW) systems, which are favored by researchers for their advantages of low power consumption, miniaturization, and high reliability.

Sample Radar Sensors Based on K-band and Below

Aiming at the K-band or even lower frequency range, researchers such as Xiaomeng Gao of the University of Hawaii used a quadrature Doppler radar sensor with a carrier frequency of 2.4 GHz, the radar working in the ISM band to collect gesture information and extract zero-crossing from its baseband signal. The features are drawn into strip-shaped feature graphics to distinguish different gestures [12]. In 2016, Fan T et al. constructed a radar sensor working at 5.8 GHz with a coherent zero-IF architecture with symmetrical subcarrier modulation, responsible for the collection of hand movement information [13]. In the same period, Youngwook Kim et al. of California State University used a Doppler radar with a frequency of 5.8 GHz and an antenna beam width of 60° to detect 10 common gestures [14]. The following year, the team used low-power radar to identify three gestures, and achieves an average accuracy rate of 99% on the test sets by using deep learning network [15].

Qifan Pu of the University of Washington and others designed a gesture recognition system named as Wise. The system used the 2.4 GHz wireless signal in daily life, and it can realize the perception of gesture information in the whole room through the difference of Doppler frequency shift of echo [16]. In addition, Matthew Ritchie of the University of London, UK, used the 24 GHz Ancortek radar system to repeatedly detect four different gestures of six people for up to 3000 times [17]. The radar signal frequency was also 24 GHz. Pavlo Molchanov of NVIDIA research center and others used the frequency modulated continuous wave (FMCW) with four antennas (one transmit and three receive antennas) radar sensor, combined with optical camera and time of flight (TOF) depth sensor and other sensors to carry out gesture recognition research in the car [18].

Radar Sensor Based on Millimeter Wave

Compared with gesture recognition systems based on low-frequency wireless signals, millimeter-wave radar sensors are easier to miniaturize and can be embedded in the device, greatly improving the integration and reliability of the device. At the same time, the millimeter wave signal has a stronger ability to capture tiny movements due to its extremely short wavelength, it has good Doppler resolution, and it can effectively identify the subtle movements of the finger. Researchers such as Ismail Nasr use a 60 GHz transmission frequency, two-transmit antennas and four-receive antennas FMCW radar sensor, used SiGe technology to achieve the detection of target gestures [19]. Xuhao Zhang used a frequency modulated continuous wave radar sensor with a working frequency of 77 GHz to provide dynamic gesture detection for the driver's gesture recognition assistance system [20].

Choi Jae-Woo of the KAIST School of Electrical Engineering in South Korea used the FMCW radar with a frequency of 60 GHz developed by Google to perceive 10 kinds of gesture information, and cooperate with the Long Short-Term Memory (LSTM) algorithm of deep learning. The recognition rate is as high as 99.10% [21]. Both Li Chuyang and Wang Yalong of the University of Electronic Science and Technology of China used a millimeter-wave radar with one transmit antenna and four receive antennas to construct a sample database of 3,200 gestures [22, 23], providing a large number of samples for later data processing. In addition, Google's Soli project publicly demonstrated the use of FMCW millimeter-wave radar chips in the 60 GHz frequency band to realize close-range micro-motion gesture recognition [24, 25].

2.2 Gesture Data Preprocessing and Feature Extraction

The main purpose of radar gesture signal preprocessing is to convert a one-dimensional gesture signal into a two-dimensional signal containing both time domain and frequency domain information. Considering that in addition to the information of the target gesture action, the received radar echo signal also contains other irrelevant objects and environmental noise and other irrelevant interference information. For the interference signal in the echo signal, a certain signal processing algorithm is used for the received data preprocessing is performed to ensure that interference information is removed while maximizing the retention of key gesture information [26]. The purpose of feature extraction is to remove redundant information in gesture data, and extract as much as possible the feature quantity that can meet the discrimination of different gestures. It is used as a feature that can well characterize the motion characteristics of dynamic gestures from the preprocessed data. The vector process is an important part of the good work of the gesture recognition system. Because the excessive redundancy of the gesture data will seriously affect the difficulty of training the classifier and extend the training time, which is a very unnecessary time loss for practical applications.

Based on Classical Time-Frequency Analysis

The classical time-frequency analysis method mainly includes Fourier Transform (FT) and a series of transform methods derived therefrom, which has an important position in the field of signal processing. Yu Chenhui used Short-Time Fourier Transform (STFT) to process the gesture echo signal, and extracts the envelope characteristics of the time-frequency map, including the maximum value, average value and variance of the time-frequency envelope curve. The time-frequency characteristics of wave signals are classified [27]. Wang W and Liu Ax proposed a human motion recognition model based on Channel State Information (CSI), using WiFi frequency band signals, analyzed the impact of multipath effects on human motion, and proposed the use of wavelet transform or time frequency analysis is used to recognize human gestures to obtain characteristic parameters [28].

WiGest is a product made by Khaled A. Harras team of Carnegie Mellon University. It used Received Signal Strength Indication (RSSI) information received by WiFi, then they used wavelet transform to identify dynamic gestures with the rising edge, falling edge, pulse and other features of the gesture signal [29]. Zhang used the Short-Time Fourier Transform to analyze the time-frequency of radar echo signal, and then constructed the positive and negative ratio of Doppler frequency offset and the duration of gesture action as features [30]. Kim used STFT to calculate the Doppler spectrum of radar signal, and used the Doppler spectrum image as the input data of convolution neural network for gesture recognition research [14]. Wang draw lessons from arctangent algorithm mentioned in communication signal processing, it can linearly solve the Doppler phase shift, thus obtaining the motion information of dynamic gesture. However, there are also problems with the pure arctangent algorithm. When the target motion amplitude is too large, the demodulation signal truncation and phase ambiguity will occur. In view of this, the team extended the arc tangent algorithm and proposed the extended differentiate and cross-multiply (DACM) algorithm. By introducing differential and integral operators, the truncation problem in phase demodulation was solved, and the complex baseband signal phase was phase continuous change, thus realizing linear demodulation of large dynamic range motion [31].

Based on Radar Signal Processing and Analysis Method

For the shortcomings of the classic time-frequency analysis method, the researchers derived a series of methods for radar signal processing on this basis. For radar signals, Schmidt R used 2-Dimensional Fast Fourier Transform (2D-FFT) to estimate the range of gesture targets and Doppler parameters, and used Multiple Signal Classification (MUSIC) algorithm [32] estimated the angle parameters of the gesture target, and the gesture action data is presented in the form of distance, speed and angle parameter changes. LIN discussed the mixing and modulation principles of FMCW signals in the article [33]. Li G used the Orthogonal Matching Pursuit (OMP) algorithm to analyze the FMCW radar echo signal, and obtained the micro-Doppler time-frequency trajectory as the characteristic input vector [34]. Wang used FMCW radar to collect gesture signals, obtained distance and speed through radar signal processing, and mapped the corresponding signal amplitude into a parameter map. Finally, the parameter graph is used to represent the gesture at each moment, and the parameter map is input into the deep learning network for feature extraction and classification [35]. However, this method is only sensitive to the radial change of the gesture, which limits the angle feature extraction sensitive to the lateral change, thus greatly limiting the application range of gesture recognition. Pavlo Molchanov introduced the estimation method of the range-Doppler map of the FMCW radar for vehicle gesture recognition, and the method of fusing radar sensor and depth sensor data [18]. Dekker processed the one-dimensional signal of the gesture into a Doppler-time spectrum and used the real and imaginary parts of the Doppler-time spectrum as the two-channel classification and recognition input data [36].

Based on radar signal processing and analysis, Wang Yong constructed a Range-Time Map (RTM), Doppler-Time-Map (DTM) and an Angle-Time-Map (ATM), synchronized the three kinds of data of the same gesture action and construct a gesture action multi-dimensional parameter data set [37]; in the same year, the team proposed a Two-Stream Fusion Neural Network (TS-FNN) based on multi-parameter images of FMCW radar signals gesture recognition method, which used gesture signals to generate distance-speed parameter graphs and angle-time parameter graphs and establishes TS-FNN for feature extraction and fusion, retaining the gesture lateral and longitudinal motion parameters [38]. Zhou Zhi used wireless channel state information and terahertz radar signals to obtain data sources, and used gesture radial velocity to characterize gesture behavior [39], but the author directly calculated a distance scalar value to represent gesture feature information at every moment. This makes the feature extraction incomplete (lack of speed and angle information), thereby reducing the accuracy of gesture recognition. Wang Jun et al. performed matched filtering, Fast Fourier Transform (FFT) of LFMCW radar echo of gesture target in turn Transform (FFT) and coherent integration processing are used to obtain the two-dimensional distribution in Range-Doppler (RD) domain, which is used as the input feature vector of subsequent deep learning network to realize automatic feature extraction and recognition of gesture actions [40].

2.3 Gesture Classification and Recognition Algorithm

The classification and recognition algorithm of gestures is the last and most important step in the research of gesture recognition. Radar gesture signals are processed by data preprocessing and feature extraction, and the echo signals are converted into abstract representations. Then, machine learning technology is used to realize the gesture classification and identification. Machine learning can not only quickly process data, predict and classify problems, but also has great development potential in the field of pattern recognition including gesture classification.

The vision-based dynamic gesture recognition algorithm is relatively mature. For the gesture recognition algorithm based on radar technology, researchers are also improving step by step. For the radar system gesture recognition technology, the current mainstream recognition algorithms include template matching-based methods, methods based on statistical learning and methods based on deep learning.

Radar Gesture Recognition Algorithm Based on Template Matching

Dynamic Time Warping (DTW) is the most commonly used template matching method in radar gesture recognition. The DTW algorithm is an algorithm commonly used in speech matching, and currently has certain applications in image processing. The algorithm is based on the nonlinear regularization method of dynamic programming, and it uses regularization functions to describe the time correspondence between the template to be tested and the reference template, so as to solve the similarity of the two time series [41]. When using DTW algorithm to recognize gestures, a series of reference templates need to be recorded in advance, and then the similarity between the gesture to be tested and the reference template is matched and calculated, and the template gesture with the highest similarity is recorded as the recognition result. The DTW algorithm has the characteristics of less training sample demand and high accuracy. Zhou Zhi used DTW method to classify multi-modal signals. Taking 10 gesture signals collected by terahertz radar as an example, the effectiveness of the analysis and recognition system is verified. The experimental results show that the recognition accuracy is above 91% [39]. Plouffe et al. used DTW algorithm to recognize dynamic gestures, and the recognition rate reached 96.25% [42].

However, the DTW algorithm also has certain limitations, including problems such as high computational complexity and poor robustness. Especially when dynamic gestures are more complex and the number of training samples is large, the recognition rate will drop significantly. Therefore, the optimization research of the original DTW algorithm came into being. Ruan X et al. improved the search path and the matching process, and then used the distortion threshold algorithm to control the process of matching the gesture to be measured with the reference gesture in real time. Compared with the traditional DTW algorithm, the improved DTW algorithm reduces the processing time 15% [43].

Radar Gesture Recognition Algorithm Based on Statistical Learning
Statistical learning is a theory that abstracts probability statistical models from data and uses the models to analyze and predict new data. For the field of radar gesture recognition technology, statistical learning-based methods are widely used in Support Vector Machine (SVM), K-Nearest Neighbor (KNN), and Hidden Markov Model (HMM) and other methods.

SVM is a supervised learning model based on statistical learning theory. Its principle is to map the input sample feature vector to a high-dimensional feature space through a kernel function, perform linear classification in the high-dimensional space, find the optimal separation hyperplane, so that the training samples can be most separable. Zhang et al. extracted two kinds of micro-Doppler features from the time-frequency spectrum of the echo signal, and used the SVM algorithm to classify these four gestures. The experimental results on the measured data showed that the classification accuracy of this method was higher than 88.56% [30]. Liu Zhao et al. used the segmented FFT signal processing method to convert the one-dimensional gesture signal into a two-dimensional gesture image, combined with the SVM algorithm to learn and classify the two-dimensional gesture signal, with an accuracy rate of 90.25% [44].

KNN is based on the training data set and calculates the k training instances that are closest to the new input instance, and the category with the largest number of k is the prediction result for the new input instance [45]. Xu Xian et al. applied traditional K-nearest neighbor algorithm and capacitive sensor to realize gesture recognition, which effectively improved the recognition success rate compared with ordinary threshold recognition methods [46]. However, in the process of gesture recognition, if there are a large number of gestures that need to be recognized, the traditional KNN algorithm training group data will be too large, which will affect the recognition efficiency. Chen Jiawei used the improved KNN algorithm to extract the features of the gestures that need to be recognized in the gesture recognition, and encode the gestures according to the extracted feature values, so that each gesture has a unique code, effectively reducing the amount of data in the training group, and it improved the success rate of gesture recognition by 5% [47].

Hidden Markov model is a typical probability and statistics model, which is usually used to describe Markov processes with hidden states. HMM can effectively capture the correlation in time series. Since gesture action is a time sequence, HMM has been widely used in the field of gesture recognition. When using HMM to recognize dynamic gestures, a separate HMM model needs to be trained in advance for each gesture, and then the probability of each HMM model producing the gesture to be measured is solved. The gesture corresponding to the HMM model with the highest probability is the recognition result.

Based on the self-developed 77 GHz millimeter-wave radar chip, Texas Instruments used the Hidden Markov Method to achieve an average classification accuracy of 83.3% for 6 medium motion amplitude gestures [48]. However, the speed energy distribution feature used in this paper is not enough to accurately represent a variety of gestures with many categories and similar features. Using the features extracted from the CSI, Wei Wang et al. proposed to use the Hidden Markov Model to build a CSI activity model containing multiple motion states, thereby completing the recognition of dynamic gestures [49]. Wang X et al. used the AdaBoost classifier to detect the user's hand, then they used particle filtering for tracking, and finally completed gesture recognition based on HMM [50]. This method has a significant improvement in recognition accuracy, but the computational complexity is extremely large and cannot meet the real-time requirements.

Radar Gesture Recognition Algorithm Based on Deep Learning

In recent years, deep learning has achieved remarkable results in computer vision, speech recognition, natural language processing and other fields. It has become one of the research fields that scholars pay attention to. Now it has become one of the most successful combination of big data and artificial intelligence. Deep learning is a kind of unsupervised learning. It can not only automatically extract features in images, but also automatically learn higher-level features, which overcomes the subjectivity and limitations of manual feature extraction [51]. The algorithms involved in deep learning methods in radar gesture recognition technology mainly include Convolutional Neural Networks (CNN) and Recurrent Neural Networks (RNN).

According to different dimensional processing methods, convolutional neural networks can be divided into two-dimensional convolutional neural networks (2-D CNN) and three-dimensional convolutional neural networks (3-D CNN). CNN is also called the two-dimensional convolutional neural network. It is one of the most basic network structures in deep learning. It usually consists of a convolutional layer, a pooling layer, and a fully connected layer. At present, CNN has achieved great success in the fields of face recognition and target detection, and it is also rapidly emerging in the field of radar gesture recognition technology.

Wang Yong used the method of convolutional neural network classification based on the distance, Doppler and angle multi-dimensional parameter features of gesture actions, which can achieve an average classification accuracy of 95.3% for 6 kinds of large motion amplitude gestures [37]. However, in this article, the multi-dimensional parameter features combined with distance spectrum, Doppler spectrum and angle spectrum are used. Among them, distance spectrum and angle spectrum are difficult to accurately characterize the micro-motion gestures with insignificant distance and angle changes. Therefore, the combination of range spectrum and angle spectrum features will deteriorate the representation ability, which is only suitable for gesture representation with large motion range. Researchers such as Dekker B used the real and imaginary parts of the Doppler-time spectrum as the input data of the two channels. Then, the author used the network structure of the convolutional layer, the pooling layer and the fully connected layer to extract the features of the image, and they used softmax classifier completes the gesture classification, the classification accuracy rate of 99% is reached on the test set [36]. Sruthy used the two receiving antennas of continuous wave Doppler radar to generate the in-phase and quadrature components of the beat signal, and they mapped the two beat signals to the CNN model, so that the accuracy of gesture classification exceeded 95% [52]. Pavlo Molchanov used optical, depth and radar sensor data fusion to recognize typical driving gestures through heterogeneous image registration and three-dimensional convolutional neural networks [53].

Recurrent Neural Network (RNN) is a type of recurrent neural network that takes sequence data as input, recursively in the evolution direction of the sequence, and all nodes are connected in a chain. In recent years, some scholars have adopted the combination of CNN and sequence model to recognize dynamic gestures. Infineon, a chip supplier of Soli, uses a long cycle full convolution neural network method based on sequence based range Doppler feature image training for gesture recognition, which can achieve an average classification accuracy of 94.34% for five micro-motion gestures [25]; Soli team adopt end-to-end cyclic neural network method based on distance Doppler features to classify gestures, which can achieve an average classification accuracy of 87% for 11 kinds of small and medium motion amplitude gestures [26], but the above two recognition systems based on RNN use the original range Doppler features, which is difficult to characterize the micro hands. Choi J used the distance Doppler sequence generated by data processing as the input of the Long Short-Term Memory (LSTM) network, enabling the gesture recognition system to successfully recognize 10 gestures with a classification accuracy rate of 99.10%. At the same time, the recognition accuracy rate of new participants' gestures was 98.48% [21]. However, this method has the best performance on small data sets. When the number of training samples is large, the computational efficiency will be greatly reduced.

From the current radar gesture recognition algorithm based on deep learning, the commonly used methods have their own advantages and disadvantages. In order to facilitate comparison and selection, Table 1 summarizes the different gesture recognition methods.

Table 1. Comparison of common radar gesture recognition algorithms

Common algorithms	Advantage	Disadvantage
DTW	Less training sample needs, high recognition accuracy	High computational complexity and poor stability
SVM	Can effectively solve small sample, high-dimensional, non-linear problems, and has strong generalization ability	When the number of training samples is large, the efficiency is low
KNN	The algorithm is simple and easy to understand	Need to take up a lot of storage space, time complexity is high
HMM	Can effectively capture the correlation in timing	The training process is more complicated, the training time is long, and the amount of calculation is large
CNN	No need to extract features manually, weight sharing	Requires a lot of training data and high computational cost
RNN	Can be used to describe the output of continuous state in time, with memory function, weight sharing	Poor parallel computing power and large amount of calculation

3 Existing Problems and Development Direction

The radar-based gesture recognition system does not require the human body to wear additional sensors and it is not affected by light conditions, it can also work normally around the clock. And the radar signal has good penetrability, even if the radar sensor is embedded inside the electronic device, the system can perform gesture recognition stably. This technology is becoming more and more mature. In addition, it is closely related to the fields of medical assistance, smart cars, and robots. It is an inevitable trend that radar technology gesture recognition will develop toward intelligence. The use of radar technology for dynamic gesture recognition has achieved certain results, but there are still some problems to be solved.

3.1 Radar Gesture Recognition in Complex Scenarios

Aiming at the K-band or even lower frequency range Compared with the complex application environment, the experimental scenes of the existing research are relatively simple, and there are not too many interfering objects within the transmitting range of the radar. In practical applications, due to the complexity of the actual application environment, the detected gesture signal sample data is often mixed with various noises. Therefore, more complex experimental scenarios can be considered in future research work to ensure that the system can still maintain stable recognition performance in the presence of other active human bodies and body movements.

3.2 Multi-view Data Fusion of Multiple Radar Sensors

Among the existing research results, most researchers use separate radar sensors for dynamic gesture information perception and have achieved good results. It is a potential direction worthy of research in this field to design a recognition system with higher accuracy and stability by fusing multiple radar sensors with multi-view and multi-scale data.

3.3 Radar Gesture Recognition Under Multiple Users

Most existing gesture recognition systems cannot recognize gestures performed by multiple users simultaneously. Different from recognizing a single user's gestures, this type of problem needs to consider the extraction of different gesture features and the construction of gesture models. In addition, due to differences in personal habits, proficiency and time, different users perform specific gestures differently. Therefore, the development of a multi-user gesture recognition technology that considers these individual differences is another future research direction in this field.

3.4 Higher Requirements for Training Data

Radar gesture recognition algorithms have high requirements for the quality and quantity of training samples. Sample data with quality and quantity is the premise and basis for effective recognition of dynamic gestures. In the process of radar collecting gesture signals, some data is lost. Aiming at the problem of poor data quality, studying how to improve the robustness of the model and maintaining a high recognition rate is one of the research directions to be carried out. Aiming at the problem of insufficient number of samples, on the one hand, the researchers should improve the learning efficiency of the model, capture the effective information in the sample and improve the classification and recognition ability; on the other hand, the researchers should enhancement and simulation data generation to expand the sample library.

4 Conclusion

The development of radar gesture recognition technology has brought a new way of human-computer interaction. The user can complete the interactive function without direct contact with the computer. This allows the user to successfully get rid of the shackles of external wearable devices such as data gloves, so as to improve the flexibility and naturalness of human-computer interaction. With the advancement of science and technology, intelligent radar gesture recognition system is an important development direction in the future. Machine learning is the core technology of artificial intelligence. Combining machine learning with radar gesture recognition is a good solution. In this paper, the methods of gesture recognition based on radar technology are summarized, sorts out the problems that need to be solved, and lays the foundation for the next step of research. It has certain reference significance for promoting the gesture recognition methods and research based on radar technology.

References

1. Cheng, N.H., Dai, N.Z., Liu, N.Z.: Image-to-class dynamic time Warping for 3D hand gesture recognition. In: IEEE International Conference on Multimedia & Expo. IEEE Computer Society (2013)
2. Nasr-Esfahani, E., et al.: Hand gesture recognition for contactless device control in operating rooms (2016)
3. Gupta, H.P., Chudgar, H.S., Mukherjee, S., Dutta, T., Sharma, K.: A continuous hand gestures recognition technique for human-machine interaction using accelerometer and gyroscope sensors. IEEE Sens. J. **16**(16), 6425–6432 (2016)
4. Mistry, C., Maes, P., Link, C.: WUW - Wear ur world - A wearable gestural interface. In: International Conference Extended Abstracts on Human Factors in Computing Systems (2009)
5. Ohn-Bar, E., Trivedi, M.M.: Hand gesture recognition in real time for automotive interfaces: a multimodal vision-based approach and evaluations. IEEE Trans. Intell. Transp. Syst. **15**(6), 2368–2377 (2014)
6. Ren, Z., Yuan, J., Meng, J., Zhang, Z.: Robust part-based hand gesture recognition using kinect sensor. IEEE Trans. Multimedia **15**(5), 1110–1120 (2013)
7. Grégory Rogez, III, J.S.S., Ramanan, D.: Understanding everyday hands in action from RGB-D images. In: IEEE International Conference on Computer Vision. IEEE (2015)
8. Hasan, H., Abdul-Kareem, S.: Retraction note to: human–computer interaction using vision-based hand gesture recognition systems: a survey. Neural Comput, Appl. (2017)
9. Rautaray, S.S., Agrawal, A.: Vision based hand gesture recognition for human computer interaction: a survey. Artif. Intell. Rev. **43**(1), 1–54 (2012). https://doi.org/10.1007/s10462-012-9356-9
10. Fragkiadaki, K., Levine, S., Felsen, P., Malik, J.: Recurrent network models for human dynamics (2015)
11. He, J., Zhang, C., He, X., Dong, R.: Visual recognition of traffic police gestures with convolutional pose machine and handcrafted features. Neurocomputing (2019)
12. Gao, X., Xu, J., Rahman, A., Yavari, E., Boric-Lubecke, O.: Barcode based hand gesture classification using AC coupled quadrature Doppler radar. In: 2016 IEEE/MTT-S International Microwave Symposium - MTT 2016. IEEE (2016)
13. Fan, T., Ma, C., Gu, Z., Lv, Q., Chen, J., Ye, D., et al.: Wireless hand gesture recognition based on continuous-wave doppler radar sensors. IEEE Trans. Microw. Theory Tech. **64** (11), 4012–4020 (2016)
14. Kim, Y., Toomajian, B.: Hand gesture recognition using micro-doppler signatures with convolutional neural network. IEEE Access **4**, 7125–7130 (2016)
15. Kim, Y., Toomajian, B.: Application of Doppler radar for the recognition of hand gestures using optimized deep convolutional neural networks. In: 2017 11th European Conference on Antennas and Propagation (EUCAP). IEEE (2017)
16. Pu, Q., Gupta, S., Gollakota, S., Patel, S.: Whole-home gesture recognition using wireless signals. In: Proceedings of the ACM SIGCOMM 2013 Conference on SIGCOMM. ACM (2013)
17. Ritchie, M., Jones, A.M.: Micro-Doppler gesture recognition using Doppler, time and range based features. In: 2019 IEEE Radar Conference (RadarConf19). IEEE. (2019)
18. Molchanov, P., Gupta, S., Kim, K., Pulli, K.: Short-range FMCW monopulse radar for hand-gesture sensing. Radar Conference, vol. 2015, pp. 1491–1496. IEEE (2015)

19. Nasr, I., Jungmaier, R., Baheti, A., et al.: A Highly integrated 60 GHz 6-Channel transceiver with antenna in package for smart sensing and short-range communications. IEEE J. Solid-State Circuits **51**(9), 2066–2076 (2016)
20. Zhang, X., Wu, Q., Zhao, D.: Dynamic hand gesture recognition using FMCW radar sensor for driving assistance. In: 2018 10th International Conference on Wireless Communications and Signal Processing (WCSP). IEEE (2018)
21. Choi, J.W., Ryu, S.J., Kim, J.H.: Short-range radar based real-time hand gesture recognition using LSTM encoder. IEEE Access **7**, 33610–336181 (2019)
22. Li, C.: Research on gesture recognition algorithm based on millimeter wave radar. University of Electronic Science and Technology of China (2020)
23. Wang, Y.: Research on gesture recognition based on millimeter wave radar. University of Electronic Science and Technology of China (2020)
24. Lien, J., Gillian, N., Karagozler, M.E., Amihood, P., Poupyrev, I.: Soli: ubiquitous gesture sensing with millimeter wave radar. ACM Trans. Graphics **35**(4), 1–9 (2016)
25. Hazra, S., Santra, A.: Robust gesture recognition using millimetric-wave radar system. IEEE Sens. Lett. **2**(4), 1–4 (2018)
26. Wang, S., Song, J., Lien, J., Poupyrev, I., Hilliges, O.: Interacting with soli. In: Proceedings of the 29th Annual Symposium on User Interface Software and Technology- UIST 2016 (2016)
27. Yu, C.: Research on Gesture Recognition Technology Algorithm Based on UWB Radar. Nanjing University of Science and Technology (2017)
28. Wang, W., Liu, A.X., Shahzad, M., Ling, K., Lu, S.: Understanding and modeling of WiFi signal based human activity recognition. In: International Conference on Mobile Computing & Networking. ACM (2015)
29. Abdelnasser, H., Harras, K.A., Youssef, M.: A ubiquitous wifi-based fine-grained gesture recognition system. IEEE Trans. Mob. Comput. 11, 2478-2487 (2019)
30. Zhang, S., Ritchie, M.A., Fioranelli, F., Griffiths, H., Li, G.: Dynamic hand gesture classification based on radar micro-Doppler signatures. In: 2016 CIE International Conference on Radar. IEEE. (2016)
31. Wang, J., Wang, X., Chen, L., et al.: Noncontact distance and amplitude-independent vibration measurement based on an extended DACM algorithm. IEEE Trans. Instrum. Meas. **63**(1), 145–153 (2014)
32. Schmidt, R., Schmidt, R.O.: Multiple emitter location and signal parameter estimation. IEEE Trans. Antennas Propag. **34**(3), 276–280 (1986)
33. Lin, J., Li, Y., Hsu, W.C., et al.: Design of an fmcw radar baseband signal processing system for automotive application. Springerplus **5**(1), 42 (2016)
34. Li, G., Zhang, R., Ritchie, M., Griffiths, H.: Sparsity-driven micro-doppler feature extraction for dynamic hand gesture recognition. IEEE Trans. Aerospace Electr. Syst. 1–1 (2017)
35. Saiwen, W., Jie, S.: Interacting with soli: exploring fine-grained dynamic gesture recognition in the radio-frequency spectrum. In: Proceedings of the 29th Annual Symposium on User Interface Software and Technology, pp. 851–860. ACM, Tokyo (2016)
36. Dekker, B., Jacobs, S., Kossen, A.S., Kruithof, M.C., Geurts, M.: Gesture recognition with a low power FMCW radar and a deep convolutional neural network. In: 2017 European Radar Conference (EURAD) (2017)
37. Wang, Y., Wu, J., Tian, Z., et al.: Multi-dimensional parameter gesture recognition algorithm based on FMCW radar. J. Electron. Inf. **41**(4), 822–829 (2019)
38. Wang, Y., Wang, S., Tian, Z., et al.: Two-stream fusion neural network gesture recognition method based on FMCW radar. Chinese J. Electron. **47**(7), 1408–1415 (2019)
39. Zhi, Z., Zongjie, C., Yiming, P.: Dynamic gesture recognition with a terahertz radar based on range profile sequences and Doppler signatures. Sensors **18**(2), 10 (2018)

40. Wang, J., Zheng, T., Lei, P., et al.: Gesture action radar recognition method based on convolutional neural network. J. Beijing Univ.ersity of Aeronautics Astronautics **44**(006), 1117–1123 (2018)
41. Hang, C., Zhang, R., Chen, Z., Li, C., Li, Z.: Dynamic gesture recognition method based on improved DTW algorithm. In: International Conference on Industrial Informatics-computing Technology. IEEE Computer Society (2017)
42. Plouffe, G., Cretu, A.M.: Static and dynamic hand gesture recognition in depth data using dynamic time warping. IEEE Trans. Instrum. Meas. **65**(2), 305–316 (2016)
43. Ruan, X., Tian, C.: Dynamic gesture recognition based on improved DTW algorithm. In: IEEE International Conference on Mechatronics & Automation. IEEE (2015)
44. Liu, Z., Cui, H., Wu, J., et al.: Gesture signal recognition algorithm based on continuous wave Doppler radar. Inf. Technol. Network Secur. **038**(003), 30–34 (2019)
45. Cover, T.M., Hart, P.E.: Nearest neighbor pattern classification. IEEE Trans. Inf. Theory **13** (1), 21–27 (1953)
46. Xu, X., Zhao, J., Zhang, H.: Design and production of gesture recognition device based on capacitive sensor. Inf. Comput. (Theoretical Edition) **417**(23), 134–136 (2018)
47. Jiawei, C., Jing, H., Ruiling, H., et al.: Research on dynamic gesture recognition based on improved KNN algorithm. J. North Univ. China: Nat. Sci. Ed. **2020**(3), 232–237 (2020)
48. Malysa, G., Wang, D., Netsch, L., Ali, M.: Hidden Markov model-based gesture recognition with FMCW radar. Signal & Information Processing. IEEE (2017)
49. Wang, W., Liu, A.X., Shahzad, M., Ling, K., Lu, S.: Device-free human activity recognition using commercial wifi devices. IEEE J. Sel. Areas Commun. 1 (2017)
50. Wang, X., Xia, M., Cai, H., Gao, Y., Cattani, C.: Hidden-markov-models-based dynamic hand gesture recognition. In: Mathematical Problems in Engineering, (2012–04–24), 2012 (PT.5), 137–149 (2012)
51. Zhang, R., Li, W., Mo, T.: A review of deep learning research. Inf. Control. 47(04): 385–397 +410. (2018)
52. Skaria, S., Al-Hourani, A., Lech, M., Evans, R.J.: Hand-gesture recognition using two-antenna doppler radar with deep convolutional neural networks. IEEE Sens. J. **19**(8), 3041–3048. (2019)
53. Molchanov, P., Gupta, S., Kim, K., Pulli, K.: Multi-sensor system for driver's hand-gesture recognition. In: IEEE International Conference & Workshops on Automatic Face & Gesture Recognition. IEEE (2015)

Analysis of the Influence of Convolutional Layer in Deep Convolutional Neural Network on SAR Target Recognition

Wei Qu[1(✉)], Gang Yao[2], and Weigang Zhu[1]

[1] Space Engineering University, Beijing, People's Republic of China
quweistar@163.com
[2] Beijing Institute of Tracking and Telecommunication Technology, Beijing, People's Republic of China

Abstract. As a frontier hot spot in the current image processing field, deep learning has unparalleled superiority in feature extraction. Deep learning uses deep network structure to perform layer-by-layer nonlinear transformation, which can achieve the approximation of complex functions. From low-level to high-level, the representation of features becomes more and more abstract, and the more essential the original data is described. Aiming at the problem of SAR image target detection, this paper studies the influence of the number of convolution kernels, the size of the convolution kernel and the number of convolution layers in the deep convolutional neural network on SAR target recognition.

Keywords: Deep convolutional neural network · SAR · Target recognition

1 Introduction

Synthetic Aperture Radar (SAR), as a branch of microwave remote sensing, obtains ground object information through the interaction between electromagnetic waves and various media. Compared with optical remote sensing, it has all-weather and all-weather detection. Reconnaissance capabilities. In addition, SAR obtains large-area high-resolution radar images through the use of range-wise pulse compression technology and azimuth synthetic aperture technology, which can provide strong support for military intelligence acquisition, precision guidance, and strike effect evaluation [1, 2].

From the perspective of SAR image target detection and recognition algorithms, its essence is based on the acquisition of image features and the design of feature usage rules. For example, the most commonly used target detection algorithm based on Constant False Alarm Rate (CFAR) [3] mainly uses the gray-scale features of SAR images [6]. The feature extraction rules of these methods are often manually designed. In practical applications, when the amount of data is too large and the data is complex, the features extracted in this way are usually not representative and cannot represent the uniqueness between different types of data. This limits the accuracy of detection and recognition.

S. Shi et al. (Eds.): AICON 2020, LNICST 356, pp. 404–413, 2021.
https://doi.org/10.1007/978-3-030-69066-3_35

2 SAR Image Target Detection Method Based on Deep Convolutional Neural Network

The SAR target detection method based on the deep convolutional neural network (CNN) constructs a structure containing multiple deep neural networks to transform the original data into a higher-level and more abstract expression using a combination of nonlinear mapping relationships. Compared with the method based on pattern classification Methods such as support vector machines and traditional neural networks have the ability to extract features autonomously. Compared with deep neural network architectures such as restricted Boltzmann machines and autoencoders, CNN has more advantages in image classification and recognition, and has achieved more achievements. With the widespread recognition of CNN in optical image processing in recent years [7], it has also set off a research boom in the field of SAR target recognition. Morgan et al. [8] directly used the typical CNN structure to obtain good recognition performance, but did not consider the impact of the network structure design on the SAR image target recognition performance, and its accuracy was not ideal. In order to further improve the recognition performance of CNN, scholars have made improvements in the design of the network structure. Tian Zhuangzhuang [9] and others first improved the classification ability of CNN by introducing a category separability measure in the error cost function, and finally used SVM to classify features. Zhao et al. [10] used the highway convolutional layer to reduce the data requirements of the network, and the recognition rate was further improved. Xu Feng [11] and others use a sparse convolutional architecture instead of a fully connected layer, which can avoid the overfitting problem caused by a small training set.

Deep Convolutional Neural Network (CNN) due to its powerful feature extraction capabilities, not only has made great achievements and widespread recognition in optical image processing, but also achieved a better recognition rate than traditional recognition methods in SAR target recognition. However, the accuracy, timeliness, and generalization of deep CNN in SAR target recognition need to be further explored, mainly as follows: 1) In terms of recognition accuracy, different activation functions, regularization functions, number of convolution kernels and Factors such as size have different effects on the recognition rate; 2) In terms of recognition timeliness, changing the convolutional layer structure parameter settings within a certain range is beneficial to the improvement of the recognition rate, and it also often increases the amount of network calculation and training time. It is not conducive to the timeliness of recognition; 3) In terms of recognition generalization, since there are multiple variants of the same type of target and the SAR image target has attitude sensitivity, the recognition network needs to extract more robust and generalized features.

In response to the above problems, this paper studies the SAR target recognition method based on deep CNN. First, based on the classic convolutional network Lenet-5, it is improved to construct the basic structure of SAR target recognition network (SAR-Lenet), and the influence of activation function and regularization function on recognition performance is analyzed; then, each volume The effect of the number and size of the convolution kernel of the buildup layer on the recognition performance is comparatively analyzed; again, in order to explore the impact of the number of network convolutional layers on the recognition performance, the basic network is deformed by the layer replacement idea, and the number of convolutional layers is increased. While keeping the training time of the network basically unchanged; finally, optimize the SAR-Lenet network structure for the generalization of the network, use the dense block network structure to merge multiple convolutional layer features, and use the convolutional layer to replace the fully connected layer to increase The sparseness of the network, and at the same time, the optimized network (SAR-Net) is trained with data augmentation and expansion training set to further improve the generalization of the recognition network.

The convolutional layer is the main component of CNN, and different settings of its structural parameters will have different effects on the recognition network. The convolutional layer structure parameters refer to the number of convolution kernels, the size of convolution kernels, and the number of convolution layers. From the SAR-Lenet convolution layer structure parameters, it follows the parameter settings of Lenet-5, and these parameter settings are based on handwritten numbers. The recognition requirements may not meet the SAR target recognition requirements. Therefore, through comparative experiments, this paper explores the influence of the number of convolution kernels, the size of convolution kernels and the number of convolution layers in the SAR-Lenet convolution layer on recognition performance, and provides a basis for the subsequent parameter setting of the convolution layer in the target recognition network.

3 The Effect of the Number of Convolution Kernels on Recognition Performance

In the setting of the number of SAR-Lenet convolution kernels, keep the size of the convolution kernels of the three layers unchanged, keep the number of convolution kernels of the three layers the same, and set the number of convolution kernels to 8, 16, 32, 64, 128, respectively and 256, where A1–A6 represent the network with six different convolution kernel configurations. The time loss of the network convolution layer increases by 4 times. At the same time, A7, A8 and A9 are set on the basis of six orders of magnitude. The network parameters of nine different convolution kernels are set as shown in Table 1. In the experiment, the SOC data set is used for training and testing. During the training process, the recognition rate curve of the test set is shown in Fig. 1, and the network training time consumption is shown in Fig. 2.

Table 1. Networks with different numbers of convolution kernels

Layer	Kernel number	Step	A1	A2	A3	A4	A5	A6	A7	A8	A9
Conv1	7 × 7	2	8	16	32	64	128	256	64	128	32
Conv2	5 × 5	1	8	16	32	64	128	256	128	64	64
Conv3	3 × 3	1	8	16	32	64	128	256	128	128	128

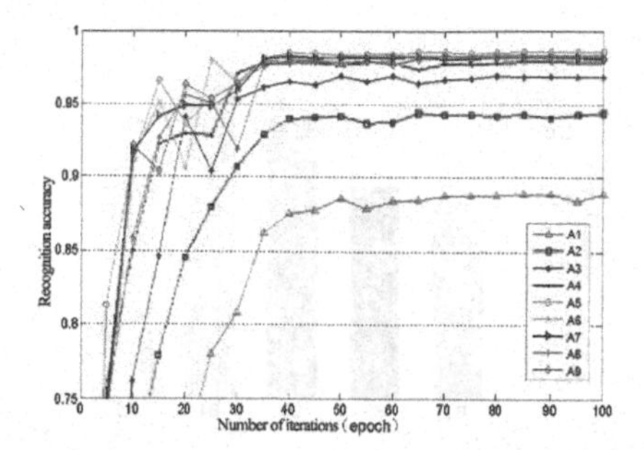

Fig. 1. Network recognition rate curve

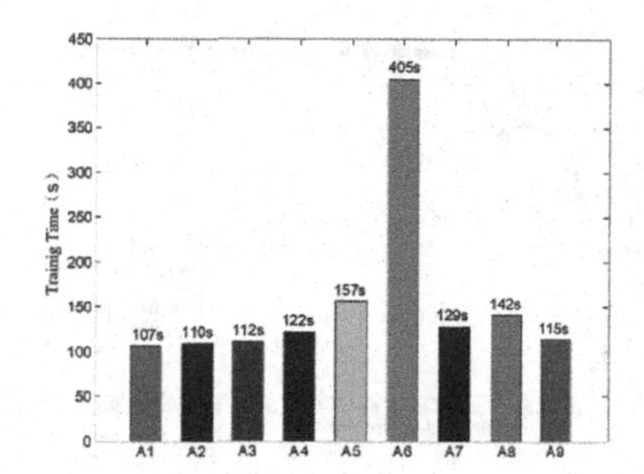

Fig. 2. Network training time consumption graph.

Comparing the recognition rate curves of A1, A2, and A3, it can be seen that as the number of convolution kernels increases, the recognition rate of the network increases greatly. From A3 to A5, the recognition rate increases less and less, and the accuracy increases by 1. % Or less, and the recognition rate from A5 to A6 has dropped. The results of the six networks show that the number of convolution kernels in the A1, A2, and A3 networks is not enough to meet the network's demand for the number of feature maps. The increase in the number of feature maps in the A3, A4, and A5 networks weakens the recognition rate. Recognition performance tends to be saturated. A6, due to the excessive number of feature maps, easily leads to network overfitting, and the recognition rate drops instead.

Comparing the A7 and A8 networks, the total number of convolution kernels is the same, but the number of conv1 and conv1 feature maps is reversed. Because SAR-Lenet adds a pooling layer to each convolutional layer to reduce the dimensionality, the size of the conv2 feature map It is half of conv1. Conv1 consumes more time than conv2 under the same number of convolution kernels. Therefore, the training time of A8 network is 13 s longer than that of A7 network, and the recognition rate of the two is basically the same. Comparing A9 and A4, the number of A9 convolution kernels is 32 more than that of A4, but the training time is 7 s less.

4 The Effect of Convolution Kernel Size on Recognition Performance

The size of the convolution kernel represents the size of the receptive field. In the performance study of the size of the convolution kernel, based on the A5 network with the highest recognition rate and the A1 network with the lowest recognition rate, the convolution kernel size is selected from 3, 5, 7, 9 Four sizes, four networks are obtained based on the A5 network, namely B1, B2, B3, B4; based on the A1 network, six networks are obtained, namely B5, B6, B7, B8, B9 and B10; The size of the convolution kernel for each network is shown in Table 2 and 3. In the experiment, the SOC data set was used for training and testing. During the training process, the recognition rate curves of the test set are shown in Fig. 3 and Fig. 5. The network training time consumption is shown in Fig. 4 and Fig. 6.

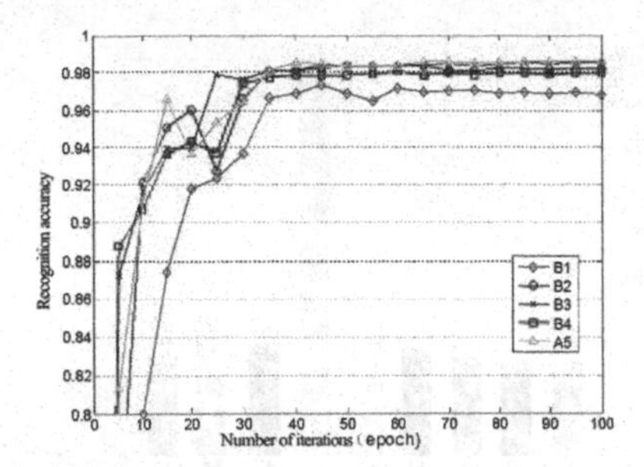

Fig. 3. Network recognition rate curve

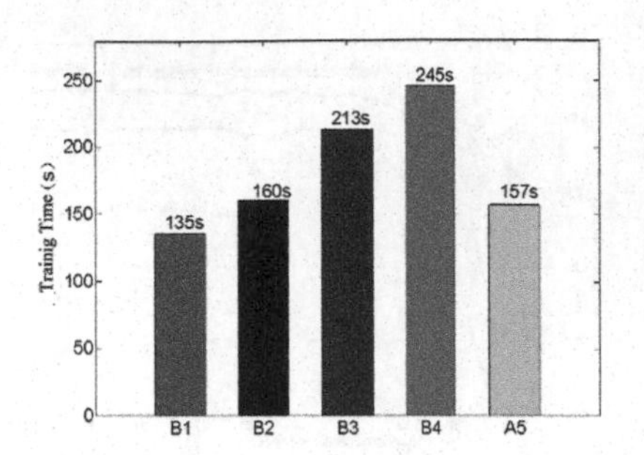

Fig. 4. Network training time consumption graph

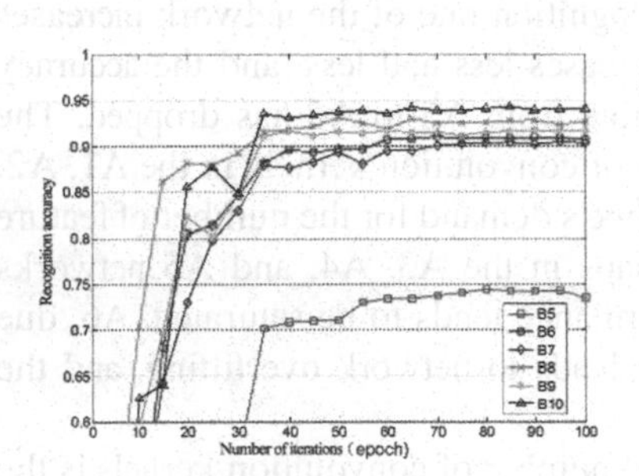

Fig. 5. Network recognition rate curve

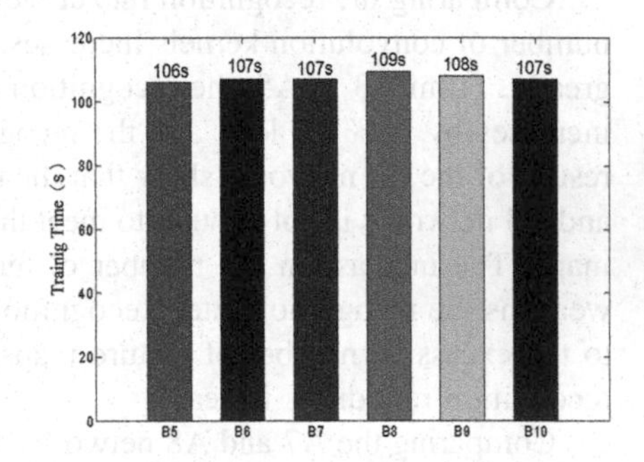

Fig. 6. Network recognition rate curve

Table 2. Networks with different convolution kernel sizes

Layer	Kernel size	Step	B1	B2	B3	B4
Conv1	128	2	3×3	5×5	7×7	9×9
Conv2	128	1	3×3	5×5	7×7	9×9
Conv3	128	1	3×3	5×5	7×7	9×9

Table 3. Networks with different convolution kernel sizes

Layer	Kernel size	Step	B5	B6	B7	B8	B9	B10
Conv1	8	2	3×3	5×5	7×7	9×9	5×5	9×9
Conv2	8	1	3×3	5×5	7×7	9×9	7×7	7×7
Conv3	8	1	3×3	5×5	7×7	9×9	9×9	5×5

From the above experimental results, when the number of convolution kernels is set to 128, since the number of network convolution kernels is basically saturated, the size of the convolution kernel has little effect on recognition performance, and the network training time increases as the size of the convolution kernel increases.; Comparing the five networks of A5 and B1–B4, A5 has the best recognition performance, indicating that different convolution kernel size combinations are more conducive to the network extracting features of different scales, which is conducive to the improvement of network recognition rate, and A5 also accounts for timeliness excellent.

When the number of convolution kernels is set to 8, the number of network convolution kernels is under-saturated, and the size of the convolution kernel has a more obvious impact on network recognition performance. Compare B5, B6, B7 and B8, as the size of the convolution kernel increases the recognition rate of the network is improved. It can be considered that increasing the scale of the convolution kernel to extract the features can effectively compensate for the impact of the insufficient number of feature maps; compare B10 and B5–B8 five networks, B10 improves the recognition rate by combining different convolutions and sizes; Comparing B9 and B10, the two networks both use different convolution kernel size combinations, but the recognition rate of B10 is higher. It can be considered that shallow features are suitable for large-size convolution kernel extraction, and high-level features need to be extracted by small-size convolution kernel; It can be seen from Fig. 6 that the training time consumption of the six networks is relatively small. The main reason is that the number of network convolution kernels is small. The calculation difference between the networks is smaller than the total calculation, but it can still be seen the size of the convolution kernel increases and the training time of the network increases.

5 The Effect of the Number of Convolutional Layers on Recognition Performance

In general, increasing the number of convolutional layers can improve the accuracy of recognition and easily increase the risk of network overfitting. This section continues to use SAR-Lenet as the basis to explore the influence of different numbers of convolutional layers on network recognition performance. In the experiment, the SOC data set is used for training and testing.

Using the number of convolution kernels of the A9 network, the network has a higher recognition rate and better timeliness. Therefore, the number of convolution kernels of the three convolution layers of SAR-Lenet is set to 32, 64, and 128 in sequence. Since the network's demand for feature maps is basically satisfied with this number of convolution kernels, the size of the convolution kernel has a small impact on the recognition performance. To facilitate subsequent research, the size of the convolution kernel adopts two scales of convolution kernel size. At the same time, in accordance with the principle of setting the larger size of the first layer, the convolution kernel sizes of the three convolution layers are set to 7, 5, and 5 in sequence. The SAR-Lenet corresponding to the number and size of the convolution kernel is referred to as network C here. The experimental results of network C in Table 4–7 show that the above parameter settings have a better recognition rate and further reduce network training time overhead.

In order to avoid the increase of network time training while increasing the number of convolutional layers, the idea of layer replacement is adopted to split a convolutional layer into two convolutional layers, and at the same time change the size of the convolution kernel or the number of convolution kernels in the replacement layer, and other the convolutional layer remains unchanged. Since only the number or size of the convolution kernel of the replacement layer has been changed, without considering its impact on the network recognition performance, the network overhead can be kept basically unchanged. There are two main aspects to increase the number of convolutional layers:

Increase the number of convolutional layers while reducing the size of the convolution kernel of the replacement layer, and the number of convolution kernels remains unchanged. Specifically, one convolutional layer in convolution block 3 of the C network can be replaced with two layers, and then the size of the convolution kernel can be reduced. The number of convolution kernels of the two convolutional layers is 128. The specific replacement process is as follows:

$$64 \times 5^2 \times 128 \Rightarrow 64 \times 3^2 \times 128 + 128 \times 3^2 \times 128 \tag{1}$$

So as to get the C1 network. The convolution block 2 of the C network can also be replaced as follows:

$$32 \times 5^2 \times 64 \Rightarrow 32 \times 3^2 \times 64 + 64 \times 3^2 \times 64 \tag{2}$$

To get the C2 network. The C3 network can be obtained by combining the C1 network and the replaced convolution block of C2. Then continue to replace the convolution block 2 and convolution block 3 of the C3 network as follows:

$$32 \times 3^2 \times 64 + 64 \times 3^2 \times 64 \Rightarrow 32 \times 3^2 \times 64 + 64 \times 2^2 \times 64 + 64 \times 2^2 \times 64 \tag{3}$$

$$64 \times 3^2 \times 128 + 128 \times 3^2 \times 128 \Rightarrow 64 \times 3^2 \times 128 + 128 \times 2^2 \times 128 + 128 \times 2^2 \times 128 \tag{4}$$

To get the C4 network. In this way, C1 and C2 are obtained from C, C3 is obtained from C1 and C2, and C4 is obtained from C3.

Increase the number of convolutional layers while reducing the number of convolution kernels in the replacement layer, and the size of the convolution kernel remains unchanged. In this way, the number of convolution kernels of the convolution layer added to the convolution block 3 in the C network can be reduced from 128 to 44, and the number of convolution kernels of the second convolution layer is still 128, which ensures the input of the convolution block 3 And the number of output feature maps remains unchanged, and the specific replacements are as follows:

$$64 \times 5^2 \times 128 \Rightarrow 64 \times 5^2 \times 44 + 44 \times 5^2 \times 128 \tag{5}$$

In this way, D1 is obtained; also the convolution block 2 of the C network is replaced as follows:

$$32 \times 5^2 \times 64 \Rightarrow 32 \times 5^2 \times 22 + 22 \times 5^2 \times 64 \tag{6}$$

In this way, D2 is obtained, and D3 is also obtained by combining D1 and D2 networks. At the same time, the convolution block 3 and convolution block 2 of the C1 network can also be replaced as follows:

$$64 \times 3^2 \times 128 + 128 \times 3^2 \times 128 \Rightarrow 64 \times 3^2 \times 88 + 88 \times 3^2 \times 88 + 88 \times 3^2 \times 128 \quad (7)$$

$$32 \times 3^2 \times 64 + 64 \times 3^2 \times 64 \Rightarrow 32 \times 3^2 \times 44 + 44 \times 3^2 \times 44 + 44 \times 3^2 \times 64 \quad (8)$$

Get the D4 and D5 networks respectively, and combine the two to get the D6 network. See Table 4 for details of the above deformation in the C network. For the layer replacement of the above network, in order to control the size of the output feature map after each convolution operation, the sliding step size of convolution block 2 and convolution block 3 is S = 1, and for the 5 × 5 convolution kernel zero padding P = 2. For 3 × 3 convolution kernel zero-filling P = 1, for two consecutive 2 × 2 convolution kernels, the first convolution kernel zero-filling P = 0, the second convolution kernel zero-filling P = 1. This makes the size of the feature map output by each convolution block unchanged. The parameters in the convolution block in Table 4 are the number and size of the convolution kernel. The sliding step size of the first convolution block is 2, and the step size of the remaining two convolution blocks is 1. The pooling layer in the network remains unchanged, and there are three maximum pooling layers after three convolutional blocks.

Table 4. The relationship between convolutional layer parameters and recognition performance

Network	Convolution block 1	Convolution block 2	Convolution block 3	Number of convolutional layers	Training time (s)	Average recognition rate (%)
C	32 × 7 × 7/2	64 × 5 × 5/1	128 × 5 × 5/1	3	123	98.04
C1	32 × 7 × 7/2	128 × 5 × 5/1	(128 × 3 × 3) × 2/1	4	121	98.57
C2	32 × 7 × 7/2	(64 × 3 × 3) × 2/1	128 × 5 × 5/1	4	122	98.89
C3	32 × 7 × 7/2	(64 × 3 × 3) × 2/1	(128 × 3 × 3) × 2/1	5	121	99.30
C4	32 × 7 × 7/2	64 × 3 × 3 + (64 × 2 × 2) × 2/1	128 × 3 × 3 + (128 × 2 × 2) × 2/1	7	123	99.07
D1	32 × 7 × 7/2	128 × 5 × 5/1	44 × 5 × 5/1 + 128 × 5 × 5/1	4	124	98.54
D2	32 × 7 × 7/2	22 × 5 × 5/1 + 64 × 5 × 5/1	128 × 5 × 5/1	4	125	98.51
D3	32 × 7 × 7/2	22 × 5 × 5/1 + 64 × 5 × 5/1	44 × 5 × 5/1 + 128 × 5 × 5/1	5	124	99.13
D4	32 × 7 × 7/2	128 × 5 × 5/1	(88 × 3 × 3) × 2/1 + 128 × 3 × 3/1	5	125	98.81
D5	32 × 7 × 7/2	(44 × 3 × 3) × 2/1 + 64 × 3 × 3/1	128 × 5 × 5/1	5	122	99.04
D6	32 × 7 × 7/2	(44 × 3 × 3) × 2/1 + 64 × 3 × 3/1	(88 × 3 × 3) × 2/1 + 128 × 3 × 3/1	7	120	98.96

It can be seen from the network training time that the layer replacement idea basically keeps the network time overhead unchanged. From the point of view of recognition rate, the recognition rate of the network after increasing the number of convolutional layers is higher than that of the C network, indicating that the C network is not saturated in the network depth. Increasing the network depth can effectively improve the recognition rate. Among them, the C3 network has a higher recognition rate. The C network has increased by 1.26%, which is the largest improvement. Comparing C3, C4 and D6, although the number of C4 and D6 convolutional layers is two more than C3, due to the saturation of recognition performance, the network has a certain degree of overfitting, and the recognition rate is instead Decrease; for networks with the same number of convolutional layers, the recognition rate of the C3 network is higher than that of the D3, D4, and D5 networks, and the recognition rate of the C1 and C2 networks is higher than that of D1 and D2, indicating that the combination of the number and size of different convolution kernels can identify the network It will also bring a certain impact; in the combination of convolutional layers, the recognition rate of C3 and D3 networks is higher than other networks, indicating that the matching method of convolutional blocks in this network structure is more suitable for target feature extraction.

6 Conclusion

Based on the above analysis, it can be seen that the number of different convolution kernels not only affects the recognition rate of the network, but also affects the timeliness of the network. When setting the number of convolution kernels, when the total number of convolution kernels in the network remains unchanged, the number of shallow convolution kernels can be appropriately reduced, and the number of deep convolution kernels can be appropriately increased, which will help improve the timeliness of the network. Combining the influence of the number and size of convolution kernels on the recognition performance of SAR-Lenet, it can be concluded that: 1) The number of convolution kernels has a greater impact on the recognition network performance than the size of the convolution kernel. Only when the number of convolution kernels is met, the recognition the rate is high; 2) When the number of convolution kernels is saturated, the size of the convolution kernel has little effect on the network performance. When the number of convolution kernels is insufficient, increasing the size of the convolution kernel is conducive to a very high recognition rate; 3) Different convolution kernels the size combination is more conducive to network feature extraction, and the scale of the first layer of convolution kernel is relatively large; 4) Increasing the number and size of the convolution kernel will increase the training time of the network. The number of SAR-Lenet convolutional layers is set to 5 and the C3 network structure has the highest recognition rate; when the number of convolution kernels tends to be saturated, increasing the number of convolution kernels does not improve the recognition rate, increasing the number of convolutional layers, the recognition rate can be It is further improved, but it is not that the more the number of convolutional layers, the better, it needs to be controlled.

References

1. Licheng, J., Xiangrong, Z., Biao, H., et al.: Intelligent SAR image processing and interpretation (2008)
2. Jianshe, S., Yongan, Z., Lihai, Y.: Synthetic Aperture Radar Image Understanding and Application. Science Press, Beijing (2008)
3. Performance analysis of CA-CFAR two approximate methods under linear detection
4. Gangyao, K., Gui, G., Yongmei, J.: Synthetic Aperture Radar Target Detection Theory, Algorithm and Application (2007)
5. Novak, L.M., Hesse, S.R.: On the performance of orderstatistics CFAR detectors. In: IEEE 25th Asilomar Conference on Signals, Systems and Computers, Pacific Grove, California, USA, November 1991, pp. 835–840 (1991)
6. Wenqing, S., Yinghua, W., Hongwei, L.: Automatic region screening target detection algorithm for high-resolution SAR images. J. Electron. Inf. Technol. **38**(5), 1017–1025 (2016)
7. Krizhevsky, A., Sutskever, I., Hinton, G.E.: Imagenet classification with deep convolutional neural networks. In: Advances in Neural Information Processing Systems, pp. 1097–1105 (2012)
8. Morgan, D.A.E.: Deep convolutional neural networks for ATR from SAR imagery. In: Proceedings of the Algorithms for Synthetic Aperture Radar Imagery XXII, Baltimore, MD, USA, 2015, vol. 23:94750F (2015)
9. Zhuangzhuang, T., Ronghui, Z., Jiemin, H., et al.: Research on SAR image target recognition based on convolutional neural network. J. Radars **5**(3), 320–325 (2016)
10. Lin, Z, Ji, K., Kang, M., et al.: Deep Convolutional Highway Unit Network for SAR target classification with limited labeled training data. IEEE Geosci. Remote Sens. Lett. (2017)
11. Chen, S., Wang, H., Xu, F., et al.: Target classification using the deep convolutional networks for SAR images. IEEE Trans. Geosci. Remote Sens. **54**(8), 4806–4817 (2016)

A Real-Time Two-Stage Detector for Static Monitor Using GMM for Region Proposal

Yingping Liang[1], Yunfei Ma[3], Zhengliang Wu[1], and Mingfeng Lu[1,2](✉)

[1] Beijing Institute of Technology, Beijing 100081, China
lumingfeng@bit.edu.cn
[2] Beijing Key Laboratory of Fractional Signals and Systems, Beijing Institute of Technology, Beijing 100081, China
[3] Zaozhuang University, Shandong 277100, China

Abstract. CNN-based object detectors have been widely exploited for vision tasks. However, for specific real-time tasks (e.g. object detection on static monitor), the enormous computation cost makes it difficult to work. To reduce the computation cost for object detection on static monitor while inheriting high accuracy of CNN-based networks, this paper proposals a method with a two-stage detector using Gaussian mixture model for region proposal. We test our method on MOT16 datasets. Compared with original models, the two-stage detectors equipped with Gaussian region proposal achieve a better performance with the mAP increased by 0.20. We also design and train a light-weight detector based on our method, which is much faster and more suitable for mobile and embedded device with little drop in accuracy.

Keywords: Deep learning · Computer vision · Intelligent monitoring

1 Introduction

VID (object detection from video) has become a challenge task in recent years. Compared to image object detection, static videos are highly redundant, containing a large amount of temporal locality (that is, similar at different times) and spatial locality (that is, they look similar in different scenes), and frame by frame processing is time-consuming and computationally expensive. So making full use of the timing context can solve the problem of a large amount of redundancy between consecutive frames in the video and improve the detection speed and the detection quality. Although deep learning methods excel on some very large datasets for image detection, there is still a big gap for the application of specific tasks.

The one-stage method (e.g. SSD [1], YOLOv3 [2]) has achieved high efficiency by densely sampling on feature maps over different scales and ratios but suffers from low accuracy; the two-stage method (e.g. Faster-RCNN [3], FPN [4]) has achieved high accuracy by using two-stage structures to describe features and regress bounding box

Beijing Natural Science Foundation (L191004).

S. Shi et al. (Eds.): AICON 2020, LNICST 356, pp. 414–423, 2021.
https://doi.org/10.1007/978-3-030-69066-3_36

parameters but suffers from low efficiency. And one of the key points to improve the detection speed of two-stage methods is to improve the structure for extracting regions of interest.

Traditional methods [5, 6] for video tracking have fewer parameters and low latency, suitable for mobile and embedded vision applications, but lack the ability to recognize categories and have some other problems (see Fig. 4). Some CNN methods utilize optical flow vectors for local area to detect moving target, which is heavily influenced by background. For methods using Convolutional LSTM [7] Network, in order to integrate features across time, ConvLSTM is used to obtain temporal information, which is computationally expensive. The mainstream idea is to combine the context information and tracking information between frames. However, lack of datasets is one of the obstacles for training an end-to-end detector for applications.

It inspires us to think: can we use an efficient mixture model for region proposal and an lightweight convolutional neural network trained on image detection datasets for classification and cascade regression?

Thus we use a pixel to pixel rather than region to region approach by an efficient adaptive Gaussian mixture model for the static video background subtraction. The mixture model presents a set of N models for each pixel. As the pixel value at each point follows a separate non-stationary temporal distribution, the mixture model generates the parameters using an adaptive learning rate for each model at every frame and selects the number of components (N) simultaneously [8]. The upgrade process is achieved by a variant of the maximum a posterior (MAP) solution. While updating, the Gaussian mixture model also produces the region of interest for moving objects as the foreground, which is highly memory and time efficient.

In the CNN part, we follow the framework of recent two-stage detection network with an extra Context Enhancement Module [9] to combine global context with local information and use a lightweight mobilenet as backbone. In the mixture model, we have got the separation of foreground and background and coarse boxes for locations. Thus the RPN module for filtering and adjusting anchors in some modules can be abandoned in the detection part.

In the next section we will review some of the prior work in background subtraction and video object detection. Section 3 describes the upgrade algorithm for Gaussian mixture model and the structure of CNN-based detection part. Section 4 shows the training details and Sect. 5 describes the test results on the MOT dataset without extra training on the training video.

2 Related Work

2.1 Moving Object Tracking

A moving object can be detected if the background of the video is known. To build the model of the background, a well-known process is to use a Gaussian mixture model (GMM) to estimate the probability distribution of pixels in real time [5]. Thus the foreground is detected by observing the local regions which do not fit the distributions. The standard equation for GMM estimation uses a fixed number of components at each pixel.

In [6], a effective recursive unsupervised learning method of finite mixture models is proposed. This method suggests an efficient method for maximizing probability distribution and parameter estimation. By using a prior as a bias for maximally mixture models, the method enables us to estimate the parameter and select the number of components. This method is also applicable to Gaussian mixture models in version tasks, which is much more efficient than RPN method for region proposal.

2.2 CNN-Based Detectors

CNN-based two-stage detectors gains excellent performance on large datasets in recent years. And the two-step methods have a more accurate detection than one-step method, by producing more features to describe object and two-step cascade regression. R-CNN uses selective search method for searching candidate region. Faster-RCNN [3] uses Region Proposal Network (RPN) to generate region of interest. And R-FCN [11] designs a fully convolutional structure and a position sensitive RoI pooling module for region proposal. And in the detection part, an extra subnet is usually added to achieve more accurate regression of object boxes and prediction of object categories.

There are also some real-time detectors with less parameters. Light-Head R-CNN [10] designs a lightweight detection network for lower latency and less computation cost, but still reserves the problems of mismatch between a fast backbone and a computational expensive detection network. SSDLite for mobile device adopts the MobileNet as backbone and a one-stage method similar to SSD. By widely using depth separable convolution with residuals, SSDLite generates a larger receptive field and reduces the computation cost but makes poor performance on accuracy. ThunderNet [9] utilizes DWConv to reduce the computation complexity in RPN net and a Special Attention Module (SAM) to enhance the foreground feature.

Fully utilizing the context between frames is also one of the keys for the improvement of state of the art detectors on video tasks. As superior models perform well on GPUs, their performance on mobile and other resource-restrained platforms will be greatly reduced. However, few work focuses on a particular application for target detection in static surveillance video. In this paper, we propose a method combining moving object tracking method and CNN-based detector for the surveillance of static scene.

3 Approach

3.1 Moving Region Proposal

The basic algorithm of Moving Region Proposal follows the formulation of a recursive unsupervised learning method of finite mixture model (see Fig. 1). For every frame in the video, the background frame can not be directly captured as there appears foreground beyond the background. And the background is not always the same on time series (e.g. A car can be either the foreground while moving or a part of background).

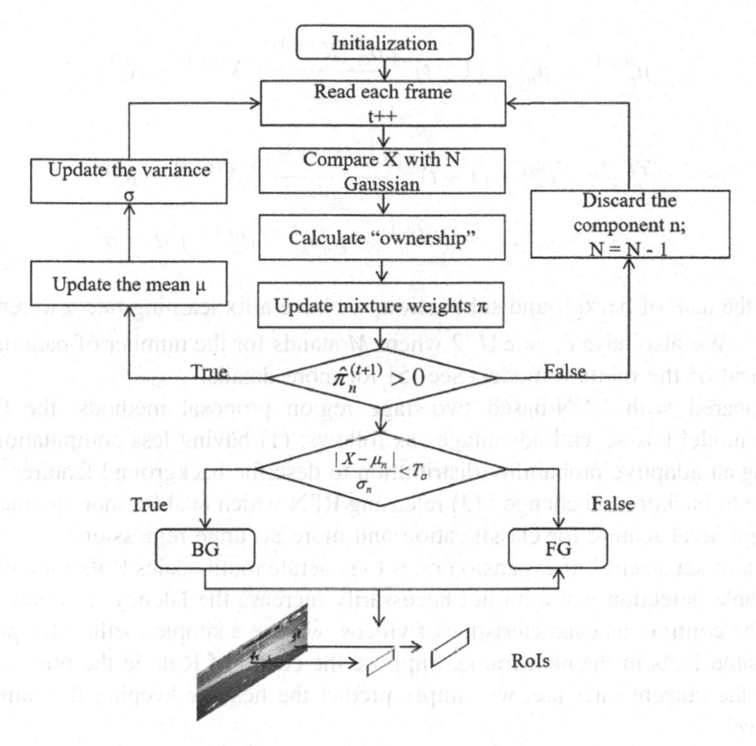

Fig. 1. The program frame and flow chart of moving region proposal.

A Gaussian mixture model with N is defined as:

$$\vec{\theta} = \{\pi_1, \pi_2, \ldots, \pi_N, \vec{\theta}_1, \vec{\theta}_2, \ldots, \vec{\theta}_N\} \tag{1}$$

where π_n is the weight of the n th component in mixture model and $\vec{\theta}_n = \{\mu_n, \sigma_n\}$ is the mean value and standard deviation.

Suppose a set of sampled data X and the vector $\vec{Y} = [y_1, y_2, \ldots, y_N]^T$ with only 0 and 1, which stands for the subordinate component of data X, the joint probability density is:

$$p(X, \vec{Y}; \vec{\theta}) = \prod_{n=1}^{N} \{\pi_n p_n(X; \vec{\theta}_n)\}^{y_n} \tag{2}$$

Thus when a new sample of pixel value is added at iteration *t+1*, we get the update equation for estimation of π, μ and σ:

$$\hat{\pi}_n^{(t+1)} = \hat{\pi}_n^{(t)} + (1+t)^{-1}\left(\frac{o_n^{(t)}(X^{(t+1)})}{1 - Nc_T} - \hat{\pi}_n^{(t)}\right) - (1+t)^{-1}\frac{c_T}{1 - Nc_T} \tag{3}$$

$$\hat{\mu}_n^{(t+1)} = \hat{\mu}_n^{(t)} + (1+t)^{-1}\frac{o_n^{(t)}(X^{(t+1)})}{\hat{\pi}_n^{(t)}}(X^{(t+1)} - \hat{\mu}_n^{(t)}) \tag{4}$$

$$\hat{d}_n^{(t+1)} = \hat{d}_n^{(t)} + (1+t)^{-1}\frac{o_n^{(t)}(X^{(t+1)})}{\hat{\pi}_n^{(t)}}((X^{(t+1)} - \hat{\mu}_n^{(t)}) \\ (X^{(t+1)} - \hat{\mu}_n^{(t)})^T - \hat{d}_n^{(t+1)}), d = \sigma^2 \tag{5}$$

For the task of background subtraction, we have a fix learning rate α to replace the $(1+t)^{-1}$. We also have $c_T = \alpha M/2$ where M stands for the number of parameters per component of the mixture model. See [5] for more details.

Compared with CNN-based two-stage region proposal methods, the Gaussian mixture model has several advantages as follows: (1) having less computational cost; (2) using an adaptive probability distribution to describe background features which is sensitive to background change.; (3) releasing RPN which enables more gradients flow from high-level feature for classification and more accurate regression.

We also set a series of expansion rates to generate mutil-scales RoIs for a better and more stable detection while do not necessarily increase the latency. In order to better utilize the continuous characteristics of videos, we use a simple method for predicting the possible RoIs in the next frame. Suppose the center of RoIs in the previous frame are and the current ones are, we simply predict the next by keeping the same vector difference.

3.2 Light-Weight Detector

To accurate the detection speed and decrease the storage and computation load, we adopt an input with a minimum side length of 300 pixels on region only with mask. As the Gaussian mixture model is sensitive to the moving region, a smaller backbone have little effect on the accuracy of the model while reducing the computational complexity. The main task of the detector is to adjust the coordinates of candidate areas and identify the classification, solving the problems of detection mismatch for local region in Gaussian mixture model (see Fig. 2).

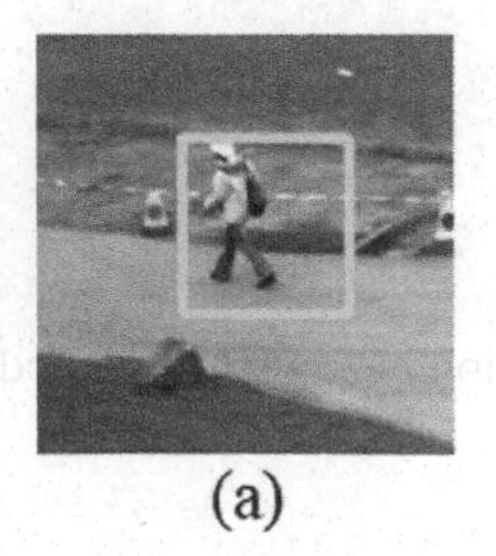

(a)

(b)

(c)

Fig. 2. Some problems in moving objection tracking by GMM method: (a) Expansion of detection frame caused by residual image of moving object. (b) Mismatch for some local statics regions. (c) Discontinuous detection for a single target.

We build a light-weight CNN backbone from MobileNets [12] for its superior performance on mobile and embedded device. By widely using depth-wise separate convolution, the model makes a further compression of the parameters.Generally, a larger receptive field is able to capture more context information which is essential for regression of coordinate values. Thus we set the initial convolutions be 7×7 convolutions with 64 channels.

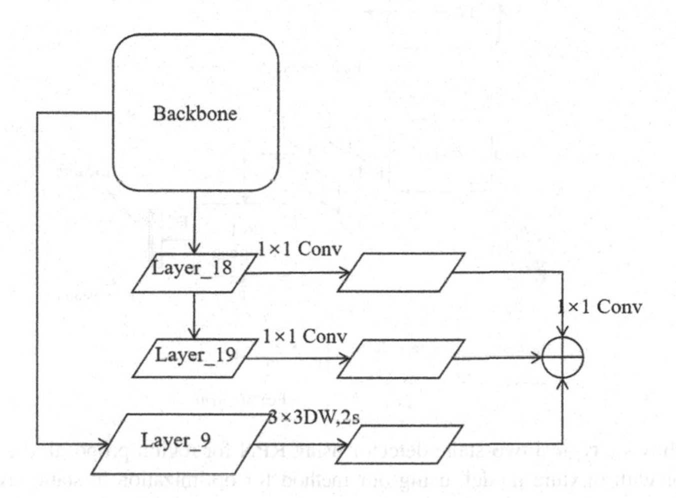

Fig. 3. Context enhance module for mixing up low-level and high-level features.

Besides, low-level features play an import role in regression while high-level features are significant to classification. To better combine the high-level features with low-level features, we design and optimize the Context Enhance Module (see Fig. 3). The key idea of CEM is to merge multi-level context and further enlarge the receptive field. Based on the MobileNet, we preserve the layer 7 as the low-level feature map and use a 3×3 depth wise convolutions and a point wise convolutions to enlarge receptive field and expand channels. For layer 12 and 18, we use a point wise convolution to compress the channels to 128. By stacking these there layers, the base feature map for classification and regression comes to a better balance between low-level and high-level features.

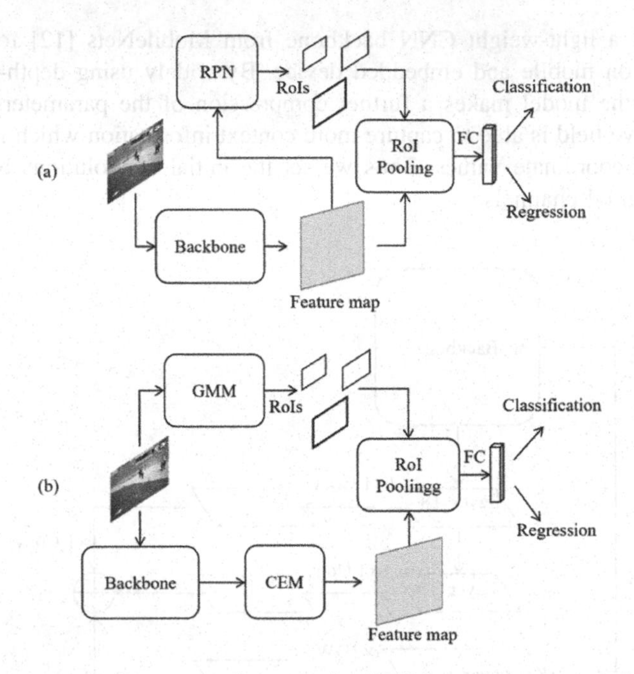

Fig. 4. (a) shows a typical two-stage detector using RPN for region proposal. (b) shows a two-stage detector with mixture model, using our method for optimization in static tasks.

3.3 Classifier and Regressor

A typical two-stage objection detection model uses RPN for region proposal (see Fig. 4). Compared with the RPN with a pre-trained model, we utilize the detection results from mixture model which is more adaptive and sensitive to the foreground. As the coordinate values of RoIs are relatively more accurate and require less adjustment, extra gradients is supposed to flow from the classification module. Thus we also present two adaptive hyper-parameters ω, ω' to estimate the learning weights of classification module and regression module in training losses. Assuming the number of epoches is t, the ω, ω' is defined as follow:

$$\begin{aligned} \omega &= 1 - 0.5 \times e^{-t/T} \\ \omega' &= 0.5 \times e^{-t/T} \end{aligned} \tag{6}$$

For train the classification, we assign a softmax label with a 0.5 weight for background and a 2.0 weight for objects in loss function. For the regerssor we use smooth L1 loss.

4 Implement Details

The moving region proposal using Gaussian mixture model has the same function with RPN, providing regions of interest and making a rough regression forecast. Therefore, we can train our model on conventional object detection datasets (e.g. VOC2012, COCO). Different scales and ratios are used to generate anchors while training. To put weights more on the CNN-based detector for classification and the second regression, we use the hyper-parameter ω to balance the losses while training. As the first regression for roughly adjusting RoIs while training can be achieved by Gaussian mixture model, we also set the weights of RPN losses (both RPN score loss and RPN regression loss) finally converge to a low value by $\omega' = 1 - \omega$. This method generates different RoIs with adjusted scales and ratios, which is closer to scene in reality (Fig. 5).

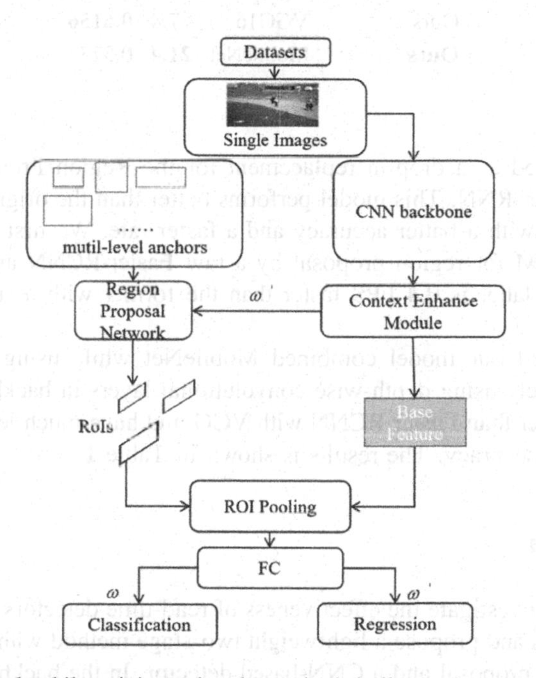

Fig. 5. The framework while training using the method similar to two-stage network. The value next to the arrow line represents the weight in training losses.

Our models are trained on VOC2012 using SGD with a momentum of 0.9. For anchors, we use 4 scales of 32,64,128, and 256, and 3 ratios of 1:1, 1:2, 2:1. By default we set the positive threshold 0.6 and the negative threshold 0.3. Note that unlike large detectors, we use less regularization such as weight decay and random dropout. As mobilenet and the light-weight subnet are small enough that there are few parameters and have less trouble with overfitting. Too much regularization may make it hard to converge. We also use the pre-trained MoblieNet on ImageNet to help the model converge faster. The network is trained for 40K iterations on VOC2012 dataset. The learning rate drops from initialized 0.001 to 0.0001 by a factor of 0.1 at 30K iterations.

5 Experiments

The MOT15 dataset [13] consists of a set of videos labeled with track boxes. 5 of 11 are static videos which are suitable for the performance testing of out model. The results of accuracy and stability are shown in Table 1.

Table 1. Performance of backbone on complexity, comparison for different networks.

Method	Backboke	FPS	mAP
Faster-RCNN	VGG16	7.0	0.4083
Ours	VGG16	7.4	**0.6156**
Ours*	MobileNet	**21.4**	0.5737

We use method as a drop-in replacement for the Region Proposal Network and backbone in Faster-RNN. This model performs better than the original model in static monitoring tasks with a batter accuracy and a faster rate. We first compared the performance of GMM for region proposal by a raw Faster-RCNN and a Faster-RCNN with GMM. The latter is 0.4 FPS faster than the former with a superior increase of mAP.

We also tested our model combined MobileNet while using GMM for region proposal. By widely using depth-wise convolutional layers in backbone, our model is almost 200% faster than Faster-RCNN with VGG and have much less parameters with little decrease in accuracy. The results is shown in Table 1.

6 Conclusion

In this paper we investigate the effectiveness of read-time detectors in object detection on static monitors and propose a lightweight two-stage method with Gaussian Mixture Model for region proposal and a CNN-based detector. In the backbone, we utilize the MobileNet for its superior performance on embedded device and low latency. For the detection part, we propose a Context Enhance Module to enlarge the receptive field and merge low-level feature and high-level feature. For RoIs, we abandon the Region Proposal Network which is computational expensive and utilize the online training GMM for region proposal, making the detector practical for real-time applications.

References

1. Liu, W., et al.: SSD: single shot MultiBox detector. In: Leibe, Bastian, Matas, Jiri, Sebe, Nicu, Welling, Max (eds.) ECCV 2016. LNCS, vol. 9905, pp. 21–37. Springer, Cham (2016). https://doi.org/10.1007/978-3-319-46448-0_2
2. Redmon, J., Farhadi, A.: Yolov3: an incremental improvement. arXiv:1804.02767 (2018)
3. Ren, S., He, K., Girshick, R., et al.: Faster R-CNN: towards real-time object detection with region proposal networks. In: Advances in Neural Information Processing Systems, pp. 91–99 (2015)
4. Lin, T.Y., Dollár, P., Girshick, R., et al.: Feature pyramid networks for object detection. In: Proceedings of the IEEE Conference on Computer Vision and Pattern Recognition, pp. 2117–2125 (2017)
5. Lee, D.-S.: Effective gaussian mixture learning for video background subtraction. IEEE Trans. Pattern Anal. Mach. Intell. **27**(5), 827–832 (2005)
6. Zivkovic, Z., van der Heijden, F.: Recursive unsupervised learning of finite mixture models. IEEE Trans. Pattern Anal. Mach. Intell. **26**(5), 651–656 (2004)
7. Chen, X., Wu, Z., Yu, J.: TSSD: temporal single-shot detector based on attention and LSTM for robotic intelligent perception (2018)
8. Zivkovic, Z., van der Heijden, F.: Efficient adaptive density estimation per image pixel for the task of background subtraction. Pattern Recogn. Lett. **27**(7), 773–780 (2006)
9. Qin, Z., Li, Z., Zhang, Z.: ThunderNet: Towards Real-time Generic Object Detection (2019). https://arxiv.org/pdf/1903.11752.pdf
10. Li, Z., Peng, C., Yu, G., et al.: Light-head R-CNN: in defense of two-stage object detector. arXiv preprint arXiv:1711.07264 (2017)
11. Dai, J., Li, Y., He, K., et al.: R-FCN: object detection via region-based fully convolutional networks. In: Advances in Neural Information Processing Systems, pp. 379–387 (2016)
12. Howard, A.G., Zhu, M., Chen, B., et al.: Mobilenets: efficient convolutional neural networks for mobile vision applications. arXiv preprint arXiv:1704.04861 (2017)
13. Milan, A., Leal-Taixé, L., Reid, I., et al.: MOT16: a benchmark for multi-object tracking. arXiv preprint arXiv:1603.00831 (2016)

An Optimized Lee Filter Denoising Method Based on EIP Correction

Yipeng Liu[1], Guoxing Huang[1](✉), Weidang Lu[1], Hong Peng[1], and Jingwen Wang[2]

[1] College of Information Engineering, Zhejiang University of Technology, Hangzhou 310023, China
hgx05745@zjut.edu.cn

[2] College of Information Engineering, China Jiliang University, Hangzhou 310023, China

Abstract. The speckle noise inherent in synthetic aperture radar (SAR) images seriously affects the visual effect of the image and brings difficulties to the subsequent parameter inversion and interpretation. However, the existing SAR image filtering methods are not effective in preserving the image edge details. In this paper, an exponential image processing (EIP) correction based lee filter denoising method is proposed to solve this problem. This method carried out a reasonable fuzzy division on the image gray histogram, and extracted its statistical characteristics from it. Such feature is used to correct the image based on the mathematical structure of EIP, and divide the filter area of the image to avoid the loss of edge information and dark details of the image. Simulation results have shown that the proposed method outperform the traditional methods in suppressing noise and protecting edge details.

Keywords: SAR · Image filter · Gray histogram · Lee filter

1 Introduction

Synthetic Aperture Radar (Synthetic Aperture Radar, SAR) is an advanced microwave remote sensing radar with good multi-polarization measurement and interference measurement capability, as well as all-day, all-weather, penetrating and other advantages. in remote sensing applications, synthetic aperture radar has better measurement accuracy than optical remote sensing. At the same time, synthetic aperture radar can be carried out in a variety of platforms, and drone, airborne and on-board SAR radar can meet the requirements of many monitoring scenarios, which makes SAR radar more popular in the field of defense and military. The application of SAR's final product, SAR image, is more diverse, for example, in the field of military reconnaissance, SAR can be used to identify or track ground objects and suspected aircraft, and in the civilian field, the information of SAR image is used in the prediction of natural disasters, forestry exploration, and surface temperature observation [1–3].

This work was supported by the National Natural Science Foundation of China (NSFC, No. 61871348) and the Natural Science Foundation of Zhejiang Province (No. LQ20F030017).

S. Shi et al. (Eds.): AICON 2020, LNICST 356, pp. 424–436, 2021.
https://doi.org/10.1007/978-3-030-69066-3_37

Due to the backscatter characteristics of the radar, this will produce speckle noise, which is represented by random changes in the brightness of the area contaminated by the speckle noise in the image, which seriously affects the visual effect of the image, and makes it difficult to extract the information of the SAR image [4]. In order to extract effective information through processing of SAR images, it is necessary to has a preprocessing process to suppress the influence of speckle noise and reduce the loss of edge information while smoothing the image as far as possible. Therefore, the research of SAR speckle reduction is a research focus in SAR imaging processing and the analysis of SAR image. At present, the commonly used methods to suppress speckle noise include multi-view processing; spatial filtering, such as Lee filtering algorithm [5], frost filtering algorithm [6]; transform domain filtering [7, 8], which is such as the algorithms based on wavelet transform. Because the spatial filtering method obtains the local statistical characteristics of the image through the sliding window, it is relatively simple to implement and has a wide range of applications compared with other methods. However, the local structure information around the pixel is not taken into account when the image is operated through a fixed window, and the edge and detail features are not well preserved.

Although the multi-view processing improves the radiation resolution of the SAR image, it reduces the utilization rate of the signal bandwidth. With the expansion of SAR image applications, the requirements for spatial resolution are also increasing, and the multi-view processing can no longer meet the higher requirements. The transform domain filtering denoising method can remove high-frequency noise well, but also needs the conversion in the spatial domain and the transform domain, the decomposition estimation and reconstruction of coefficients [8], so it has the disadvantages of large amount of calculation, high complexity and loss of details. Overall, The above methods can all have certain effects in coherent speckle noise suppression, but they all have certain defects.

In view of the above problems, this paper based on the image fuzzy modeling method of literature [9, 10], and taking into account the inherent fuzziness of the image, a reasonable fuzzy division is carried out on the image gray histogram, and its statistical characteristics which can be used to correct the image through the mathematical structure of exponential image processing (EIP) are extracted from divided gray histogram. Under the spatial network of the EIP framework, the characteristic of the exponential function is used to perform nonlinear operations on image pixels to improve the visual contrast. At the same time, the processing area of the speckle noise filtering method is subdivided based on the local mean and variance of the image as the basis for regional division in order to avoid the loss of edge information and dark details of the image as much as possible. In this paper, through the SAR image noise suppression experiment, the proposed method of this paper not only effectively suppresses speckle noise, but also maintains the edge information well, and the proposed method has outstanding performance compared with similar methods and the original algorithms.

This article is organized as follows: Firstly, in the Sect. 2, introducing the mathematical model of speckle noise and the basic principle of Lee filtering algorithm. Then, in the Sect. 3, the proposed method of this paper is given. Finally, through some simulation results, we have a conclusion.

2 The Basic Principle of SAR Image Denoising

The key to synthetic aperture radar technology is to synthesize an antenna with a larger equivalent size using an antenna moving along the track, as shown in Fig. 1. By letting a small antenna, which is a single radiating individual, "moves" back and forth in a straight line, the echo signals reflected from the same ground object are collected, demodulated and compressed at different positions. In this way, a small "moving" antenna can be combined into a "large antenna" with the same effect, which can significantly improve the azimuth resolution of the acquired radar image and because the azimuth resolution is irrelevant with the distance between object and radar, so the SAR radar can be installed on the satellite platform to get a relatively large synthetic hole [1]. Due to the coherence of the SAR system imaging method, speckle noise will inevitably be generated. More precisely, the image is contaminated due to the fading of the echo reflection signal of the ground target, which causes the brightness of some areas of the image to change randomly. In general, the basic principle of SAR system imaging: using a single small antenna to receive the echo signal of the target by constantly moving to be equivalent to a large antenna, which improves the resolution while it also brings the interference of speckle noise, this is also the focus of this article.

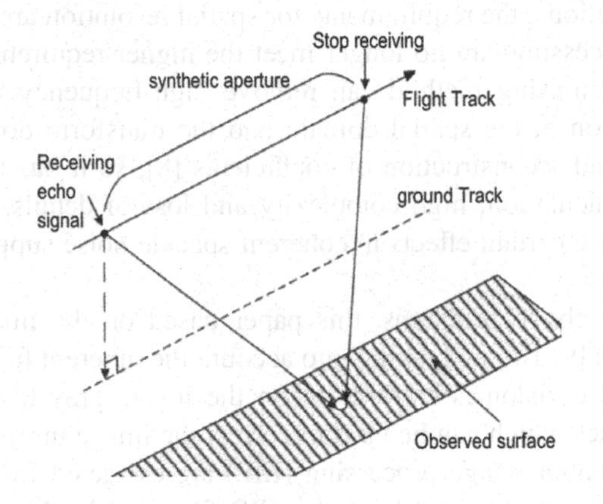

Fig. 1. SAR imaging method

2.1 The Mathematical Model of Speckle Noise in SAR Images

The speckle noise in the SAR image is caused by the fading phenomenon of echo signal and the fading process is formed as follows: There are multiple scatters illuminated at the same time, and when there is relative motion between the local object and the radar station, the multiple ground objects and the radar have different distances and different propagation speeds, which produces random fluctuations and the measurement of the scattering coefficient of the ground object by SAR have a large deviation, The interference of speckle noise makes the pixels of the image area change randomly, so it belongs to random noise. The speckle noise in SAR images is essentially different from the noise in optical images, This is because the physical processes that they generate are fundamentally different [11]. The speckle noise in the SAR image is caused by the fading of the radar echo signal, which is the inherent shortcoming of all imaging systems based on the principle of coherent light, including SAR radar. Unlike additive noise, speckle noise in SAR images is a model of multiplicative noise, The expression is as follows:

$$R(x, y) = I(x, y)V(x, y) \tag{1}$$

Where (x, y) represents the azimuth and range coordinate values of the SAR image pixel space, $I(x, y)$ represents the noise-free intensity, $R(x, y)$ represents the observed SAR image intensity and $V(x, y)$ represents the independent noise data. Usually the mean value of noise is 1, the standard deviation of noise is ρ, and its value is related to the sight numbers of SAR images.

2.2 Principle of Lee Filter Denoising

Lee filter is one of the commonly used spatial filtering methods based on the minimum mean square error for noise suppression. It does not need to establish an accurate statistical model and is designed based on a fully developed multiplicative noise model [5]. The formula is as follows:

$$\hat{I}(m, n) = \bar{R}(m, n) + W(m, n)(R(m, n) - \bar{R}(m, n)) \tag{2}$$

Where m = 1, 2 ··· M, n = 1, 2 ··· N; $\hat{I}(m, n)$ is the center pixel value of the filtered window; $\bar{R}(m, n)$ is the mean value of pixels covered by this filter window; $R(m, n)$ is the center pixel value of the window covered pixel before filtering; $W(m, n)$ is a weighting factor, which is calculated as follows:

$$W(m, n) = 1 - \frac{(\sigma_u/\bar{u})^2}{(\sigma_R(m, n)/\bar{R}(m, n))^2} \tag{3}$$

Where, σ_u is the noise standard deviation; $\bar{u}$ is the noise mean value; $\varsigma_R(m, n)$ is the standard deviation of the pixels covered by the local window in the translation process.

It can be seen that the classic Lee filtering algorithm processes the noise under the premise of obtaining the statistical parameters of the noise, which assumes that the noise is fully developed. For partially developed regions, the result of filtering directly using the Lee algorithm is not ideal. At the same time, due to the inadaptability of local window sliding filtering [11], the filtered image details and edge information cannot be effectively retained is also one of the defects of this algorithm.

3 SAR Image Denoising Based on Adaptive Lee Filtering Algorithm

Because the traditional Lee filter has the disadvantages of noise suppression and insufficient protection of the edge details of the SAR image. To this end, the mathematical structure of exponential image processing (EIP) is introduced and SAR images are corrected based on EIP structure. Using the characteristics of the exponential function to define a series of non-linear calculations for the gray value of the image, and through a reasonable fuzzy division of the gray histogram, the gray statistical characteristics of the image are obtained according to the result of the division, and the EIP is guided according to this characteristic, the parameters of the nonlinear operation are selected, and the corrected SAR image is finally obtained. The calculation result shows that the visual effect of SAR image can be effectively improved, especially the "dark part" detailed area while the local mean and variance of the image are used as the basis for dividing the filtering area to subdivide the processing area of speckle noise filtering method. to avoid the loss of edge information and image details as much as possible, it also can achieve high efficiency suppression of speckle noise.

3.1 Image Correction Method Based on EIP

From the visual experience of human eyes, the visual effects of images are generally divided into three situations: too dark, too bright, and gray values concentrated in middle value area. This also corresponds to the use of different correction functions for images in different situations [9, 10], and there are three types of correction functions corresponding to this. The gray distribution characteristics of the obtained image are divided and the appropriate correction function is adopted according to the characteristics. The three types of gray correction function curves are shown in Fig. 2.

The non-linear transformation process given in Fig. 2(a) can expand the dynamic range of the dark area of the image and enhance the details and contrast of the dark area of the image. The change in the gray value of the image before and after the change in Fig. 2(b) shows that the transformation The process can expand the dynamic range of the bright area of the image, and enhance the details and contrast of the bright area of the image. Similarly, the correction transformation method in Fig. 2(c) can expand the dynamic range of the middle area of the image and enhance the detail and contrast of the middle area of the image. The method proposed in this paper firstly divides the image gray histogram 3-fuzzy, and the divided areas are dark, middle and bright areas. Then the statistical characteristics of the three parts are extracted, and the correction function is selected adaptively according to the gray value distribution characteristics of the image, so as to achieve the purpose of improving the overall image details and edge information.

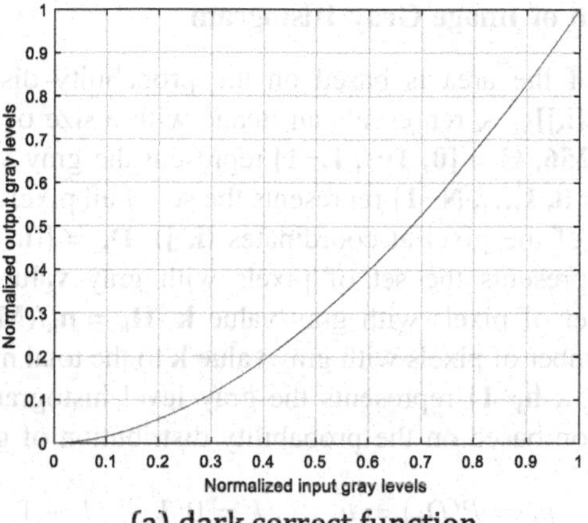

(a) dark correct function

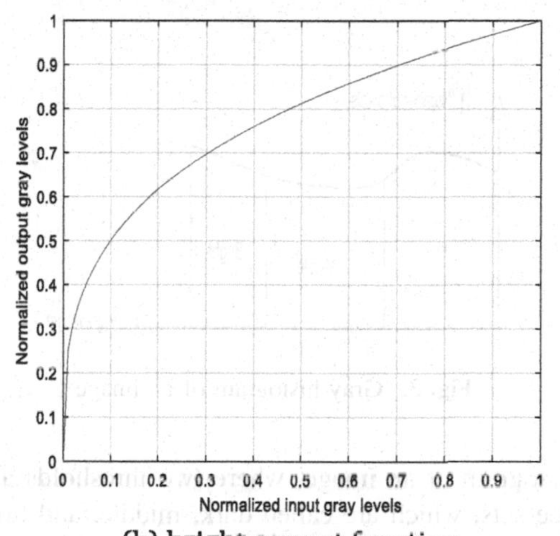

(b) bright correct function

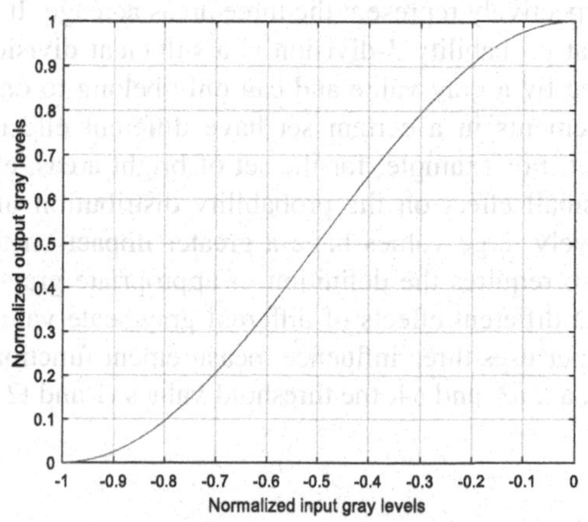

(c) middle correct function

Fig. 2. Calibration curve

3.2 Fuzzy Division of Image Gray Histogram

Since the division of the area is based on the probability distribution of the gray histogram, let $\mathbf{Y} = \mathbf{y[i,j]_{M\times N}}$ represents an image with a size of $\mathbf{M} \times \mathbf{N}$, and its gray level is **L**, Let $\mathbf{L = 256, G = [0, 1\cdots, L-1]}$ represent the gray value set, $\mathbf{D = \{(i, j) : i = 0, 1..., M-1; j = 0, 1,..., N-1\}}$ represents the set of all pixels in the image, $\mathbf{x(i, j)}\in$ **G** is the gray value of the pixel at coordinates **(i, j)**, $\mathbf{D_k = \{(i, j): x(i,j) = k(i, j)\in D\}}$ $\mathbf{(k = 0,1,...,L-1)}$ represents the set of pixels with gray value k in the image, $\mathbf{n_k}$ represents the number of pixels with gray value **k**, $\mathbf{H_k = n_k/(M \times N)}$ represents the percentage of the number of pixels with gray value **k** to the total number of pixels in the image, $\mathbf{H = \{h_0, h_1..., h_l-1\}}$ represents the gray level histogram of the image, then constructing a division based on the probability distribution of gray values:

$$p_k = P(D_k) = h_k \qquad k = 0, 1 \cdots, L - 1 \tag{4}$$

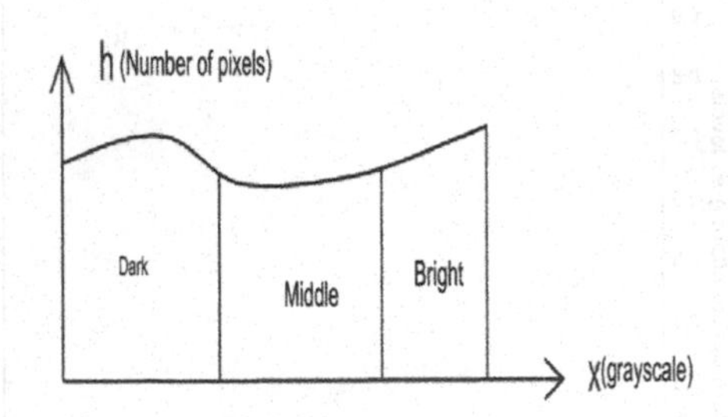

Fig. 3. Gray histogram of an image

Figure 3 is a histogram of an image, where two thresholds are used to divide its gray levels into three sets, which are called dark, middle, and bright. The gray value area of the image represents a probability 3-division, $P_d = P(E_D)$, $P_m = P(E_M)$, $P_b = P(E_B)$, E_D, E_m, E_D respectively represent the three areas acreage, It can be seen from the above expression that probability 3-division is a sufficient division. The pixels in the image are represented by a gray value and can only belong to one of the three region sets. In addition, elements in a certain set have different effects on the probability distribution of the set. For example, for the set of bright areas, elements with a small gray value have a small effect on the probability distribution of the set, while gray elements with relatively large values have a greater impact on the probability distribution [12]. This also requires the definition of appropriate gray-scale influence principles to describe the different effects of different gray-scale values on the probability distribution. This paper uses three influence measurement function curves as shown in Fig. 4, by setting a1, a2, a3, and a4, the threshold values t1 and t2 are more in line with

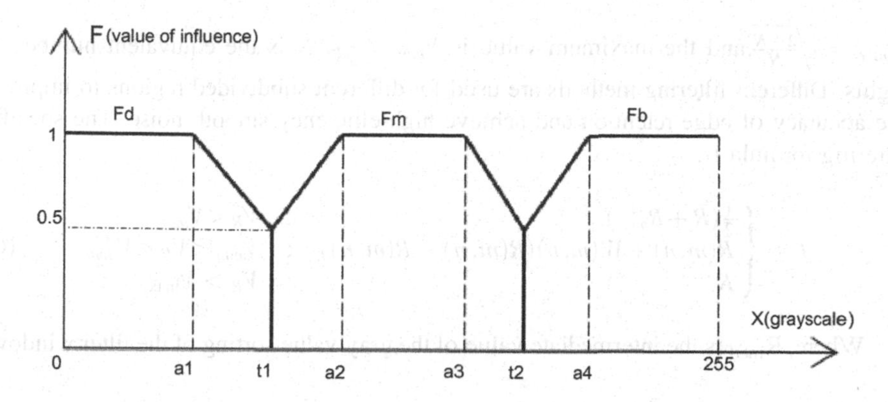

Fig. 4. Gray scalc influence curve

people's visual habits of dark, middle and bright areas. In this way, a new probability 3-partition is obtained, and its probability distribution is:

$$\cdot\begin{cases} p_d = \sum_{0}^{255} P_k \cdot F_d(k) \\ p_m = \sum_{0}^{255} P_k \cdot F_m(k) \\ p_b = \sum_{0}^{255} P_k \cdot F_b(k) \end{cases} \tag{5}$$

The gray level of the whole image is divided into three equal parts, and the statistical properties of the three parts P_d, P_m and P_b are calculated respectively. If P_d is the largest, the image is dark, if P_m is the largest, the gray value of the image is concentrated in the middle area, if P_b is thc largest, the image is bright. After mastering the characteristics of the image, different correction function forms for different situations can be selected to achieve the purpose of adaptively enhancing the image to avoid the loss of edge detail information.

3.3 Division of Filtering Area

Aiming at the problem of how to achieve high-efficiency noise suppression, it can be seen from the above that for some areas where the noise is not fully developed, the result of directly using the Lee algorithm to filter is not ideal. To improve the denoising effect of SAR images, Coefficient of variation $V_R = \frac{\sigma_R(m,n)}{\bar{R}(m,n)}$ is introduced to Subdivide the processing area [13, 14], the maximum value of the coefficient of variation is

$V_{max} = \sqrt{\frac{2+N}{N}}$,and the maximum value is $V_{min} = \frac{1}{\sqrt{N}}$, N is the equivalent number of sights. Different filtering methods are used for different subdivided regions to improve the accuracy of edge retention and achieve high-efficiency smooth noise. The specific filtering formula is:

$$\hat{I} = \begin{cases} \frac{1}{2}(\bar{R} + R_{med}) & ; \quad V_R < V_{min} \\ \bar{R}(m,n) + W(m,n)(R(m,n) - \bar{R}(m,n)) & ; \quad V_{min} < V_R < V_{\max} \\ R & ; \quad V_R > V_{\max} \end{cases} \quad (6)$$

Where, R_{med} is the intermediate value of the gray value sorting of the filter window.

3.4 Process of the Method

Step1, Initialization. According to the probability distribution of the gray value of the SAR image, its gray scale histogram is divided into dark, middle and bright areas, and the statistical characteristics of the gray scale distribution of the image are obtained as (4).

Step 2, Define the principle of the influence of different gray values on the probability distribution, and obtain a fuzzy 3-division that conforms to vision, as in (5).

Step 3, After obtaining the characteristics of the image, different correction functions can be selected for different situations to achieve the purpose of adaptively enhancing image details.

Step 4, By calculating the statistical characteristic parameters (mean, standard deviation, etc.) of the filter window, the coefficient of variation of the divided area is obtained by using the statistical characteristics.

Step 5, The coefficient of variation is introduced to subdivide the filtering area, and different filtering methods are selected for different areas, as shown in (6).

Step 6, Finally, the filtered image is obtained.

4 Simulation Results

In this section, we provide several experiments to evaluate the effectiveness of the proposed method in this paper and compare it with other methods. In order to verify the effectiveness of the method in this paper, the following simulation experiments are performed: Two SAR image data are selected for filtering experiments, and the original SAR image is shown in Fig. 5. At the same time, in order to compare the advantages of the method in this paper, Frost filtering algorithm, mean filtering algorithm, median filtering algorithm and original Lee filtering algorithm are selected for comparative analysis. The experimental results are shown in Fig. 6, 7, 8, 9 and 10.

In order to evaluate the effect of the method, two indicators, equivalent look number (ENL) and edge retention index (EPI), are selected as the quality evaluation criteria after image denoising. The larger the value of ENL, the richer the image

Fig. 5. Original image

Fig. 6. Lee filter algorithm

Fig. 7. Frost filtering algorithm

Fig. 8. Median filtering algorithm

Fig. 9. Mean filter algorithm

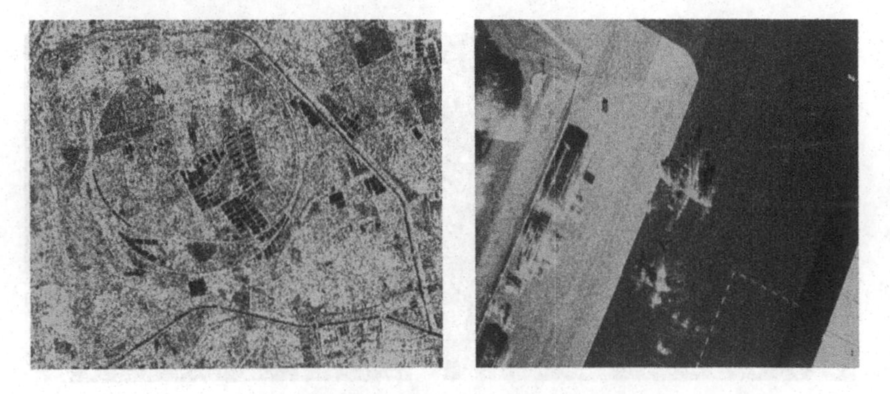

Fig. 10. The algorithm of this paper

information and the better the denoising effect. The closer the EPI is to 1, the higher the edge protection accuracy is. The calculation formula is as follows:

$$\mathrm{ENL} = \frac{\mu^2}{\sigma^2} \tag{7}$$

$$\mathrm{EPI} = \frac{\sum_{i=1}^{m} |I_h(i) - I_o(i)|}{\sum_{i=1}^{m} \left|I_h^p(i) - I_o^p(i)\right|} \tag{8}$$

Where μ and σ represent the mean and standard deviation of the filtered image respectively, $I_h(i)$ and $I_o(i)$ represent the gray values of adjacent pixels on both sides of the filtered edge respectively, $I_h^p(i)$ and $I_o^p(i)$ represents the gray value of adjacent pixels on both sides of the edge before filtering.

Judging from subjective vision, the Frost filter algorithm is better than the Lee filter algorithm, but the overall image is blurry and the visual effect is poor. It filters out noise at the cost of losing edge detail information. The noise suppression effect of median filter and average filter is not good enough, and the edge protection effect is not ideal. The original Lee filtering algorithm has a good effect in maintaining edge information, but the noise suppression ability is poor. The methods proposed in this paper are better than the selected contrast algorithms in terms of noise suppression, and are significantly better than several types of contrast algorithms in terms of maintaining edge information.

In order to quantitatively evaluate the denoising ability of various filtering methods, Equivalent Look Number (ENL) and Edge Preservation Index (EPI) are used as evaluation indicators. The results are shown in Table 1. It can be seen from the table that the equivalent look number (ENL) of the method in this paper is the highest among similar algorithms, which is better than comparison algorithms, and the edge preservation index (EPI) is much higher than similar comparison algorithms. Especially compared with the original Lee filter algorithm, the EPI index of the original Lee filter algorithm is 0.5774, and the EPI value of the algorithm proposed in this paper is 0.9196. Similarly, the ENL index of Lee filtering is 1.8382, and the ENL index of the algorithm proposed in this paper is 5.1651. Therefore, the algorithm proposed in this paper has improved noise removal effect and edge retention compared with the original Lee filter and other methods.

Table 1. Comparison of noise suppression and edge retention effects of the first picture and the second picture.

Methods	EPI	ENL	EPI	ENL
Original	1	1.7452	1	1.1682
Lee filter	0.5774	1.8382	0.5132	1.7234
Frost filter	0.5658	2.2763	0.4920	1.8337
Median filter	0.2851	2.1343	0.2508	1.8239
Mean filter	0.2844	2.8872	0.2358	1.8303
The proposed method	0.9196	5.1651	0.8096	2.6295

5 Conclusion

This paper proposes a new method of SAR image adaptive correction based on gray-scale fuzzy division combined with optimized area filtering to remove speckle noise. First, the gray scale histogram of the image is divided into 3-fuzzy to guide the selection of correction methods, the filtering area is divided according to the coefficient of variation, and finally, the optimized Lee filtering method is used to denoise the SAR image. Experimental results show that the image quality obtained by this method is better. Compared with Frost algorithm, traditional Lee filter and other algorithms, it achieves filtering and denoising while improving the accuracy of maintaining edge information and image texture details. So it is worthy of further study.

References

1. Chen, S.X., Gao, L., Li, Q.Y.: SAR image despeckling by using nonlocal sparse coding model. Circ. Syst. Signal Process. **37**, 3023 (2018)
2. Parrilli, S., Poderico, M., Angelino, C.V., Verdoliva, L.: A nonlocal SAR image denoising algorithm based on LLMMSE wavelet shrinkage. IEEE Trans. Geosci. Remote Sens. **50**(2), 606–616 (2012)
3. Jeong, H., Park, J.H., Ryu, H.Y., Kwon, J.B., Oh, Y.: VLSI architecture for SAR data compression. IEEE Trans. Aerosp. Electron. Syst. **38**(2), 427–440 (2002)
4. Martnez, A., Marchand, J.L.: SAR image quality assessment. IEEE Trans. Image Process. **7**, 54–59 (2011)
5. Wang, Y., Ainsworth, T.L., Lee, J.S.: Application of mixture regression for improved polarimetric SAR speckle filtering. IEEE Trans. Geosci. Remote Sens. **55**(1), 453–467 (2017)
6. Ma, X., Shen, H., Zhao, X., Zhang, L.: SAR image despeckling by the use of variational methods with adaptive nonlocal functionals. IEEE Trans. Geosci. Remote Sens. **54**(6), 3421–3435 (2016)
7. Ma, X., Wu, P., Shen, H.: A nonlinear guided filter for polarimetric SAR image despeckling. IEEE Trans. Geosci. Remote Sens. **57**(4), 1918–1927 (2019)
8. Ren, Y., Yang, J., Zhao, L., Li, P., Shi, L.: SIRV-based high-resolution PolSAR image speckle suppression via dual-domain filtering. IEEE Trans. Geosci. Remote Sens. **57**(8), 5923–5938 (2019)
9. Chen, G., Li, G., Liu, Y., Zhang, X., Zhang, L.: SAR image despeckling based on combination of fractional-order total variation and nonlocal low rank regularization. IEEE Trans. Geosci. Remote Sens. **58**(3), 2056–2070 (2020)
10. Martino, G.D., Simone, A.D., Iodice, A., Riccio, D.: Scattering-based nonlocal means SAR despeckling. IEEE Trans. Geosci. Remote Sens. **54**(6), 3574–3588 (2016)
11. Jiang, Y., Wang, X., Xu, X., Ye, X.: Speckle noise filtering for sea SAR image. In: 2009 WRI Global Congress on Intelligent Systems, Xiamen, pp. 523–527 (2009)
12. Nakai, K., Hoshi, Y., Taguchi, A.: Color image contrast enhancement method based on differential intensity/saturation gray-levels histograms. In: International Symposium on Intelligent Signal Processing and Communication Systems, Naha, pp. 445–449 (2013)
13. Lee, J.S., Ainsworth, T.L., Wang, Y., Chen, K.S.: Polarimetric SAR speckle filtering and the extended sigma filter. In: IEEE Transactions on Geoscience and Remote Sensing, vol. 53, no. 3, pp. 1150–1160 (2015)
14. Jeon, B.K., Jang, J.H., Hong, K.S.: Road detection in spaceborne SAR images using a genetic algorithm. IEEE Trans. Geosci. Remote Sens. **40**(1), 22–29 (2002)

Research on Azimuth Measurement Method of CCD Camera Based on Computer 3D Vision System

Yixiong He(✉), Yiqun Zhang, Weizhi Wang, and Su Ma

Peng Cheng Laboratory, Research Center of Networks and Communication, Shenzhen, China
heyx@pcl.ac.cn

Abstract. In the field of artificial intelligence (AI), three-dimensional (3D) vision system is increasingly used to obtain 3D information of targets. In this paper, a method for measuring the azimuth of CCD camera's apparent axis with high precision applied to 3D vision system is proposed. The azimuth angle of the apparent axis of CCD camera can be easily and accurately measured by using the laser projection transfer method through the horizontal two-dimensional turntable and the linear laser, which provides a method for 3D vision system calibration. This paper introduced the principle of measuring angles of the system, deduced the equation of coordinate transformation of the system, and made systematic error analysis. The results show that the measurement accuracy and reliability of this method meet the needs of 3D vision system calibration, which is much higher than the measurement accuracy and stability of geomagnetic sensors.

Keywords: Artificial intelligence · Three-dimensional vision system · CCD camera · Azimuth measurement

1 Introduction

Since the 21st century, artificial intelligence technology has developed rapidly and has been widely used in medical treatment, security, autopilot, robotics, industrial intelligent manufacturing [1] and other fields. Computer vision [2], as an important branch of artificial intelligence technology, automatically receives and processes the image of a real object through optical devices and non-contact sensors, to obtain the required information or devices for controlling robot or mechanical movement. As one of the hot topics in the field of computer vision, 3D vision system is an important means of 3D perception and measurement of complex environment. Compared with traditional 2D vision system, 3D vision system can perceive and measure shapes-related features, such as object flatness, surface area and volume, etc. 3D vision systems are generally divided into three categories: binocular vision system [3], structured light system [4], and time of flight (TOF) system [5].

CCD camera is the core sensor of binocular vision system and structural light system. The azimuth of CCD camera has a decisive influence on the measurement

S. Shi et al. (Eds.): AICON 2020, LNICST 356, pp. 437–446, 2021.
https://doi.org/10.1007/978-3-030-69066-3_38

precision of the whole 3D vision system. Therefore, accurate measurement of the azimuth angle of the camera's visual axis can effectively reduce the installation error of the camera's azimuth axis and improve the measurement accuracy of 3D vision system. There are mainly two methods in azimuth measurement: geomagnetic sensor measurement and dual-antenna GPS measurement. Geomagnetic sensor is widely used (such as the traditional compass), but it is greatly affected by the magnetic environment and cannot work normally in some occasions. Dual-antenna GPS needs to measure the position information of the two antennas respectively, and then calculate the azimuth information of the baseline, but it cannot be used because the indoor GPS signal quality does not meet the requirements. In addition, the system takes up a large space (about 2 m of the baseline length). In response to the above needs and problems, this paper presents a method to measure the azimuth angle of light source optical axis or camera optical axis in the room by laser projection method, and carries out precision analysis and test. This paper proposes a method for measuring the azimuth of the apparent axis of CCD camera, which is applied to 3D vision system with high precision. The azimuth of the apparent axis of CCD camera can be easily and accurately measured by using the laser projection transfer method with the cooperation of the horizontal 2D turntable and the linear laser.

2 Angle Measurement Principle

In order to accurate measurement and calibration the azimuth angle of machine vision CCD, laser projection transfer method is used in this paper (see Fig. 1).

First of all, the base of the 2D turntable is fixed above the reference object, and the linear laser A is fixed on the pitch axis of the turntable, which requires to be perpendicular to the azimuth axis of the turntable. The projection plane A on which the laser from the linear laser A located is parallel to the pitch axis of the turntable. The linear laser A select the 532 nm semiconductor laser.

Secondly, the CCD mounting plate is fixed on the tripod. The CCD and linear laser B is fixed on the CCD mounting plate. It is required that the projection plane B where the laser from linear laser B located is parallel to the CCD visual axis, and the linear laser B select the 650 nm semiconductor laser.

Thirdly, set up the tripod to ensure the level of the cloud platform. Turn on the linear laser A, and generate a red light on the horizontal ground, that is the projection line A of plane A on the horizontal ground. Turn on the linear laser B, and a green light is generated on the horizontal ground, that is the projection line B of plane B on the horizontal ground.

Finally, adjust the azimuth axis of the turntable to make the projection line B (green light) parallel to the projection line A (red light); Adjust the pitch axis of the turntable to make the projection line A coincide with the projection line B. In this case, the angle value output by the azimuth axis angle encoder of the turntable is the azimuth angle of the CCD visual axis.

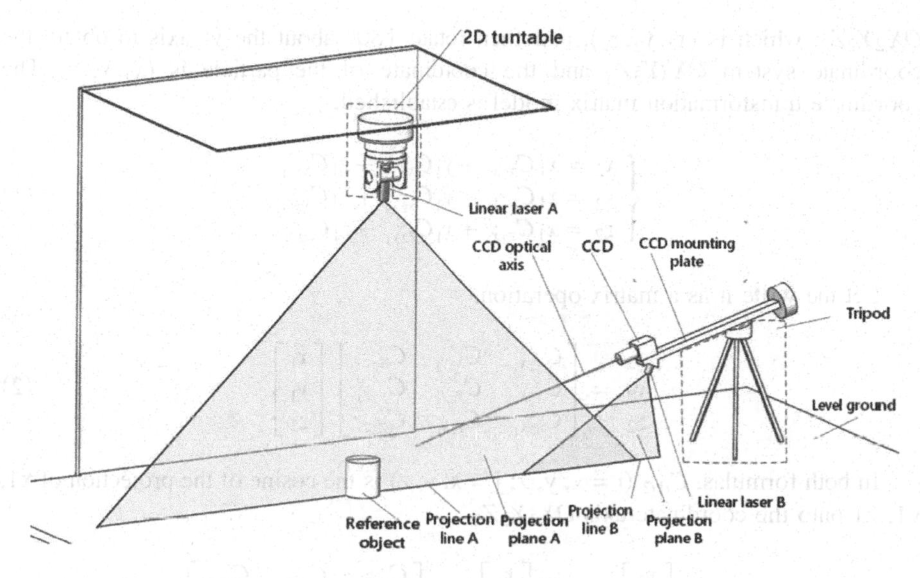

Fig. 1. System diagram.

3 Transformation of Coordinates

Let the coordinate system of the 2D turntable be $OX_1Y_1Z_1$, and the turntable be hung upside down in the room. $OX_3Y_3Z_3$ is the CCD coordinate system. In order to realize the transformation from 2D rotary coordinate system to CCD coordinate system, the transition coordinate system is established. The coordinate system definition is shown in the Fig. 2.

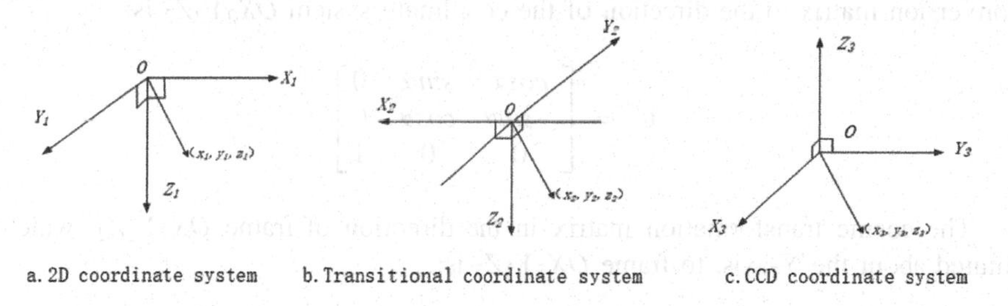

a. 2D coordinate system
b. Transitional coordinate system
c. CCD coordinate system

Fig. 2. Transformation coordinate system.

The vector measured by the 2D turntable needs to be converted to the CCD coordinate system through multiple coordinate transformations to obtain the azimuth Angle parameters for CCD modeling. Suppose the coordinate of the particle in $OX_1Y_1Z_1$ is (x_1, y_1, z_1), rotate 180° about the z_1 axis to obtain the coordinate system

$OX_2Y_2Z_2$, which is (x_2, y_2, z_2), and then rotate 180° about the y_2 axis to obtain the coordinate system $OX_3Y_3Z_3$ and the coordinate of the particle is (x_3, y_3, z_3).The coordinate transformation matrix model is established.

$$\begin{cases} x_2 = x_1 C_{x_2x_1} + y_1 C_{x_2y_1} + z_1 C_{x_2z_1} \\ y_2 = x_1 C_{y_2x_1} + y_1 C_{y_2y_1} + z_1 C_{y_2z_1} \\ z_2 = x_1 C_{z_2x_1} + y_1 C_{z_2y_1} + z_1 C_{z_2z_1} \end{cases} \tag{1}$$

Let me write it as a matrix operation.

$$\begin{matrix} x_2 \\ y_2 \\ z_2 \end{matrix} = \begin{bmatrix} C_{x_2x_1} & C_{x_2y_1} & C_{x_2z_1} \\ C_{y_2x_1} & C_{y_2y_1} & C_{y_2z_1} \\ C_{z_2x_1} & C_{z_2y_1} & C_{z_2z_1} \end{bmatrix} \begin{bmatrix} x_1 \\ y_1 \\ z_1 \end{bmatrix} \tag{2}$$

In both formulas, C_{i1j2} (i = x, y, z; j = x, y, z) is the cosine of the projection of x1, y1, z1 onto the coordinate axis $OX_2Y_2Z_2$.

$$\mathbf{r}^2 = \begin{bmatrix} x_2 \\ y_2 \\ z_2 \end{bmatrix}, \mathbf{r}^1 = \begin{bmatrix} x_1 \\ y_1 \\ z_1 \end{bmatrix}, C_1^2 \begin{bmatrix} C_{x_2x_1} & C_{x_2y_1} & C_{x_2z_1} \\ C_{y_2x_1} & C_{y_2y_1} & C_{y_2z_1} \\ C_{z_2x_1} & C_{z_2y_1} & C_{z_2z_1} \end{bmatrix} \tag{3}$$

The above equation has been simplified.

$$\mathbf{r}^2 = C_1^2 \mathbf{r}^1 \tag{4}$$

Where r_1 is the projection of the r vector onto the x_2 axis, r_1 is the projection of the r vector onto the x_1 axis. C_1^2 is the transformation direction cosine matrix.

Then the coordinate system $OX_1Y_1Z_1$ rotates about the Z axis, and the cosine conversion matrix to the direction of the coordinate system $OX_2Y_2Z_2$ is:

$$C_1^2 = \begin{bmatrix} cos\alpha & sin\alpha & 0 \\ -sin\alpha & cos\alpha & 0 \\ 0 & 0 & 1 \end{bmatrix} \tag{5}$$

The cosine transformation matrix in the direction of frame $OX_2Y_2Z_2$, which is rotated about the Y-axis, to frame $OX_3Y_3Z_3$ is:

$$C_2^3 = \begin{bmatrix} cos\beta & 0 & -sin\beta \\ 0 & 1 & 0 \\ sin\beta & 0 & cos\beta \end{bmatrix} \tag{5}$$

The conversion matrix from coordinate $OX_1Y_1Z_1$ to coordinate $OX_3Y_3Z_3$ is:

$$C_1^3 = C_2^3 C_1^2 \tag{6}$$

Let r^3 be the final converted vector of the coordinate system $OX_3Y_3Z_3$, and the coordinate transformation equation is obtained.

$$r^3 = C_1^3 r^1 \tag{7}$$

Substitute in the rotation Angle, which is represented by matrix operation.

$$\begin{bmatrix} x_3 \\ y_3 \\ z_3 \end{bmatrix} = \begin{bmatrix} 1 & 0 & 0 \\ 0 & -1 & 0 \\ 0 & 0 & -1 \end{bmatrix} \begin{bmatrix} x_1 \\ y_1 \\ z_1 \end{bmatrix} \tag{8}$$

Arrange to get:

$$\begin{bmatrix} x_3 \\ y_3 \\ z_3 \end{bmatrix} = \begin{bmatrix} 1 & 0 & 0 \\ 0 & -1 & 0 \\ 0 & 0 & -1 \end{bmatrix} \begin{bmatrix} x_1 \\ y_1 \\ z_1 \end{bmatrix} = \begin{bmatrix} x_1 \\ -y_1 \\ -z_1 \end{bmatrix} \tag{9}$$

After deriving the coordinate transformation, the azimuth and the pitch Angle are respectively:

$$\alpha = -\arctan\frac{x_1}{y_1} \tag{10}$$

$$\beta = -arctan\frac{z_1}{\sqrt[2]{x_1^2 + y_1^2}} \tag{11}$$

4 Error Analysis

4.1 Installation Error

If the azimuth axis deflects α angle around the X axis, the angle error of projection on the XOY plane was analyzed.

According to the parametric equation of the circle in the space coordinate system:

$$\begin{cases} x(\theta) = c_x + r(a_x cos\theta + b_x sin\theta) \\ y(\theta) = c_y + r(a_y cos\theta + b_y sin\theta) \\ z(\theta) = c_z + r(a_z cos\theta + b_z sin\theta) \end{cases} \tag{12}$$

Where, $C(c_x, c_y, c_z)$ is the center of the circle, r is the radius of the circle, vector $a = (a_x, a_y, a_z)$ and vector $b = (b_x, b_y, b_z)$ are orthogonal unit vectors on the plane of the circle.

To simplify the calculation, considering that the center of the circle $C(c_x, c_y, c_z)$ is the origin of coordinates $(0, 0, 0)$, the radius $r = 1$. Let's think about the rotation angle α of the circle $x^2 + y^2 = 1$ around the X-axis, so the normal vector $n = (0, -\sin\alpha, \cos\alpha)$ of the circle, the two orthogonal vectors perpendicular to the normal vector can be expressed as vector $a = (1, 0, 0)$ and vector $b = (0, cos\alpha, \sin\alpha)$. Therefore, the circle considered here is shown in Fig. 3. The parametric equation is specifically expressed as:

$$\begin{cases} x(\theta) = cos\theta \\ y(\theta) = cos\alpha \cdot sin\theta \\ z(\theta) = \sin\alpha \cdot sin\theta \end{cases} \tag{13}$$

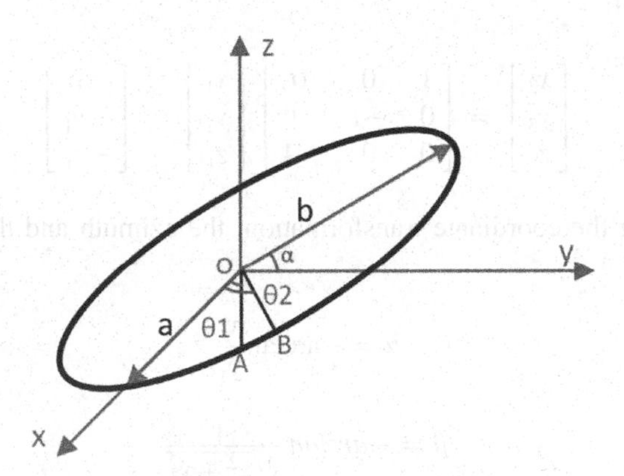

Fig. 3. Azimuth axis installation error model.

The initial angle θ_1 is defined as the included angle between the initial measurement angle pointing $(\overrightarrow{OA})$ and the rotation axis (X axis) of the installation rotation error angle α, and the measurement angle θ_2 is defined as the included angle between the secondary measurement angle pointing $(\overrightarrow{OB})$ and the rotation axis (X axis) of the installation rotation error angle α. Therefore, the coordinates of point A can be expressed as $A(\cos\theta_1, \cos\alpha\sin\theta_1, \sin\alpha \cdot \sin\theta_1)$, the coordinates of point B as $B(cos\theta_2, \cos\alpha\sin\theta_2, \sin\alpha \bullet sin\theta_2)$, and the projection of points A and B on the XOY plane as $A'(cos\theta_1, \cos\alpha\sin\theta_1, 0)$ and $B'(cos\theta_2, \cos\alpha\sin\theta_2, 0)$. According to the included angle formula of vectors, the included angle of vector $\overrightarrow{OA}$ and $\overrightarrow{OB}$ is $\varphi = \cos^{-1}\left(\frac{\overrightarrow{OA} \cdot \overrightarrow{OB}}{\left|\overrightarrow{OA}\right| \cdot \left|\overrightarrow{OB}\right|}\right) = \theta_2 - \theta_1$, and the included Angle of vector $\overrightarrow{OA'}$ and $\overrightarrow{OB'}$ is φ', which can be expressed as:

$$\begin{aligned}\varphi' &= \cos^{-1}\left(\frac{\overrightarrow{OA'}\cdot\overrightarrow{OB'}}{\left|\overrightarrow{OA'}\right|\cdot\left|\overrightarrow{OB'}\right|}\right)\\ &= \cos^{-1}\left[\frac{cos\theta_1 cos\theta_2 + (\cos\alpha)^2\sin\theta_1\sin\theta_2}{\sqrt{(cos\theta_1)^2+(\cos\alpha)^2(\sin\theta_1)^2}\cdot\sqrt{(cos\theta_2)^2+(\cos\alpha)^2(\sin\theta_2)^2}}\right]\end{aligned} \tag{14}$$

Based on the above error model, when $\alpha = 1$ mrad, the initial angle $\theta_1 \in (0, \pi)$ and the code rotation $\varphi \in (0, \pi)$, the value range of angle measurement error was analyzed. The numerical analysis results are shown as Fig. 4.

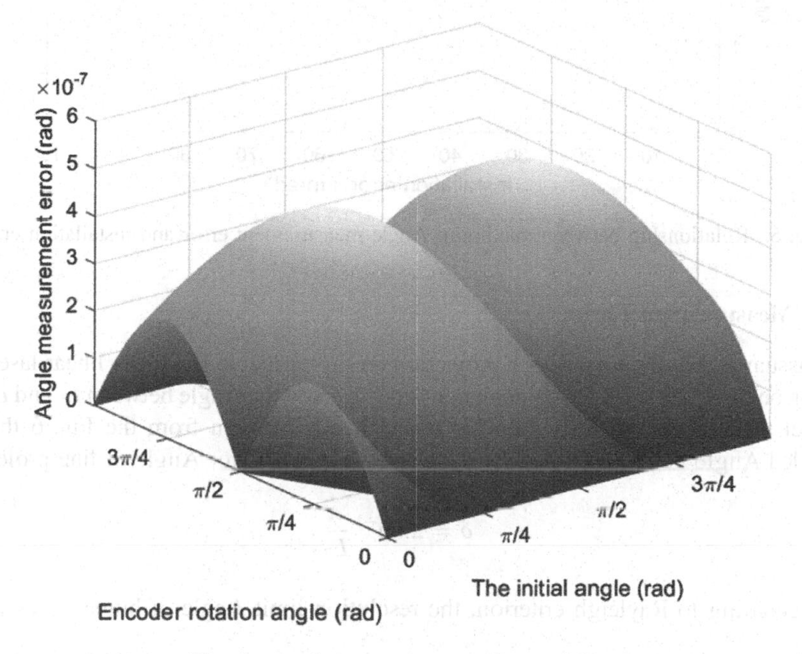

Fig. 4. Angle measurement error analysis.

When there is 1mrad installation error in the azimuth axis and horizontal plane of the turntable, the actual angular measurement error is between 0 μrad to 0.5 μrad. When the initial azimuth angle was $\pi/4$ or $3\pi/4$ and the azimuth displacement was $\pi/2$, the angular error δ_1 was the maximum, with a maximum of 0.5 rad.

Analyze the relationship between the maximum error value of angle measurement 1 mad and the installation error Angle under the conditions of 1 mard–100 mrad, and the results are shown in Fig. 5.

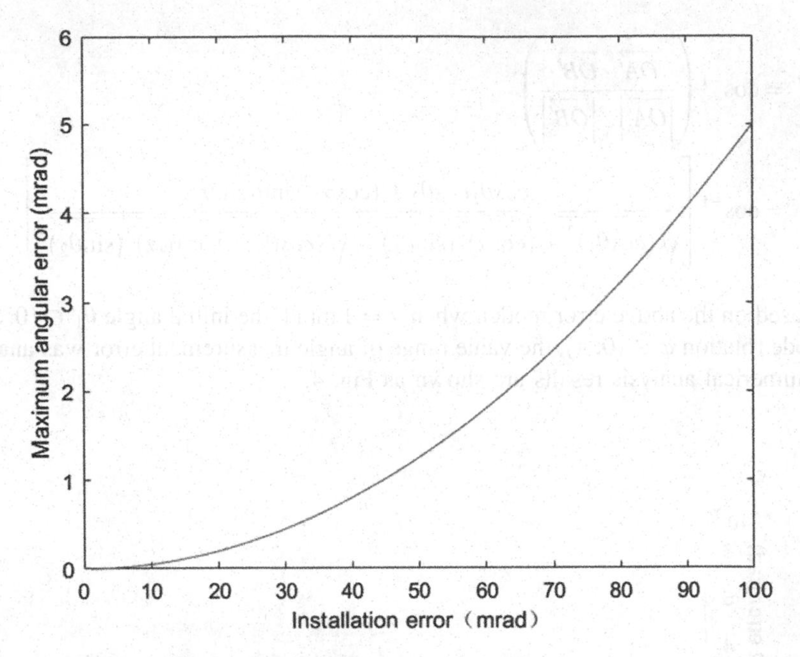

Fig. 5. Relationship between maximum Angle measurement error and installation error.

4.2 Measurement Error

It is assumed that the linear laser projection on the turntable is L_1, the linear laser projection on the CCD is L_2, the coincidence part is L_3, and the Angle between L_1 and L_2 is α.

Let the human eye resolution be a, and it can be seen from the Fig. 6 that the included Angle between L_1 and L_2 is the measurement error Angle of line projection.

$$\delta = 2tan^{-1}\frac{a}{L_3} \tag{15}$$

According to Rayleigh criterion, the resolution limit Angle of human eyes is:

$$\theta = 1.22 \cdot \frac{\lambda}{D} \tag{16}$$

Where D is the pupil size of the human eye, and λ is the working wavelength of light. When D = 3 mm, then the limit resolution Angle of human eyes θ is 0.264 mrad. Therefore, when the observation distance is 0.5 m, the human eye resolution limit distance B is 0.132 mm. In this paper, we take the human eye resolution distance A under normal conditions as:

$$a = n \cdot b(n = 1.5) = 0.198 \approx 0.2\,\text{mm} \tag{17}$$

Based on the above error model, the human eye resolution a is 0.2 mm respectively, and the curve of coincidence length L3 and measurement error Angle δ is established as Fig. 6.

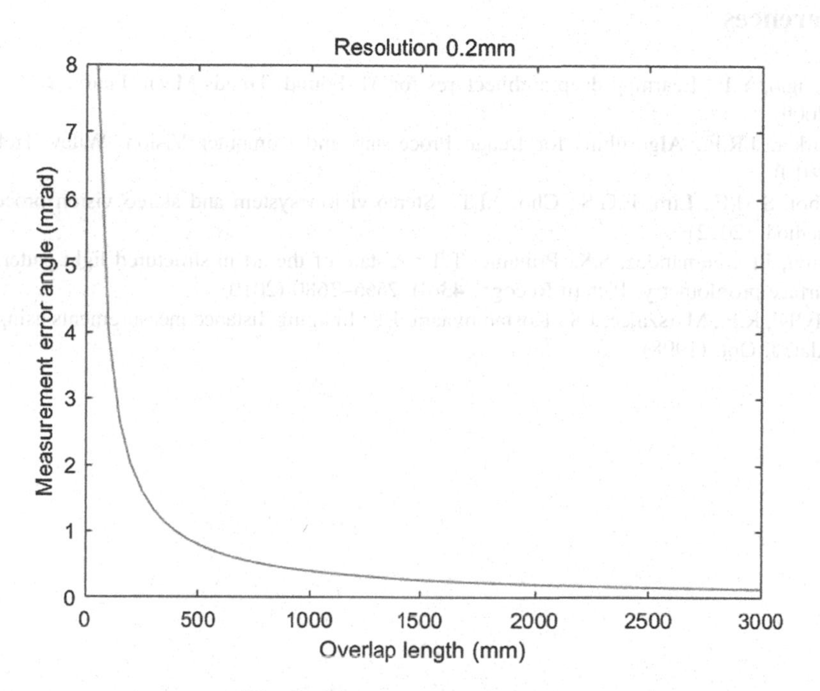

Fig. 6. Measurement error angle.

It can be concluded that the measurement error angle decreases with the increase of coincidence length and resolution. When the human eye resolution is 0.2 mm and coincidence length is 3 m, the error angle δ is 0.13 mrad. Therefore, the thinner the projection line, the clearer the boundary, the more obvious the contrast, the larger the coincidence area, and the smaller the measurement error angle.

5 Conclusion

In this paper, the linear laser alignment measurement method is applied to the 3D vision system, the projection characteristics and measurement methods of the two-dimensional platform are systematically analyzed, and the measurement mathematical model is obtained through coordinate transformation. Of two-dimensional horizontal rotary table with CCD camera produces in the process of laser projection alignment error factors are analyzed and the simulation, the main error is divided into the installation error and measurement error, the overall error is less than 0.13 mrad, the device has simple and reliable, the advantages of high precision and reliability, to improve the measurement precision of 3D-vision system has important significance.

Acknowledgement. This work is supported by the project "The Verification Platform of Multi-tier Coverage Communication Network for oceans (LZC0020)".

References

1. Bengio, Y.F.: Learning deep architectures for AI. Found. Trends Mach. Learn. **2**(1), 1–127 (2009)
2. Parker, J.R.F.: Algorithms for Image Processing and Computer Vision. Wiley, Hoboken (2010)
3. Choi, S.M.F., Lim, E.G.S., Cho, J.I.T.: Stereo vision system and stereo vision processing method. (2012)
4. Salvi, J.F., Fernandez, S.S., Pribanic, T.T.: A state of the art in structured light patterns for surface profilometry. Pattern Recogn. **43**(8), 2666–2680 (2010)
5. Myllyl, R.F., Marszalec, J.S., Kostamovaara, J.T.: Imaging distance measurements using TOF lidar. J. Opt. (1998)

An Overspeed Capture System Based on Radar Speed Measurement and Vehicle Recognition

Long Bai[1], Jiayi Yang[1], Jie Wang[1], and Mingfeng Lu[2](✉)

[1] School of Optics and Photonics, Beijing Institute of Technology, Beijing 100081, China

[2] School of Information and Electronics, Beijing Key Laboratory of Fractional Signals and Systems, Beijing Institute of Technology, Beijing 100081, China

lumingfeng@bit.edu.cn

Abstract. Overspeed has always been a very dangerous behavior for people. This may cause a variety of bad consequences such as car accidents and casualties. We need to be able to obtain the relevant information of the car while detecting the speeding now, so that subsequent punishments can be made, otherwise the perpetrators may commit the crime again. This paper proposes a high-precision, efficient method for taking photos of speeding vehicles and vehicle recognition. We directly connect the radar speed measurement module with the camera module, so that we only have one terminal for the whole system. When the radar module detects that the vehicle is speeding, it will send it directly to the camera module, so that it can capture the overspeed vehicle. This accelerates the response speed of the camera module. Therefore, when we design imaging devices, we can lower the requirements without reducing the accuracy. We can timely capture the image even if we choose the camera with low price and low quality. At last, we use image processing and support vector machines to identify the license plate. The whole system has not much equipment and can be installed in a narrow space.

Keywords: Overspeed · Speed measurement · License plate recognition · Photographing

1 Introduction

All over the world, overspeed is a very serious violation of the law. If overspeed, the driver may not be able to fully and correctly perceive the changes inside and outside the car, and at the same time the ability to recognize the space and reduce the ability to judge will be weakened. The driver will also have difficulty diverting attention, resulting in an inability to operate in a timely and accurate manner. This will undoubtedly lead to serious car accidents such as rear-end collisions and rollovers. Overspeed will also increase the strength and load of the vehicle and affect the safety performance of the vehicle. We plan to design a system that can capture speeding vehicles and identify their license plates to help manage the vehicles.

This system mainly includes imaging equipment and vehicle recognition equipment, the most important of which is the license plate recognition system. The speed

S. Shi et al. (Eds.): AICON 2020, LNICST 356, pp. 447–456, 2021.
https://doi.org/10.1007/978-3-030-69066-3_39

measurement system we add on the basis of license plate recognition allows this system to be used to monitor overspeed vehicles. Vehicle License Plate Recognition (VLPR) is a technology that can detect vehicles on monitored roads and automatically extract vehicle license plate information (including Chinese characters, English letters, Arabic numerals, and plate colors) for processing. It is an application of computer video image recognition technology in vehicle license plate recognition. License plate recognition is one of the important components of modern intelligent transportation systems. It has a wide range of applications. It requires the ability to extract and recognize vehicle license plates in motion from complex backgrounds through license plate extraction, image preprocessing, feature extraction, and license plate character recognition and other technologies to identify information such as vehicle license plate and color.

Mullot R et al. [11] developed a system that can be used for both container recognition and license plate recognition. The system mainly uses the commonness of text textures in vehicle images for positioning and recognition. License plate recognition is shared with container recognition. A hardware system. Lee E R et al. [12] used the color components in the image to locate and recognize vehicle license plates. He used three methods in a sample set of 80 graphic plates: 1. Edge detection and location recognition based on Hough transform; 2. Recognition algorithm based on gray value transformation; 3. License plate recognition system based on HLS color mode, the recognition rate reached 81.25%, 85%, 91.25%. Tindall D W [13] analyzed the significance of the whole-day work of the license plate recognition system, and pointed out the difficulty of the license plate recognition system in Europe. There are more than ten countries in Europe, and each country has multiple license plates, and there should be no obstruction between recognizing license plates from different countries. Therefore, if the license plate recognition system is to be applied in Europe, the system must be able to recognize license plates in multiple formats at the same time. Tindail has developed a license plate recognition system using the principle of license plate reflection, which can recognize all five British format license plates. The Japanese have done a lot of research on the acquisition of license plate images and have done a lot of work for the industrialization of the system. Among them, a set of the license plate recognition system developed by Sirithinaphong T et al. [14] has a full-day recognition rate of 84.2%. The system devel.

Oped by Luis [15] has an all-day recognition rate of over 90%, and 70% in bad weather. The application environment of his system is a highway toll station.

Based on previous research on license plate recognition, we have added modules for overspeed capture. The direct connection between the camera and the speed measuring module makes only one MCU between the two modules. This can reduce the requirements for equipment and improve the system's effectiveness. At the same time, this allows us to have lower requirements for imaging equipment, so that we can buy cheap equipment with a lower frame rate to build our system. This can save a lot of costs. We hope that our research can bring great help to road monitoring and vehicle management.

2 Device Structure

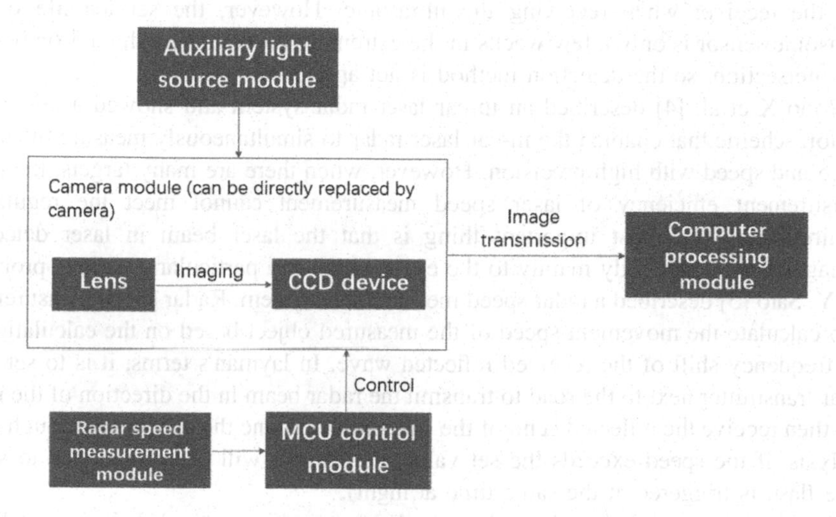

Fig. 1. Device structure

The device is mainly composed of five basic structures: speed measurement module, camera module, auxiliary lighting module, microcontroller unit module and computer processing module. It can be seen from Fig. 1 that the radar speed measurement module detects the speeding vehicle and sends the signal to the preset microcontroller unit. After the microcontroller unit receives the signal, it controls the camera to take pictures. The photo information is then transmitted to the computer for processing such as license plate number reading. Due to the requirement to realize all-weather monitoring, we choose to add an auxiliary light source to supplement light when taking pictures. Figure 1 shows the composition of our entire device.

2.1 Speed Measurement Module

We have found many vehicle speed measurement methods currently in use, mainly the following five.

Pumrin S et al. [1] used video detection to measure speed. Their work was in support of using un-calibrated traffic management roadside cameras for automated speed estimates. They constructed an activity region using moving vehicle edges, and small differences in the activity region in consecutive images were used as a decision criterion for recalibrating the camera.

Xinyi Jiang [2] studied the speed measurement capability and speed resolution capability of radar. For high-speed roads with a single direction, microwave radar is the best partner for high-speed cameras. The high-speed cameras receive the high-speed moving vehicles detected by the microwave radar, and quickly enter the state of rapid capture, and cooperate with the high-speed shutter for illegal evidence collection. The international mainstream product is radar and high-speed cameras to shoot speeding.

Lobur M et al. [3] proposed a method to measure vehicle speed using sound waves. It can know the distance through the time difference between the ultrasonic transmitter and the receiver when receiving the ultrasonic. However, the service life of the ultrasonic sensor is only a few weeks in the extremely dusty and harsh environment of the intersection, so the detection method is not applicable.

Mao X et al. [4] described an in-car laser radar system and showed a new modulation scheme that enabled the in-car laser radar to simultaneously measure the target range and speed with high precision. However, when there are many targets, the point measurement efficiency of laser speed measurement cannot meet the regulatory requirements. The most important thing is that the laser beam in laser detection damages the human body mainly to the eyes, which is a particularly serious problem.

Y. Sato [5] described a radar speed measurement system. Radar speed measurement is to calculate the movement speed of the measured object based on the calculation of the frequency shift of the received reflected wave. In layman's terms, it is to set up a radar transmitter next to the road to transmit the radar beam in the direction of the road, and then receive the reflected echo of the car, and determine the car speed through echo analysis. If the speed exceeds the set value, the camera will be commanded to shoot (The flash is triggered at the same time at night).

In summary, we choose the radar speed measurement method. It is more reliable and has a longer service life. Moreover, it can be adapted to various environments, even in bad weather conditions.

2.2 Camera Module

Field of View Calculation

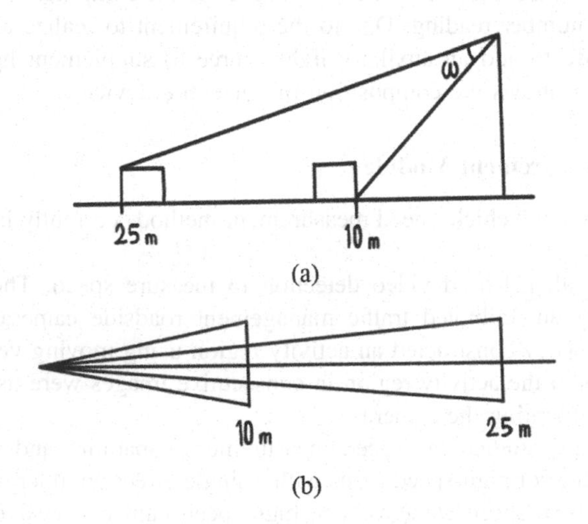

Fig. 2. Field of view calculation

It is estimated that the height of the overpass is about 5–7 m, the length of the vehicle is 3 m, the height of the vehicle is 1.5 m, and the height of the overpass is x, as shown in Fig. 2(a):

$$\omega = arctan\frac{2.5}{x-1.5} - \arctan\frac{10}{x}$$

Solution is

$$\omega_{1max} = \omega(7) = arctan\frac{2.5}{7-1.5} - arctan\frac{10}{7} \approx 22.5846^{\circ}$$

As shown in Fig. 2(b), we set the road width to 3 m:
Take the field of view angle at a distance of 10 m as:

$$\omega_2 = 2arctan\frac{1.5}{\sqrt{10^2+3.5^2}} \approx 16.1166^{\circ}$$

Take the field of view angle at a distance of 25 m as:

$$\omega_2 = 2arctan\frac{1.5}{\sqrt{25^2+3.5^2}} \approx 6.8011^{\circ}$$

In summary, the angle of view should be at least 23° × 17°

Focal Length Calculation
According to the survey of related imaging devices, the CCD pixel size is selected as 4.8 μm.

Considering that the camera angle is tilted when taking pictures, there is a certain degree of compression in the vertical direction of the license plate shooting. The smallest detail on the license plate is 1 cm, which we will use for calculation. Considering that there is almost no compression when the shooting position is at infinity, the closer the shooting position is, the greater the degree of compression, so 25 m is used for calculation.

According to the three sides and angle formula of a triangle:

$$b^2 + c^2 - a^2 = 2bc\cos A$$

Find:

$$\cos A = \frac{b^2+c^2-a^2}{2bc} = \frac{(7^2+25^2) + \left((7-0.01)^2+25^2\right) - 0.01^2}{2\sqrt{7^2+25^2}\sqrt{(7-0.01)^2+25^2}}$$

$$h^2 = (7^2 + 25^2) + \left((7 - 0.01)^2 + 25^2\right) - 2\sqrt{7^2 + 25^2}\sqrt{(7 - 0.01)^2 + 25^2}\cos A$$
$$\approx 9.63\,\text{mm}$$

According to the object image relation:

$$\frac{h}{h'} = \frac{u}{v} = \frac{u}{f'} \Rightarrow f'_{min} = \frac{u \times h'}{h} = \frac{25\,\text{m} \times 4.8\,\mu\text{m}}{9.63\,\text{mm}} \approx 12.4611\,\text{mm}$$

Therefore, the focal length of the lens is about 12 mm.

Image Size Calculation
The fixed image surface size under the field of view angle is calculated according to the focal length:

$$d_1 = 2f'\tan\frac{\omega_1}{2} = 2 \times 12\,\text{mm} \times \tan 11.2923^\circ \approx 4.7923\,\text{mm}$$

$$d_2 = 2f'\tan\frac{\omega_2}{2} = 2 \times 12\,\text{mm} \times \tan 8.0583^\circ \approx 3.3979\,\text{mm}$$

Therefore, the selected CCD image size should be larger than this value.

Exposure Time and Frame Rate Calculation. According to the exposure time calculation formula:

$$t_1 = \frac{10\,\text{mm}}{1 \times 22222\,\text{mm/s}} = 0.45\,\text{ms}$$

According to the survey, the reading time of CCD with a high frame rate is about 6.5 ms.
The frame frequency can be approximately:

$$fps = \frac{1}{0.00045\,\text{s} + 0.0065\,\text{s}} \approx 143.88\,\text{Hz}$$

Therefore, we select a CCD device with a frame frequency of about 150 Hz.

2.3 Auxiliary Lighting Module

Because the lighting environment is very different during the day and night, and different application sites also have different lighting environments, we add an auxiliary constant light source, and at the same time, use the flash to fill the light to achieve a better photo effect. We choose the LED [6] as the constant light source because it has the advantages of low energy consumption, long life, high luminous efficiency and so on. Besides, LEDs can be made into various shapes to meet our needs.

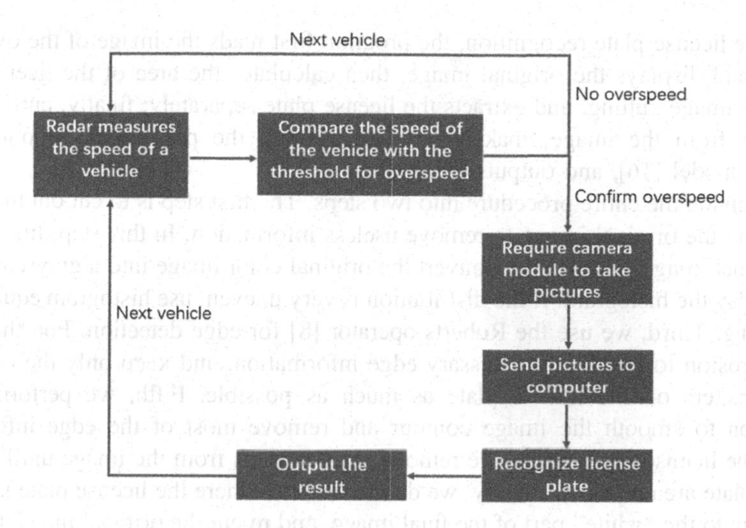

Fig. 3. Data transfer process

2.4 Microcontroller Unit Module

This module receives the vehicle speed information sent from the speed measurement module to determine whether the vehicle is overspeed. If the microcontroller determines that the vehicle is overspeed, it will send the information to the camera module to remind the camera to take pictures of the vehicle for subsequent processing. Figure 3 shows the entire data transfer process.

2.5 Computer Processing Module

This module processes the license plate information after reading the photos of the overspeed car. The entire process can be seen in Fig. 4.

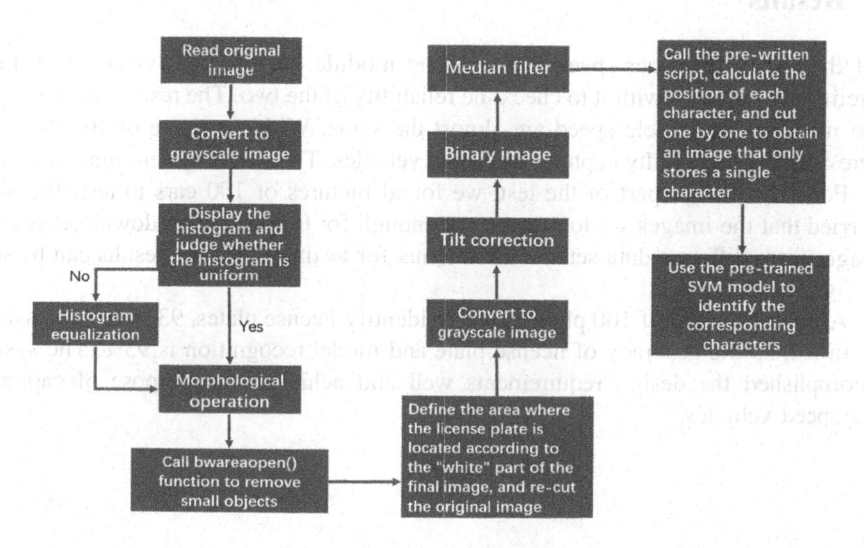

Fig. 4. Image processing flow

In the license plate recognition, the program first reads the image of the overspeed vehicle and displays the original image; then calculates the area of the license plate, performs image cutting, and extracts the license plate separately; finally, cuts out each character from the image, makes them go through the pre-trained support vector machine model [16], and outputs the final result.

We divide the entire procedure into two steps. The first step is to cut out the license plate from the original image to remove useless information. In this step, first we read the original image. Second, we convert the original color image into a grayscale image and display the histogram. If the distribution is very uneven, use histogram equalization processing. Third, we use the Roberts operator [8] for edge detection. Fourth, we use image erosion to remove unnecessary edge information, and keep only the outline of the characters on the license plate as much as possible. Fifth, we perform image expansion to smooth the image contour and remove most of the edge information except the license plate. Sixth, we remove small objects from the image until only the license plate area remains. Finally, we define the area where the license plate is located according to the "white" part of the final image, and re-cut the original image to obtain the image of the license plate.

The second step of license plate recognition is to cut out each character from the license plate image obtained earlier. First, we convert the license plate image to a grayscale image and draw a histogram. Second, in order to enhance the contrast, we perform histogram equalization on the gray image. Third, we correct the tilt of the image and binarize it. Fourth, we perform median filtering [9] to filter out some useless information. Fifth, we calculate the position of each character and cut them one by one to obtain an image that only stores a single character. Finally, we use pre-prepared character templates and train them with support vector machines [16], and then classify the single characters separated from the license plate by the trained model to obtain the license plate recognition result.

3 Results

For the test of the radar speed measurement module, we used the video speed measurement to compare with it to check the reliability of the two. The results show that the two measures of vehicle speed are almost the same. Within a range of 10–25 m, we were able to successfully capture overspeed vehicles. The whole system runs smoothly.

For the software part of the test, we found pictures of 100 cars to test. We were worried that the images we took were not enough for testing, so we downloaded some images from different data sets on the Internet for testing [10]. The results can be seen in Fig. 5 and Table 1.

After testing, out of 100 photos used to identify license plates, 93 were successfully identified, so the accuracy of license plate and model recognition is 93%. The system accomplished the design requirements well and achieved the purpose of capturing overspeed vehicles.

Fig. 5. Part of the images used for license plate recognition [10] (The images are 1,2,3,4,5,6,7,8,9 from left to right and top to bottom)

Table 1. Part of results of license plate recognition

	Test1	Test2	Test3	Test4	Test5	Test6	Test7	Test8	Test9
Result	桂A C3692	桂A V6388	桂A YG299	桂A F2830	桂A W7566	桂A E0886	桂A 72668	桂F 02235	桂A M9678

4 Conclusion

In previous studies, people's equipment is usually relatively large, and this causes them to take up a lot of space. In this paper, we proposed a vehicle photographing and recognition system which has high precision. We directly connect the camera module to the speed measurement module, which improves efficiency and reduces money costs. At the same time, we also use image processing and support vector machines [16] to complete the license plate recognition. Under normal conditions, this system has high accuracy and good practicability.

However, we still need to improve the system. Under bad weather conditions such as rain, snow, fog, etc., the recognition accuracy of the system will be greatly reduced. Therefore, there is still room for improvement in accuracy.

In the subsequent work, we can upgrade the imaging equipment so that it has high penetration under bad weather conditions and still takes clear pictures. Additionally, if we can have more data sets involved in training, we can also get higher accuracy.

Acknowledgement. This paper is sponsored by Beijing Natural Science Foundation (L191004).

References

1. Pumrin, S., Dailey, D.J.: Roadside camera motion detection for automated speed measurement. In: Proceedings of the IEEE 5th International Conference on Intelligent Transportation Systems, pp. 147–151. IEEE (2002)
2. Jiang, X.: Research on millimeter wave radar velocity measurement. Nongjia Staff **597**(19), 242 (2018)
3. Lobur, M., Darnobyt, Y.: Car speed measurement based on ultrasonic Doppler's ground speed sensors. In: 2011 11th International Conference the Experience of Designing and Application of CAD Systems in Microelectronics (CADSM), pp. 392–393. IEEE (2011)
4. Mao, X., Inoue, D., Kato, S., et al.: Amplitude-modulated laser radar for range and speed measurement in car applications. IEEE Trans. Intell. Transp. Syst. **13**(1), 408–413 (2011)
5. Sato, Y.: Radar speed monitoring system. In: Proceedings of VNIS'94–1994 Vehicle Navigation and Information Systems Conference, pp. 89–93. IEEE (1994)
6. Shailesh, K.R., Kini, S.G., Kurian, C.P.: Summary of LED down light testing and its implications. In: 2016 10th International Conference on Intelligent Systems and Control (ISCO), pp. 1–5. IEEE (2016)
7. Brunelli, R.: Template Matching Techniques in Computer Vision: Theory and Practice. Wiley, Chichester (2009)
8. Wang, A., Liu, X.: Vehicle license plate location based on improved Roberts operator and mathematical morphology. In: 2012 Second International Conference on Instrumentation, Measurement, Computer, Communication and Control, pp. 995–998. IEEE (2012)
9. George, G., Oommen, R.M., Shelly, S., et al.: A survey on various median filtering techniques for removal of impulse noise from digital image. In: 2018 Conference on Emerging Devices and Smart Systems (ICEDSS), pp. 235–238. IEEE (2018)
10. Zhu, Y.: License plate recognition based on Matlab (2018). https://download.csdn.net/download/weixin_42618564/10533369?utm_medium=distribute.pc_relevant_download.none
11. Mullot, R., Olivier, C., Bourdon, J.L., et al.: Automatic extraction methods of container identity number and registration plates of cars. In: Proceedings IECON'91: 1991 International Conference on Industrial Electronics, Control and Instrumentation, pp.: 1739–1744. IEEE (1991)
12. Lee, E.R., Kim, P.K., Kim, H.J.: Automatic recognition of a car license plate using color image processing. In: Proceedings of 1st International Conference on Image Processing, vol. 2, pp. 301–305. IEEE (1994)
13. Tindall, D.W.: Deployment of automatic licence plate recognition systems in multinational environments (1997)
14. Sirithinaphong, T., Chamnongthai, K.: Extraction of car license plate using motor vehicle regulation and character pattern recognition. In: IEEE. APCCAS 1998. 1998 IEEE Asia-Pacific Conference on Circuits and Systems. Microelectronics and Integrating Systems. Proceedings (Cat. No. 98EX242), pp. 559–562. IEEE (1998)
15. Salgado, L., Menendez, J.M., Rendon, E., et al.: Automatic car plate detection and recognition through intelligent vision engineering. In: Proceedings IEEE 33rd Annual 1999 International Carnahan Conference on Security Technology (Cat. No. 99CH36303), pp. 71–76. IEEE (1999)
16. Cortes, C., Vapnik, V.: Support-vector networks. Mach. Learn. **20**(3), 273–297 (1995)

Study on Elevation Estimation of Low-Angle Target in Meter-Wave Radar Based on Machine-Learning

Di Chen(✉) and Chengyu Hou

Harbin Institute of Technology,
NO. 92, Xidazhi Street, Nangang District, Harbin, Heilongjiang, China
dchen@hit.edu.cn

Abstract. In these years, meter-wave radar has gotten more and more attention from the researchers all over the world for its advantages in anti-stealth. However, the beam width of meter-wave radar is wider because of the size of radar antenna's vertical aperture, and it makes the low-angle targets detecting and tracking more difficult, which has become one of the urgent problems in radar field. In this paper, the multipath effect in low-angle target detecting will be researched, and an elevation estimation algorithm of low-angle target in meter-wave radar based on machine-learning will be proposed.

Keywords: Elevation estimation · Meter-wave radar · Reflected echoes cancelling

1 Introduction

In recent years, with the rapid development of stealth technology, the ability of microwave radar to detect targets is declining day by day. Meanwhile, meter-wave radar has attracted more and more attention because of its advantages in anti-stealth. However, due to the long wave length, the beam width of receiving antenna is wider, which leads to the low elevation resolution of meter-wave radar. In the detecting and tracking low-angle targets, the reflected echoes from the ground or sea surface will enter the receiving antenna together with the direct ones, which cause the multipath effect, and seriously affects the targets real location estimation, resulting in the decline of radar elevation estimation and tracking performance [1, 2]. The multipath effect has become an important factor affecting the accuracy of elevation estimation of low-angle targets [3]. Nowadays, the problem of the multipath effect of low-angle targets has become an important issue in meter-wave radar field. In this paper, the elevation estimation algorithm which can effectively eliminate the influence of the multipath effect is studied, and the real elevation estimation of the target is obtained.

S. Shi et al. (Eds.): AICON 2020, LNICST 356, pp. 457–464, 2021.
https://doi.org/10.1007/978-3-030-69066-3_40

2 The Multipath Effect Model

The elevation estimation of low-angle targets is influenced by multipath effect. So in order to solve the problem in elevation estimation of low-angle targets in meter wave radar, it is necessary to understand the cause of multipath effect, and the establishment of multipath model is a more effective means [4]. Therefore, when establishing the multipath effect model, the actual situation of the multipath effect should be reflected as much as possible, and give a model that can accurately describe the actual situation of the multipath effect.

Radar can detect, track and sometimes recognize targets because of the existence of the target's echoes. And the echo characteristics depend largely on the size and properties of the target surface exposed in the radar beam, and are sensitive to the changes of the target scattering characteristics. For the meter-wave radar, the geometric size of the aerial target (such as aircraft) is equivalent to the radar wavelength, which is usually of multiple wavelength orders of magnitude, so its scattering characteristics have resonance characteristics. In order to simplify the analysis, the total electromagnetic scattering of the target is simplified as the sum of the electromagnetic scattering of multiple isolated scattering centers. Therefore, the target in the multipath effect model is no longer a simple point target, but a set of equivalent multiple scattering centers. There are the direct echo path and the reflected echo path between each scattering center and the antenna, so the combination form of echoes is complex. Figure 1 shows the multipath effect model based on multiple scattering centers.

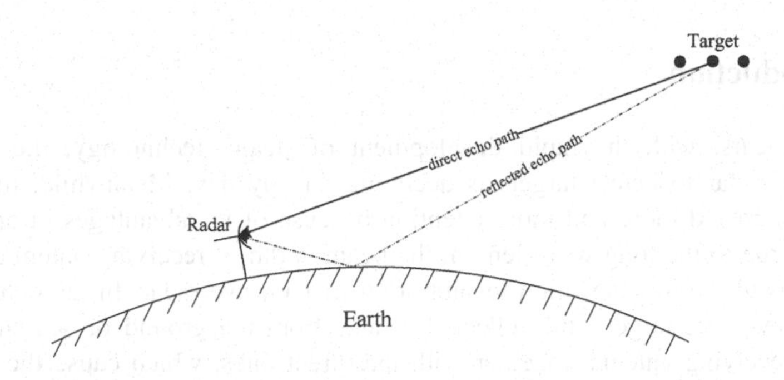

Fig. 1. The multipath effect model based on multiple scattering centers

In the Fig. 1, the distance of the direct path d_d and the distance of the reflection path d_f are respectively:

$$d_d = \sqrt{(r_{earth} + h_r)^2 + (r_{earth} + h_t)^2 - 2(r_{earth} + h_r)(r_{earth} + h_t)\cos(\beta_r + \beta_t)} \quad (1)$$

$$d_f = \sqrt{2(r_{earth}^2 + r_{earth} \cdot h_t)(1 - \cos\beta_t) + h_t^2} + \sqrt{2(r_{earth}^2 + r_{earth} \cdot h_r)(1 - \cos\beta_r) + h_r^2} \quad (2)$$

where, h_r is the height of the radar, h_t is the height of the target, r_{earth} is the radius of curvature of the earth's prime vertical circle, β_r is the sphere center angle between the radar and the ground reflection point, β_t is the sphere center angle between the target and the ground reflection point.

3 Elevation Estimation Algorithm

At present, in array signals processing, some classic techniques are usually used. Such as the super-resolution technology, it is very mature for array signal processing, and can obtain good angular resolution and estimation accuracy [5]. But this is based on the premise that the radar echoes only contains the direct echoes of the target, which is not for meter-wave radar. So this paper will analyze the characteristics of the reflected echoes, study the method of extracting the reflected components based on machine learning, and use the obtained reflected components to construct a cancellation vector to cancel the reflected echoes to eliminate the multipath effect, which lets the direct echoes dominate the whole echoes, makes the application of super-resolution algorithm possible, and the true elevation estimation of the target can be obtained.

3.1 Preprocessing of Elevation Estimation Credibility

For the meter-wave radar, due to the influence of multipath effects, the angle estimation of the low-angle targets is neither credible nor accurate. The meter-wave radar echoes of the low-angle target is actually the composite echoes of the direct ones and the reflected ones reflected by the ground. Among them, the direct ones is no longer dominant in the whole echoes. The target elevation estimation is influenced by the multipath effects. However, the influence of the multipath effects on elevation estimation presents a credibility phenomenon, that is, when the estimation is within a certain range, the credibility is very high, and it can be used as a true estimation, while it is incredibility in some range, and the estimation needs to be corrected. Therefore, after receiving the echo data, it is necessary to preprocess the elevation estimation credibility of the data by the elevation estimation interval.

In the process of space propagation, electromagnetic wave is blocked by undulating terrain surface and ground obstacles, forming terrain blind area, resulting in electromagnetic wave cannot be transmitted. In the evaluation credibility prediction process, if the elevation estimation obtained falls in the terrain blind area, the evaluation estimation credibility is lower, so the estimation should be corrected for the multipath effects.

For a radar, the terrain blind area is mainly related to the position of setting up the antenna and the terrain undulations within the radar coverage. Without considering its propagation influence, the radar's coverage blind area is mainly determined by the earth curvature blind area and terrain blocking blind area. The relationship between blind angle α_z and terrain is as follows:

$$\alpha_z = \cos^{-1}\left(\frac{d^2 - 2h \cdot r_{earth} - h^2}{2d \cdot r_{earth}}\right) - \frac{\pi}{2} \tag{3}$$

where, d is the distance of undulating terrain, h is the height of undulating terrain, r_{earth} is the radius of curvature of the earth's prime vertical circle where the radar is erected.

The flow chart of the calculation of terrain blind area is shown.

Firstly, according to the performance index of radar, the azimuth, elevation coverage and elevation resolution of the radar are input.

Secondly, a certain elevation value is set which needs to be calculated, and calculate the effective coverage distance d after propagation attenuation according to the working parameters and environmental parameters of the radar.

Thirdly, the latitude and longitude range of the radar coverage is calculated according to the latitude, longitude and altitude information, azimuth β, elevation α and coverage distance d. According to the latitude and longitude information, the terrain data of coverage area is indexed, and the effective elevation α' and its corresponding effective distance d' are calculated when the coverage distance is d and the azimuth is β.

Then, compare the values of α and α', if the former is greater than the latter, it means that there is terrain occlusion, at this time d is updated to d'; otherwise, d does not need to be updated.

Finally, traverse all the ranges of the elevation, and the output d is the coverage distance under the specified azimuth.

3.2 The Elevation Estimation Algorithm Based on the Reflected Components Cancellation

Therefore, the key to improving the elevation estimation performance of the targets is to eliminate the influence of the reflected echoes. To eliminate the reflected echoes, the prerequisite is to accurately grasp the reflected echoes components information. Therefore, the acquisition of the reflected components is very important. According to the multipath effect model of complex targets with multiple scattering centers, the target should be regarded as a group of multiple scattering centers that obey a specific distribution. Since different types of targets have different distributions of multiple scattering centers and the scattering characteristics of each scattering center, and the target type cannot be known in advance in actual situations, the data classification method can be used to extract the reflected echo vectors corresponding to each scattering center.

According to the multipath effect model of the multi scattering centers, the echoes of the low-angle target in meter-wave radar are composed of two parts: the direct echoes which returns to the receiving antenna after the target scattering and the reflected ones arriving at the receiving antenna after the ground reflection. Echoes received at the radar antenna $E(t)$ shall be expressed as:

$$\begin{aligned} E(t) &= \sum_{k=1}^{K} E_{dk}(t) + \sum_{k=1}^{K} E_{fk}(t) \\ &= \sum_{k=1}^{K} A_{dk} e^{j\phi_k} + \sum_{k=1}^{K} A_{fk} e^{j\varphi_k} \end{aligned} \quad (4)$$

where, $E_{dk}(t)$ and $E_{fk}(t)$ are the direct echoes and reflected echoes generated by the kth scattering center of the target, A_{dk} and A_{fk} are the amplitude of direct echoes and reflected echoes from the kth scattering center respectively, and φ_k and ϕ_k are the phases of direct echoes and reflected echoes respectively.

Among them, reflected echoes are the main reason for inaccurate measurement of low-angle target in meter-wave radar. Therefore, the key to solve the problem of the elevation estimation of low-angle target is how to eliminate the influence of reflected echoes. According to the influence of each equivalent reflection components on the estimation results, the most appropriate combination of reflection echoes cancellation components is extracted based on machine-learning feature matching. Then, according to the statistics and analysis of the cancellation results of different combinations, which is the estimation is classified, and then the estimation results of different combination cancellation are mapped into the classes one by one, and the estimations in one class are combined, and the target track and target motion trend information are combined to obtain the more accurate target elevation estimation.

Through the analysis of the multipath effect model, it is known that each reflected component corresponds to a phase and amplitude. Therefore, searching for the reflected component is actually to search for the phase and amplitude of the reflected component.

Therefore, the steps for searching the reflected components are as follows:

1) The amplitude of the reflected component is set, and then the angle of the reflected component is adjusted to construct the cancellation vector phase.
2) After cancellation, the cancellation result is judged, and those that meet the judgment conditions are output as the extraction result.
3) By adjusting the amplitude of the reflected component, the phase search is carried out again, and finally a set of cancellation components is obtained.

In the process of searching cancellation components, how to determine the cancellation components is an important part. Through the analysis of the cancellation results, it can be found that the peak energy of super-resolution spectrum after cancellation is higher than that corresponding to the adjacent search phase when the reflected components extraction is more accurate. The reason is that the cancellation components is sensitive to the phase, and some difference in the phase will lead to different cancellation results. Once the phase is aligned, the cancellation result will be significantly improved compared with that without alignment, and the reflected components will be restrained to some extent, thus the super-resolution spectrum energy will be improved. At the same time, because the search step is very small, many cancellation vectors with small phase difference often obtain similar cancellation results, which makes the same reflected components be extracted many times, resulting in overlapping of extracted reflected components. Therefore, the validity of the reflected components can be judged and extracted according to the cancellation result of the reflected components. The extraction of the reflected components can be equivalent to the two-classification problem of the good or bad cancellation effect of the reflected components. In this paper, BP neural network is used to extract the characteristic samples of the reflected components with prior information to train the network, so as to realize the classification and extraction of the reflected components.

The specific steps of cancelling component extraction based on BP network are as follows:

1) Continuous weight/threshold initialization, using a small random number for assignment, and the weight and threshold cannot be equal to avoid network errors.
2) Provide the training samples, test samples and the network parameters of the reflected cancellation component to the neural network for normalization preprocessing.
3) Each sample in the training set is calculated: the output of each neuron in the hidden layer and output layer is calculated in the forward direction; the error between the expected output and the network output is calculated; the weights and thresholds of the modified network are calculated backward.

3.3 The Solution of the Problem of Multi-value Suppression

In the process of meter-wave radar signal processing, since the elevation estimation is obtained by cancelling the searched reflected components, the cancellation results directly affect the estimation results. However, the reflection components used in the cancellation is obtained by searching, which cannot be completely equivalent to the real multipath effect. Therefore, there will be residual components after cancellation, which will produce multi value phenomenon in the estimation results. The multiple estimation generated are regarded as multiple observation values of the current batch. The plot-track association method based on data classification and recognition is used, and the training samples are obtained through the analysis of historical data, and the batch of plots containing multi-value information are input into the trained model for target association classification, so as to realize the association of target plots and tracks and solve the problem of multi value suppression.

The plot-track association algorithm based on data classification adopts the support vector machine (SVM) based on posterior probability. In the process of track association, the association threshold is set based on the previous batch of plots of the target's existing track, and the association area is established based on the threshold value. If the value of multiple estimates of the current batch falls within the association area, the estimation and the existing track of the target form an association hypothesis, And whether the hypothesis is true can be obtained by the classifier. When constructing the target track sample for training classifier, it is obtained by extracting the track data, including a specified number of batch data and error data affected by the multipath effect. The training data set **C** is composed of the classifier training samples whose characteristic parameters include information such as speed, acceleration, etc., which is obtained from each sample as features:

$$\mathbf{C} = \{(c_{11}, c_{12}, \cdots, c_{1j}), \cdots, (c_{i1}, c_{i2}, \cdots, c_{ij})\}, i = 1, 2, \cdots, M \quad j = 1, 2, \cdots, N \tag{5}$$

where, c_{ij} is the characteristic parameter of the sample, M is the number of samples in the training data set, N is the number of extracted features of each sample.

By inputting the characteristic parameters of the track association hypothesis sample into the trained posterior probability SVM classifier, the probability of

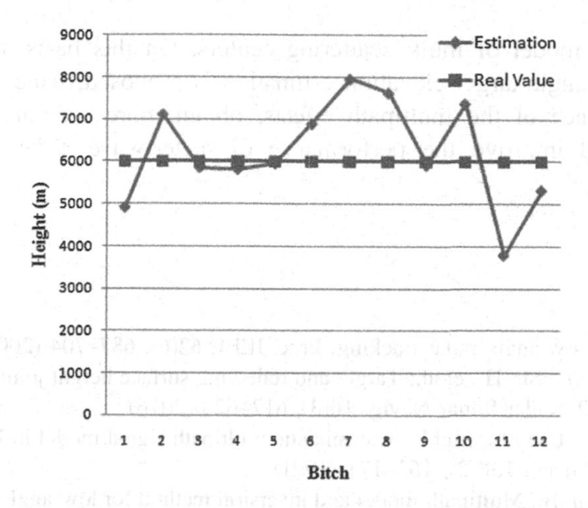

Fig. 2. Elevation estimation of target without reflected components cancellation

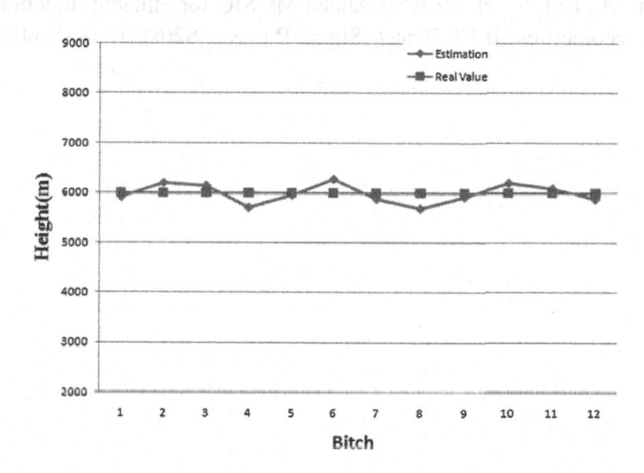

Fig. 3. Elevation estimation of target based on reflected components cancellation

interconnection between the associated track and the batch of plots can be obtained under the condition that the associated track in the current hypothesis which the characteristic parameters correspond to is tenable. After classification, the probability of each hypothesis can be obtained. Finally, the multi value problem is solved and the elevation of low-angle target is estimated correctly. Figure 2 and Fig. 3 show the processing results of elevation estimation of target without reflected components cancellation and based on reflected components cancellation, respectively.

4 Conclusion

In this paper, the elevation estimation algorithm of low-angle target in meter-wave radar based on reflected components cancellation is studied. The characteristics of multi scattering centers and the multipath effects are analyzed by establishing the multipath effect model of multi scattering centers. On this basis, a solution to the problem of low-angle target elevation estimation is proposed, which can better overcome the influence of the multipath effects, obtain more accurate target elevation information, and improve the performance of meter-wave radar low-angle target tracking.

References

1. Barton, D.K.: Low-angle radar tracking. Proc. IEEE **62**(6), 687–704 (2005)
2. Wang, S., Cao, Y., Su, H., et al.: Target and reflecting surface height joint estimation in low-angle radar. IET Radar Sonar Navig. **10**(3), 617–623 (2016)
3. Lo, T., Litva, J.: Use of a highly deterministic multipath signal model in low-angle tracking. Radar Signal Process **138**(2), 163–171 (1991)
4. Zheng, Y., Chen, B.: Multipath model and inversion method for low-angle target in very high frequency radar. J. Electron. Inf. Technol. **38**, 1468–1474 (2016)
5. Yan, F.G., Jin, M., Liu, S., et al.: Real-valued MUSIC for efficient direction estimation with arbitrary array geometries. IEEE Trans. Signal Process. **62**(6), 1548–1560 (2014)

Target Registration Based on Fusing Features of Visible and Two Wave Bands Infrared Images

Junhua Yan[1,2(✉)], Kai Su[2], Xuyang Cai[2], Tianxia Xie[2], Yin Zhang[1,2], and Kun Zhang[2]

[1] Key Laboratory of Space Photoelectric Detection and Perception, Nanjing University of Aeronautics and Astronautics, Ministry of Industry and Information Technology, No. 29 Jiangjun Street, Nanjing 211106, China
yjh9758@126.com

[2] College of Astronautics, Nanjing University of Aeronautics and Astronautics, Nanjing 211106, Jiangsu, China

Abstract. In order to register the same target in images from different sources to improve the accuracy of target recognition of multi-source images, based on the principle that the same target has the highest similarity among targets in these images, this paper proposes a new target registration algorithm by fusing features of Visible (VIS), Long Wave Infrared (LWIR) and Middle Wave Infrared (MWIR) images, which registers the same target in these images by calculating the targets similarity in different source images. Firstly, the similarity between targets in LWIR and MWIR images is calculated by using the improved structural similarity. Then, the similarity between targets in VIS and LWIR images is calculated by using Hu invariant moment feature and cosine similarity. Finally, the similarity among targets in VIS, MWIR and LWIR images is calculated by fusing these two kinds of target similarity, so that target registration of these three-source images is realized. Experimental results show that the proposed algorithm has high correct rate and accuracy of target registration. Specifically, the correct rate of target registration is 83.87% and the accuracy of target registration is higher than 0.95.

Keywords: Target registration · Fusion · VIS image · LWIR image · MWIR image · Improved structural similarity

1 Introduction

The expansion of human activities in time and space demands recognition of interested targets from complex background under all weather conditions, hence multi-source images are needed to realize this demand. In order to improve the accuracy of target recognition of multi-source images, the same target in these images must be registered firstly. Because of different imaging principles, images from different sources have different image features and the same target in these images has different target features. VIS images have rich color information and texture information to capture the details of the target. However, adverse weather conditions render VIS images

S. Shi et al. (Eds.): AICON 2020, LNICST 356, pp. 465–483, 2021.
https://doi.org/10.1007/978-3-030-69066-3_41

ineffective. Infrared images, by contrast, are not affected by the weather. Not many details are captured in infrared images, yet the edge information of the target, which provides the complete structural information about the target, is well preserved. Specifically, MWIR images are richer in texture information than LWIR images, especially in high thermal radiation regions. LWIR images, however, are brighter than MWIR images. Although the same target has different target features in images from different sources, it has the highest similarity among targets in these images. In order to recognize targets effectively under all weather conditions and at all times, the target features of the visible and two wave bands infrared images are fused in this paper based on the targets similarity in different source images, realizing target registration of multi-source images.

J. Ma et al. [1] realized the registration of corresponding regions in the VIS and LWIR images. This method combines edge enhancement and normalized correlation coefficient, and the registration rate reaches 83.33%. This algorithm highlights the complete edge information of LWIR images through edge enhancement. However, the normalized correlation coefficient is only related to the gray information of the image. Moreover, the different imaging principles of VIS and LWIR images cause significant differences in the gray of these images, which renders the algorithm ineffective when the image is complex. J. Jiang et al. [2] registered the VIS and LWIR images based on wavelet transform and mutual information, with a registration rate of 80%. This method extracts the complete edge information of VIS and LWIR images with wavelet transform. However, the computation of mutual information is entirely based on the statistical information of the image gray, leaving out the spatial information of the image gray. Thus, similarly, this algorithm is ineffective when the image is complex. L. L. Mao et al. [3] combined the mutual information and the gradient information to realize the registration of VIS and LWIR images, and the registration rate is higher than those of registration algorithms which are only based on mutual information. This algorithm uses the gradient information to represent the complete edge information of VIS and LWIR images. However, this algorithm do not have high robustness, because the mutual information is only related to the statistical information of the image gray, and the spatial distribution of the image gray is ignored. Z. Li [4] put forward an algorithm to register VIS and LWIR images using the normalized covariance correlation coefficient. Although the normalized covariance correlation coefficient is strongly adaptable to gray transformations between images, it belongs to gray correlation. Therefore, this algorithm is not suitable for a variety of images from different sources. G.A. Bilodeau et al. [5] used the histogram of oriented gradient to register VIS and LWIR images. The registration rate is around 50% when the registration window is small and can reach more than 90% when the registration window becomes larger. Although the histogram of oriented gradient can represent the common edge information of VIS and LWIR images, this algorithm is not robust to the sizes of registration windows. M. An et al. [6] presented a robust image registration and localization algorithm based on SURF features to register remote sensing video images. This algorithm can compute camera poses under different conditions of scale, viewpoint and rotation so as to precisely localize object's position. However, the algorithm almost based on the land-marks, it only works when landmarks are available in the scene. J.Y. Ma [7] proposed a regularized Gaussian fields criterion for non-rigid registration of visible and infrared face images, which represents an image by its edge map and align the edge maps by a robust criterion with a non-rigid model. This algorithm is suitable for non-rigid bodies such as face images, but it is not robust to images lacking of rich edge information. Y. Zhuang [8] registered visible and infrared face images based on mutual information, which combines the PSO algorithm and Powell search method to find the most appropriate registration parameters. However, mutual information is entirely based on the statistical information of the image gray so this method is not suitable for complex images.

The above mentioned registrations algorithms of multi-source images make good use of the complete edge information of VIS and LWIR images. However, they only use the statistical information of the gray or the edge orientation to characterize the edge information of these images, and do not utilize the complete shape and structural information. Hence they are not suitable for the registration of a variety of complex images from different sources. The Hu invariant moment feature is a typical shape feature that can represent the shape information of the target in the image. J.F. Dou et al. [9] proposed a registration algorithm of VIS and LWIR images, in which the Hu invariant moment feature of the neighboring area around every feature point is used to match feature points. This algorithm makes stable use of the shape feature of the neighboring area around every feature point, and has a good registration performance in the presence of changes in rotation, scale, translation, field, etc. X. Wen et al. [10] put forward an improved structural similarity algorithm. For the overall structural feature of the target, it uses the images and the corresponding edge images of different scales in Gaussian scale space to calculate the multi-scale edge structural similarity, realizing the target registration of two wave bands infrared images. The correct rate and accuracy of the target registration of this algorithm is high. F. Wu [11] presented an image registration method improving the precision of visible and infrared (VIS/IR) image registration based on visual salient (VS) feature detector. This detector can detect the VS correspondences between images with higher repeatability score. However, this method is not applicable to target with big scale change. St-Charles et al. [12] presented an online multimodal video registration method that relies on the matching of shape contours to estimate the parameters of a planar transformation model. They used foreground-background segmentation on each video frame to obtain shape contours from targets present in the scene, then described and matched contour points using the iterative shape context approach. However, the algorithm is not suitable for targets with approximate shape. Y.J. Chen [13] presented an image registration method for visible and infrared images based on stable region features and edginess, which uses Zernike moments to describe salient region features for a coarse registration, and uses an entropy optimal process based on edginess to refine the registration to achieve a more accurate result. However, Zernike invariant moments are effective for images in which the target shape is dominant, and are less effective for texture-rich images.

Hence, the edge information, shape information and structure information in VIS, LWIR and MWIR images are fully utilized in this paper. Based on the principle that the same target has the highest similarity among targets in images from different sources, target features in VIS, LWIR and MWIR images are fused by calculating the targets similarity, which improves the registration accuracy of the same target in these images. Firstly, the similarity between targets in LWIR and MWIR images is calculated by using the improved structural similarity. Then, the similarity between targets in VIS and LWIR images is calculated by using Hu invariant moment feature and cosine similarity, which utilizes the complete shape information of VIS and LWIR images. Finally, the similarity among targets in VIS, MWIR and LWIR images is calculated by fusing these two kinds of target similarity, so that target registration based on fusing features of VIS and two wave bands infrared images is realized. In this paper, the proposed algorithm is tested on multiple sets of VIS, MWIR and LWIR images, and the superiority of the proposed algorithm is verified.

2 The Target Registration Algorithm Based on Fusing Features of VIS and Two Wave Bands Infrared Images

The framework of the target registration algorithm is shown in Fig. 1.

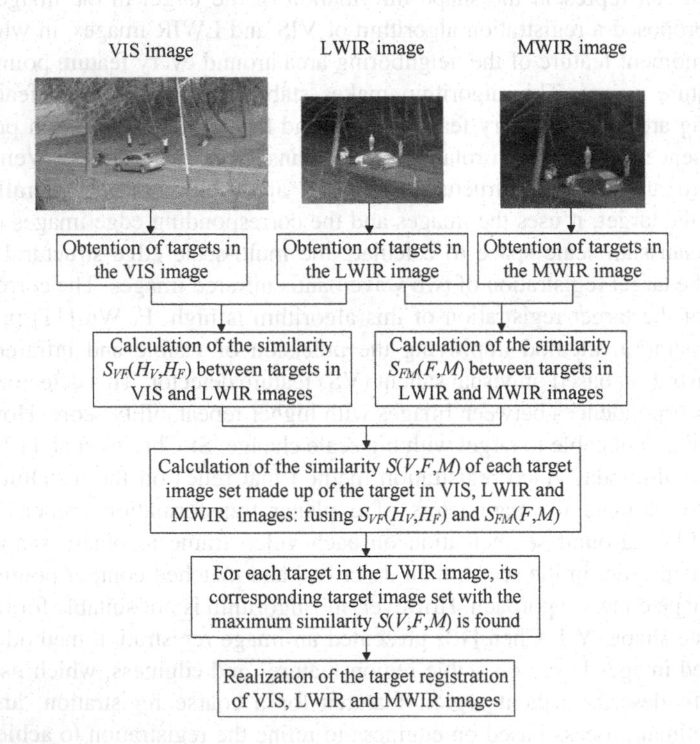

Fig. 1. The framework of the target registration algorithm based on fusing features of VIS and two wave bands infrared images

Based on the principle that the same target has the highest similarity among targets in images from different sources, target features in VIS, LWIR and MWIR images are fused by calculating the targets similarity, which realizes the registration of the same target in these images. Firstly, since LWIR and MWIR images contain rich edge information and complete structure information, the similarity $S_{FM}(F,M)$ between targets in LWIR and MWIR images is calculated by using the improved structural similarity; since VIS and LWIR images contain rich edge information and complete geometric shapes information, the similarity $S_{VF}(H_V,H_F)$ between targets in VIS and LWIR images is calculated by using the Hu invariant moment feature and cosine similarity. Then, the similarity $S(V,F,M)$ among targets in VIS, LWIR and MWIR images is calculated by fusing $S_{FM}(F,M)$ and $S_{VF}(H_V,H_F)$ in the same proportion. Finally, for each target in LWIR image, the corresponding target image set with the maximum value of similarity $S(V,F,M)$ is identified by searching, which realizes the target registration of VIS, LWIR and MWIR images.

3 Calculation of the Similarity Among Targets in VIS and Two Wave Bands Infrared Images

3.1 Calculation of the Similarity Between Targets in LWIR and MWIR Images

LWIR and MWIR images contain rich edge information and complete structure information [14]. The improved structural similarity uses these images and their edge images of different scales in Gaussian scale space to calculate the multi-scale structural similarity. It combines the complete edge information with the structural information of the target in LWIR and MWIR images [10]. Therefore, the similarity between targets in LWIR and MWIR images is calculated by using the improved structural similarity. The specific computational formula is as follows:

$$S_{FM}(F,M) = [l_{MSG}(F,M)]^{\alpha}\cdot[c_{MSG}(F,M)]^{\beta}\cdot[s_{MSeG}(F,M)]^{\gamma},\ 0<\alpha,\beta,\gamma\leq 1 \quad (1)$$

where $l_{MSG}(F,M) = [l(F_N,M_N)]^{w_N}$, representing the value of multi-scale luminance comparison.

$c_{MSG}(F,M) = \prod_{i=1}^{N}[c(F_i,M_i)]^{w_i}$, representing the value of multi-scale contrast comparison.

$s_{MSeG}(F,M) = \prod_{i=1}^{N}[s_e(F_i,M_i)]^{w_i}$, representing the value of multi-scale structure comparison.

Where $F_i = F*G(x,y,t_i)$, $M_i = M*G(x,y,t_i)$, $i = 1,2,\ldots,N$, N is the number of scales, F_i and M_i represent the Gaussian scale space of LWIR image F and MWIR image M respectively. Here, $N = 5$, and the values of w_i $(i = 1,2,\ldots,5)$ are 0.0448, 0.2856, 0.3001, 0.2326, 0.1333 respectively.

3.2 Calculation of the Similarity Between Targets in VIS and LWIR Images

VIS and LWIR images contain rich edge information and complete geometric shapes [15]. However, camera motion causes changes of scale, translation and rotation between images. Hu invariant moment feature can describe the shape information of the image, and it is invariant to scale, translation and rotation [6]. Hence, it can serve as the effective feature of the target in VIS and LWIR images to realize target registration. Therefore, Hu invariant moment feature vector of 7 dimensions is extracted in this paper. The computational formula is as follows:

$$Hu_0 = \eta_{20} + \eta_{02}; \tag{2}$$

$$Hu_1 = (\eta_{20} - \eta_{02})^2 + 4\eta_{11}^2; \tag{3}$$

$$Hu_2 = (\eta_{30} - 3\eta_{12})^2 + (3\eta_{21} - \eta_{03})^2; \tag{4}$$

$$Hu_3 = (\eta_{30} + \eta_{12})^2 + (\eta_{21} + \eta_{03})^2; \tag{5}$$

$$\begin{aligned} Hu_4 = {} & (\eta_{30} - 3\eta_{21})(\eta_{30} + \eta_{12})[(\eta_{30} + \eta_{12})^2 - 3(\eta_{21} + \eta_{03})^2] \\ & + (3\eta_{21} - \eta_{03})(\eta_{21} + \eta_{03}) \cdot [3(\eta_{30} + \eta_{12})^2 - (\eta_{21} + \eta_{03})^2]; \end{aligned} \tag{6}$$

$$Hu_5 = (\eta_{20} - \eta_{02})[(\eta_{30} + \eta_{12})^2 - (\eta_{21} + \eta_{03})^2] + 4\eta_{11}(\eta_{30} + \eta_{12})(\eta_{21} + \eta_{03}); \tag{7}$$

$$\begin{aligned} Hu_6 = {} & (3\eta_{21} - \eta_{03})(\eta_{30} + \eta_{12})[(\eta_{30} + \eta_{12})^2 - 3(\eta_{21} + \eta_{03})^2] \\ & + (3\eta_{12} - \eta_{03})(\eta_{21} + \eta_{03}) \cdot [3(\eta_{30} + \eta_{12})^2 - (\eta_{21} + \eta_{03})^2]; \end{aligned} \tag{8}$$

Where η_{pq} (p, q = 0,1,2,3) is the normalized central moment of (p + q) order of the image.

After extracting the Hu invariant moment feature of the target in VIS and LWIR images respectively, cosine similarity [16] is used to measure the similarity between targets in VIS and LWIR images. The computational formula of cosine similarity is as follows:

$$S_{VF}(H_V, H_F) = \frac{\sum_i H_V(i) \cdot H_F(i)}{\sqrt{\sum_i (H_V(i))^2 \cdot \sum_i (H_F(i))^2}} \tag{9}$$

Where HV and H_F represent the Hu invariant moment feature vector of the target in VIS and LWIR images respectively; i represents the dimension of the Hu invariant moment feature vector, and i = 1,2,…,7.

3.3 Calculation of the Similarity Among Targets in VIS and Two Wave Bands Infrared Images

Based on the similarity among targets in images from different sources, target features in VIS, LWIR and MWIR images are fused to realize the registration of the same target in these images. Firstly, the similarity $S_{VF}(H_V, H_F)$ between targets in VIS and LWIR images is calculated by using the Hu invariant moment feature and cosine similarity; the similarity $S_{FM}(F, M)$ between targets in LWIR and MWIR images is calculated by using the improved structural similarity. Then, the similarity $S(V,F,M)$ among targets in VIS, LWIR and MWIR images is calculated by fusing $S_{FM}(F,M)$ and $S_{VF}(H_V,H_F)$ in the same proportion. The calculation formula is as follows:

$$S(V,F,M) = 0.5 \times S_{VF}(H_V,H_F) + 0.5 \times S_{FM}(F,M) \tag{10}$$

4 Realization of Target Registration of VIS and Two Wave Bands Infrared Images

The flow chart of target registration of VIS and two wave bands infrared images is shown in Fig. 2.

In Fig. 2, N_0 represents the number of targets in the VIS image; N_1 represents the number of targets in the LWIR image; N_2 represents the number of targets in the MWIR image; n_V represents the index of the target in the VIS image; n_M represents the index of the target in the MWIR image; n_F represents the index of the target in the LWIR image.

Realization steps:

(1) Target regions of VIS, LWIR and MWIR images are extracted.
(2) The similarity between targets in VIS and LWIR images is calculated by using Hu invariant moment feature and cosine similarity.
(3) The similarity between targets in LWIR and MWIR images is calculated by using the improved structural similarity.
(4) The similarity among targets in VIS, LWIR and MWIR images is calculated by fusing these two kinds of target similarity abovementioned.
(5) For the target n_F (the initial value of n_F is 1) in the LWIR image, the similarity between the target n_F and the target n_V (the initial value of n_V is 1) as well as the similarity between the target n_F and the target n_M (the initial value of n_M is 1) are calculated. Then, determine the similarity of this target image set is the maximum similarity of the target n_F or not. If yes, registration of the target n_F with targets in VIS and MWIR images is realized; if no, the value of n_M is increased by 1 and the process aforementioned is repeated. If the maximum similarity of the target n_F is not found after all targets in the MWIR image are tested, the value of n_V is increased by 1, the value of n_M is reset to 1 and the process aforementioned is repeated.
(6) Determine whether n_F is the terminate value or not. If yes, the program ends; if no, the value of n_F is increased by 1, the values of n_V and n_M is reset to 1, and the step (5) is repeated.

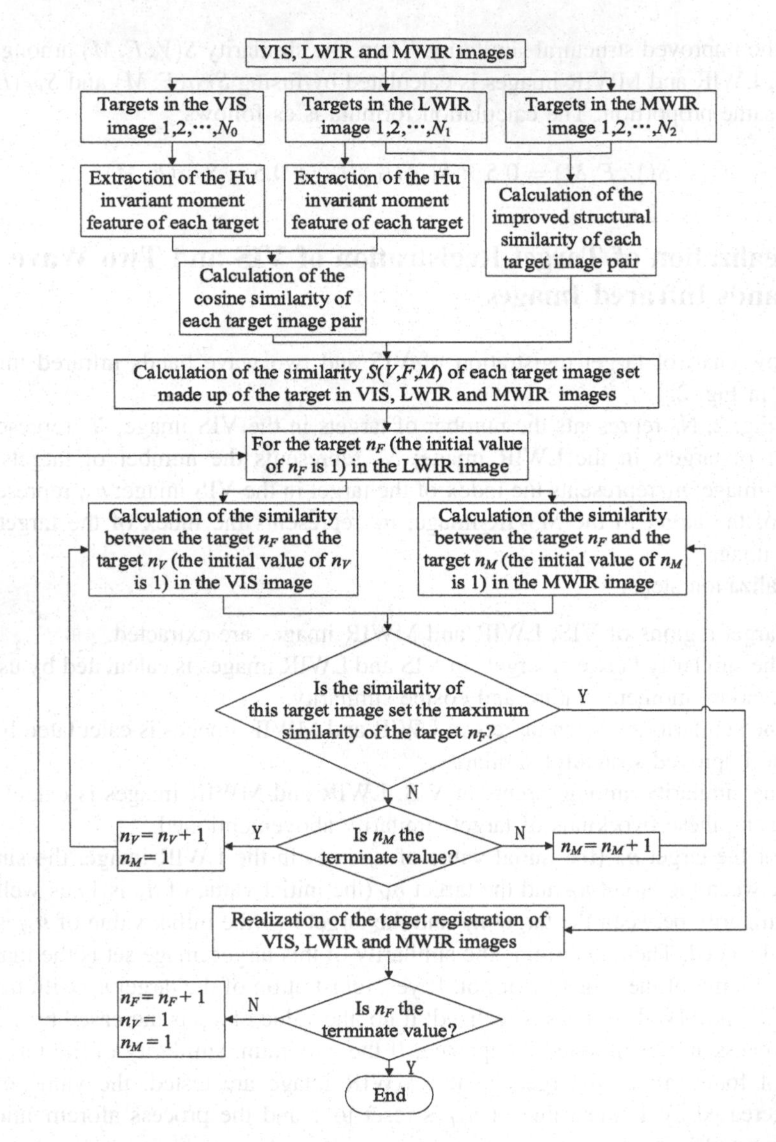

Fig. 2. The flow chart of target registration of VIS and two wave bands infrared images

5 Experimental Results and Discussion

The experimental images in this paper are VIS, LWIR and MWIR images of the same scene. With extensive research, no open data set was found to provide all these three types of source images for a same scene. Therefore, 9 sets of experimental images were taken in the field. VIS images were taken by a digital single lens reflex camera, and the resolution is 1920 × 1080 pixels; LWIR images were taken by a handheld infrared thermal imager with a single eyepiece (type specification: IR518, working band: 8–14 μm), and the resolution is 384 × 288 pixels; MWIR images were taken by a cooled thermal imager (type specification: IR300, working band: 3–5 μm), and the resolution is 320 × 256 pixels. Due to the long shooting distance, the targets occupy a small percentage of pixels in the entire images, which shows that the targets are small.

The proposed algorithm in this paper is named TRA_VFM algorithm. The proposed algorithm and the target registration algorithms of VIS and LWIR images mentioned in literatures [1–5] are applied to 9 sets of VIS, LWIR and MWIR images. The experimental purpose is to compare the accuracy of target registration of various registration algorithms, in order to verify the superiority of the proposed algorithm. Each image in experimental image sets 1–5 includes 3 targets while each image in experimental image sets 6–9 includes 4 targets. Images containing the same target with the same attitude in VIS, MWIR and LWIR images are considered as a set of target images, adding up to 31 sets of target images; images containing the same target with the same attitude in VIS and LWIR images are considered as a pair of target images, adding up to 31 pairs of target images.

All experiments were carried out on ASUS's G11 flight fortress desktop computer with the Intel Core i7–7700 processor and 8G memory. The operating system is Win7 and the experimental software isVS2010 + OpenCV2.4.6.

5.1 Target Registration Experiment of VIS and Two Wave Bands Infrared Images

The six registration algorithms are TRA_VFM algorithm, NorCroCor algorithm [1], NormMI algorithm [2], NormMI_Gradient algorithm [3], NorCovCor algorithm [4] and HOG algorithm [5] respectively. Experimental results are shown in Fig. 3, Fig. 4, and Fig. 5. Targets in boxes of the same color (red, green, yellow, or pink) represent the same target. Specifically, targets in red boxes, green boxes, yellow boxes and pink boxes are target 1, target 2, target 3 and target 4 respectively. For each image set processed by the TRA_VFM algorithm, the upper left image is the LWIR image, the upper right image is the MWIR image and the lower image is the VIS image. For each image pair processed by the other 5 algorithms, the upper image is the LWIR image and the lower image is the VIS image.

There are changes in image background, target attitude, shooting angle, etc. among the 9 image sets. Experimental results shown in Fig. 3, Fig. 4 and Fig. 5 indicate that the correct rate of target registration of the TRA_VFM algorithm is higher than that of the other 5 target registration algorithms. Moreover, the TRA_VFM algorithm realizes the registration of the target from three different image sources, which are VIS, LWIR and MWIR images. However, the other 5 algorithms can only realize target registration in VIS and LWIR images. For the TRA_VFM algorithm, target 1 and target 2 are registered erroneously in image set 2, target 1 is registered erroneously in image sets 3 and 5, and target 2 is registered erroneously in image set 6. For the NorCroCor algorithm, the NormMI algorithm and the HOG algorithm, some targets are registered erroneously in image sets 1 to 9. For the NormMI_Gradient algorithm, all targets are registered correctly in image sets 1, 3, and 9, but some targets are registered erroneously in the rest of the image sets. For the NorCovCor algorithm, all targets are

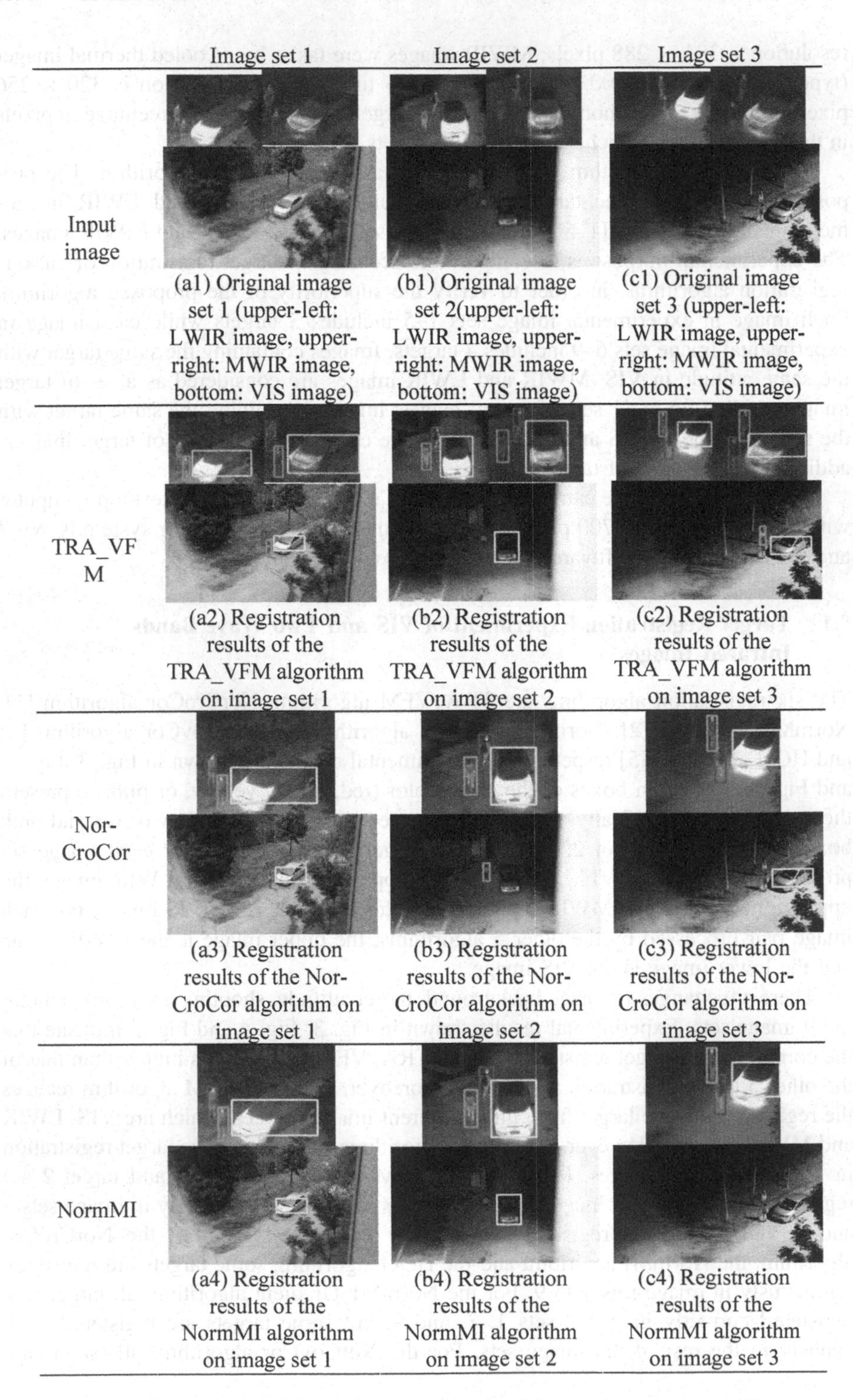

Fig. 3. Target registration results of six algorithms on image sets 1 to 3

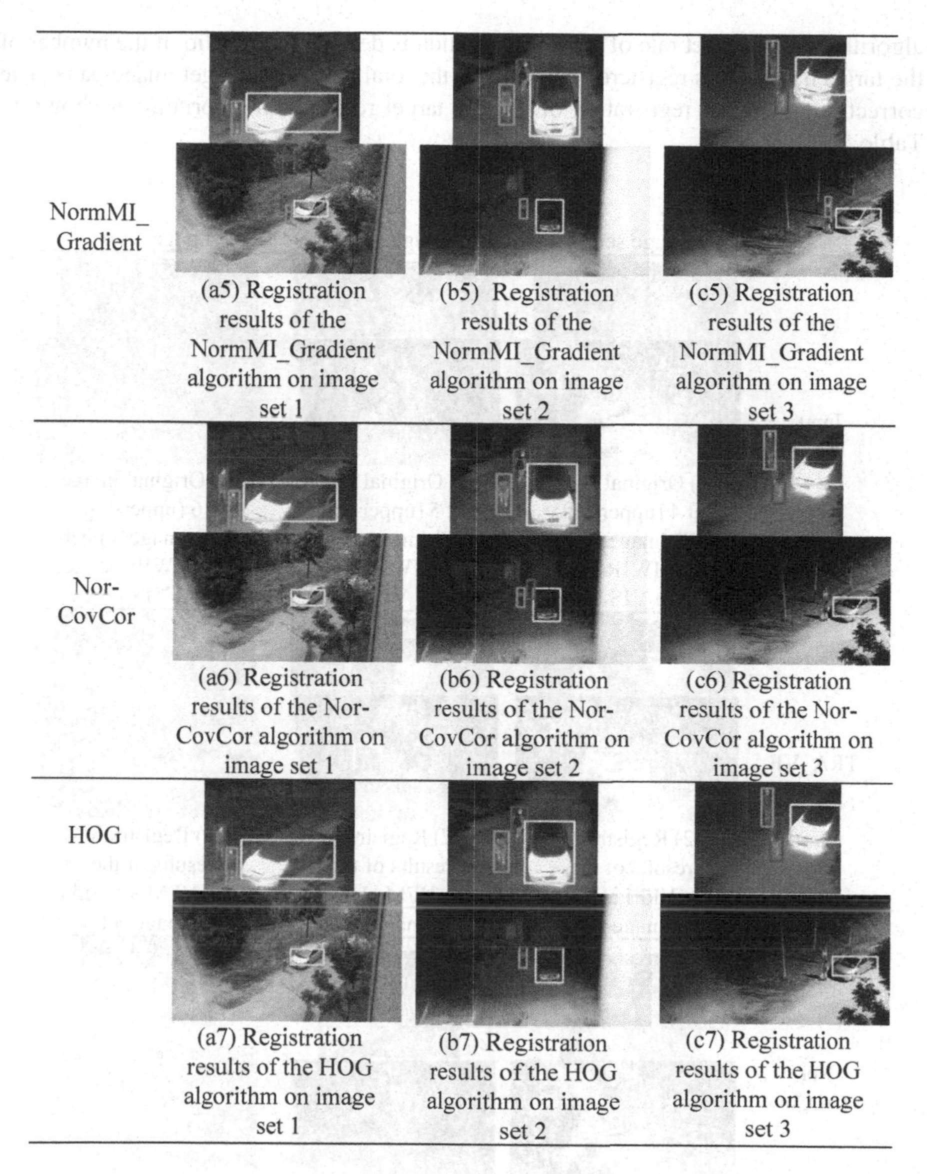

(a5) Registration results of the NormMI_Gradient algorithm on image set 1
(b5) Registration results of the NormMI_Gradient algorithm on image set 2
(c5) Registration results of the NormMI_Gradient algorithm on image set 3

(a6) Registration results of the Nor-CovCor algorithm on image set 1
(b6) Registration results of the Nor-CovCor algorithm on image set 2
(c6) Registration results of the Nor-CovCor algorithm on image set 3

(a7) Registration results of the HOG algorithm on image set 1
(b7) Registration results of the HOG algorithm on image set 2
(c7) Registration results of the HOG algorithm on image set 3

Fig. 3. (*continued*)

registered correctly in image set 2, but some targets are registered erroneously in the rest of the image sets.

In order to compare the correct rates of target registration of the 6 tested target registration algorithm quantitatively, an evaluation index is put forward in this paper: the correct rate of target registration. For the TRA_VFM algorithm, the correct rate of target registration is defined as the ratio of the number of the target image sets registered correctly to the total number of target image sets; for the other 5 target registration algorithms, the correct rate of target registration is defined as the ratio of the number of the target image pairs registered correctly to the total number of target image pairs. The correct rate of target registration of these 6 target registration algorithms is shown in Table 1.

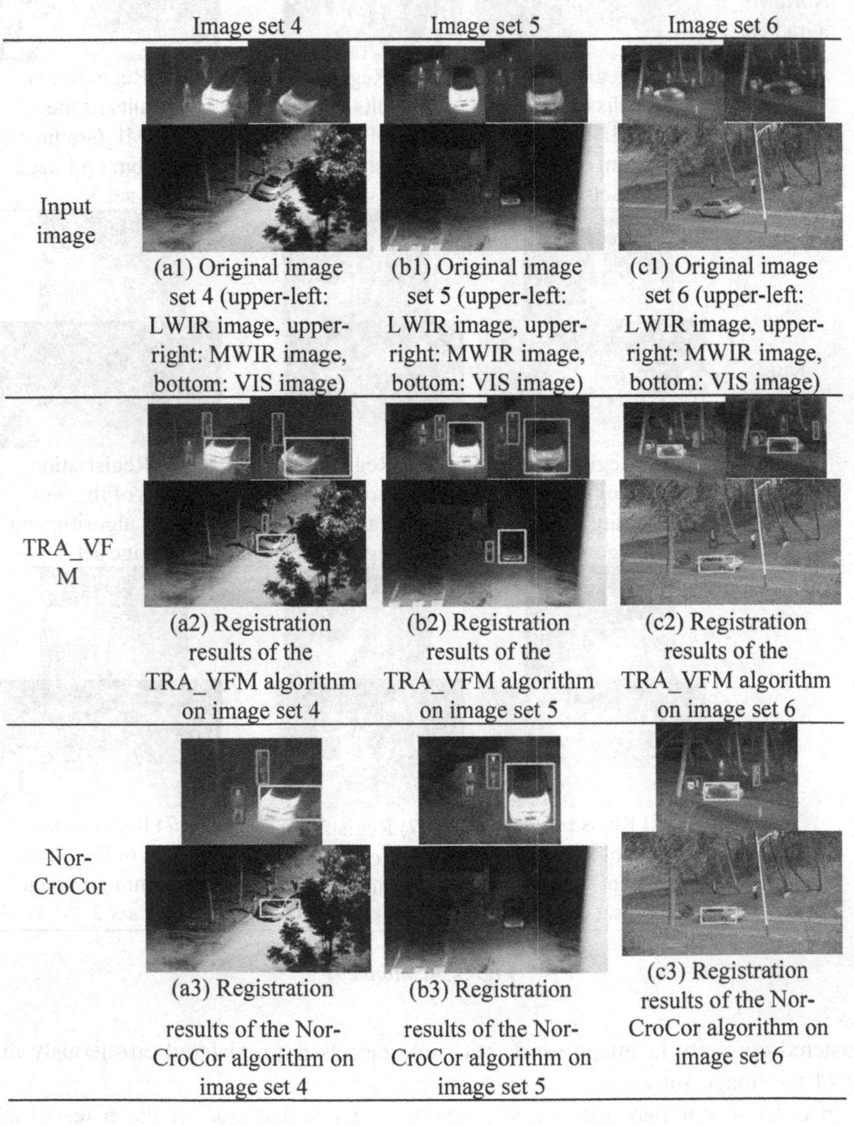

Fig. 4. Target registration results of six algorithms on image sets 4 to 6

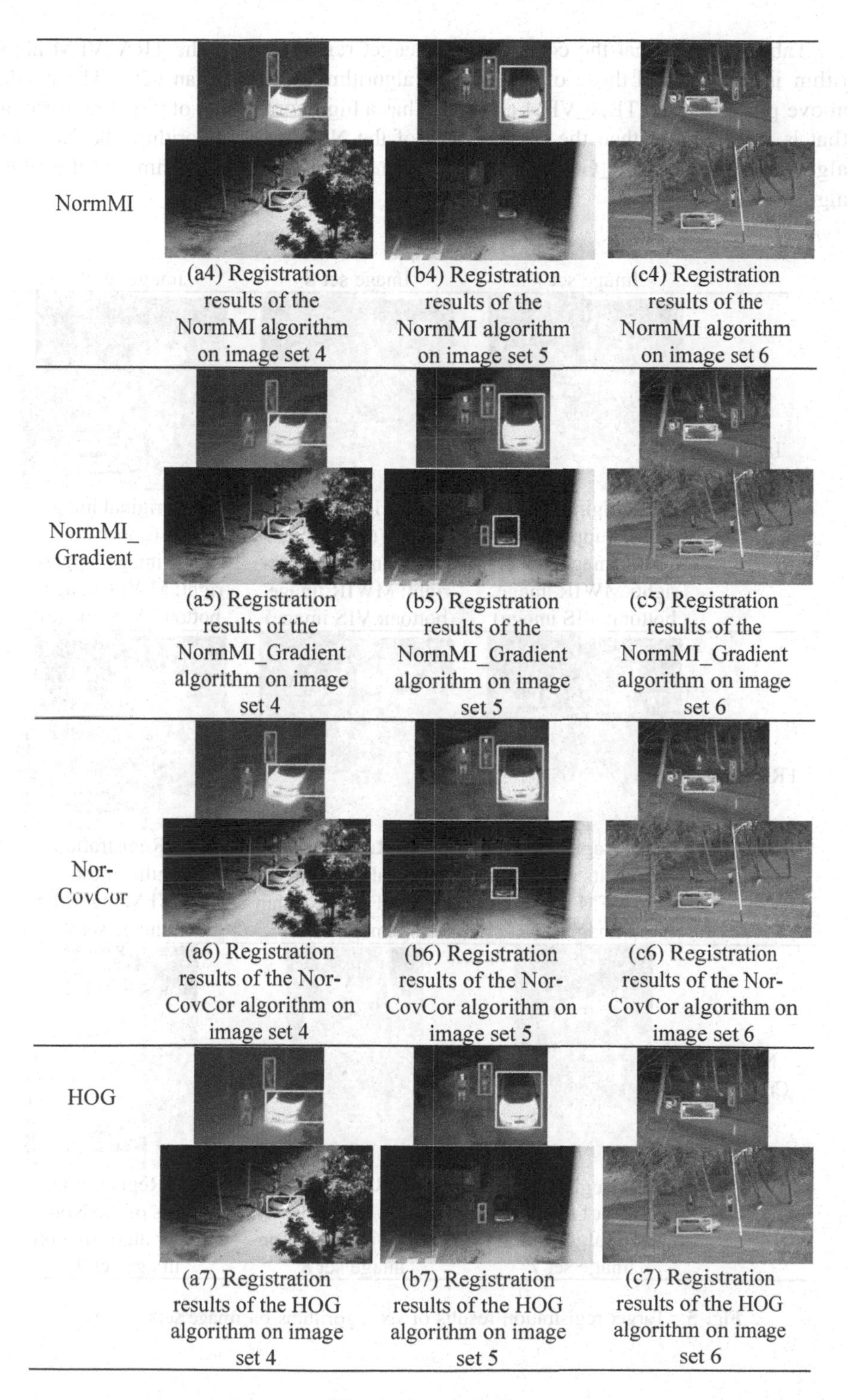
NormMI
NormMI_
Gradient
Nor-
CovCor
HOG
(a4) Registration results of the NormMI algorithm on image set 4
(b4) Registration results of the NormMI algorithm on image set 5
(c4) Registration results of the NormMI algorithm on image set 6
(a5) Registration results of the NormMI_Gradient algorithm on image set 4
(b5) Registration results of the NormMI_Gradient algorithm on image set 5
(c5) Registration results of the NormMI_Gradient algorithm on image set 6
(a6) Registration results of the Nor-CovCor algorithm on image set 4
(b6) Registration results of the Nor-CovCor algorithm on image set 5
(c6) Registration results of the Nor-CovCor algorithm on image set 6
(a7) Registration results of the HOG algorithm on image set 4
(b7) Registration results of the HOG algorithm on image set 5
(c7) Registration results of the HOG algorithm on image set 6

Fig. 4. *(continued)*

Table 1 shows that the correct rate of target registration of the TRA_VFM algorithm is 83.87% and those of the other 5 algorithms are less than 62%. The results above prove that the TRA_VFM algorithm has a high correct rate of target registration that is much higher than the correct rates of the NorCroCor algorithm, the NormMI algorithm, the NormMI_Gradient algorithm, the NorCovCor algorithm and the HOG algorithm.

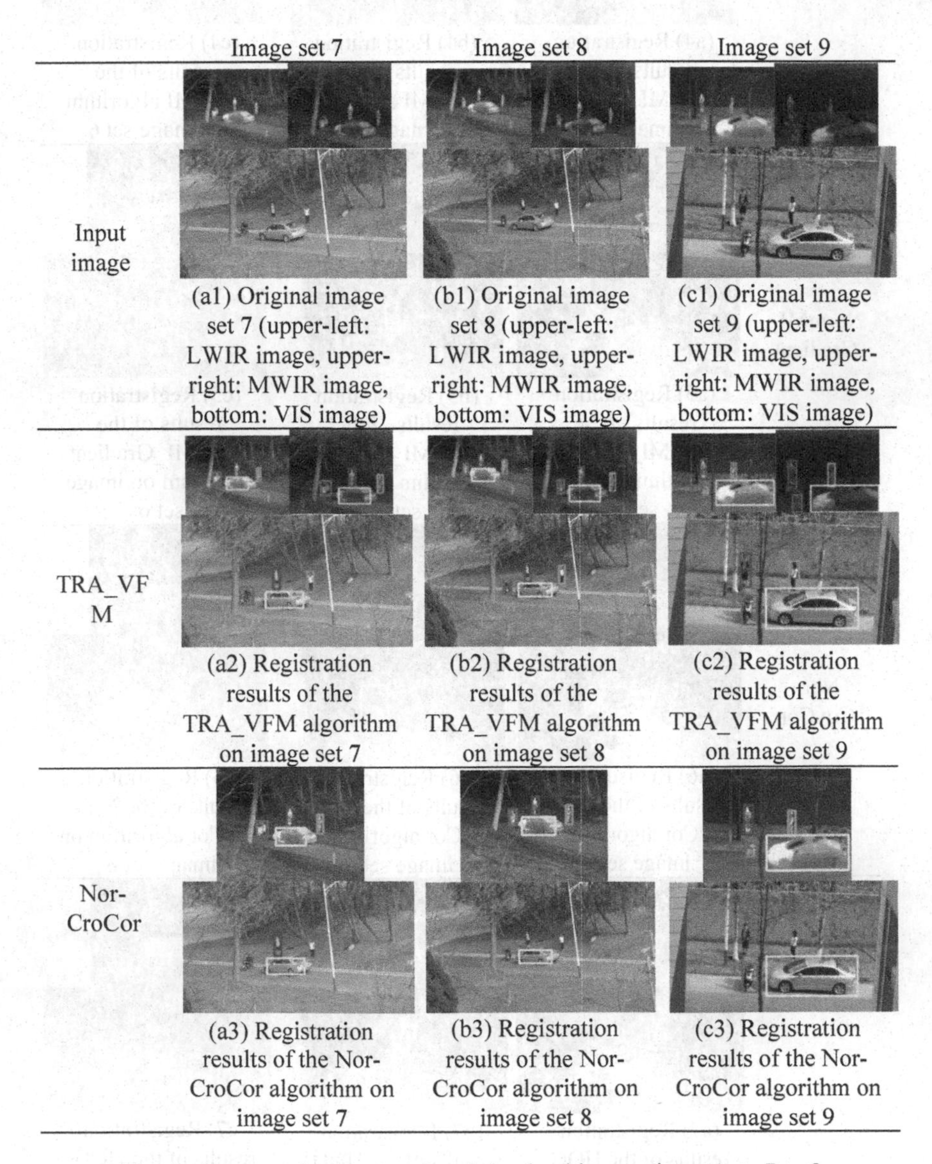

Fig. 5. Target registration results of six algorithms on image sets 7 to 9

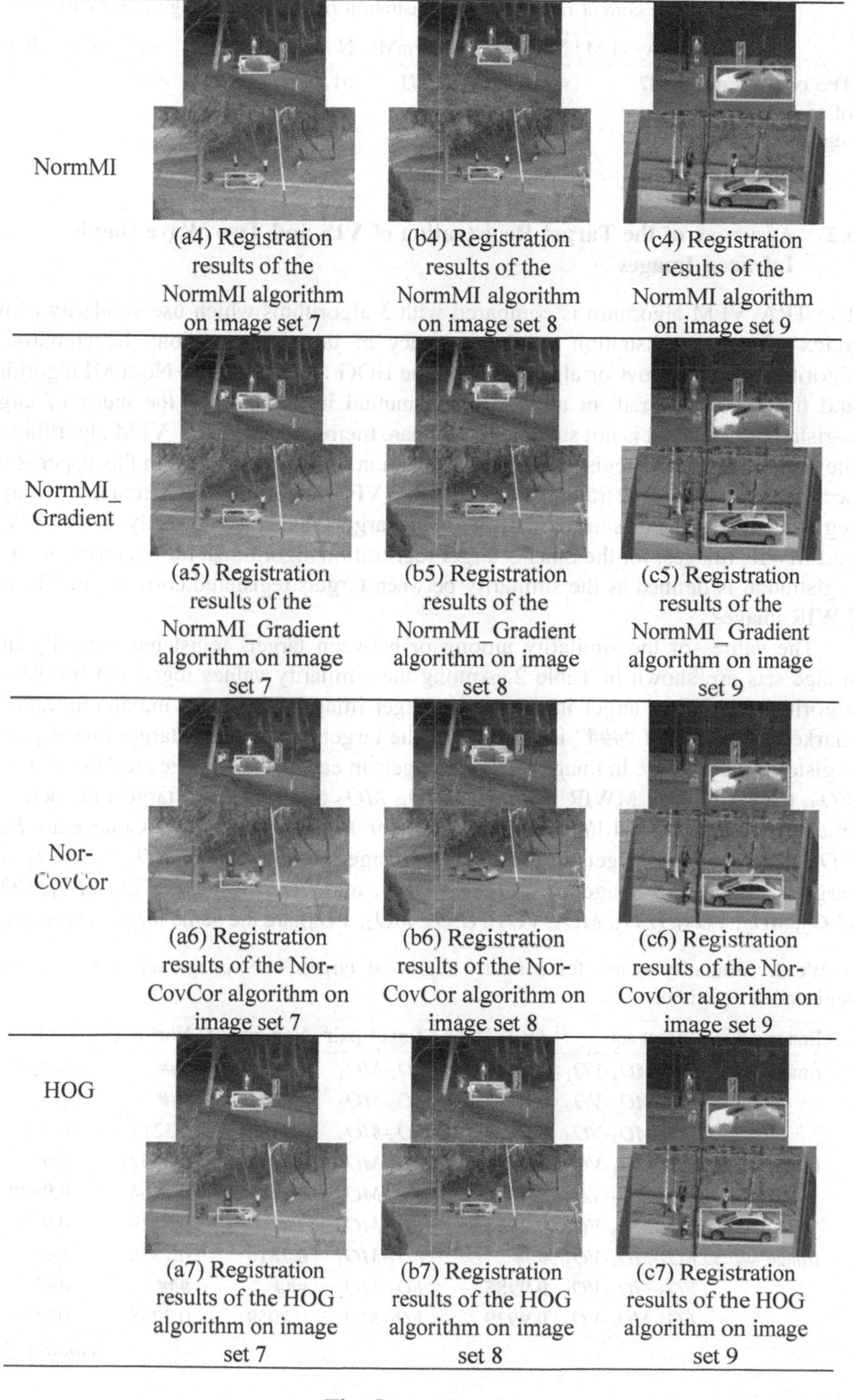
NormMI
(a4) Registration results of the NormMI algorithm on image set 7
(b4) Registration results of the NormMI algorithm on image set 8
(c4) Registration results of the NormMI algorithm on image set 9
NormMI_ Gradient
(a5) Registration results of the NormMI_Gradient algorithm on image set 7
(b5) Registration results of the NormMI_Gradient algorithm on image set 8
(c5) Registration results of the NormMI_Gradient algorithm on image set 9
Nor-CovCor
(a6) Registration results of the Nor-CovCor algorithm on image set 7
(b6) Registration results of the Nor-CovCor algorithm on image set 8
(c6) Registration results of the Nor-CovCor algorithm on image set 9
HOG
(a7) Registration results of the HOG algorithm on image set 7
(b7) Registration results of the HOG algorithm on image set 8
(c7) Registration results of the HOG algorithm on image set 9

Fig. 5. (*continued*)

Table 1. The correct rate of target registration of the 6 tested algorithms (%)

	TRA_VFM	NorCroCor	NormMI	NormMI_Gradient	NorCovCor	HOG
The correct rate of target registration	**83.87**	45.16	38.71	61.29	41.94	61.29

5.2 Accuracy of the Target Registration of VIS and Two Wave Bands Infrared Images

The TRA_VFM algorithm is compared with 3 algorithms which use similarity as the index of target registration on the accuracy of target registration: the NorCroCor algorithm, the NorCovCor algorithm and the HOG algorithm. The NormMI algorithm and the NormMI_Gradient algorithm use mutual information as the index of target registration. Hence it is not suitable to compare them with the TRA_VFM algorithm on the accuracy of target registration. The evaluation index put forward in this paper is the accuracy of target registration. For the TRA_VFM algorithm, the accuracy of target registration is defined as the similarity among targets registered correctly in VIS, LWIR and MWIR images; for the other 3 target registration algorithms, the accuracy of target registration is defined as the similarity between targets registered correctly in VIS and LWIR images.

The values of the similarity among or between targets registered correctly in 9 image sets are shown in Table 2. Among the similarity values registered by these 4 algorithms in every target image set or target image pair, every maximum value is marked in bold, and "###" indicates that the target image set or target image pair is registered incorrectly. In image sets 1–5, targets in each LWIR image are FO_1, FO_2 and FO_3, targets in each MWIR image are MO_1, MO_2 and MO_3, and targets in each VIS image areVO_1, VO_2 and VO_3. In image sets 6–9, targets in each LWIR image are FO_1, FO_2, FO_3 and FO_4, targets in each MWIR image are MO_1, MO_2, MO_3 and MO_4, and targets in each VIS image are VO_1, VO_2, VO_3 and VO_4. Besides, (FO_1, MO_1, VO_1), (FO_2, MO_2, VO_2), (FO_3, MO_3, VO_3), (FO_4, MO_4, VO_4) are the same target respectively.

Table 2. Similarity values for correctly registered targets in 9 image sets processed by 6 registration algorithms

Image set	Target set	TRA_VFM	Target pair	NorCroCor	NorCovCor	HOG
Image set 1	FO_1-MO_1-VO_1	**0.9995**	FO_1-MO_1	###	###	0.9882
	FO_2-MO_2-VO_2	**0.9997**	FO_2-MO_2	###	###	###
	FO_3-MO_3-VO_3	**0.9950**	FO_3-MO_3	0.8163	0.3213	0.9832
Image set 2	FO_1-MO_1-VO_1	###	FO_1-MO_1	**0.8542**	0.2552	###
	FO_2-MO_2-VO_2	###	FO_2-MO_2	0.8685	0.2464	**0.9939**
	FO_3-MO_3-VO_3	**0.9660**	FO_3-MO_3	###	0.5129	0.9163
Image set 3	FO_1-MO_1-VO_1	###	FO_1-MO_1	**0.7816**	0.1338	###
	FO_2-MO_2-VO_2	**0.9985**	FO_2-MO_2	###	###	###
	FO_3-MO_3-VO_3	**0.9929**	FO_3-MO_3	0.7059	0.2248	0.9658

(*continued*)

Table 2. (*continued*)

Image set	Target set	TRA_VFM	Target pair	NorCroCor	NorCovCor	HOG
Image set 4	FO_1-MO_1-VO_1	**0.9970**	FO_1-MO_1	0.8007	###	0.9943
	FO_2-MO_2-VO_2	**0.9984**	FO_2-MO_2	###	###	###
	FO_3-MO_3-VO_3	**0.9983**	FO_3-MO_3	0.6827	0.0219	0.9747
Image set 5	FO_1-MO_1-VO_1	###	FO_1-MO_1	**0.8305**	0.1895	###
	FO_2-MO_2-VO_2	**0.9923**	FO_2-MO_2	###	###	0.9922
	FO_3-MO_3-VO_3	**0.9681**	FO_3-MO_3	###	0.4501	###
Image set 6	FO_1-MO_1-VO_1	**0.9931**	FO_1-MO_1	###	###	0.9875
	FO_2-MO_2-VO_2	###	FO_2-MO_2	###	###	**0.9646**
	FO_3-MO_3-VO_3	**0.9894**	FO_3-MO_3	###	###	###
	FO_4-MO_4-VO_4	**0.9776**	FO_4-MO_4	0.8960	0.2061	0.9744
Image set 7	FO_1-MO_1-VO_1	**0.9936**	FO_1-MO_1	0.8879	0.4441	0.9863
	FO_2-MO_2-VO_2	**0.9559**	FO_2-MO_2	###	###	###
	FO_3-MO_3-VO_3	0.9744	FO_3-MO_3	###	###	**0.9764**
	FO_4-MO_4-VO_4	0.9695	FO_4-MO_4	0.9119	0.2311	**0.9799**
Image set 8	FO_1-MO_1-VO_1	**0.9939**	FO_1-MO_1	###	###	###
	FO_2-MO_2-VO_2	**0.9803**	FO_2-MO_2	###	###	0.9594
	FO_3-MO_3-VO_3	**0.9899**	FO_3-MO_3	###	###	###
	FO_4-MO_4-VO_4	0.9676	FO_4-MO_4	0.8925	###	**0.9967**
Image set 9	FO_1-MO_1-VO_1	**0.9882**	FO_1-MO_1	###	###	###
	FO_2-MO_2-VO_2	**0.9829**	FO_2-MO_2	###	###	0.9820
	FO_3-MO_3-VO_3	**0.9929**	FO_3-MO_3	0.8877	###	0.9909
	FO_4-MO_4-VO_4	**0.9960**	FO_4-MO_4	0.8702	0.1502	0.9889

It is shown in Table 2 that, in all 31 target sets (target pairs), the TRA_VFM algorithm has the maximum similarity value in 23 target sets; the HOG algorithm has the maximum similarity value in 5 target pairs; the NorCroCor algorithm has the maximum similarity value in 3 target pairs; and the NorCovCor algorithm has no maximum similarity value in any target sets. Overall, the accuracy of target registration of the TRA_VFM algorithm is higher than 0.95. It is demonstrated that the accuracy of target registration of the TRA_VFM algorithm is higher than that of the HOG algorithm, the NorCroCor algorithm and the NorCovCor algorithm.

5.3 Analysis of the Target Registration of VIS and Two Wave Bands Infrared Images

Based on the principle that the same target has the highest similarity among targets in images from different sources, the TRA_VFM algorithm fuses target features in VIS, LWIR and MWIR images by calculating the targets similarity, which improves the registration accuracy of the same target in these images. VIS and LWIR images contain rich edge information and complete geometric shapes information, which can be described by the Hu invariant moment feature. The effect of changes of scale, translation and rotation between images caused by camera motion can be mitigated because

Hu invariant moment feature is invariant to scale, translation and rotation. Hence, the TRA_VFM algorithm calculates the similarity between targets in VIS and LWIR images based on the Hu invariant moment feature and cosine similarity. LWIR and MWIR images contain rich edge information and complete structure information. The improved structural similarity uses images of different scales and their edge images in the Gaussian space to calculate the multi-scale structural similarity, thus it can combine the complete edge structure information of targets in LWIR and MWIR images. Hence, the TRA_VFM algorithm calculates the similarity between targets in LWIR and MWIR images based on the improved structural similarity. The TRA_VFM algorithm calculates the similarity among targets in VIS, LWIR and MWIR images by fusing the two similarities, which realizes the target registration of images of these three sources. Based on the similarity among targets in images from different sources, the TRA_VFM algorithm fuses shape features and structural features of targets in VIS, MWIR and LWIR images. The multi-source information such as edge information, shape information and structural information in these three-source images is fully utilized, which improves the correct rate and the accuracy of target registration. For the proposed TRA_VFM algorithm, the correct rate of target registration is 83.87%, and the accuracy of target registration is higher than 0.95. Both of them are superior to other algorithms. Compared with our algorithm, the investigated references methods only utilize the target information in VIS and LWIR images, lacking of the rich edge information and complete structural information in MWIR images, and comprehensive information of the target is unavailable. Therefore, the target registration result is not as good as our algorithm.

6 Conclusion

In this paper, the shape and structural features of targets in VIS, LWIR and MWIR images are fused based on the similarity among targets in images from different sources, which improves the registration accuracy of the same target in these images. The proposed algorithm and 5 registration algorithms with good performances are tested for target registration performance on 9 sets of VIS, LWIR and MWIR images. Experimental results show that the correct rate and the accuracy of target registration of the proposed algorithm are both superior than those of the other algorithms. In this paper, the target registration of VIS, LWIR and MWIR images is realized, and the correct rate and accuracy of target registration are improved by the registration of multi-source images. In the future, features that better represent target information will be researched in order to realize target registration of images from more than three sources.

Acknowledgments. This work was supported by the National Natural Science Foundation of China (61471194 and 61705104), the Fundamental Research Funds for the Central Universities (NJ2020021), the Natural Science Foundation of Jiangsu Province (BK20170804), National Defense Science and Technology Special Innovation Zone Project.

Conflicts of Interest. None.

References

1. Ma, J., Cao, G.Z.: Image matching technology of infrared and visible images based on edge information. Comput. Digit. Eng. **34**, 30–32 (2006)
2. Jiang, J., Zhang, X.S.: Visible and infrared image automatic registration algorithm using mutual information. In: Control and Decision Conference, pp. 1322–1325. IEEE (2010)
3. Mao, L.L., Xu, G.F., Chen, X.B.: Research on multi-source image matching based on fusion of mutual information and gradient. Infrared Technol. **31**(9), 532–536 (2009)
4. Li, Z.: Research on the key technology of heterogeneous image matching, Master Thesis, National University of Defense Technology (September 2011)
5. Bilodeau, G.A., Torabi, A., Charles, P.L., Riahi, D.: Thermal-visible registration of human silhouettes: A similarity measure performance evaluation. Infrared Phys. Technol. **64**, 79–86 (2014)
6. An, M., Jiang, Z., Zhao, D.: High speed robust image registration and localization using optimized algorithm and its performances evaluation. J. Syst. Eng. Electronics **21**(3), 520–526 (2010)
7. Ma, J.Y., Zhao, J., Ma, Y.: Non-rigid visible and infrared face registration via regularized Gaussian fields criterion. Pattern Recogn. **48**(3), 772–784 (2015)
8. Zhuang, Y., Gao, K., Miu, X.: Infrared and visual image registration based on mutual information with a combined particle swarm optimization–Powell search algorithm. Optik-Int. J. Light Electron Optics **127**(1), 188–191 (2016)
9. Dou, J.F., Li, J.X.: Automatic registration of visible and infrared images based on corner and Hu moment invariants. Infrared **32**(7), 23–27 (2011)
10. Wen, X., Xie, T.X., Yan, J.H., Zhang, Y., Huang, W., Chen, X.: Target registration for two wave bands infrared images based on the improved structural similarity. Chinese J. Sci. Instrum. **38**(12), 3112–3120 (2017)
11. Wu, F., Wang, B., Yi, X.: Visible and infrared image registration based on visual salient features. J. Electron. Imaging **24**(5), 053017 (2015)
12. St-Charles, P.L., Bilodeau, G.A., Bergevin, R.: Online multimodal video registration based on shape matching. In: Proceedings of the IEEE Conference on Computer Vision and Pattern Recognition Workshops (2015)
13. Chen, Y., Zhang, X., Zhang, Y., Maybank, S.J., Fu, Z.: Visible and infrared image registration based on region features and edginess. Mach. Vis. Appl. **29**(1), 113–123 (2017). https://doi.org/10.1007/s00138-017-0879-6
14. Zhang, T., Liu, L., Gao, T.C., Hu, S.: The actuality and progress of whole sky infrared cloud remote sensing techniques. Instrumentation **2**(3), 65–74 (2015)
15. Ma, J.Y., Zhao, J., Ma, Y., Tian, J.W.: Non-rigid visible and infrared face registration via regularized Gaussian fields criterion. Pattern Recogn. **48**(3), 772–784 (2015)
16. Liu, B., Li, W.S.: Indoor location method of fingerprint matching algorithm based on cosine similarity. Bull. Sci. Technol. **33**(3), 198–202 (2017)

Deep Learning Based Target Activity Recognition Using FMCW Radar

Bo Li[1], Xiaotian Yu[2(✉)], Fan Li[3], and Qiming Guo[2]

[1] School of Information Science and Engineering, Dalian Polytechnic University, Dalian 116034, China
libo_15@dlpu.edu.cn

[2] School of Information Science and Technology, Dalian Maritime University, Dalian 116026, China
{yuxiaotian,gqm}@dlmu.edu.cn

[3] Faculty of Electronic Information and Electrical Engineering, Dalian University of Technology, Dalian 116023, China
lifanfl@mail.dlut.edu.cn

Abstract. Target activity recognition has many potential applications in the fields of human-computer interaction, smart environment, smart system, *etc.* Recent years, due to the miniaturized design of the frequency modulated continuous wave (FMCW) radar, it has been widely utilized to realize target activity recognition in our daily life. However, the activity recognition accuracy is usually not high due to the surrounding noise and variation of the activity. To realize high accuracy activity recognition, one feasible way is to extract discriminative features from the weak radar signals reflected by the activity. Inspired by the successful application of deep learning in computer vision, in this paper, we try to explore leveraging deep learning to solve the target activity recognition task. Specifically, based on the characteristics of the FMCW signals, we design the Doppler radio images suitable for the deep network to deal with. Then, we develop a deep convolutional network to extract discriminative activity features from the Doppler radio images. Finally, we feed the features into a Softmax classifier to recognize the activity. We carry out extensive experiments on a 77 GHz FMCW radar testbed. The experimental results show the excellent target activity recognition performance.

Keywords: Deep learning · Activity recognition · FMCW

1 Introduction

Target activity recognition has many potential applications in our daily life. Due to the availability of pervasive wireless signals, it becomes popular to sense our world by leveraging wireless signals [1–6]. Therefore, target activity recognition using wireless signals has drawn considerable attention in recent years. Compared with traditional target activity recognition techniques, such as vision or wearable

S. Shi et al. (Eds.): AICON 2020, LNICST 356, pp. 484–490, 2021.
https://doi.org/10.1007/978-3-030-69066-3_42

devices based techniques, target activity recognition using wireless signals does not need the target equipped with any devices, could work under dark or smoky conditions, does not concern privacy disclosure, *etc.* These advantages make it an ideal technique to realize human-computer interaction, smart environment, smart system, *etc.*

Researchers have conducted valuable exploration on this technique. Kim and Moon [7] leverage the micro-Doppler signatures to recognize the activity of a person. Wang *et al.* [8–10] explore how to recognize activity gesture under cross-scenario conditions. Gao *et al.* [11] extract the coherence histogram features to characterize different activities. Ma *et al.* [12] solve the activity recognition problem under small sample set. Huang and Dai [13] leverage the link quality as metric to realize activity recognition. However, due to the surrounding noise and variation of the activity, the activity recognition accuracy is still not very high.

To realize high accuracy activity recognition, one feasible way is to extract discriminative features from the weak radar signals reflected by the activity. Inspired by the successful application of deep learning in computer vision, in this paper, we try to explore leveraging deep learning to solve the target activity recognition task. The main contributions of this paper can be summarized as follows:

1. We design a deep network based target activity recognition architecture, which could realize high performance activity recognition.
2. We design the Doppler radio images as the input to the deep network, and develop a deep convolutional network to extract discriminative activity features from the radio images.
3. We develop a 77 GHz frequency modulated continuous wave (FMCW) hardware based prototype system, and carry out extensive evaluations to evaluate the proposed deep learning based target activity recognition method.

The rest of the paper is structured as follows. Section 2 presents the architecture of proposed system. Section 3 introduces the detailed implementation of the proposed system, presents the Doppler radio image construction method, and gives the deep network architecture. Section 4 presents the experimental evaluation. Finally, the conclusion is drawn in Sect. 5.

2 System Architecture

The developed deep learning based target activity recognition system is shown in Fig. 1. It mainly consists of three function modules, *i.e.*, Doppler radio image construction module, deep convolutional network module, and softmax classifier module. The Doppler radio image construction module acquires the reflection wireless signals from the target, build the Doppler radio image of the target, so as to provide measurement information for realizing the activity recognition task. The deep convolutional network module tries to extract discriminative activity features from the Doppler radio images by performing a series of convolution, pooling, and ReLU operations. The softmax classifier module recognizes the target activity by projecting the activity features to the class space. We will present the detailed implementation of each module in the next section.

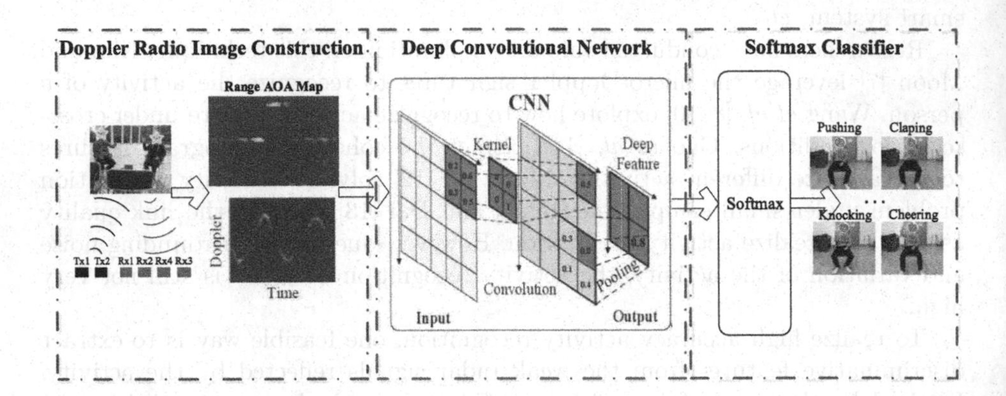

Fig. 1. Architecture of the deep learning based target activity recognition system.

3 System Implementation

3.1 Doppler Radio Image Construction

The FMCW radar transmits a frequency modulated wireless signal to the target, and then receives the reflection signal from the target. When the target locates at different distances with the receiver or performs different activities, the reflection signals will be different. The distance can be identified by the frequency of the reflection signal, and the activities determines the Doppler of the reflection signal. We firstly detect the target in a polar coordinate system by building the Range angle of arrival (AOA) map (RAM). We estimate the range and AOA of the target by performing 2-D Fast Fourier Transform (FFT) Algorithm along the fast time axis and receiver axis, respectively, as follows

$$\mathbf{RAM} = \mathbf{2D} - \mathbf{FFT}\left(\mathbf{S}_{1,1}, \ldots, \mathbf{S}_{i,j}, \ldots, \mathbf{S}_{I,J}\right), \tag{1}$$

where I and J denote the total number of samples on the fast time axis and the number of receivers, respectively. As shown in Fig. 1, we can detect the two targets clearly from the RAM.

With the range and AOA information of the target, we focus on the target by firstly performing beamforming on the expected target. Then, we perform 2-D FFT along the fast time axis and slow time axis to get the range and Doppler information, respectively. As we have know the range of the expected target, we can further filter out the noise by using a narrow range filter. Finally, we integrate the Doppler information within a narrow range and get the expected

Doppler radio image, as shown in Fig. 1. The Doppler radio image characterizes the Doppler information change over time, which depicts the movement of the activity over time. We will leverage it as the measurement information to realize target activity recognition.

3.2 Deep Convolutional Network Based Feature Extraction

Deep Network has been widely utilized to realize many computer vision tasks in recent years. It has a powerful ability to learn latent discriminative features from the data. There are many types of deep networks, such as deep fully connected networks, deep convolutional neural networks, deep recurrent neural networks, generative adversarial networks, *etc.* Different types of deep networks are suitable for different types of tasks. As for the target activity recognition task, deep fully connected networks and deep convolutional neural networks are ideal choice due to their excellent ability of extracting discriminative features. Meanwhile, since of use 2-D Doppler radio images as the measurement, thus, the 2-D deep convolutional neural network is the best choice for the target activity recognition task. Therefore, we leverage it in this paper.

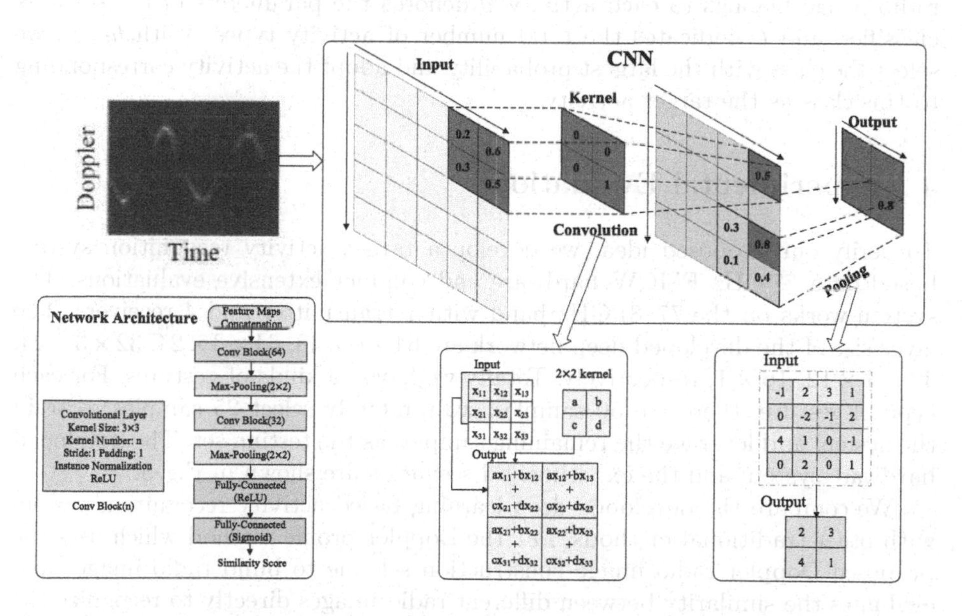

Fig. 2. Working principle implementation of the deep convolutional network.

The working principle and implementation of the developed deep convolutional network is illustrated in Fig. 2. The developed deep convolutional network is made up of a series of convolution operations, ReLU nonlinear operations, and pooling operations. The convolution operation perform convolution on the kernel and input image, and the ReLU operation adds nonlinear information into the network, they jointly capture the informative information of the image. The pooling operation downsampling the radio image. The above three key operations execute in turn for many times in the network. For simplicity, we term a convolution operation and a ReLU operation as a Conv Block. The developed deep convolutional network is made up of two Conv Blocks, two max pooling layers, two fully connected layers, as shown in Fig. 2.

3.3 Softmax Classifier

With the extracted deep features x, we feed it into a softmax classifier to recognize the target activity y as follows

$$h_\theta(x) = \begin{bmatrix} p(y=1|x;\theta) \\ p(y=2|x;\theta) \\ \vdots \\ p(y=C|x;\theta) \end{bmatrix}, \tag{2}$$

where $h_\theta(x)$ is a $C \times 1$ vector which indicates the probabilities that the input radio image belongs to each activity, θ denotes the parameters of the softmax classifier, and C indicates the total number of activity types. With $h_\theta(x)$, we select the class with the largest probability and adopt the activity corresponding to this class as the target activity.

4 Experimental Evaluation

To verify our proposed idea, we develop a target activity recognition system based on a 77 GHz FMCW hardware and conduct extensive evaluations. The system works on the 77–81 GHz band with 1 transmitter and 4 receivers. The layer size of the developed deep network are $64 \times 6 \times 48$, $64 \times 3 \times 24$, $32 \times 3 \times 24$, $32 \times 1 \times 12$, 16×1, respectively. Totally we have 16 kinds of gestures. For each type of gesture, there are 50 samples, we randomly select 25 samples to train the system and leverage the remaining samples as the testing set. The developed hardware system and the experimental scenarios are shown in Fig. 3.

We compare the developed deep learning based activity recognition system with other traditional methods, *i.e.*, the Doppler profile method which uses the proposed Doppler radio image construction scheme to build radio images and evaluates the similarity between different radio images directly to recognize the activity, the raw Doppler method which use the Doppler information without any range and AOA filtering operation. The results are summarized in Table 1. From the results, we can discover that our proposed deep learning based method achieves the best performance, which confirms the effectiveness of the developed deep network. Meanwhile, we also discover that the accuracy of Doppler profile method is better than the traditional raw Doppler method as well, which confirms that the developed Doppler radio image construction method is valid.

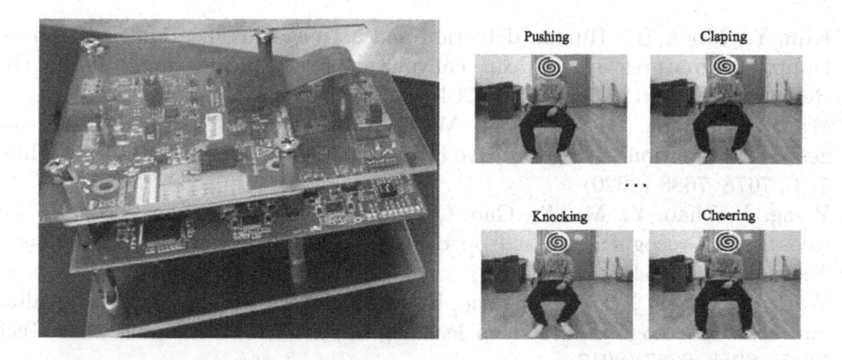

Fig. 3. Developed hardware system and the experimental scenarios.

Table 1. Activity recognition accuracy with different methods.

	Deep Learning	Doppler Profile	Raw Doppler
Accuracy(%)	95.8	91.5	86.5

5 Conclusion

In this paper, we develop a novel deep learning based target activity recognition system using FMCW hardware. We design a scheme to build the informative Doppler radio images, and develop a deep convolutional network to extract discriminative activity features from the Doppler radio images. These strategies guarantee a good activity recognition accuracy. Experiment conducted on a 77 GHz FMCW radar indicates that the developed system could achieve an accuracy of 95.8% when there are 16 types of activities.

References

1. Li, C., Lubecke, V.M., Boric-Lubecke, O., Lin, J.: A review on recent advances in Doppler radar sensors for noncontact healthcare monitoring. IEEE Trans. Microwave Theory Technol. **61**(5), 2046–2060 (2013)
2. Savazzi, S., Sigg, S., Nicoli, M., Rampa, V., Kianoush, S., Spagnolini, U.: Device-free radio vision for assisted living: Leveraging wireless channel quality information for human sensing. IEEE Signal Process. Mag. **33**(2), 45–58 (2016)
3. Chen, X., Chen, L., Feng, C., Fang, D., Xiong, J., Wang, Z.: Sensing our world using wireless signals. IEEE Internet Comput. **23**(3), 38–45 (2019)
4. Wang, J., Gao, Q., Pan, M., Fang, Y.: Device-free wireless sensing: challenges, opportunities, and applications. IEEE Network **32**(2), 132–137 (2018)
5. Wang, J., Gao, Q., Ma, X., Zhao, Y., Fang, Y.: Learning to sense: Deep learning for wireless sensing with less training efforts. IEEE Wirel. Commun. **27**(3), 156–162 (2020)
6. Li, X., He, Y., Jing, X.: A survey of deep learning-based human activity recognition in radar. Remote Sensing **11**(9), 1–22 (2019)

7. Kim, Y., Moon, T.: Human detection and activity classification based on micro-Doppler signatures using deep convolutional neural networks. IEEE Geosci. Remote Sens. Lett. **13**(1), 8–12 (2016)
8. Wang, J., Zhang, L., Wang, C., Ma, X., Gao, Q., Lin, B.: Device-free human gesture recognition with generative adversarial networks. IEEE Internet Things J. **7**(8), 7678–7688 (2020)
9. Wang, J., Zhao, Y., Ma, X., Gao, Q., Pan, M., Wang, H.: Cross-scenario device-free activity recognition based on deep adversarial networks. IEEE Trans. Veh. Technol. **69**(5), 5416–5425 (2020)
10. Wang, J., Zhang, X., Gao, Q., Yue, H., Wang, H.: Device-free wireless localization and activity recognition: a deep learning approach. IEEE Trans. Veh. Technol. **66**(7), 6258–6267 (2017)
11. Gao, Q., Wang, J., Zhang, L., Yue, H., Lin, B., Wang, H.: Device-free activity recognition based on coherence histogram. IEEE Trans. Industr. Inf. **15**(2), 954–964 (2019)
12. Ma, X., Zhao, Y., Zhang, L., Gao, Q., Pan, M., Wang, J.: Practical device-free gesture recognition using WiFi signals based on meta-learning. IEEE Trans. Industr. Inf. **16**(1), 228–237 (2020)
13. Huang, X., Dai, M.: Indoor device-free activity recognition based on radio signal. IEEE Trans. Veh. Technol. **66**(6), 5316–5329 (2017)

Recent Advances in AI and Their Applications in Future Electronic and Information Field

A Visible Light Indoor Location System Based on Lambert Optimization Model RSS Fingerprint Database Algorithm

Xiaoqian Ding[1], Shuo Shi[1,2(✉)], Xuemai Gu[1,3], and Shihang Chen[4]

[1] School of Electronic and Information Engineering, Harbin Institute of Technology, Harbin 150001, Heilongjiang, China
crcss@hit.edu.cn

[2] Peng Cheng Laboratory, Network Communication Research Centre, Shenzhen 518052, Guangdong, China

[3] International Innovation Institute of HIT in Huizhou, Huizhou 516000, Guangdong, China

[4] Guangxi Communication Planning and Design Consulting Co., Ltd., Nanning 530001, Guangxi, China

Abstract. In order to further improve the positioning accuracy of indoor positioning based on visible light communication, this paper proposes an RSS fingerprint database location algorithm based on Lambert optimization model, which significantly improves the positioning accuracy. The Internet of Things technology is used to realize the connection between devices and mobile phones, and the positioning information is transmitted to mobile phone clients in real time. To test this system, a visible light indoor positioning system based on STM32F407 development platform was built, which verified that the positioning error of the system was basically stable within 30 mm in the area of 800 mm * 800 mm. This system has stable signal transmission, high precision, small time consuming and low cost of power consumption, which has a good development and application value.

Keywords: Visible light communication · Indoor location · Location algorithm · Internet of Things · High precision

1 Introduction

In recent years, visible light communication technology has been increasingly mature, and the scale of investment in the industry has also gradually increased. It has effectively made up for the deficiencies of existing wireless communication technologies, such as low positioning accuracy, high power consumption and slow speed, via its prominent advantages, such as rich spectrum resources, unrestricted use of sensitive areas, high confidentiality and low cost [1]. As a new indoor positioning technology, visible light communication has become a hot research and application field. However, most existing visible indoor positioning systems have low positioning accuracy, long time delay and high-power consumption, which reduce the availability of the system.

S. Shi et al. (Eds.): AICON 2020, LNICST 356, pp. 493–501, 2021.
https://doi.org/10.1007/978-3-030-69066-3_43

Visible light communication technology is transmitting digital signals in the visible light band to achieve communication between sending and receiving points via using light-emitting diodes. Compared with traditional wireless communication, visible light communication technology has both lighting and communication functions, and has its unique advantages, that is, rich spectrum resources, unlimited use of sensitive areas, green safety, high confidentiality, and low-cost. Therefore it is more suitable for application in the field of communications, with very considerable application prospects.

At present, the commonly used indoor positioning algorithms [2] based on visible light communication mainly include geometric measurement method, fingerprint identification method [3], approximate perception method, image sensor imaging method and so on. Geometric analysis location algorithm mainly includes RSS, TOA, and TDOA algorithms [1]. These location algorithms have their own advantages and disadvantages. Although indoor positioning can be achieved theoretically [4], it is difficult to achieve accurate and efficient positioning when used alone due to the complex actual situation. And it is difficult to obtain accurate parameters of the lighting model in the actual environment [5]. Among them, RSS algorithm, the received light intensity algorithm [5], is the most widely used visible light indoor location algorithm at present. And the key point is to estimate the distance between the measured point and the luminous spot. The accuracy of distance estimation directly determines the positioning accuracy. Generally, the distance can be estimated according to the phase detection method and the intensity estimation method. However, the phase detection method requires high synchronicity and is difficult to realize positioning [6], while the intensity estimation method can estimate the distance of optical transmission by directly detecting the optical power [6], which is usually used in combination with the three-side algorithm. Although this algorithm is easy to implement, it has certain defects [7].

To solve the existing problems, this paper proposes an RSS fingerprint database location algorithm based on Lambert optimization model, and builds a visible light indoor location model device. The spectrum of visible band is used for high-speed data transmission in order to improve the positioning accuracy up to the magnitude of mm, then comprehensively improve the positioning accuracy and speed of the system. Through software coding realize signal filtering, data storage and other functions, and the location information can be transmitted to the client in real time.

2 RSS Fingerprint Database Location Algorithm Based on Lambert Optimization Model

2.1 Lambert Optimization Model

In general, we regard the lighting model of a single LED as the Lambert radiation model [8], that is, the light intensity of the received signal conforms to the law of cosine of light intensity between the light intensity and the light propagation distance. In lambert radiation model, if you want to determine the value of the parameter *m*, you need to convention half the range of power Angle $\theta_{1/2}$ at first, thus lambert parameter *m* also has a scope of constraint.

However, most LED do not contain only one lamp bead, but may contain multiple lamp beads. In this case, the upper half power Angle of the same LED in different radiation directions may be different, and its Lambert parameter may also deviate to some extent, then results a positioning error. Therefore, in practice, lambert parameters need to be optimized to obtain high precision positioning. A new estimation method of Lambert parameters is proposed in literature [9]. Based on this method, a geometric diagram of the lambert radiation parameter optimization model proposed in this design for its existing problems.

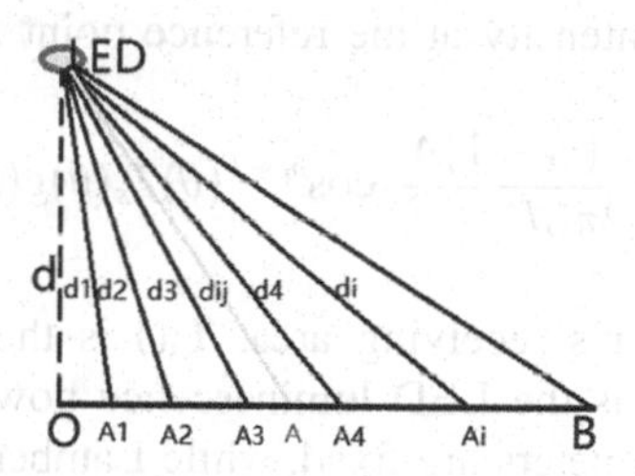

Fig. 1. Lambert optimization model diagram

As shown in Fig. 1, if the matched object point is at A position between the reference point O and B, then A[*i*] of several known coordinate positions is selected from the direction O to B, and the measurement value of the direct vision distance between the reference point *i* and the light source is assumed to be d_i. At present, the discrete sequence *m*[] about Lambert parameters can be obtained with a certain step distance, and each element *m*[*j*] in the sequence can be used as the Lambert radiation coefficient to conduct ranging for the reference point A*i* with a known distance of d_i, and the measured value can be obtained as *dij*.

The *m* value is solved as follows: suppose the receiving optical power at O is P_o, and the receiving optical power at reference point A_i is P_i, then:

$$m = \log_{d_i/d} {P_o/P_i} - 3 \tag{1}$$

Where d_i is the direct distance from the reference point *i* to the light source, and its value can be directly retrieved from the fingerprint database. Then, a sequence *m*[] about *m* should be obtained. And named the maximum value of this sequence as m_{max}, the minimum value as m_{min}.

$$m_j = m_{\min} + \frac{m_{\max} - m_{\min}}{n} \times j; j = 0, 1, 2, \cdots n \tag{2}$$

As shown in Eq. (2), n is the number of steps, and the greater the value, the more accurate estimation of m_j. When m_j is taken as Lambert coefficient, we can obtain the theoretical distance *dij* between the reference point A*i* and the LED source, via substituting each m_j in Eq. (2) back into Eq. (1).

Then extend the above theory to N reference points, and use variance as error measurement standard. When the minimum value of formula (3) is reached, that is, the error of d_i and *dij* is the minimum, and the corresponding *m*[*j*] is the reliable estimate value of Lambert parameter in this direction.

$$e^j = \sum_{i}^{n} \sqrt{(d_i - d_{ij})^2} \quad (3)$$

Then substitute this reliable estimate value into the classic single-lamp Lambert model, the received light intensity at the reference point A*i*, as shown in Eq. (4).

$$P_A = \frac{(m+1)A}{2\pi(d^2 + r^2)} \cos^{m+1}(\theta) T_s(\theta) g(\theta) \times P_t \quad (4)$$

Where, A is the sensor's receiving area, $T_s()$ is the filter gain, g() is the light accumulation gain, and P_t is the LED luminescence power [8]. In general, when the system is stable, these parameters are fixed, while Lambert parameter *m* represents the directivity of luminescence, and its value is related to the half power Angle. According to the geometric relationship, it can be seen that:

$$\cos(\theta) = \frac{d}{\sqrt{d^2 + r^2}} \quad (5)$$

In combination Eq. (4) and (5), the linear distance between point A*i* and point O in the projection plane can be obtained, as shown in Eq. (6).

$$r = d(\sqrt[m+3]{\frac{P_o^2}{P_A^2}} - 1)^{\frac{1}{2}} \quad (6)$$

2.2 RSS Fingerprint Database Location Algorithm

The RSS fingerprint database location algorithm based on The Lambert optimization model is adopted. Firstly, collecting the light intensity information of reference points then establishing a database. Then introducing an optimization algorithm of the Lambert model in the real-time location stage. Afterwards, the precise estimation value of the obtained Lambert parameter was used for RSS measurement, and the coordinate position of the point to be measured was calculated by the three-side algorithm and sent to the computer and LCD display.

This optimization hybrid algorithm makes up for the disadvantages and limitations of the low accuracy of RSS algorithm and the huge database of fingerprint algorithm. It can not only ensure the accuracy of positioning, but also not introduce a large amount of work. When the positioning points is reference points or within a very small range of their vicinity, it can achieve minimal or even zero errors. Algorithm model is shown in Fig. 2.

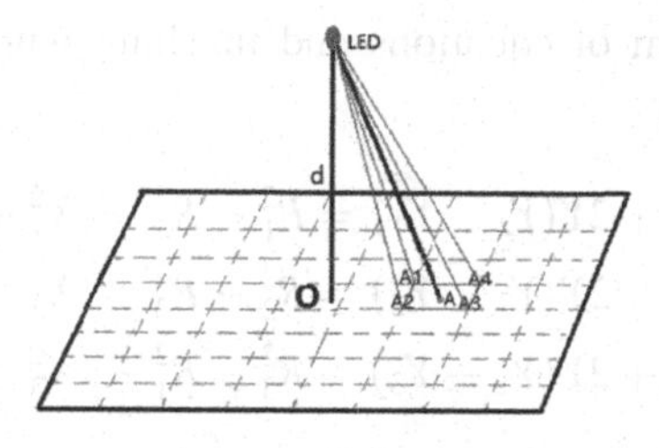

Fig. 2. Model diagram of hybrid positioning algorithm

Where A is the point to be measured and its coordinate position A(X, Y) is unknown; *d* is the vertical distance between point O and the light source. Assuming that the nearest probable point for point A to be detected by fingerprint database matching is A1(X_1, Y_1), then if the system continues to match, the nearest neighbor points to be found as A2(X_2, Y_2), A3(X_3, Y_3), A4(X_4, Y_4). The four reference points will mark out a small illumination region, and there must be an optimal Lambert radiation coefficient *m* in this illumination region, so that the positioning error of all reference points in this illumination region can be minimized when calculating by formula (4).

Then using the Lambert optimization model to calculate the distance between the object point and the three sources, as R1, R2 and R3. Finally, using the Three-side measurement algorithm RSS to figure out the coordinate of the object point.

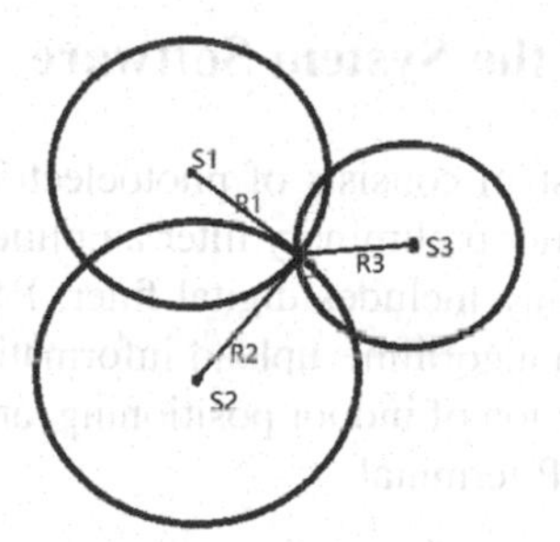

Fig. 3. Three-sided measurement

As shown in Fig. 3. Where, S1, S2, S3 respectively represent the three LEDs' vertical projection on the lighting underside, then set their coordinates as (X_1, Y_1), (X_2, Y_2), (X_3, Y_3), set the coordinates for point O as (X, Y). And the distance of the point of S1, S2, S3 respectively are R1, R2, R3, then using simultaneous system of equations by geometric relations. As shown in formula (7).

$$\begin{cases} (X - X_1)^2 + (Y - Y_1)^2 = R_1^2 \\ (X - X_2)^2 + (Y - Y_2)^2 = R_2^2 \\ (X - X_3)^2 + (Y - Y_3)^2 = R_3^2 \end{cases} \tag{7}$$

Subtracting the system of equations and finishing with each other, then we can obtain Eq. (8).

$$\begin{cases} 2X(X_2 - X_1) + 2Y(Y_2 - Y_1) = R_1^2 - R_2^2 - (X_1^2 + Y_1^2 - X_2^2 - Y_2^2) \\ 2X(X_3 - X_2) + 2Y(Y_3 - Y_2) = R_2^2 - R_3^2 - (X_2^2 + Y_2^2 - X_3^2 - Y_3^2) \\ 2X(X_1 - X_3) + 2Y(Y_1 - Y_3) = R_3^2 - R_1^2 - (X_3^2 + Y_3^2 - X_1^2 - Y_1^2) \end{cases} \quad (8)$$

Making that, $A_i = \begin{cases} A_1 = X_2 - X_1 \\ A_2 = X_3 - X_2 \\ A_3 = X_1 - X_3 \end{cases}$, $B_i = \begin{cases} B_1 = Y_2 - Y_1 \\ B_2 = Y_3 - Y_2 \\ B_3 = Y_1 - Y_3 \end{cases}$,

And $C_i = \begin{cases} C_1 = R_1^2 - R_2^2 - (X_1^2 + Y_1^2 - X_2^2 - Y_2^2) \\ C_2 = R_2^2 - R_3^2 - (X_2^2 + Y_2^2 - X_3^2 - Y_3^2) \\ C_3 = R_3^2 - R_1^2 - (X_3^2 + Y_3^2 - X_1^2 - Y_1^2) \end{cases}$

Finally, we can obtain formula (9), and we can obtain coordinates location (X, Y) of object point.

$$A_i X + B_i Y = C_i, \; i = 1, 2, 3 \quad (9)$$

3 The Realization of the System Software

As shown in Fig. 4, the system consists of photoelectric detector receives the signal sent to the microprocessor after preliminary filter amplifier treatment STM32 algorithm processing. The system mainly includes digital filter, RSS fingerprint database established, optimization location algorithm, upload information to cloud server. Then user can enable the system's function of indoor positioning, and access to the location of the detector, just by mobile APP terminal.

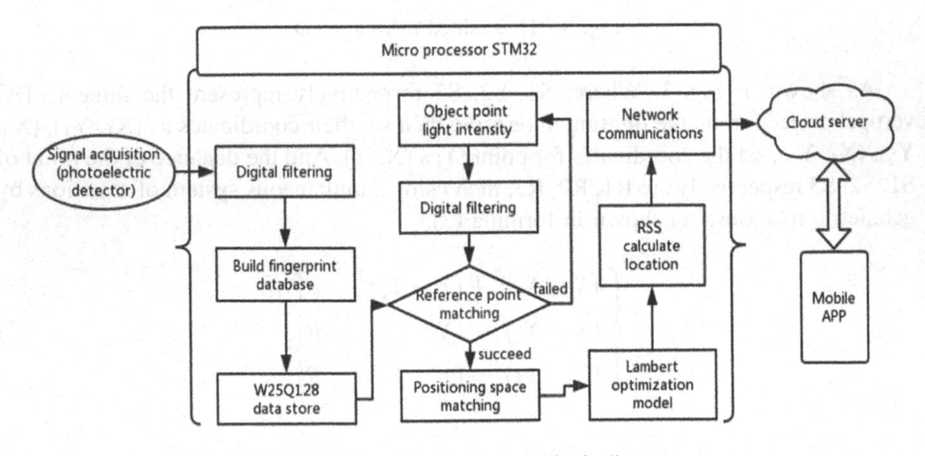

Fig. 4. The software design block diagram

3.1 Optimization Localization Algorithm Module

The receiver control processor STM32F407 communicate with external memory W25Q128 using SPI. When system running localization algorithm, it will match reference point data from memory firstly. Take the variance between object point's signal strength and fingerprint data as the matching error measure, and minimum variance of the corresponding fingerprint information is the closest reference point data information.

As shown in Fig. 2, assuming that the matched results of closest reference point from the fingerprint database is A1(X1, Y1), then system will continue to find the nearest neighbor points respectively, as A2(X_2, Y_2), A3(X_3, Y_3), A4(X_4, Y_4). Finally, four reference points can box out a small piece of illumination area.

Since then, the system will calculate corresponding lambert parameters of four reference points respectively, and deposit them into the array *m*[] following the growing up order. Then get the corresponding lambert parameters in the direction of extreme value by doing 400 times step by step, coexist in the array *m*[]. Then calculate the light power of this point using each element of array *m*[] via plugging in formula (4), and match with object point information once again to find the best Lambert radiation parameter *m*. And then calculate the distance between the object point and three sources respectively, as R1, R2 and R3, via plugging in formula (6). Finally, using RSS to figure out the coordinate of the object point.

3.2 Remote Control and Information Interaction Module

This system use ATK-ESP8266 module (WIFI devices) with STM32F407 controller to develop Gizwits Agent, it realized system device terminal docking with the cloud server and mobile phone user interaction.

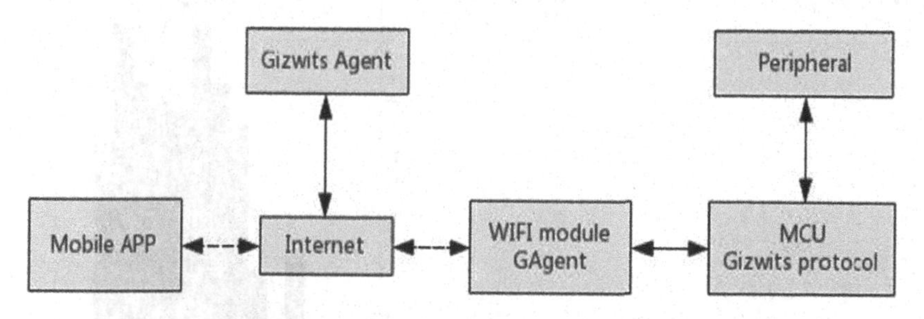

Fig. 5. Gizwits-Agent product control process

The working process of device terminal and client information interaction is shown in Fig. 5, after WIFI equipment receiving instructions sent by users, those instruction will send to MCU through the protocol frame format, and processor STM32F407will send receiving data into buffer. Then processor will catch bag from buffer at set intervals, and begin depth analysis if it was right, then push those data into event processing (the action execution), and MCU will implement its own logic according to the data point corresponding events.

The MCU will send data sampled by sensor to WIFI devices according to the protocol stack frame format packaging, and WIFI devices is responsible for sending those data to cloud server. Finally, the mobile client can receive object-point's location information after sending control command in APP interface.

4 Testing Results and Analysis

In order to test the stability of this system and avoid the accident of test results, we tested this system many times and recorded 192 effective data points. Use the positioning error as precision standard, and analyze its positioning accuracy. Where, the positioning error is defined as the absolute gap between the actual location of the object point (x0, y0) and the detected location by this system (x, y), and it also use the standard gap sigma σ and the average error μ, as shown in Eq. (10) and Eq. (11).

$$\delta = \sqrt{\delta_x^2 + \delta_y^2} = \sqrt{(x - x_0)^2 + (y - y_0)^2} \tag{10}$$

$$\sigma = \sqrt{\frac{1}{N}\sum\nolimits_{i=1}^{N}(\sigma_i - \mu)^2}, \mu = \frac{1}{N}\sum\nolimits_{i=1}^{N}\delta_i \tag{11}$$

The positioning error statistical analysis diagram is shown in Fig. 6. Take these positioning error of test data for statistical analysis, we find that the maximum position error is less than 50 mm, the average error is 25.92 mm, and the standard error is 10.195 mm.

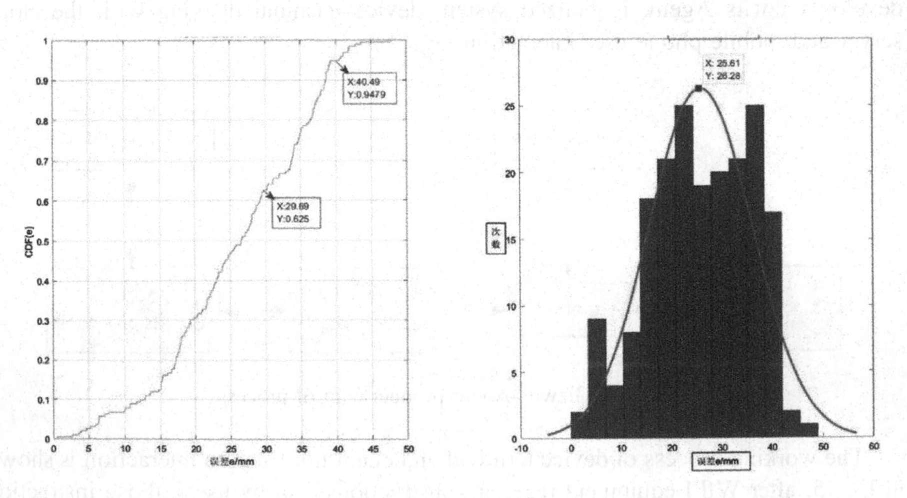

Fig. 6. (a) Positioning error of the cumulative distribution, (b) Gauss curve fitting

Here, the positioning error cumulative distribution function is mainly used for observation the current error value percentage of all error, and Gauss curve is mainly used for observation the statistical distribution of all error values. As shown in Fig. 6 (a), the test point which positioning error are under 30 mm occupy 62.5%, and the test point which positioning error are under 40 mm occupy 94.8%. And as shown in Fig. 6 (b), we find that the positioning error distribution between 20 mm to 40 mm basically, the center value of Gaussian fitting curve is 25.61 mm.

5 Conclusion

The system can only obtain basic error in 70.7 mm through using traditional fingerprint algorithm, which take the nearest fingerprint data as the object points coordinate, while this system can realize the average positioning error less than 30 mm through using RSS fingerprint database location algorithm based on Lambert optimization model proposed by this paper.

Compared with the traditional visible light indoor positioning system, this system has its advantages, such as ideal speed, high positioning accuracy, low latency, low consumption, and without any complicated sensors and other synchronization devices. Combining the device terminal with mobile phone via Internet of things technology, it achieves the user's information interaction with the cloud server. Within the scope of network coverage, users only need to send control commands in the APP, then he can get the location of the object in real time. To summary, this visible light indoor positioning has high feasibility, high positioning accuracy and strong practicability.

In future, we can achieve free or portable positioning via improving the hardware design of the receiving end and adopting a mobile power supply, so that it can be miniaturized for easy wearing. For the last mile of navigation, we can consider using visible light indoor positioning technology to achieve seamless connection between indoor and outdoor positioning.

References

1. Jiaqi, Z., Nan, C.: Comparative study on several key technologies of indoor LED visible light positioning. Lamp Light. **39**(01), 34–41 (2015)
2. Wang, K.: Design of visible indoor positioning system based on received signal intensity. Huazhong University of Science and Technology (2017)
3. Wang, H.: Research on indoor visible light positioning method based on location fingerprint. Xidian University (2018)
4. Guan, W., Wu, Y., Wen, S., et al.: A novel three-dimensional indoor positioning algorithm design based on visible light communication. Optics Commun. **392**, 282–293 (2017)
5. Tang, Q.: Research on indoor positioning method based on LED communication. University of Electronic Science and Technology (2016)
6. Hu, Q.: Indoor positioning algorithm based on LED visible light. Dalian Maritime University (2016)
7. Gu, J.: An improved three-point indoor light positioning scheme and design considering occlusion. Nanjing University of Posts and Telecommunications (2017)
8. Chen, X.: Simulation and experimental study of indoor visible multi-parameter location fingerprint database location method. Nanjing University of Posts and Telecommunications (2015)
9. Shun, C., Kun, Y., Zhuo, L., et al.: Visible indoor positioning technology with lambert model parameter optimization. Optical Commun. Res. **05**, 69–73 (2018)

A Target Detection Algorithm Based on Faster R-CNN

XinQing Yan(✉), YuHan Yang, and GuiMing Lu

North China University of Water Resources and Electric Power, Zhengzhou, People's Republic of China
1161547277@qq.com

Abstract. Target detection is one of the hotspots of image processing research. In the image, due to factors such as distance or light, it will affect the target detection result and increase the error detection rate. Moreover, the existing network training time is too long to meet the actual needs. In order to reduce the lack of light or shadow interference and other factors, based on the Faster R-CNN framework, this paper innovatively proposes a method to improve its feature network ResNet-101 to extract deep features of images.In order to shorten the running time, this paper introduces the region number adjustment layer to adaptively adjust the number of candidate regions selected by RPN during the training process. This paper conducts experiments on the PASCAL VOC data set. The experimental results show that the improved feature network model proposed has an accuracy improvement of 2% compared with the original feature network model. The results show that the target detection algorithm proposed in this paper has higher recognition accuracy than the original algorithm.

Keywords: Target detection · ResNet-101 · Faster R-CNN

1 Introduction

Object detection is an important research direction in the field of computer vision, which is very important for extracting and mining regions of interest. Target detection is a central problem in computer vision [1], and has important research value in the fields of pedestrian tracking, license plate recognition and unmanned driving. The features extracted by traditional algorithms are basically low-level and simple features manually selected. These features are more targeted at specific objects and better represent multiple targets. In addition, some prior knowledge needs to be manually set. In recent years, deep learning has greatly improved the accuracy of image classification, so target detection algorithms based on deep learning have gradually become mainstream. Traditional target detection algorithms are mainly based on the matching of traditional frames or feature points in sliding windows. AlexNet won the championship in the 2012 ImageNet Large-scale Visual Recognition Challenge [2]. Its role goes far beyond traditional algorithms, and brings the general public's vision into the convolutional neural network. Target detection methods based on deep learning have just appeared in the field of computer vision. Target detection based on deep learning

S. Shi et al. (Eds.): AICON 2020, LNICST 356, pp. 502–509, 2021.
https://doi.org/10.1007/978-3-030-69066-3_44

can be roughly divided into two categories: 1) Regression-based detection algorithms; 2) Classification-based detection algorithms. The former is represented by the YOLO system algorithm. Reference [3] proposes the YOLO model. YOLO divides the image into SxS grids, and each grid is responsible for detecting targets centered on the grid. YOLO uses a cell-centered multi-scale area to replace the regional target proposal network, thus giving up some accuracy in exchange for faster detection speed. This method is directly trained on the original image. This method is faster and less accurate. The latter's typical algorithm is Faster-RCNN. Reference [4] proposed a regional convolutional neural network model, which uses a deep convolutional neural network to select some candidate regions in the image to be detected by a selective search method, so as to perform advanced feature extraction, and then use multiple SVM pairs Functions are classified and target discovery tasks are completed. In the article [5], the Faster-RCNN (Faster-RCNN) model is proposed to improve the detection accuracy and speed of the RCNN model. His central idea is to use RPN network to exclude areas of interest and train based on it. The method is characterized by high accuracy and slow speed. Therefore, if the accuracy requirements are high, the Faster-RCNN series algorithm should be adopted based on the classification idea. However, when real-time requirements are high, fast RCNN is slower and not easy to apply [6]. In this article, the Faster-RCNN model is improved. This article innovatively proposes a method to improve its feature network ResNet-101 to extract the deep features of the target, extract the deeper features of the image, and use the PASCAL VOC data. And in order to shorten the running time, this paper introduces the region number adjustment layer to adaptively adjust the number of candidate regions selected by RPN during the training process. The set is compared with the Faster-RCNN model experiment to verify the method proposed in this paper.

2 Related Work

The convolutional neural network obtains the features of the target by learning the data set of manually labeled features [7]. At present, algorithms based on convolutional neural networks can be roughly divided into two modes, namely, two-stage mode and one-stage mode. The detection process of the former is divided into two steps [8]. First, the algorithm generates several candidate boxes, and then classifies the candidate boxes through CNN [9]. The latter is a direct regression to the category probability and location coordinates of the target. The emergence of R-CNN has successfully applied CNN to the field of target detection, but R-CNN also has certain problems.

In order to break through the time bottleneck of the candidate region algorithm, Ren Shaoqing et al. proposed FasterR-CNN in 2016 [10]. FasterR-CNN uses RPN instead of selective search, which greatly reduces the time to extract candidate frames. This algorithm can be roughly understood as a combination of RPN and FastR-CNN. From R-CNN to FasterR-CNN, its detection speed and detection accuracy are constantly improving [11]. Such algorithms are still an important branch of target detection algorithms.

3 Faster R-CNN

The Faster R-CNN target detection network is mainly divided into two steps. The first step is to locate the target. Enter a picture into the feature extraction network, and extract the feature map of the image after a series of convolution and pooling operations. The task of target detection is not only to detect the target, but also to find and locate the candidate target on the feature map through the RPN network. The second step is to classify the specific category of the target. The range box regressor is used to modify the position of the candidate target to generate the final candidate target area. The softmax classifier is used to identify the categories of candidate targets. This paper uses a classification network to determine whether the candidate area belongs to the target of interest, so as to realize the detection of the target of interest. The structure of Faster R-CNN is shown in Fig. 1.

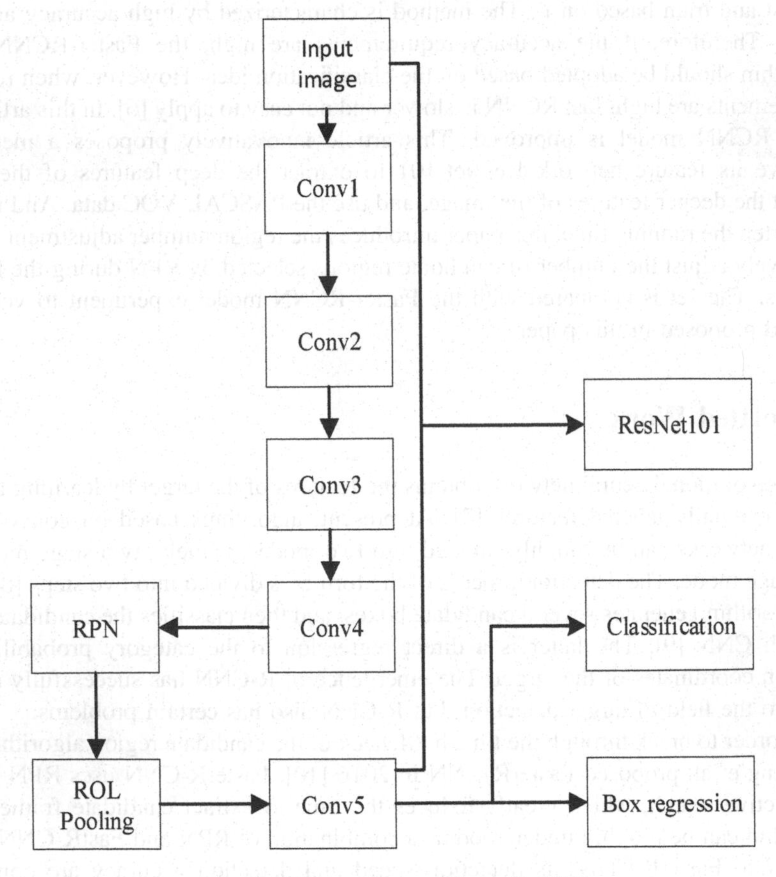

Fig. 1. The structure of Faster R-CNN

3.1 ResNet

In this paper, the feature extraction of the target is completed by improving the ResNet-101 model based on the convolutional neural network. Compared with the feature extraction network such as AlexNet and VGG16, its characteristics are as follows.

ResNet adopts the shortcut connection method of "shortcut" identity mapping network, which neither generates additional parameters nor increases computational complexity. As shown in Fig. 2, the shortcut connection simply performs identity mapping and adds its output to the output of the overlay. Through SGD back propagation, the entire network can be trained in an end-to-end manner.

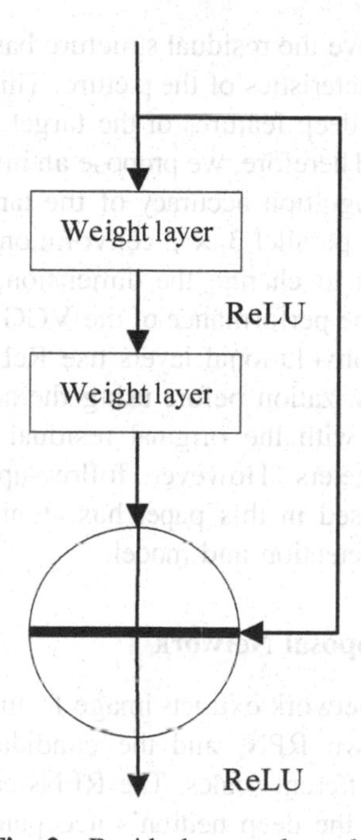

Fig. 2. Residual network model

$$X_{l+1} = X_l + F(X_l, W_l) \tag{1}$$

$$X_{l+2} = X_{l+1} + F(X_{l+1}, W_{l+1}) = X_l + F(X_l, W_l) + F(X_{l+1}, W_{l+1}) \tag{2}$$

$$X_L = X_l + \sum_{i=l}^{L-1} F(X_i, W_i) \tag{3}$$

$$\frac{\partial \varepsilon}{\partial X_l} = \frac{\partial \varepsilon}{\partial X_L}\frac{\partial X_L}{\partial X_l} = \frac{\partial \varepsilon}{\partial X_L} + \left[1 + \frac{\partial}{\partial X_L}\sum_{i=l}^{L-1} F(X_i, W_i)\right] \tag{4}$$

In the formula, the network input value is X_l, and the network output value is $X_{l+1}, X_{l+2}, \ldots, X_{:L}, \varepsilon$. The network residual block is $F(X_l, W_l)$, $F(X_{l+1}, W_{l+1})$.

ResNet adopts the "Bottleneck design" network design structure, uses 1×1 convolution to change the dimension, 3×3 convolution inherits network performance, and controls the number of input and output feature maps of 3×3 convolution. When the number of layers is high, the number of 3×3 convolutions is reduced, which greatly reduces the number of convolution parameters and the amount of calculation while increasing the depth and width of the network.

3.2 Improved Feature Extraction Network

This article attempts to improve the residual structure based on the traditional ResNet-101, combined with the characteristics of the picture. This paper increases the width of the network and extracts the deep features of the target so that the network can learn key distinguishable features. Therefore, we propose an improved deep residual network structure to improve the recognition accuracy of the target person. The residual unit used in this paper, using two parallel 3×3 convolutional layers. The function of the 1×1 convolutional layer is to change the dimension, and the two parallel 3×3 convolutional layers inherit the performance of the VGG network. The unit uses a pre-activation method, and all convolutional layers use ReLU as the activation function, That is, perform batch regularization before using the activation function and convolution operation, Compared with the original residual unit, this structure has little difference in training parameters. However, follow-up experiments show that the residual unit structure proposed in this paper has significantly improved the performance of the target person detection and model.

3.3 Improved Region Proposal Network

When the feature extraction network extracts image features, it is a process from low to deep. Each layer has its own RPN, and the candidate regions are generated by extracting feature maps of different scales. The RPNs corresponding to different proportions are different. When the deep neuron's acceptance range expands, the corresponding anchor box size also increases. The larger the candidate area, the smaller the RPN. After obtaining the candidate area, the features are converted into uniform size through RI pooling. Finally it is sent to the classifier. So as to complete the entire Faster R-CNN process.

This paper introduces the NP (number of proposals) layer to adaptively adjust the number of candidate regions selected by the RPN during the training process.

$$N_{Pi+1} = \begin{cases} N_{Pi}(1+\mu_1) & L_i \geq 2L_{i-1} \\ N_{Pi} & 0.5L_{i-1} < L_i < 2L_{i-1} \\ N_{Pi}(1-\mu_2) & L_i \leq 0.5L_{i-1} \end{cases} \tag{5}$$

In the formula, i represents the sequence number of every N training. N_{Pi} represents the number of candidate regions used from the $\mathrm{Ni}^{\mathrm{th}}$ training to the $(\mathrm{N}+1)\mathrm{i}^{\mathrm{th}}$ training. L_i represents the average regression loss from the $\mathrm{Ni}^{\mathrm{th}}$ training to the (N + 1) i^{th}training. μ_1 represents the penalty factor and μ_2 represents the reward factor.

In the training process, the NP layer is introduced to feedback and adjust the training results. Calculate the average of regression loss every N training intervals. Through the blank control group experiment (fixing $N_{p_{i+1}}$ and taking different values for experiment), every N times L_i reduces by half and self-increases by 1 time as the reasonable change jitter interval. Beyond this interval, it is considered that feedback adjustment is required. When L_i doubles or more, the number of candidate regions is increased by a multiple of $(1+\mu_1)$. When L_i is reduced by half or smaller, it is considered that the number of selection boxes can be appropriately reduced, and the number of candidate regions becomes a multiple of itself $(1-\mu_2)$.

Set the upper and lower limits of the number of candidate regions, so that the number of candidates can be adaptively changed from 300 to 2000.

4 Experiment and Analysis

4.1 Data Set and Experimental Environment

The experimental hardware adopts Z440 workstation with 32G memory and NVIDIA P2000 graphics card, and the operating system is Ubuntu 16.04. The programming environment used is as follows, the programming language used is Python language, and the deep learning framework is TensorFlow 2.0. The training samples for the experiment come from the training set in the Pascalvoc dataset.

4.2 Experimental Results and Analysis

Experiment 1. In the training phase, each target in each image in the training set needs to be marked with a rectangular box, and the occluded target should also be marked. During the test, if the overlap between the identified target detection frame and the marked rectangular frame reaches more than 90% of the marked rectangular frame, the test is recorded as successful. The PASCAL VOC 2007 data set is a classic open source data set, including 5000 training set sample images and 5000 test sample images, and a total of 21 different object categories. This article uses the training set and test set of the data set, and the experiment uses the Tensorflow framework to implement the convolutional neural network model. The parameters such as random inactivation, maximum iteration value, batch size in Faster-RCNN generate the average accuracy value (mAP) Greater impact. In order to get a better output, these parameters need to be optimized.

In the experiment, the maximum number of iterations is 70,000, the batch size of the RPN network is 256, and the random deactivation value is selected as 0.6. Experiment one is the performance of the target detection network with the adjustment layer of the number of regions in terms of average detection accuracy and speed. After many experiments, the calculation speed can be increased by 18% on average, with almost no loss of accuracy, saving a lot of overhead. The results of experiment one are shown in Table 1.

Table 1. Detection effect with different number of candidate regions

Network model	Number of candidate regions	Mean average precision	Total time (s)
VGG16	2000	81.86%	54135
VGG16	1000	81.86%	50967
VGG16	500	80.13%	47843
VGG16	100	79.22%	56901
VGG16	10	56.81%	44697
VGG16 + NP	Trained	83.14%	45631
Improved ResNet-101 + NP	Trained	85.62%	45531

Experiment 2. Experiment setup is the same as experiment one. Experiment 2 mainly compares the detection accuracy under different feature extraction models. The results are shown in Table 2.

Table 2. Detection Accuracy under different characteristic networks

Feature network	Mean average precision	Total time (s)
AlexNet	85.26%	54135
ResNet-101	86.23%	50967
VGG16	83.14%	45631
Improved ResNet-101	85.62%	45531

It can be concluded from Table 2 that the improved model has the highest accuracy rate, which is 3% higher than that of the original model. That is, for AlexNet and the original feature network, the improved feature network can better extract image features, which can be more accurate. At the same time, the improved ResNet101 network model has the highest value of mAP, that is, the target network model has relatively good performance, thereby increasing the recognition accuracy of the model. Therefore, the improved model detection effect in this paper is relatively good.

5 Conclusion

Aiming at the problem of target recognition accuracy and target positioning accuracy affecting detection, this paper proposes a method to improve its feature network ResNet-101 based on the Faster R-CNN framework, and improves the Region Proposal Network. The following conclusions can be obtained.

Improve the design of the ResNet-101 feature extraction network based on convolutional neural network in order to make the model more sensitive to the target.

This paper increases the width of the network on the basis of the original network. This improvement enables the network to learn key distinguishable features, thereby improving the accuracy of target recognition. Improvements to the Region Proposal Network greatly increase the running time. The introduction of the region number adjustment layer also improves the accuracy rate to a certain extent.

In short, compared with the original model, the improved Faster RCNN model has improved the accuracy of target recognition.

References

1. Liu, Z., Lyu, Y., Wang, L., et al.: Detection approach based on an improved Faster RCNN for brace sleeve screws in high-speed railways. IEEE Trans. Instrum. Meas. **100**(1), 39–46 (2019)
2. Girshick, R., Donahue, J., Darrell, T,. et al.: Rich feature hierarchies for accurate object detection and semantic segmentation. In: Proceedings of the IEEE Conference on Computer Vision and Pattern Recognition, pp. 580–587 (2014)
3. Girshick, R.: Fast R-CNN. In: Proceedings of the IEEE International Conference on Computer Vision, pp. 1440–1448 (2015)
4. Duan, L., Zhang, D., Xu, F., et al.: A novel video encryption method based on faster R-CNN. In: 2018 International Conference on Computer Science, Electronics and Communication Engineering, pp. 112–116 (2018)
5. Ren, S., He, K., Girshick, R., et al.: Faster R-CNN: and resnet features are equivalent with region proposal networks. In: Advances in Neural Information Processing Systems, vol. 10, no. 3, pp. 91–99 (2015)
6. McNeely-White, D., Beveridge, J.R., Draper, B.A.: Inception and resnet features are equivalent. Cogn. Syst. Res. **59**(1), 312–318 (2020)
7. Khan, S.H., Hayat, M., Bennamoun, M., et al.: Cost-sensitive learning of deep feature representations from imbalanced data. IEEE Trans. Neural Netw. Learn.g Syst. **29**(8), 3573–3587 (2017)
8. Ouyang, W., Wang, X., Zhang, C., et al.: Factors in finetuning deep model for object detection with long-tail distribution. In: Proceedings of the IEEE Conference on Computer Vision and Pattern Recognition, vol. 12, no. 1, pp. 864–873 (2016)
9. Zhang, D., Li, J., Xiong, L., et al.: Cycle-consistent domain adaptive faster RCNN. IEEE Access **12**(7), 123903–123911 (2019)
10. Ren, S., He, K., Girshick, R., et al.: Faster R-CNN: towards real-time object detection with region proposal networks. In: Advances in Neural Information Processing Systems, vol. 10, no. 4, pp. 91–99 (2015)
11. Chen, X., Zhang, Q., Han, J., et al.: Object detection of optical remote sensing image based on improved faster RCNN. In: 2019 IEEE 5th International Conference on Computer and Communications, vol. 12, no. 3, pp. 1787–1791 (2019)

A Kind of Design for CCSDS Standard GF(2^8) Multiplier

Wei Zhang(✉), Aihua Dong, Hao Zhang, and Dacheng Cao

Shandong Institute of Space Electronic Technology, Yantai 264003, China
wzzw1219@126.com

Abstract. Through theoretical analysis, the calculation method of dual basis multiplication in GF (2^8) field based on CCSDS Berlekamp is given. Based on this calculation method, a VLSI architecture for parallel multiplication and serial operation in circuits is proposed. At the same time, the hardware resource occupation and the timing performance of each VLSI architecture are analyzed in detail.

Keywords: CCSDS reed-solomon Code · Dual basis · Multiplication

1 Preface

In the Reed Solomon code specified in CCSDS standard, the codeword is located in GF (2^8) Galois domain and is represented by Berlekamp [1]. Because the basis components used in Berlekamp representation are not directly related to common polynomial bases, polynomial dual bases and normal bases, the multiplication of two elements in Galois field under the Berlekamp representation can not be applied to the more mature design methods under the representation of other bases, such as polynomial base multiplication, dual base multiplication, normal base multiplication and so on [2]. In general, we can transform the Berlekamp base representation to other common base representations, and the results are then converted to the Berlekamp basis after calculating. This calculation process is relatively complicated. In this paper, a method for computing dual basis is proposed, which can be multiplied directly on the Berlekamp basis. The product of the two elements is still expressed by the Berlekamp basis, which greatly simplifies the calculation process and is convenient to realize in practical circuits.

2 Theoretical Analysis

The generating polynomial of GF (2^8) field in CCSDS standard is shown in Formula 1:

$$F(x) = x^8 + x^7 + x^2 + x + 1 \tag{1}$$

The Berlekamp representation of any element Z in the GF (2^8) field is shown in Formula 2:

S. Shi et al. (Eds.): AICON 2020, LNICST 356, pp. 510–516, 2021.
https://doi.org/10.1007/978-3-030-69066-3_45

$$z = z_0l_0 + z_1l_1 + z_2l_2 + z_3l_3 + z_4l_4 + z_5l_5 + z_6l_6 + z_7l_7 \quad (2)$$

The basic component series $\{l_i\}$ and a group of base component series $\{u_i\}$ are dual bases, and their corresponding relationship is shown in Table 1 (a is the primitive element).

Table 1. Dual relation between $\{l_i\}$ and $\{u_i\}$

$\{l_i\}$	$\{u_i\}$
$l_0 = a^{125}$	$u_0 = a^{117\times0} = a^0$
$l_1 = a^{88}$	$u_1 = a^{117\times1} = a^{117}$
$l_2 = a^{226}$	$u_2 = a^{117\times2} = a^{234}$
$l_3 = a^{163}$	$u_3 = a^{117\times3} = a^{96}$
$l_4 = a^{46}$	$u_4 = a^{117\times4} = a^{213}$
$l_5 = a^{184}$	$u_5 = a^{117\times5} = a^{75}$
$l_6 = a^{67}$	$u_6 = a^{117\times6} = a^{192}$
$l_7 = a^{242}$	$u_7 = a^{117\times7} = a^{54}$

We know that the key to the implementation of polynomial dual base multiplier is to generate polynomials in Galois domain, which defines the iterative relationship between polynomial basis components [3]. In GF (2^8) domain, the iterative formula is shown in Formula 3:

$$a^8 = a^7 + a^2 + a + 1 \quad (3)$$

In the dual basis system $\{u_i\}$ epresented by Berlekamp, there is no iterative relationship determined by the generating polynomial, but according to the relationship between dual basis and trace [4], the following Formula 4, holds:

$$u_8 = a^{117\times8} = a^{171} = Tr(u_{8*}l_0)u_0 + \ldots + Tr(u_{8*}l_7)u_7 \quad (4)$$

Where *Tr*(.) is the trace function. Through calculation, the iterative relationship shown in Formula 1 can be obtained:

$$u_8 = u_7 + u_3 + u_1 + 1 \quad (5)$$

The following is the multiplication operation. Suppose:

$$a = bc \quad (6)$$

In the above formula,

$$\begin{aligned} a &= \sum a_il_i \\ c &= \sum c_iu_i \\ b &= \sum b_il_i \end{aligned} \quad (7)$$

There is:

$$a_i = Tr(a * u_i) = Tr(bc * u_i) = \sum_{j=0}^{7} c_j Tr(bu_i u_j) = \sum_{j=0}^{7} c_j Tr(bu_{i+j}) \quad (8)$$

When i + j $\leq$ 7,

$$Tr(bu_{i+j}) = b_{i+j} \quad (9)$$

When i + j > 7, let i + j = 8 + n, according to Formula 5 we can get the following results:

$$Tr(bu_{i+j}) = Tr(bu_{8+n}) = Tr(bu_{7+n} + bu_{3+n} + bu_{1+n} + b_n) \quad (10)$$

Formula 10 produces the following iterative relationship:

$$\begin{aligned} Tr(bu_8) &= Tr(bu_{8+0}) = b_7 + b_3 + b_1 + b_0 \\ Tr(bu_9) &= Tr(bu_{8+1}) = b_8 + b_4 + b_2 + b_1 \\ &= b_7 + b_4 + b_3 + b_2 + b_0 \\ &\cdots\cdots \\ Tr(bu_{14}) &= Tr(bu_{8+6}) = b_{13} + b_9 + b_7 + b_6 \\ &= b_7 + b_6 + b_5 + b_3 + b_2 + b_1 + b_0 \end{aligned} \quad (11)$$

It can be seen that in CCSDS Berlekamp representation, there is also an iterative relationship similar to the common polynomial base representation, which suggests that we can use a similar implementation method to polynomial basis multiplication to design multipliers.

3 Hardware Structure

In GF (2^8), "addition" corresponds to XOR operation in logic circuit, and "multiplication" corresponds to "and" operation. The multipliers used in Reed Solomon codes usually have serial structure and parallel structure.

3.1 Serial Structure

The hardware structure of the serial multiplier represented by Berlekamp is described in Fig. 1. The structure adopts the form of parallel input and serial output. Each clock cycle outputs one bit of product data. It takes 8 cycles to calculate a multiplication. The working principle of the circuit in Fig. 1 is analyzed below.

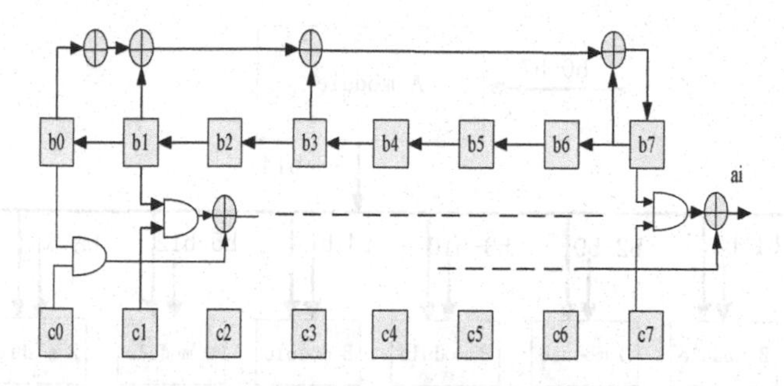

Fig. 1. Serial multiplier.

The first clock cycle: Multipliers b8 ~ b0 and multipliers c8 ~ c0 are input into corresponding registers in parallel. The output is:

$$a_i = b_0c_0 + b_1c_1 + \cdots + b_7c_7 = a_0 \tag{12}$$

The second clock cycle:

$$b_0^{'} = b_1, b_1^{'} = b_2, \cdots, b_6^{'} = b_7 \tag{13}$$

$$b_7^{'} = b_0 + b_1 + b_3 + b_7 \tag{14}$$

According to Formula 10, it can be calculated that:

$$b_7^{'} = b_8 \tag{15}$$

The output is:

$$a_i^{'} = c_0b_1 + c_1b_2 + \cdots + c_6b_7 + c_7b_8 = a_i \tag{16}$$

In this way, a2, a4,..., a7 can be calculated in turn. It can be seen that the circuit shown in Fig. 1 is essentially a pulsating structure.

3.2 Parallel Architecture

The structure of the parallel multiplier represented by Berlekamp is described in Fig. 2. The function of module A is to calculate b8–b14 from B0 to B7 according to the calculation method described in Eq. 8, and the function of module B is to calculate a bit of product according to Formula 6.

Taking B8 as an example, the circuit structure for calculating the bit in module A is shown in Fig. 3.

All B modules have the same circuit structure, as shown in Fig. 4.

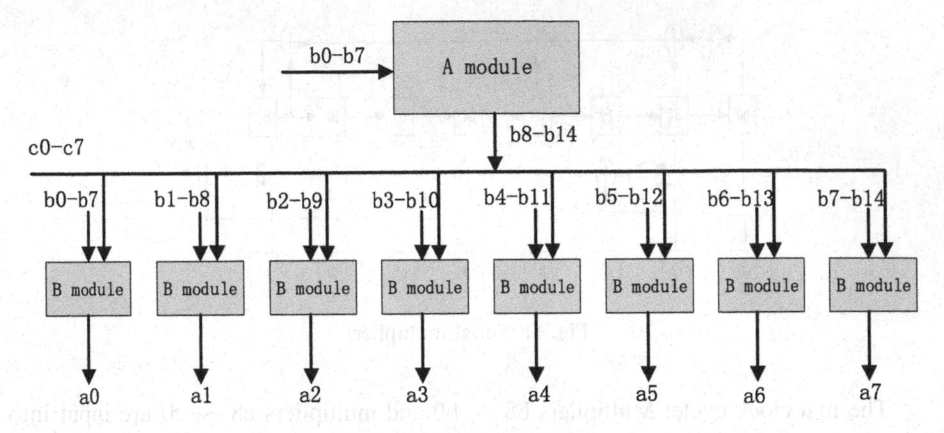

Fig. 2. Parallel multiplier

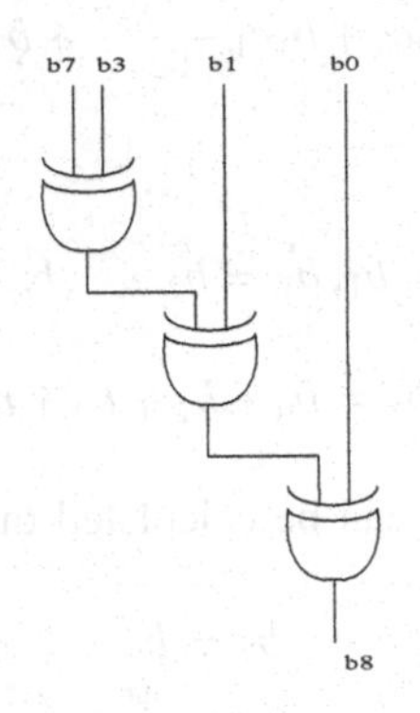

Fig. 3. Example of A module

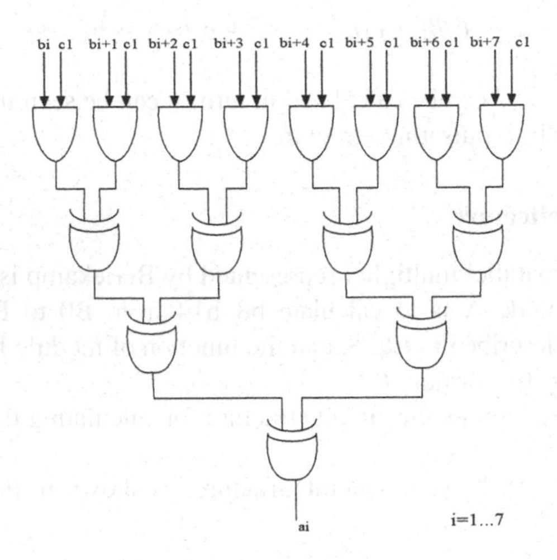

Fig. 4. Example of B module

4 Analysis of Resources and Time Sequence

Table 2 shows the resource usage and timing comparison of serial and parallel multipliers. In the table, Na is the number of two input AND gates. Da is AND gates's delay. NX is the number of two input XOR gates. DX is XOR gates's delay. Nd is the number of triggers.

Table 2. Performance analysis of serial and parallel architectures.

		Na	Nx	Nd	Delay
Serial		8	7	8	Da + 7Dx
Parallel	A module	0	26	0	6Dx
	B module	8	7	0	Da + 7Dx

The statistics of hardware resources in CCSDS dual base multiplier do not include the hardware circuits needed to convert $\{l_i\}$ basis to $\{u_i\}$ basis. This is because in CCSDS Reed Solomon code decoding scheme, the common Berlekamp Massey algorithm or Euclid algorithm uses multiple multipliers. These multipliers share the same multiplier [5] and only need one conversion.

We can see that the XOR gate resources occupied by parallel multipliers are more than 4 times that of serial multipliers. In FPGA, XOR gates are scarce resources (one XOR gate for each LE), and the overall resource consumption of parallel multipliers is relatively large. The multiplier (B0 $\sim$ B7) of serial multiplier needs to be saved for the next clock cycle, which increases the usage of 8 flip flops. In FPGA, flip flops are rich resources and will not become a limiting factor.

From the timing point of view, the parallel structure consumes more than six XOR gates' inherent delay produced by a module than the serial structure. Assuming that the delay of the parallel structure is approximately equal to that of the gate and XOR gate, it can be estimated that the signal path delay of the parallel structure is twice that of the serial structure.

5 Conclusion

The key to the design of CCSDS standard Reed Solomon encoder and decoder is the multiplier design on Galois GF (2^8). In order to reduce the resource consumption and improve the timing performance, the multiplier with simple structure and short path delay must be adopted.The design method of multiplier proposed in this paper uses the iterative relation of GF (2^8) field on the dual basis of Berlekamp basis, and adopts the structure similar to polynomial basis multiplication. Compared with the commonly used dual base multiplier, the multiplier does not need to transform from the Berlekamp base representation to the dual base representation before the multiplication. In the calculation process, the basis transformation is performed automatically, and the calculated product is directly expressed on the Berlekamp basis. The hardware structure of the base transformation is omitted and the hardware structure is greatly simplified. The serial multiplier needs 8 clock cycle processing delay, while the parallel multiplier only needs 1 clock cycle processing delay, but its resource usage is large and the timing path delay is long. In the actual using process, we can reasonably choose the serial or parallel structure according to the requirements of encoding and decoding delay clock cycle.

References

1. CCSDS 101.0-B-6: Telemetry Channel Coding, October 2002
2. Hsu, I.S., Troung, T.K., Deutsch, L.J., Reed, L.S.: A comparison of VLSI architecture of finite field multipliers using dual, normal, or standard bases. IEEE Trans. Comput. **37**(6), 735–739 (1988)
3. Fenn, S.T.J., Benaissa, M., Taylor, D.: Bit-serial Berlekamp-like multipliers for GF(2m). Electron. Lett. **31**(22), 1893–1894 (1995)
4. Wang, X., Zhen, X.: Error correcting codes - Principles and methods. Xidian University Press, Xi'an (2001)
5. Troung, T.K., Cheng, T.C.: A new decoding algorithm for correcting both erasures and errors of reed-solomon codes. IEEE Trans. Commun. **51**(3), 381–388 (2003)

Overview of Terahertz 3D Imaging Technology

Haohao Jiang(✉) and Wei Qu

Space Engineering University, Beijing 101416, China
1052290231@qq.com

Abstract. Terahertz three-dimensional imaging system can realize the detection and imaging of near-field targets with high frame rate and high resolution, and can provide more comprehensive information about the three-dimensional geometric distribution structure of the target and the imaging scene. It is suitable for the current high real-time requirements Security inspection, seeker terminal guidance, military reconnaissance and other fields. The high-resolution three-dimensional imaging technology of radar targets in the terahertz band is of great significance to the development of radar technology and the application of radar imaging. In this paper, the research background and significance of the terahertz near-field imaging technology, radar three-dimensional imaging technology, the development status of terahertz radar system and terahertz radar imaging algorithm are reviewed, and the existing problems of terahertz near-field imaging technology are summarized and prospected.

Keywords: Terahertz radar · Three-dimensional imaging · Time domain imaging algorithm · Frequency domain imaging algorithm

1 Introduction

Terahertz (THz) waves refer to electromagnetic waves with frequencies in the range of 100 GHz to 10 THz. They are between millimeter waves and infrared, called sub-millimeter waves or far-infrared light, and are in the transition zone from electronics to photonics. Compared with signals in the microwave and millimeter wave bands, the wavelength of the terahertz band is relatively short, which is more conducive to achieving large signal bandwidth and narrow antenna beams, and can obtain fine imaging of the target [1–3], which is very conducive to target identification. Compared with infrared light signals, terahertz waves have stronger transmission ability for non-polar and weak-polar media materials, can penetrate clouds and smoke, see through camouflage, and are suitable for complex battlefield environments, and the echo has better Coherence, with higher anti-interference ability.

Terahertz technology can also be used to detect dangerous goods and hidden objects and perform imaging. Because terahertz radiation is non-ionizing, it can penetrate materials that are opaque to other frequency bands, such as packaging materials, clothes and walls. These properties of terahertz port make terahertz remote sensing better used in non-destructive applications. Testing and safety inspection.

In recent years, with the continuous improvement of the performance of terahertz devices, the research and development of terahertz application systems has gradually

S. Shi et al. (Eds.): AICON 2020, LNICST 356, pp. 517–530, 2021.
https://doi.org/10.1007/978-3-030-69066-3_46

attracted the attention of countries all over the world. Terahertz science and technology have been recognized by the international scientific community as a strategically significant field and will gradually become a battleground for high-tech industries.

2 Radar 3D Imaging Technology

At present, based on traditional SAR and ISAR imaging technologies, radar systems with three-dimensional imaging capabilities mainly include: Interferometric SAR (InSAR), Interferometric SAR (InISAR), Circular SAR (CSAR), tomographic SAR (TomoSAR), array 3D SAR, and array 3D ISAR. InSAR and InISAR both use multi-antennas and interferometric processing technology to achieve 3D imaging of the observation scene based on 2D imaging. Circular SAR and tomographic SAR are The synthetic aperture formed by the radar trajectory is used to realize the resolution in the height direction, while the array 3D SAR and the array 3D ISAR use the real apertures formed by the antenna array to obtain the resolution of the third dimension. The following mainly discusses and briefly discusses these three types of imaging technologies analysis.

2.1 Interferometric 3D Imaging

The concept of interferometric processing technology is to extract the phase difference of the same resolution unit through a single observation of multiple antennas or multiple observations of a single antenna to obtain the information of the observed resolution unit in the vertical line of sight and the direction of the flight path. Image-based, to achieve three-dimensional imaging of the observation area or target. In 1974, L.C. Graham first proposed the use of interferometric SAR for three-dimensional measurement and introduced the imaging principle of interferometric SAR [4]. With the development and advancement of radar technology, interferometric SAR imaging technology has become more mature, and many interferometric SAR processing methods have been studied in depth. Zhenfang Li et al. proposed a method for estimating synthetic aperture in the presence of large co-registration errors. A new method of radar interference phase, which uses the coherent information of adjacent pixel pairs to automatically register the SAR image, and uses the joint signal subspace to project to the corresponding joint noise subspace to estimate the terrain interference phase [5]. PAN Zhou-Hao et al. improved the three-baseline phase unwinding method based on clustering analysis and solved the problem of elevation inversion in the sudden elevation region of the scene [6]. Today, these interferometric SAR processing methods have been widely used in terrain surveying and mapping, moving target detection and positioning, and ocean monitoring.

InISAR is based on ISAR imaging of multiple antennas perpendicular to each other and uses interference technology to reconstruct the three-dimensional coordinates of the scattering points on the target. In 1996, M. Soumekh et al. proposed the use of an interferometric algorithm for dual-channel ISAR image processing. The algorithm has sufficient accuracy to use the phases measured by ISAR single and dual stations for interference processing [8]. In 2007, Lincoln Laboratories built a multistatic radar interferometric imaging test system, which verified the three-dimensional imaging capabilities of interferometric ISAR through the processing of one-shot and three-recovered wave data. In 2014, the University of Pisa in Italy conducted a dual-antenna interferometric ISAR imaging experiment, which realized the interferometric three-dimensional measurement of trucks, but the imaging results only contained strong scattering points [8, 9]. Domestically, many units have also carried out research on InISAR imaging. Among them, Northwestern Polytechnical University proposed a close-range high-resolution microwave imaging method, and carried out imaging experiments on aircraft models in a microwave anechoic chamber, and realized the aircraft The interferometric 3D imaging of the model verifies the feasibility of this method [10].

The advantage of the interferometric three-dimensional imaging technology is that the number of array elements is relatively small and the interferometric measurement accuracy is high. However, the interference processing technology assumes that only one scattering point is included in the same resolution unit. Only one elevation information can be calculated through the interference phase between antennas, so there is no resolution ability for scattering points in the same distance unit.

2.2 Synthetic Aperture 3D Imaging

The synthetic aperture formed by the circular trajectory of the radar movement or the multiple trajectories that enable the radar to achieve three-dimensional resolution based on the SAR two-dimensional imaging is called the synthetic aperture three-dimensional imaging technology, including circular SAR and tomographic SAR. In 1996, M. Soumekh first proposed the circular SAR imaging mode, and then analyzed its three-dimensional imaging capabilities [11, 12]. In 2004, Per-Olov Frölind et al. carried out the first airborne CSAR test. Compared with linear tracking SAR, the detection performance has been significantly improved [13]. In 2006, the U.S. Air Force Laboratory carried out a series of CSAR airborne tests and published Gotcha data to achieve high-resolution 3D imaging of the detection target.

Tomography SAR draws on computer tomography (CT) technology. It uses the synthetic aperture formed by radar in the vertical line of sight and azimuth and combines the principle of three-dimensional tomography to achieve three-dimensional imaging. In 1998, the German Aerospace Agency first carried out the experiment of polarization airborne SAR tomography, which verified the feasibility of airborne tomographic 3D SAR imaging [14]. In 2004, G. Fornaro et al. used the data of 30 voyages of the ERS satellite to achieve 3D imaging of the São Paulo Stadium in Italy, thus verifying the feasibility of spaceborne 3D tomographic SAR [15].

Compared with interferometric SAR, synthetic aperture 3D imaging technology can distinguish multiple scattering points in the same range-azimuth resolution unit, which can realize true 3D imaging, and the system is simple to implement. However, the synthetic aperture 3D imaging technology requires high trajectory accuracy of the flight platform, and the image registration is difficult, and the system has a long accumulation time, so the image registration is very difficult.

2.3 Array 3D Imaging

Array SAR is a two-dimensional virtual array formed by cutting track array or multi-input multi-output (MIMO) antenna combined with radar motion, and the high-dimensional resolution brought by bandwidth signal to obtain three-dimensional imaging results. Work in front-view or down-view mode. In 2004, R. Giret and others of the French Aerospace Agency carried out the imaging research of the linear array 3D SAR on the UAV, and obtained the car height image in the down-view mode [16]. Beginning in 2005, the German Institute of Applied Sciences (FGAN) has carried out airborne down-view 3D SAR imaging research based on MIMO array [17], and conducted the first actual test in 2010 [18]. Domestically, the University of Electronic Science and Technology of China built a ground-based array SAR imaging experimental platform in 2008, and subsequently conducted many field experiments, and obtained the three-dimensional imaging results of the field experiments [19].

The advantage of the array 3D imaging technology is that the imaging frame rate is relatively high, but the number of array elements required by the system is very large and the cost is high.

3 Terahertz Radar Imaging System

3.1 Foreign Terahertz Radar System

The Jet Propulsion Laboratory (JPL) of the United States is an important research institution in the field of terahertz imaging and has achieved many research results. It is one of the first institutions to develop a terahertz radar system. In 2007, the laboratory successfully developed the first high-resolution terahertz radar imaging system with a working frequency of 560 GHz–635 GHz using Frequency Modulation Continuous Wave(FMCW)system, and achieved a distance resolution of 2 cm within a range of 4 m [20, 21]. In 2008, Cooper KB and others of JPL Lab successfully developed a 580 GHz active coherent terahertz radar, which uses a linear frequency modulation continuous wave system to achieve millimeter-level range resolution, and uses very narrow antenna beam scanning to achieve The centimeter-level azimuth resolution [22]. In 2011, JPL Lab improved the original experimental system and realized a three-dimensional scanning terahertz radar imaging system with a center frequency of 675 GHz and bandwidth of 29 GHz, with a maximum imaging distance of 25 m. At the same time, the system's multi-pixel scanning at the same time The method greatly shortens the imaging time. Figure 1 shows the imaging results of three PVC pipes hidden under clothes by the radar system [1]. In 2014, JPL Lab successfully developed a 340 GHz radar array transceiver and successfully applied it to a video frame rate imaging security inspection system [23]. In 2015, JPL designed and developed a radar integrated array transceiver with 8 array elements, the array size is only 8.4 cm [24, 25], which can further increase the imaging frame rate and shorten the imaging scan time.

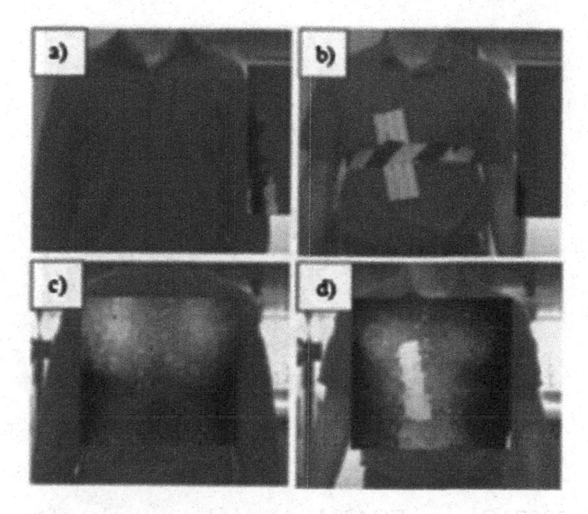

Fig. 1. 675 GHz radar imaging results of the JPL laboratory in the United States

In 2009, the Pacific Northwest National Laboratory (PNNL) developed a 350 GHz active detection imaging system. This system uses a combination of quasi-optical focusing devices and high-speed conical scanning to achieve an imaging resolution of 1 cm [26, 27]. The system can quickly detect dangerous objects hidden on cooperative targets, and can almost achieve real-time imaging. System conceptual diagram and physical diagram are shown in Fig. 2 and Fig. 3.

Fig. 2. Conceptual diagram of 350 GHz active detection imaging system

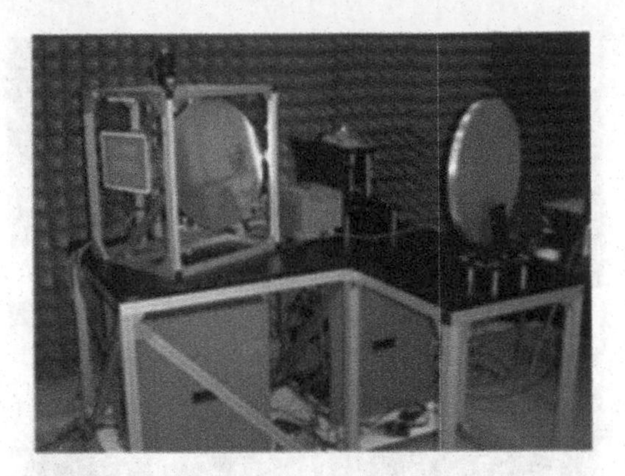

Fig. 3. 350 GHz active detection imaging system

Beginning in 2007, German FGAN has carried out a series of research on terahertz radar SAR and ISAR imaging. First, it developed the COBRA-220 radar system [28, 29], with a center frequency of 220 GHz, and carried out the development of complex targets such as bicycles and automobiles. In the high-resolution SAR and ISAR imaging experiments, the system can image a target at a distance of 135 m with a resolution of 1.8 cm. In 2013, FGAN developed the MIRANDA-300 radar system with a working frequency band of 325 GHz, with a resolution of 3.75 mm.

In 2011, & (R&S) of Germany designed a QPASS system that includes 96 transmitting antennas and 96 receiving antennas in each transceiver unit, as shown in Fig. 4. The azimuth resolution is 1.96 mm, the acquisition time of imaging data is 20 ms [30, 31].

Fig. 4. The imaging system of R&S

3.2 Domestic Terahertz Radar System

Domestic research units include the Institute of Electronics of the Chinese Academy of Sciences, the Chinese Academy of Engineering Physics, the University of Electronic Science and Technology of China, and the National University of Defense Technology. In 2011, the Institute of Electronics, Chinese Academy of Sciences designed and implemented a 0.2 THz three-dimensional holographic imaging system, which is based on optical path sector scanning and one-dimensional linear scanning to achieve the three-dimensional reconstruction of the target [32, 33]. In 2012, the Institute of Electronics of the Chinese Academy of Sciences designed and developed a three-dimensional holographic imaging system with a center frequency of 0.2 THz and a sweep bandwidth of 15 GHz [34]. The imaging resolution of the system has reached 8.8 mm, realizing the 3D image reconstruction of the human body model of hidden dangerous objects under the terahertz quasi-light Gaussian beam. Figure 5 shows the simplified structure of the system and the result of imaging the model with hidden pistol targets. In the following years, the unit continued to improve the imaging algorithm and achieved good imaging results [35–38]. In 2014, the University of Electronic Science and Technology of China established a 330 GHz radar system [39], carried out turntable imaging experiments on aircraft models and other targets, and carried out equivalent CSAR imaging experiments in the terahertz frequency band, achieving three-dimensional imaging of point targets [40, 41]. At the same time, the National University of Defense Technology built a multi-band multiplexed terahertz radar system in a darkroom environment, and carried out a series of imaging experiments on turntable targets and moving targets [42–44]. In 2017, the China Academy of Engineering Physics built a 340 GHz system, using 4 transmitters 16 receivers MIMO array and one-dimensional optical path scanning to achieve 3D imaging of the human body [45, 47].

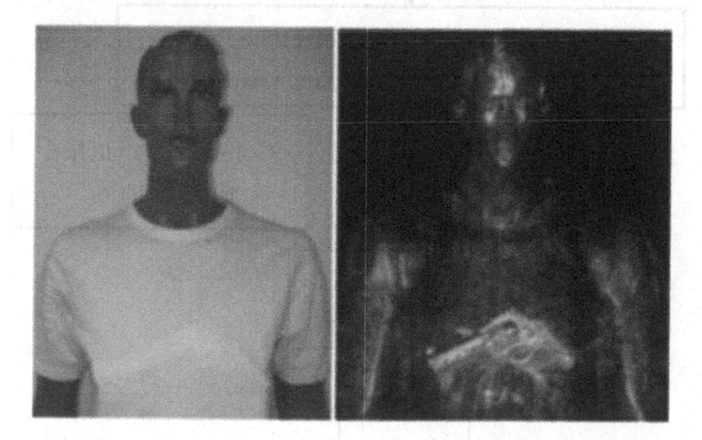

Fig. 5. 3D imaging results of a human model with a hidden pistol

4 Terahertz Radar Imaging Algorithm

4.1 THz Radar Imaging Algorithm Based on Time Domain

Time-domain imaging methods are widely used in terahertz radar imaging processing. Back-projection algorithm is the most commonly used algorithm in radar imaging technology. The principle is to back-project radar echo data to each pixel in the imaging area, and calculate the radar The distance and time delay of the echo between the radar antenna and the image pixel are cumulatively imaged. The flow diagram of the BP algorithm is shown in Fig. 6, the American JPL system in the literature [24, 25], and the early circular scanning of the PNNL in the literature [47]. The imaging system, the planar electrical scanning imaging system based on sparse matrix system of R&S company in literature [48, 49] and the 340 GHz system of China Academy of Engineering Physics in literature [45, 50] all adopt BP algorithm for imaging. The advantage of this algorithm is that it can be applied to a variety of imaging models, is not affected by imaging geometry and array form, and the imaging results are accurate, but the algorithm has a huge amount of calculation and long imaging time, which cannot meet the requirements of fast imaging.

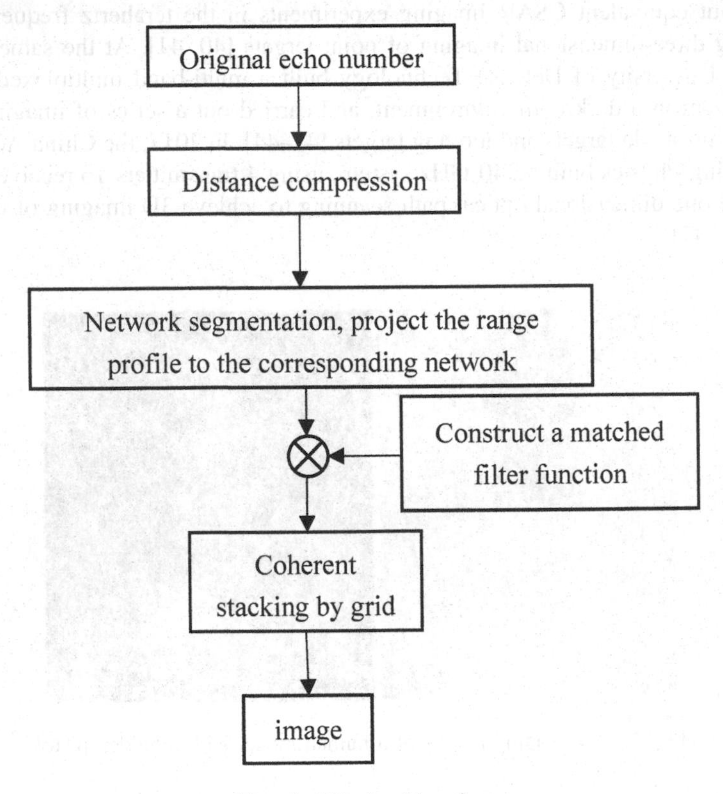

Fig. 6. BP algorithm flow

In recent years, many scientific researchers have optimized the BP imaging algorithm and promoted it to a variety of terahertz imaging systems. Among them, in the literature [51], Wang Qiong extended the BP algorithm to the MIMO linear array imaging system and realized the simulation of 5 scattering points in the scene, thus verifying the good focusing performance of the BP algorithm. Literature [52] proposed a range migration back projection (RM-BP) algorithm for near-field sparse MIMO-SAR 3D imaging. This method can achieve near-field MIMO-SAR 3D imaging focusing, and the calculation speed is greatly improved. The acceleration factor is an order of magnitude with the number of SAR-dimensional scanning arrays, but the imaging quality does not change much, which has certain advantages in practical applications.

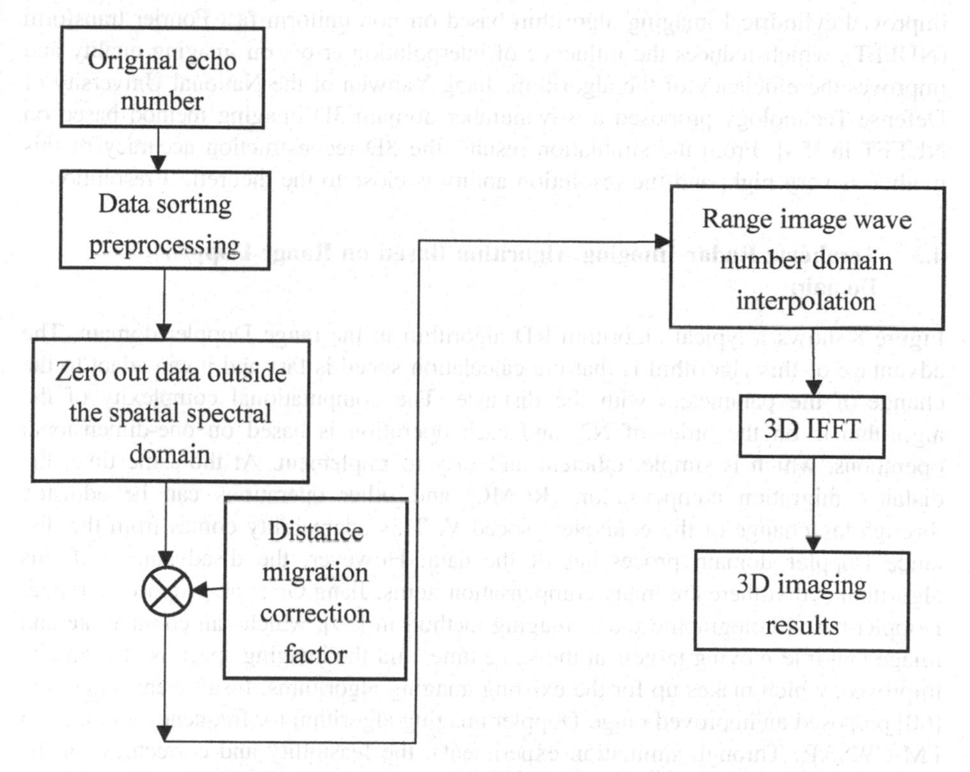

Fig. 7. 3D imaging process based on RMA algorithm

4.2 Terahertz Radar Imaging Algorithm Based on Frequency Domain

The frequency domain imaging method transforms the echo signal into the wavenumber domain space, and uses the Fourier Transform (FT) method for imaging, which greatly improves the imaging efficiency. The distance migration algorithm realizes the reconstruction of the target image in the wavenumber domain, and can achieve complete focusing without geometric deformation based on the scattering point model for the entire area without adding other approximate conditions. This method can be called terahertz near-field in principle. Optimal imaging algorithm for uniform array imaging system. The three-dimensional imaging process of near-field targets based on the RMA algorithm is shown in Fig. 7. The imaging systems in literature [53, 54] all adopt distance migration algorithms to reconstruct images of three-dimensional targets. However, the actual implementation requires the use of Stolt interpolation, which results in a very large amount of calculation and further reduces the running speed of the algorithm. The Institute of Electronics, Chinese Academy of Sciences and Tsinghua University have improved RMA [55, 56], and the processing efficiency and accuracy have been improved. n [57], Wang Youshu proposed an improved cylindrical imaging algorithm based on non-uniform fast Fourier transform (NUFFT), which reduces the influence of interpolation errors on imaging quality and improves the efficiency of the algorithm. Jiang Yanwen of the National University of Defense Technology proposed a wavenumber domain 3D imaging method based on NUFFT in [58]. From the simulation results, the 3D reconstruction accuracy of this method is very high, and the resolution ability is close to the theoretical resolution.

4.3 Terahertz Radar Imaging Algorithm Based on Range Doppler Domain

Figure 8 shows a typical algorithm-RD algorithm in the range Doppler domain. The advantage of this algorithm is that the calculation speed is fast and it can adapt to the change of the parameters with the distance. The computational complexity of the algorithm is on the order of N2, and each operation is based on one-dimensional operations, which is simple, efficient and easy to implement. At the same time, the distance migration compensation (RCMC) and other operations can be adjusted through the change of the equivalent speed V. This adaptability comes from the distance Doppler domain processing of the data. However, the disadvantage of this algorithm is that there are many compensation items. Jiang Ge et al. proposed a range-Doppler-based holographic radar imaging method in [59], which can compensate and image multiple moving targets at the same time, and the imaging speed is also greatly improved, which makes up for the existing imaging algorithms. Insufficient. Literature [60] proposed an improved range-Doppler imaging algorithm for frequency modulation FM-CWSAR. Through simulation experiments, the feasibility and correctness of the algorithm were verified, which provided a basis for further research and design of ground orbit FM-CWSAR systems. Theoretical reference basis.

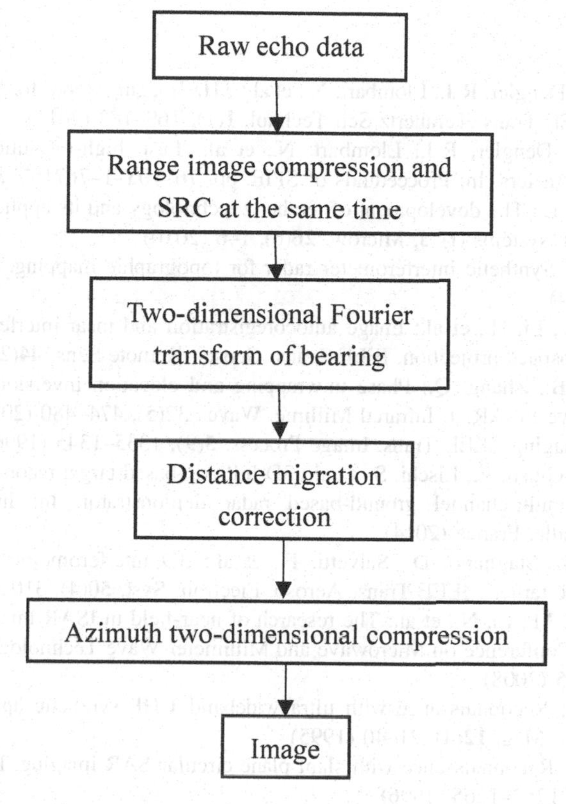

Fig. 8. RD algorithm flow

5 Summary and Outlook

Compared with traditional radars, terahertz radars have higher speed measurement accuracy, target recognition and imaging capabilities, and strong anti-jamming capabilities, as well as very good anti-stealth capabilities. Terahertz radar also has the ability to penetrate plasma to detect targets. However, judging from the development status of terahertz imaging technology, there are still many problems to be solved urgently, and serious challenges are still faced.

(1) The imaging results of the traditional BP imaging algorithm are accurate, but the algorithm has a huge amount of calculation and long imaging time. To solve this problem, we can further optimize the BP imaging algorithm to make it more suitable for terahertz three-dimensional imaging systems.

(2) The traditional distance migration imaging algorithm has the problems of long interpolation time and inaccurate interpolation, which will affect the imaging quality. It is necessary to further study the fast and accurate frequency domain imaging algorithm suitable for terahertz three-dimensional imaging system.

References

1. Cooper, K.B., Dengler, R.J., Llombart, N., et al.: THz Imaging Radar for Standoff Personnel Screening. IEEE Trans. Terahertz Sci. Technol. **1**(1), 169–182 (2011)
2. Cooper, K.B., Dengler, R.J., Llombart, N., et al.: Fast, high-resolution terahertz radar imaging at 25 meters. In: Proceedinds of SPIE, pp. 76710Y-1–76710Y-8, Orlando (2010).
3. Xin, Z., Chao, L.: The development of terahertz technology and its application in radar and communication systems (I). J. Microw. **26**(6), 1–6 (2010)
4. Graham, L.C.: Synthetic interferometer radar for topographic mapping. Proc. IEEE **62**(6), 763–768 (1974)
5. Li, Z., Bao, Z., Li, H., et al.: Image autocoregistration and insar interferogram estimation using joint subspace projection. IEEE Trans. Geosci. Remote Sens. **44**(2), 288–297 (2006)
6. Pan, Z., Liu, B., Zhang, Q.: Phase unwrapping and elevation inversion of three-baseline millimeter wave InSAR. J. Infrared Millime. Waves 32(5), 474–480 (2013)
7 Automatic, imaging: IEEE Trans. Image Process. **5**(9), 1335–1345 (1996)
8. Stagliano, D., Giusti, E., Lischi, S., et al.: 3D InISAR-based target reconstruction algorithm by using a multi-channel ground-based radar demonstrator. In: International Radar Conference, Lille, France (2014)
9. Martorella, M., Stagliano, D., Salvetti, F., et al.: 3D interferometric ISAR imaging of noncooperative targets. IEEE Trans. Aerosp. Electron. Syst. **50**(4), 3102–3114 (2014)
10. Liang, H., He, M., Li, N., et al.: The research of near-field in ISAR imaging diagnosis. In: International Conference on Microwave and Millimeter Wave Technology, Nanjing, China, pp. 1773–1775 (2008)
11. Soumekh, M.: Reconnaissance with ultra wideband UHF synthetic aperture radar. IEEE Signal Process. Mag. **12**(4), 21–40 (1995)
12. Soumekh, M.: Reconnaissance with slant plane circular SAR imaging. IEEE Trans. Image Process. **5**(8), 1252–1265 (1996)
13. Frölind, P., Gustavsson, A., Lundberg, M., et al.: Circular-aperture VHF-band synthetic aperture radar for detection of vehicles in forest concealment. IEEE Trans. Geosci. Remote Sens. **50**(4), 1329–1339 (2012)
14. Reigber, A., Moreira, A., Papathanassiou, K.P.: First demonstration of airborne SAR tomography using multibaseline L-band data. IEEE Trans. Geosci. Remote Sens. **38**(5), 2142–2152 (2000)
15. Fornaro, G., Serafino, F.: Spaceborne 3D SAR tomography: experiments with ERS Data. In: IEEE International Geoscience and Remote Sensing Symposium, pp. 1240–1243 (2004)
16. Giret, R., Jeuland, H., Enert, P.: A study of a 3D-SAR concept for a millimeter-wave imaging radar onboard an UAV. In: 2004 European Radar Conference, Amsterdam, pp. 201–204 (2004)
17. Weib, M., Ender, J.H.G.: A 3D imaging radar for small unmanned airplanes-ARTINO. In: European Radar Conference 2005 (EURAD 2005), Paris, pp. 209–212 (2005)
18. Weiß, M., Gilles, M.: Initial ARTINO radar experiments. In: EUSAR 2010, Aachen, Germany, pp. 1–4 (2010)
19. Kefei, L.: Single-stimulus three-dimensional SAR experimental system and imaging technology research. University of Electronic Science and Technology of China, ChengDu, China (2010)
20. Dengler, R.J., Cooper, K.B., Chattopadhyay, G., et al.: 600 Ghz imaging radar with 2 Cm range resolution, pp. 1371–1374 (2007)
21. Chattopadhyay, G., Cooper, K.B., Dengler, R., et al.: A 600 Ghz imaging radar for contraband detection (2008)

22. Cooper, K.B., Dengler, R.J., Chattopadhyay, G., et al.: A high-resolution imaging radar at 580 Ghz. IEEE Microwave Wirel. Compon. Lett. **18**(1), 64–66 (2008)
23. Cooper, K.B.: Performance of a 340 Ghz radar transceiver array for standoffsecurity imaging. In: International Conference on Infrared, Millimeter, and Terahertz Waves (2014)
24. Reck, T., Jung-Kubiak, C., Siles, J.V., et al.: A silicon micromachined eight-pixel transceiver array for submillimeter-wave radar. IEEE Trans. Terahertz Sci. Technol. **5**(2), 197–206 (2015)
25. Chattopadhyay, G., Reck, T., Lee, C., et al.: Micromachined packaging for Terahertz systems. Proc. IEEE **105**(6), 1139–1150 (2017)
26. Sheen, D.M., Mcmakin, D.L., Barber, J., et al.: Active Imaging at 350 Ghz for security applications. In: Proceedings of SPIE - The International Society for Optical Engineering, pp. 6948–69480M (2008)
27. Sheen, D.M., Severtsen, R.H., Mcmakin, D.L., et al.: Standoff concealed weapon detection using a 350-Ghz radar imaging system. In: Proceedings of SPIE – The International Society for Optical Engineering, vol. 7670, no. 1, pp. 115–118 (2010)
28. Essen, H., Wahlen, A., Sommer, R., et al.: Development of a 220-GHz experimental radar. In: 2008 German Microwave Conference, Hamburg, pp. 1–4 (2008)
29. Essen, H., Wahlen, A., Sommer, R., et al.: High-bandwidth 220 GHz experimental radar. Electron. Lett. **43**(20), 1114–1116 (2007)
30. Ahmed, S., Schiessl, A., Gumbmann, F., et al.: Advanced microwave imaging. IEEE Microwave Mag. **13**(6), 26–43 (2012)
31. Ahmed, S.S., Genghammer, A., Schiessl, A., et al.: Fully electronic active E-band personnel imager with 2 m2 aperture based on a multistatic architecture. IEEE Trans. Microw. Theory Tech. **61**(1), 651–657 (2013)
32. Gao, X., Li, C., Gu, S., et al.: Design, analysis and measurement of a millimeter wave antenna suitable for stand off imaging at checkpoints. J. Infrared Millim. Terahertz Waves **32**(11), 1314–1327 (2011)
33. Gu, S.M., Li, C., Gao, X., et al.: Terahertz aperture synthesized imaging with fan-beam scanning for personnel screening. IEEE Trans. Microw. Theory Tech. **60**(121), 3877–3885 (2012)
34. Gu, S., Li, C., Gao, X., et al.: Terahertz aperture synthesized imaging with fan-beam scanning for personnel screening. IEEE Trans. Microwave Theory Tech. Mtt **60**(12), 3877–3885 (2012)
35. Liu, W., Li, C., Sun, Z., et al.: A Fast Three-Dimensional Image Reconstruction with large depth of focus under the illumination of terahertz gaussian beams by using wavenumber scaling algorithm. IEEE Trans. Terahertz Sci. Technol. **5**(6), 967–977 (2015)
36. Li, C., Gu, S., Gao, X., et al.: Image reconstruction of targets illuminated by terahertz gaussian beam with phase shift migration technique. In: International Conference on Infrared, Millimeter, and Terahertz Waves, pp. 1–2 (2013)
37. Sun, Z., Li, C., Gao, X., et al.: Minimum-entropy-based adaptive focusing algorithm for image reconstruction of terahertz single-frequency holography with improved depth of focus. IEEE Trans. Geosci. Remote Sens. **35**(7), 8–93 (2015)
38. Gu, S., Li, C., Gao, X., et al.: Three-dimensional image reconstruction of targets under the illumination of terahertz gaussian beam-theory and experiment. IEEE Trans. Geosci. Remote Sens. **51**(4), 2241–2249 (2013)
39. Zhang, B., Pi, Y., Li, J.: Terahertz imaging radar with inverse aperture synthesis techniques: system structure, signal processing, and experiment results. IEEE Sens. J. **15**(1), 290–299 (2015)
40. Liu, T., Pi, Y., Yang, X.: Wide-angle CSAR imaging based on the adaptive subaperture partition method in the Terahertz Band. IEEE Trans. Terahertz Sci. Technol. **8**(2), 165–173 (2018)

41. Yang, X., Pi, Y., Liu, T., et al.: Three-dimensional imaging of space debris with space-based terahertz radar. IEEE Sens. J. **18**(3), 1063–1072 (2018)
42. Gao, J., Deng, B., Qin, Y., et al.: Efficient terahertz wide-angle NUFFT-based inverse synthetic aperture imaging considering spherical wavefront. Sensors **16**(12), 2120 (2016)
43. Jiang, Y., Deng, B., Qin, Y., et al.: Experimental results of concealed object imaging using terahertz radar. In: 8th International Workshop on Electromagnetics:Applications and Student Innovation Competition, iWEM 2017, London, United Kingdom, pp. 16–17 (2017)
44. Yang, Q., Deng, B., Wang, H., et al.: ISAR imaging of rough surface targets based on a terahertz radar system. In: 2017 Asia-Pacific Electromagnetic Week, 6th Asia-Pacific Conference on Antennas and Propagation, Xi'an, China (2017)
45. Cheng, B., Lu, B., Gao, J.K., et al.: Standoff 3-D imaging with 4Tx-16Rx MIMO-based radar at 340 GHz. In: 2017 42nd International Conference on Infrared, Millimeter, and Terahertz Waves (IRMMW-THz), Cancun, Mexico, pp. 1–2 (2017)
46. Cui, Z., Gao, J., Lu, B., et al.: Real-time 3-D imaging system with 340GHz sparse MIMO array J. Infrared Millim. Waves **36**(1), 102–106 (2017)
47. Sheen, D.M., Mcmakin, D.L., Hall, T.E., et al.: Real-Time Wideband Cylindrical Holographic Surveillance System (1999)
48. Ahmed, S.S., Genghammer, A., Schiessl, A., et al.: Fully electronic – band personnel imager of 2 M aperture based on a multistatic architecture. IEEE Trans. Microw. Theory Tech. **61**(1), 651–657 (2013)
49. Ahmed, S.S., Schiessl, A., Schmidt, L.P.: Novel fully electronic active real-time millimeter-wave imaging system based on a planar multistatic sparse array. In: IEEE MTT-S International Microwave Symposium Digest. IEEE MTT-S International Microwave Symposium 1 (2011)
50. Cui, Z., Gao, J., Lu, B., et al.: Real-time 3-D imaging system with 340GHz sparse MIMO array. J. Infrared Millim. Waves **36**(1), 102–106 (2017)
51. Qiong, W.: Implementation of GPU-based terahertz MIMO array imaging algorithm. Xidian University (2019)
52. Wen, J.: Research on terahertz radar imaging of human hidden targets and its speckle and polarization characteristics. China Academy of Engineering Physics (2019)
53. Sheen, D.M., Mcmakin, D.L., Severtsen, R.H.: Concealed explosive detection on personnel using a wideband holographic millimeter-wave imaging system. In: Proceedings of SPIE - The International Society for Optical Engineering (1996)
54. Bertl, S., Detlefsen, J.: Effects of a reflecting background on the results of active Mmw Sar imaging of concealed objects. IEEE Trans. Geosci. Remote Sens. **49**(10), 3745–3752 (2011)
55. Qiao, L., Wang, Y., Zhao, Z., Chen, Z.: Exact reconstruction for near-field three-dimensional planar millimeter-wave holographic imaging. J. Infrared Millim. Terahertz Waves **36**(12), 1221–1236 (2015). https://doi.org/10.1007/s10762-015-0207-z
56. Sun, Z., Li, C., Gu, S., et al.: Fast three-dimensional image reconstruction of targets under the illumination of terahertz gaussian beams with enhanced phase-shift migration to improve computation efficiency. IEEE Trans. Terahertz Sci. Technol. **4**(4), 479–489 (2014)
57. Wang, Y.: Research on cylindrical three-dimensional imaging algorithm of terahertz radar. University of Electronic Science and Technology of China (2016)
58. Jiang, Y.: Research on three-dimensional imaging technology of terahertz array radar. National University of Defense Technology (2018)
59. Ge, J., Jie, L., Wen, J., Binbin, C., Jianxiong, Z., Jian, Z.: Holographic radar imaging algorithm based on the concept of range Doppler. J. Infrared Millim. Waves **36**(03), 367–375 (2017)
60. Geng, S., Jiang, Z., Cheng, Z., et al.: Research on FM-CW SAR range-Doppler imaging algorithm. J. Electron. Inf. Technol. (2007)

Evaluating Recursive Backtracking Depth-First Search Algorithm in Unknown Search Space for Self-learning Path Finding Robot

T. H. Lim(✉) and Pei Ling Ng

Universiti Teknologi Brunei, Tungku Highway, Gadong, Brunei Darussalam
lim.tiong.hoo@utb.edu.bn

Abstract. Various path or route solving algorithms have been widely researched for the last 30 years. It has been applied in many different robotic systems such as bomb sniffing robots, path exploration and search rescue operation. For instance, an autonomous robot has been used to locate and assist a person trapped in the jungle or building to exit. Today, numerous maze solving algorithms have been proposed based on the some information available regarding the maze or remotely control. In real scenario, a robot is usually placed in an unknown environment. It is required for the robot to learn the path, and exhibit a good decision making capability in order to navigate the path successfully without human' assistance. In this project, an Artificial Intelligence (AI) based algorithm called Recursive Backtracking Depth First Search (RBDS) is proposed to explore a maze to reach a target location, and to take the shortest route back to the start position. Due to the limited energy and processing resource, a simple search tree algorithm has been proposed. The proposed algorithm has been evaluated in a robot that has the capability to keep track of the path taken while trying to calculate the optimum path by eliminating unwanted path using Cul-de-Sac technique. Experimental results have shown that the proposed algorithm can solve different mazes. The robot has also shown the capability to learn and remember the path taken, to return to the start and back to target area successfully.

Keywords: Maze solving · Search space · Graph theory · Search tree · Depth first search · Cul-de-sac

1 Introduction

Technology advancements in the area of robotics have made enormous contributions in both industrial and social domains. The applications of automation and

S. Shi et al. (Eds.): AICON 2020, LNICST 356, pp. 531–543, 2021.
https://doi.org/10.1007/978-3-030-69066-3_47

robotics have increased significantly from home to industrial automations, bomb sniffing robot, and search rescue operation [1,2]. For such applications to operate in a dynamic environment, the systems must exhibit autonomy and decision making properties. In a fully autonomous robot, it is important that it can perceive and adapt to the environment, and operate for an extended period without human intervention. After operating for a period of time, it should learn about its new environment and become more intelligent. In the perspective of intelligence in a robot, it can be derived from the fields of Artificial Intelligence (AI) that compromises of three separate entities namely search algorithm, knowledge representation and the extent of implementing these ideas.

In the application of search and rescue, speed and efficiency is much more concerned due to the urgency to locate the position of wounded or trapped persons in order to save them [3]. These complex situation of accident locale can be abstracted using a maze. Therefore, to test the intelligence, these robots are required to solve problem such as solving a maze. A maze is network of paths and hedges designed as a puzzle through which solver has to find a way or solution, usually from an entrance to a goal or another exit. In general, there are two types of mazes [4] namely model-based maze in which global model is available, and sensor information-based maze of which information about the maze is unknown.

Many algorithms for maze navigation and maze solving have been proposed and continue to be improved over the years [3,5–7]. Artificial Intelligence plays a vital role in defining the best possible method of solving maze effectively, such that Graph Theory appears to be an efficient tool while designing proficient maze solving techniques [8]. Primarily, the search problem includes either an explicit initial or goal node in which one has to find the path connecting them or simply find the goal which is implicitly defined [9].

It is fundamental for the robot to successfully navigate to its goal. This is usually achieved either with direct human navigation, a predefined program or set of general rules programmed in the systems. These approaches usually assume that global model or view is available in advance [7]. This assumption may not be always true as the environment to be traversed is usually unknown in advanced. Each maze is usually different and may change dynamically. Hence, the use of fixed rules or predefined path is not the best approach. In order for a robot to traverse through an unknown environment from a source position to a target, it needs to percept and act using current available information. The key information of the maze's search space, situation of branches, dead-end or passing-through crossing and current perception of the robot are all acquired locally from sensors attached to the robot during the search.

In this paper, a new Artificial Intelligence (AI) based algorithm called Recursive Backtracking Depth-first Search (RBDS) is proposed and evaluated in real robots. To the best of our knowledge, this is the first time a search tree based AI algorithm has been applied in a low processing and powered robot. The motivation behind re-using an existing simple known AI algorithm is that we do not want to design a whole new algorithm. In contrast, the simplest search algorithm can easily be applied and what is proposed in this paper can be supported by any microcontroller such as Arduino and microPy. We have also incorporated a route optimization algorithm to allow the robot use the shortest and optimal path to traverse between the goal and initial start point.

The main contributions of this paper are: 1) A novel AI-Based algorithm called Recursive Backtracking Depth-first Searh (RBDS) using an uninformed Depth First Search (DFS) to explore an unknown search space and applied an optimization algorithm called Recursive Cul-de-Sac Backtracking (RCB) to eliminate all unwanted path for route optimisation; 2) A quantitative analysis of the AI-Based RBDS algorithm on a robotic platform based on a low power, and low cost microcontroller robot to solve an unknown maze autonomously. A reinforcement learning technique is used by the robot to learn about the new environment and update its environment. Initially, the robot traverse through an unknown search space using the information received from an input sensor. Once the goal is located, the robot will identify the optimal path by eliminating the dead ends using RCB. In the subsequent run, the robot can traverse between the goal and initial position using the optimal path. The results have shown the proposed algorithm can help to solve any maze and determine the optimum path..

The rest of this paper is organized as follows. Section 2 presents related work. Section 3 provides an overview of AI based RBDS follows by the experimental setup to analyse the performance of the proposed system in Sect. 4. In Sect. 5, we summarize with our conclusions.

2 Related Works

Autonomous mobile robots need to operate safely and reliably. At the same time, they are expected to minimize energy consumption, travel time and distance. Mishra and Bande [10] conducted a comparative study on the path length and time taken to solve the maze using 3 different algorithms namely basic wall follower, Djikstra's shortest path, and flood fill algorithm [11]. Mishra and Bande show that the Djikstra's algorithm can be effectively used if both time and hardware are not critical and supported. However, both constraints need to be satisfied, flood fill would be preferable compared to the others despite the complex calculation required. The basic and simplest wall follower logic is the most supported algorithm in hardware. In another study, Sharma [12] the same results are also obtained where flood fill algorithm outperformed basic wall follower logic.

Another comparative study on wall follower, Lee's algorithm [13] and flood fill algorithm [5] were presented by Gupta and Sehgal [14]. Equally, Lee's algorithm is an application of Dijkstra's algorithm that uses Breadth First Search (BFS). It works in two phases namely the filling phase in which cells are marked while retrace phase is derived from the concept of backtracking [13]. In BFS, the tree is examined from top down such that every node at depth d is examined before any node at depth d+1. Starting from the source vertex, the entire layer of unvisited vertex is visited by some vertex in the visited set and the recently visited vertexes are added with the visited set [15]. Although floodfill is by far the most famous and efficient algorithm to solve all types of maze, it is commonly set with a preset target point. Initially, the algorithm assigns to a value of each cell representing the distance between the cell and the target cell. The robot follows the path of decreasing value and is updated in every single step.

Finally, in Depth First Search (DFS), the search starts from the root of a graph. The terminal nodes are examined from left to right. The exploration continues toward the deeper region of the tree or graph until it reach a dead-end and needs to back track. The algorithm starts searching from a specific vertex and then it navigate by branching out corresponding vertices until it reach the final goal [16]. The whole maze is mapped as graph where the nodes or vertexes are considered as maze cells. The search order may not be in specific order, it only commits the idea of always expanding a node as deep in the search tree as possible. Hence a search from right to left is also possible.

3 AI-Based Recursive Backtracking Depth-First Search (RBDS) Algorithm

In this section, an overview of design and methodology used for Artificial Intelligence Based Maze Solving robot is presented. This includes the construction and implementation of two-wheel robot that is capable of solving a line maze using an array of infra-red sensors. A line maze is used instead of the real wall maze as it is cheaper to built and modify. The robot will be used to evaluate the propose Artificial Intelligence (AI) based algorithm known as Recursive Backtracking Depth First Search (RBDS), capable of searching through an unknown search space to identify a target without any guidance.

3.1 Systems Design

The system design of the autonomous robot consists of three fundamental phases in the design process. The first involved the selection and integration of individual components of a mobile robot to operate according to the system specification required to traverse the maze. Secondly, the design of the RBDS algorithm and apply Cul-de-Sac approach to elimate any deadend from the database in order to optimize the route taken. Finally, the interface the system software with hardware components with the RBDS algorithm in order for the evaluate the robot to operate autonomously using RBDS.

The proposed path finding algorithm is based on an uninformed or blind search, Depth First Search algorithm. The operating environment is in the form of a maze with black lines representing the path and having different kinds of junctions. The starting point can be randomized with target goal differentiating from other paths. Figure 1 shows the overall system of block diagram of maze solving robot.

The working principle of robot in solving the line maze is provided by the systems program functions listed as follows;

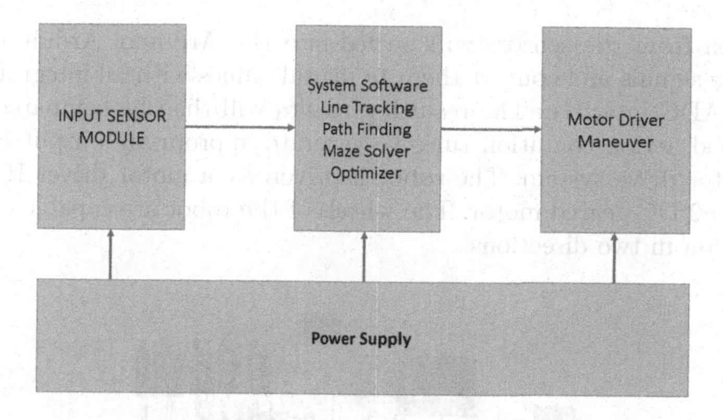

Fig. 1. Robotic systems design.

- Line tracking system In line tracking system, the robot follows a black line maze by sensing the environment using the input sensor module. The sensor values are sent to the software system, in which the microcontroller will process. The program function will instruct the robot to follow the line as it percepts and acts accordingly to the instruction sent from the system to the motor drive system.
- Depth First Search Based on the output sensory module information, the robot teaches itself to identify the route path taken using path finding algorithm. The robot will apply the Depth First Search Algorithm to explore the maze to achieve its goal and use its knowledge through the computational perception and action condition rules to learn about the route to reach the goal.
- Recursive Backtracking Path Optimisation Function After identifying the route path, the robot enters the learning phase and optimises the path. In subsequently run, the robot returns to its original source position using the optimised path. Once it returned to the original position, the robot can now operate for instance transport and transfer goods from the start to the desired position using the optimised path.

3.2 Platform Design

Figure 2 shows the circuit diagram of hardware system of the robot. It consists of 3 major hardware components namely the processing unit, the input unit and the output unit. An Arduino microcontroller processing unit is used to operate the robot. As the core component of the systems, it is responsible for processing the input data and storing the information produced during the computational process. The input and output units are responsible robot's perception and actions. Input information regarding the line tracks, dead end and turns will be obtained from the five channel infrared reflective sensor array module. The track or movement can be determined by sending an infrared signal to the line while photo-transistor receive and sense the signal correspondingly. The correspond information from the sensors will be fed into the Arduino. Arduino will then process the signals and convert them to digital values with an integrated analog to digital ADC interface. The resulting results will then be compared with the programmed action condition rules to generate appropriate output instruction to the motor drive system. The robot is driven by a motor driver H-bridge IC L298N and 2 DC geared motor. The wheels of the robot are capable of independent rotation in two directions.

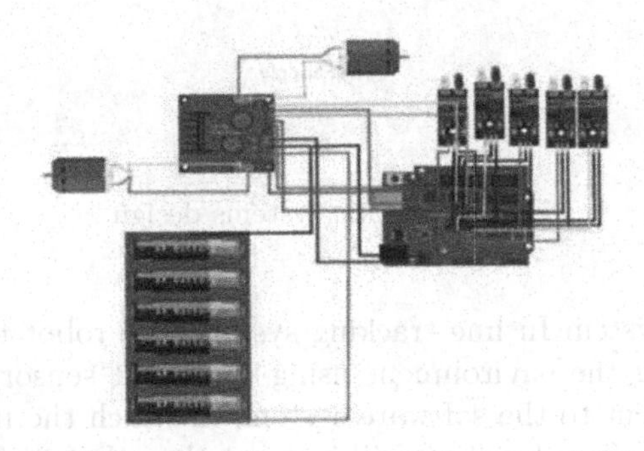

Fig. 2. Hardware circuit design for the autonomous robot

3.3 Software Design

The system application design of the AI based autonomous maze solving robot is shown in Fig. 3. C programming language is used to develop the AI based RBDS algorithm to traverse and solve a maze. The program code acts as the decision maker using Depth First Search Algorithm embedded in the microcontroller. The robot percepts and acts based on the output for particular set of combination of sensors.

The working principle of robot in solving line maze is as follows;

- Line path tracking.
  For line path tracking, the robot follows a line maze by perceiving its environment using array IR sensors. It is programmed to follow the line perfectly with both motors rotating in forward direction. Any deviation from the line will cause the motor change accordingly to ensure it is still on and within the right path. The most right and left sensors are used for taking decision on junctions while the other 3 middle sensors are for tracing the black line, for forwarding.
- Mobile robot navigation.
  For navigation, the robot is dependent on the sensors. The output from the sensor is fed into the Arduino, microcontroller which is programmed based on the action condition rules. As the input is processed, the corresponding output is sent to the H-bridge, L298N motor driver to navigate the robot by rotating it in corresponding directions.

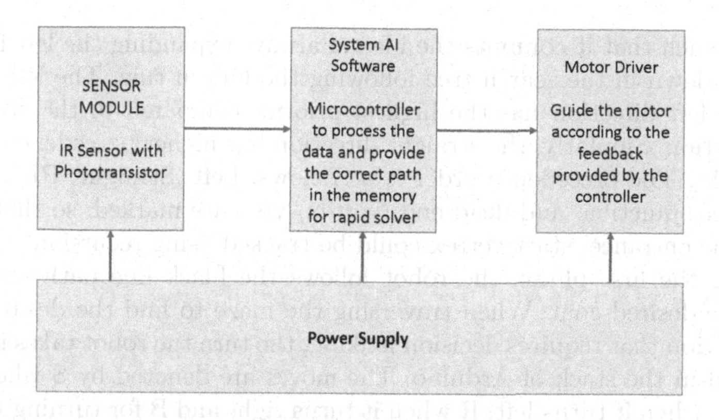

Fig. 3. Software system block diagram

- Knowledge based navigation and reinforced learning.
  The knowledge based navigation is incorporated by means of Depth First Search, left search variant rule. The first priority is given for the left, second to the straight and least to the right. Based on the knowledge, the robot navigates accordingly. In the teaching phase, it uses the rule to navigate and traverse through all paths including dead ends. When it encounter dead end, it takes a back turn or 180° turns and traverses the same path again. The turns are stored only at junctions, intersection, later used by the robot for navigation in the subsequent phase. In the second run, Depth First Search is not applied, instead the Cul-de-Sac technique is applied such that optimised path is used to traverse the robot back to its original position by eliminating unnecessary turn and dead ends. Alternatively, the robot can also go back to the goal from the start position using the optimised path learnt.
  The use of Artificial Intelligence concept is similar to the reinforcement learning applied in the second run. Where it illustrates the concept of learning from mistake. The former teaching phase, enable the learning of how to solve the maze as it proceeds through dead ends. Subsequently the next phase involves Cul-de-Sac approach by eliminating all the dead ends and remove them from the stack memory. As a result, an optimal path to the goal from the start position is identified.

3.4 Algorithms

A Depth First Search, left hand rule with Cul-de-Sac approach is used as the algorithms to solve the line maze. This algorithm is qualitative in nature, requiring no map of environment, no fundamental matrix and no assumption as used in other algorithms.

Line Tracking Depth First Search. Similar to classical strategy for exploring maze, DFS algorithm is used for exploring a maze, The DFS follows the left hand rule such that it commits the idea of always expanding the left hand node and move down in the search tree following the known rule. The left hand rule states the left direction has the highest priority compared to the straight and right direction. Similarly, the straight direction has higher percedence compared to the right. The precedence order is as follows: Left, Straight, Right. At each intersection, junction, and dead end (vertex) visit are marked, so that the path back to the entrance, start vertex could be tracked using recursion stack.

During the first phase, the robot follows the black line path and traverse to find the desired goal. When traversing the maze to find the desired goal, at every junction that requires decision making, the turn the robot takes is recorded and stored in the stack of Arduino. The moves are denoted by S when it move forward, L when it turns left, R when it turns right and B for turning back when it encounters dead end. When the desired goal is reached, the robot finishes the first phase and proceeds to the next phase of learning. The robot will then traverse from the target to the starting point and vice versa though the optimal path. The optimal path can be determined using Cul-de-Sac approach to simplify the path.

Cul-de-Sac Approach. Once the goal node is found, path simplification can be achieved using Cul-de-Sac approach to find the optimal path in the traversed maze. At every turn, the length of the recorded path increases by 1. The algorithm takes the turn stored in the stack during the first phase and applies Cul-de-Sac approach, respective reduction path rules accordingly and removes all the dead end, back turn made previously.

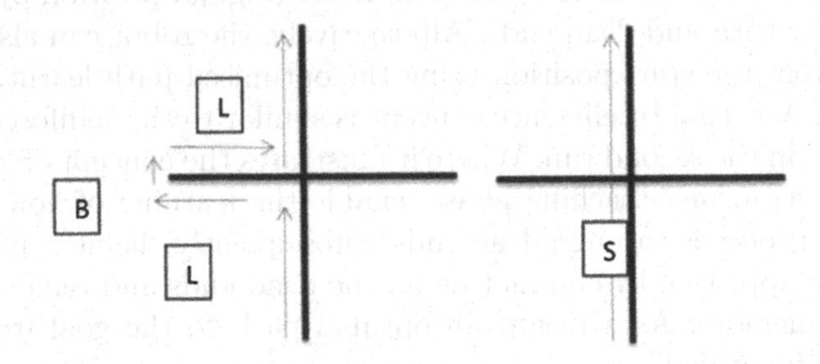

Fig. 4. Correspondence of LBL sequence and equivalent path

With path simplification, the strategy is that whenever the robot encounters a sequence xBx, the path can be simplified by eliminating the dead end and replaced with a turn corresponding to the total angle. The path can only be simplified if the second to the last turn was "B" a dead end and the path length must be greater than 3. Considering the sequence "LBL", after turning 90° to the left and 180° back and 90° to the left again, the net effect of the robot is heading back in its original direction. The path can be simplified to 0° turn, a single "S" move.

The Cul-de-Sac algorithm is based on the data in Table 1. In a traversed maze, the algorithm is applied onto the list of path moves stored in an array stack several times until all the necessary "B" moves are removed. Figure 4 exhibits two functionally equivalent paths from the start to the end.

Table 1. Path simplification

Original path sequence	Reduced path
LBL	S
LBS	R
LBR	B
SBL	R
SBR	L
SBS	B
RBR	S
RBL	B
RBS	L

L = left, R = right, B = back, S = straight

3.5 Robot Driving Control System

To allow the system to drive the robot autonomously, the robot is mounted with five channel infra-red sensors. These five sensors will be used to detect a straight line, turns, dead end or goal and the robot can use the detected information to navigate the robot. With five input values, the sensors can have numerous configuration that allow the robot to make decision whether to move straight or to turn. With digital output, there are 32 possible combinations. With that, '0' output indicates sensor sense black line and '1' output sense otherwise.

As shown in Fig. 5, the far most left and right sensors detect turning and intersection. The three sensors in the middle are evenly spaced. Chance of two sensors detecting line at the same time when it adjust itself is high. Thus, few combinations are accounted for in the action condition rule. These sensors looks directly down on the line track and processed by the program to determine the correct action. The following figures exhibit some of the possible sensor output combinations for following a line.

As the robot is expected to follow the lines and find the path from the source position to the desired goal. There are various junctions in the line maze. Given a maze with no loops, there are only 8 possible scenarios that the robot would encounter. First, it follows the line. When it reaches an intersection, the sensor will decide what type of intersection it is and make appropriate turn. These steps continue in a loop until it reach the maze end.

With the path finding algorithm implemented, the robot acts accordingly as it perceives the environment, depending on the type of turns and intersection it encounters. In the case of left turn junction and right turn junction, the robot has no alternative but to make 90° turn. As of reaching a dead end, the robot make a 180° u-turn to backtrack. For the straight or right turn junction, a sub routine is created. The robot moves forward for a bit and perceive the current state's its in. If the robot sense there is line ahead, it moves forward as the rule prioritizes straight rather than right. As of the left or straight junction, the robot will always prioritize and turn left. The junctions can be identified according to the sensor readings tabulated in Table 2.

Table 2. Truth table For direction and motor rotation

S1	S2	S3	S4	S5	Junction	Rotation
0	0	0	0	0	Left	Left
1	1	0	1	1	Following line	Straight
1	1	0	0	0	Left	Left
1	1	1	0	1	Slight left	Left
1	1	0	0	1	Slight left	Left
1	0	1	1	1	Slight right	Right
1	0	0	1	1	Slight right	Right
0	0	0	1	1	Right	Right
0	0	0	0	1	Slight left	Left
1	0	0	0	0	Slight right	Right
1	1	1	1	1	No line	U-Turn
1	0	0	0	1	End maze	Stop

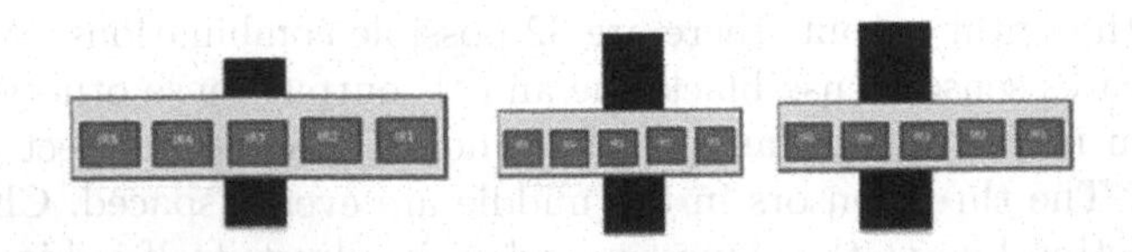

Fig. 5. Sensor combinations for following line

4 Performance Evaluation

For the evaluation of RBDF Search Algorithm, 4 line mazes are tested. Figure 6 shows the corresponding line mazes to be solve. The robot is placed at a defined source position with differentiated desired goal position to test if it is capable of solving the maze autonomously. Alternatively in each line maze with a fixed goal, the robot start position is also chosen randomly.

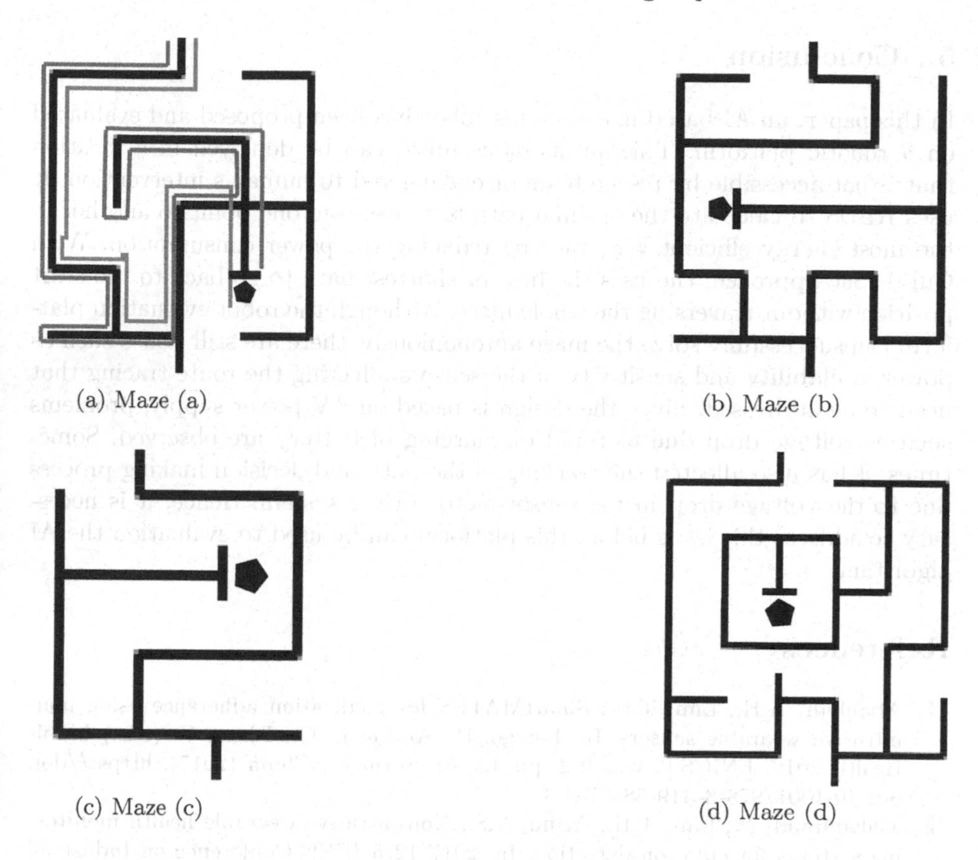

Fig. 6. Evaluation of RBDF on different mazes

4.1 Analysis

From the results obtained and observed, the algorithms are able to solve mazes in the teaching phase and learning phase. In the first phase, the learning phase indicated by the blue line in Maze (a), the robot traverse and store the paths it have taken until it reaches the goal. In the subsequent run indicated by pink line, using Cul-de-Sac approach it backtracks the path to reach the starting source position using an optimal path. Once it reaches the source position, it finishes the run by traversing through the optimal path to the desired goal. For looped maze as shown in Maze (d) Fig. 6, the fourth maze, the algorithm was not able to solve the maze. This is due to the Depth First Search algorithm, Left Hand Rule following such that the robot will always choose to move in the left direction instead of turning right to reach the goal.

5 Conclusion

In this paper, an AI based maze solving robot has been proposed and evaluated on a robotic platform. This autonomous robot can be deployed in a location that is not accessible by rescue team or endangered to human's intervention. It uses RBDS to calculate the optimal path to move from one point to another in the most energy efficient way, thereby reducing the power consumption. With Cul-de-Sac approach, the uses the best or shortest path to go back to the start position without traversing the whole maze. Although the robot evaluation platform can successfully solve the maze autonomously, there are still issues such as power availability and sensitivity of the sensors affecting the route tracing that need to be addressed. Since the design is based on 8 V power supply, problems such as voltage drop due to rapid discharging of battery are observed. Sometimes, it has also affected the tracking of the path and decision making process due to the voltage drop in the sensor-motor driver system. Hence, it is necessary to address this issue before this platform can be used to evaluation the AI algorithm.

References

1. Abdullah, A.H., Lim, T.H.: SmartMATES for medication adherence using non-intrusive wearable sensors. In: Perego, P., Andreoni, G., Rizzo, G. (eds.) MobiHealth 2016. LNICST, vol. 192, pp. 65–70. Springer, Cham (2017). https://doi.org/10.1007/978-3-319-58877-3_8
2. Muhammad, N., Lim, T.H., Arifin, N.S.: Non-intrusive wearable health monitoring systems for emotion detection. In: 2017 12th IEEE Conference on Industrial Electronics and Applications (ICIEA), pp. 985–989 (2017)
3. Stentz, A.: Optimal and efficient path planning for partially known environments. In: Hebert, M.H., Thorpe, C., Stentz, A. (eds.) Intelligent Unmanned Ground Vehicles. The Springer International Series in Engineering and Computer Science (Robotics: Vision, Manipulation and Sensors), vol. 388, pp. 79–82. Springer, Boston (1997). https://doi.org/10.1007/978-1-4615-6325-9_11
4. Jiang, H.L.: Designed of Wheeled Robot Based on Single Chip Computer, vol. 13 (2009)
5. Elshamarka, I., Saman, A.: Design and implementation of a robot for maze-solving using flood-fill algorithm. Int. J. Comput. Appl. **56**(5), 8–13 (2012)
6. Dang, H., Song, J., Guo, Q.: An efficient algorithm for robot maze-solving. In: 2010 Second International Conference on Intelligent Human-Machine Systems and Cybernetics, vol. 2, pp. 79–82 (2010)
7. Cahn, D.F., Phillips, S.R.: ROBNAV: a range-based robot navigation and obstacle avoidance algorithm. IEEE Trans. Syst. Man Cybern. SMC **5**(5), 544–551 (1975)
8. Sadik, A.M.J., Dhali, M.A., Farid, H.M.A.B., Rashid, T.U., Syeed, A.: A comprehensive and comparative study of maze-solving techniques by implementing graph theory. In: International Conference on Artificial Intelligence and Computational Intelligence, vol. 1, pp. 52–56 (2010)
9. Cai, J., Wan, X., Huo, M., Wu, J.: An algorithm of micro mouse maze solving. In: 2010 10th IEEE International Conference on Computer and Information Technology, pp. 1995–2000 (2010)

10. Mishra, S., Bande, P.: Maze Solving Algorithms for Micro Mouse, pp. 86–93, January 2009
11. Wang, H., Yu, Y., Yuan, Q.: Application of Dijkstra algorithm in robot path-planning. In: Second International Conference on Mechanic Automation and Control Engineering, pp. 1067–1069 (2011)
12. Mishra, S., Bande, P.: Maze solving algorithms for micro mouse. In: 2008 IEEE International Conference on Signal Image Technology and Internet Based Systems, pp. 86–93 (2008)
13. Lee, C.Y.: An algorithm for path connections and its applications. IRE Trans. Electron. Comput. EC **10**(3), 346–365 (1961)
14. Gupta, B., Sehgal, S.: Survey on techniques used in autonomous maze solving robot. In: 2014 5th International Conference - Confluence The Next Generation Information Technology Summit (Confluence), pp. 323–328 (2014)
15. Ginsberg, M.: Essentials of Artificial Intelligence. Elsevier Science (2011)
16. Adil, M.J.S., Maruf, A.D., Hasib, M.A.B.F.: A comprehensive and comparative study of maze solving techniques by implementing graph theory. In: International Conference on Artificial Intelligence and Computational Intelligence (2010)
17. Lim, T.H., Lau, H.K., Timmis, J., Bate, I.: Immune-inspired self healing in wireless sensor networks. In: Coello Coello, C.A., Greensmith, J., Krasnogor, N., Liò, P., Nicosia, G., Pavone, M. (eds.) ICARIS 2012. LNCS, vol. 7597, pp. 42–56. Springer, Heidelberg (2012). https://doi.org/10.1007/978-3-642-33757-4_4
18. Lim, T.H., Bate, I., Timmis, J.: A self-adaptive fault-tolerant systems for a dependable wireless sensor networks. Des. Automat. Embedded Syst. **18**(3–4), 223–250 (2014)

FTEI: A Fault Tolerance Model of FPGA with Endogenous Immunity

Jie Wang, Shuangmin Deng(✉), Junjie Kang, and Gang Hou

School of Software Technology, Dalian University of Technology, Dalian 116023, China
wang_jie@dlut.edu.cn, ku_nan_xi@163.com, 18840837856@163.com, Hg.dut@163.com

Abstract. FPGA emerges as a very promising AI chip and algorithm hardware accelerator. However, the FPGA is susceptible to complex and changeable environment, which leads to circuit configuration information faults. To address this issue, we propose FTEI, a fault tolerance model of FPGA with endogenous immunity. At fault detection phase, we put forward a fault detection models based on optimized logistic regression classification and use it to establish a fault model matching library. During fault recover stage, we use fault configuration library and online evolution to recover faults. In order to improve the success ratio of online evolution, we propose RLAGA, an adaptive genetic algorithm based on reinforcement learning. Experiments on typical functional circuits, 8-bit parity verifier and 2-bit multiplier, demonstrate that the fault detection accuracy rates reach 94.4% and 93.2%, and the fault recover success rates of RLAGA are 100% and 90%, which significantly improves FPGA errors detection and recover effectiveness.

Keywords: FPGA · Fault tolerance · Fault detection · Fault recover

1 Introduction

The extraordinary advantages of FPGA are high concurrency, low delay, and reconfiguration. Taking these advantages FPGA emerges as a very promising AI chip and algorithm hardware accelerator. However, a non-negligible issue of FPGA is that complex and changeable environment, high temperature, high pressure and high radiation, will change the configuration information on FPGA, which causes Single Event Upset (SEU). Once the FPGA chip occurs SEU errors that cannot be ruled out timely, the output results of FPGA will be wrong. More seriously, it will lead to equipment stagnation. Hence, it is a core issue to improve the reliability of FPGA chips.

In literature, flourishing researches have been achieved on fault tolerance technology. In general, approaches can be broadly grouped into three categories:

S. Shi et al. (Eds.): AICON 2020, LNICST 356, pp. 544–557, 2021.
https://doi.org/10.1007/978-3-030-69066-3_48

redundancy technology [1,2], reconfigurable technology [3,4] and evolvable hardware technology [5]. Although much effort has been devoted to redundancy technology because of its simple design ideas, this method increases the complexity of circuit design rapidly and large resource consumption. Reconfigurable technology refreshes the system by reconfiguring configuration information, saving lots of resource. However, the fault detection and location methods [6] are required to highly accurately locate the fault circuit of the system, which greatly increases the difficulty of system design. The recover technology based on evolvable hardware, using the characteristics of self-organization, self-adaption and self-recovery, can recover FPGA faults with less hardware circuit resources and has better fault tolerance performance.

In order to achieve efficient and real-time recover of SEU errors, our comprehensive analysis of the above methods found that the following three challenges need to be solved. The first challenge is (1) how to improve the real-time and accuracy of fault location. The second challenge is (2) how to ensure the normal operation of the circuit system while recovering faults. The third challenge is (3) how to improve the efficiency of fault recover under strict space-time constraints.

To tackle all the challenges mentioned above, in this paper we propose FTEI, a Fault tolerance model of FPGA with endogenous immunity. It endows FPGA fault perception, fault memory, and environment adaptation to improve the reliability of the FPGA platform. We realize fault perception through fault detection and location mechanism. The realization of fault memory and environment adaptation is achieved through fault recover mechanism. In general, the main contributions of this paper are summarized as below:

- To realize real-time fault location, we propose to take the FTRL-optimized logistic regression classification algorithm as the fault detection model.
- We propose pre-setting chromosome of known faults in fault configuration library to save recover time and truth table of circuits in circuit truth table module to guarantee the normally operation of fault circuits.
- To realize faults recover of unknown faults we recover circuit by online evolution and RLAGA algorithm is proposed to raise the success rate of online evolution.

2 Related Work

FPGA is susceptible to complex and changeable environment. Therefore, enhancing the reliability of FPGA has won a lot of research interest. These researches can be classified into two categories: fault-tolerant methods with fault detection capabilities and fault-tolerant methods with fault shielding capabilities.

2.1 Fault-Tolerant Methods with Fault Detection Capabilities

The fault-tolerant methods with fault detection capability achieve troubleshooting by adding additional fault detection resources to the system design. Wang et al. [7] took deep learning algorithms as a fault detection model to monitor runtime data. Reorda et al. [8] utilized additional logic of carrying chains and hard links to perform error detection to implement fault detection and correct single and two errors that affected FPGA configuration memory due to configuration bit flips. Du et al. [9] exploited the bitstream analysis tool readback bitstream to obtain the current status and absolute addresses of D flip-flops (DFFs) and storage units. By analyzing the above information, the fault location is obtained. Ranjbar et al. [10] devised a method which could transfer the effects of faults occurring in the LUT (Look-Up-Table) to triggers facilitating fault detection. Although the above methods can be used to locate circuit faults, it cannot guarantee the real-time performance of circuit fault detection and increase the difficulty of circuit design and realization.

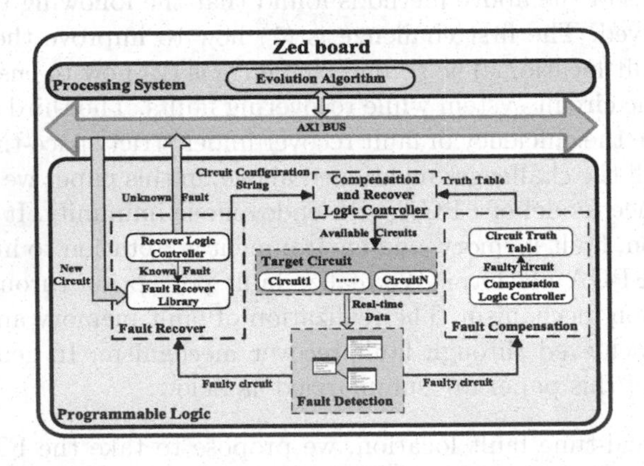

Fig. 1. Chip real-time self-healing system architecture

2.2 Fault-Tolerant Methods with Fault Shielding Capabilities

Triple Modular Redundancy (TMR) can directly shield circuit faults when a single modular circuit fails, which can effectively protect the system from circuit faults and maintain the normal operation of the circuit system [11]. Given the hardware resource consumption problem of TMR, Schweizer et al. [12] proposed a strategy using unused functional units for redundant calculation on a coarse-grained reconfigurable architecture and realized a low-cost TMR strategy. Experiments showed that this strategy reduced the area of hardware resources by 12.8% compared with the traditional TMR method. Burdyshev et al. [13] reduced a large number of extra hardware resources consumed by TMR technology to achieve fine-grained TMR by calculating the combination of channel redundancy and transistor redundancy. Although the above methods can save hardware resource, it makes the circuit more complex and difficult.

To overcome these above shortcomings, we develop FTEI, a fault tolerance model of FPGA with endogenous immunity. Besides real-time fault perception, the fault memory and environment adaptation are realized.

3 The Proposed Model

In this section, we generalize FTEI architecture firstly. Secondly, we introduce fault detection and location mechanism. Finally, the circuit fault recover mechanism are depicted.

3.1 Fault Tolerance Model of FPGA with Endogenous Immunity (FTEI)

The system structure is shown in Fig. 1. These circuits are divided into modules according to their functions in the design stage to facilitate detection and recover. The target circuit module runs on FPGA which needs to be detected and recovered when faults occur to it. The detection module is exploited to supervise the real-time data of the system. Once the detection module detects circuit faults, the circuit needs to be recovered in servery clock time. In order to buy time for fault recover and ensure the normal operation of the circuit system, the compensation module is mentioned. We pre-store truth table data of target circuits in the circuit truth table and the compensation logic controller is utilized to control data transfer of the compensation module. The fault recover module consists of two parts: fault recover library and evolution algorithm. We preset some circuit configuration chromosome in the fault configuration library whose faults are known. For unknown faults, we use evolution algorithm to recover and then store obtained configuration chromosome into the fault configuration library. The process is controlled by recover logic controller.

3.2 Fault Detection and Location Mechanism

In order to realize the precise location of fault circuit, this paper proposes a fault detection model, FTRL-optimized logistic regression, and uses it to establish a fault model matching library.

Fault Detection Mechanism. The fault detection process is divided into two stages: fault detection model acquisition and fault online detection. The fault detection model is obtained through offline training. The offline training of fault detection model needs to obtain a large number of circuit fault data. In this paper, a software tool is utilized to randomly inject faults into the Cartesian Genetic Programming (CGP) [16,17] code string running on the Virtual Reconfigurable Circuit (VRC) [7,15]. Considering that the circuit data can be divided into two categories: with fault and without fault, therefore, circuit fault detection is actually a binary classification problem. To get the fault detections model, this paper proposes FTRL-optimized logistic regression.

FTRL (Follow-the-regularized-Leader) online learning algorithm integrates the advantages of FOBOS algorithm and RDA algorithm, which can better guarantee the accuracy of model parameters and the sparsity of feature items in the training process. At the same time, after adding non smooth regular items, FTRL can get better sparsity value. So we utilized FTRL to optimize the parameters of the fault circuit logistic regression model [18].

The main iteration formula of FTRL can be expressed as Eq. (1):

$$w_{t+1} = \arg\min_{w}\{g_{1:t}w + \frac{1}{2}\sum_{s=1}^{t}\delta_s||w - w_s||_2^2 + \lambda_1||w||_1\} \quad (1)$$

The above formula is divided into three parts: $(\sum_{r=1}^{t} G^r)w$ is the cumulative gradient sum, which indicates the direction of loss function decline; $\frac{1}{2}\sum_{s=1}^{t}\delta_s||w - w_s||_2^2$ indicates that the new result should not deviate too far from the existing result; $\lambda_1||w||_1$ is the regular term, which is used to generate the sparse solution.

Let $\delta_{1:t} = \frac{1}{\sigma_t}$, $z_{t-1} = \sum_{r=1}^{t-1} G^r - \sum_{s=1}^{t-1}\delta_s - w_s$ we derive the following Eq. (2):

$$z_t = z_{t-1} + g_t - (\frac{1}{\sigma_t} - \frac{1}{\sigma_{t-1}})w_t \quad (2)$$

Substituting Eq. (2) into Eq. (1), the iteration formula can be rewritten as Eq. (3).

$$w_{t-1} = \arg\min_{w}\{(g_{1:t}w - \sum_{s=1}^{t}\delta_s w_s)w\frac{1}{2\sigma_t}||w - w_s||_2^2 + \lambda_1||w||_1 + c\} \quad (3)$$

Thus, Eq. (4) can be described as:

$$w_{t+1,i} = \begin{cases} 0, & |z_{t,i}| < \lambda_1 \\ -\sigma_t(z_{t,i} - \lambda_1 sgn(z_{t,i})), & z_{t,i}| \geq \lambda. \end{cases} \quad (4)$$

To improve the sparsity and accuracy of parameters, L2 regular term is added to the regular term of FTRL in this paper, and the method of mixed regular term is adopted to make the solution result of the fault model smoother, and the accuracy of model prediction is also increased.

The feature weight iterative formula of the optimized FTRL can be expressed as Eq. (5):

$$w_{t+1} = \arg\min_{w}\{g_{1:t}w + \frac{1}{2}\sum_{s=1}^{t}\delta_{1:t}||w - w_s||_2^2 + \lambda_1||w||_1 + \frac{1}{2}\lambda_1||w||_2^2\} \quad (5)$$

Here, g_s is the standard gradient; δ_s is learning an update strategy; $\delta_{1:t} = \frac{1}{\sigma_t}$ and σ_t is described as Eq. (6):

$$\sigma_{t,i} = \frac{\alpha}{\beta + \sqrt{\sum_{s=1}^{t} g_{s,t}^2}} \quad (6)$$

The formula (5) is expanded to obtain the optimal solution problem represented by Eq. (7):

$$w_{t+1} = \arg\min_{w}\{(g_{1:t}w - \frac{1}{2}\sum_{s=1}^{t}\delta_s w_s)w + \lambda_1||w||_1 + \frac{1}{2}(\lambda_2 + \sum_{s=1}^{t}\sigma_s)||w||_2^2 + \frac{1}{2}\sum_{s=1}^{t}\sigma_s)||w_s||_2^2\} \quad (7)$$

Where $\frac{1}{2}\sum_{s=1}^{t}\sigma_s)||w_s||_2^2$ for w is a constant, let:

$$z^{(t)} = g_{1:t}w - \sum_{s=1}^{t}\delta_s w_s \tag{8}$$

The Eq. (7) is equivalent to Eq. (9):

$$w_{t+1} = \arg\min_{w}\{z_t w + \lambda_2||w||_1 + \frac{1}{2}(\lambda_2 + \sum_{s=1}^{t}\sigma_s)||w||_2^2\} \tag{9}$$

According to each dimension of feature weight, the above formula is decomposed into n independent scalar minimization problems.

$$w_{t+1} = \arg\min_{w}\{z_{t,i}w_i + \lambda_1||w||_1 + \frac{1}{2}(\lambda_2 + \sum_{s=1}^{t}\sigma_s)w_i^2\} \tag{10}$$

Therefore, the weight formula shown in Eq. (11) can be obtained.

$$w_{t+1,i} = \begin{cases} 0, & |z_{t,i}| < \lambda_1 \\ -(\lambda_2 \sum_{s=1}^{t}\sigma_s)^{-1}(z_{t,i} - \lambda_1 sgn(z_{t,i})), & |z_{t,i}| \geq \lambda_1. \end{cases} \tag{11}$$

The pseudo code of circuit fault detection based on FTRL-optimized logistic regression is as Algorithm 1.

Algorithm 1. OPTIMAL SOLUTION OF LOGISTIC REGRESSION BASED ON FTRL-OPTIMIZED ALGORITHM

Input: $\alpha, \beta, \lambda_1, \lambda_2$;
1: **for** $i = 0$ to d **do**
2: Initialize $z_i = 0, n_i = 0$;
3: **end for**
4: **for** $i = 1$ to t **do**
5: The eigenvector of fault data set is X_t, Let $I = \{i|x_i \neq 0\}$;
6: **for** $i \in I$ **do**
7: $w_{t+1,i} = \begin{cases} 0, & |z_i| < \lambda_1 \\ -(\lambda_2 + \frac{\beta+\sqrt{n_i}}{\alpha})^{-1}(z_i - \lambda_1 sgn(z_i)), & |z_i| \geq \lambda_1. \end{cases}$
8: **end for**
9: Calculate estimated probability $p_t = \sigma(X_t \cdot W)$ by $w_{t,i}$ according to Equation 4;
10: Tags from training set $y_t \in \{0, 1\}$, indicates if the circuit is faulty;
11: **for** $i \in I$ **do**
12: Calculate gradient loss $g_i = (p_t - y_t)x_i$;
13: Calculate $\sigma_i = \frac{1}{\alpha}(\sqrt{n_i - g_i^2} - \sqrt{n_i})$;
14: Update $z_i = z_i + g_i - \sigma_i W_{t,i}$;
15: Update $n_i = n_i + g_i^2$;
16: **end for**
17: **end for**

3.3 Fault Recovery Mechanism

RLAGA: Adaptive Genetic Algorithm Based on Reinforcement Learning. When an unknown circuit fault occurs, in order to ensure the recover of the fault, the evolutionary recover algorithm is used to obtain an alternative circuit. Therefore, this paper proposes RLAGA. In this algorithm, the crossover operators and mutation operators of genetic algorithm are dynamically and adaptively adjusted by the reward feedback mechanism of reinforcement learning, so the diversity of population in the iterative process can be maintained which avoids the algorithm falling into the local optimal solution, and improves the evolution efficiency of genetic algorithm. Q-learning algorithm [19] is adopted as the learning algorithm to strengthen the learning agent. Through the crossover operators and mutation operators of the population to reducing the probability of local optimum. Therefore, it is necessary to consider the impact of crossover operators and mutation operations on population diversity, when designs the learning process of reinforcement learning.

The population diversity is measured by the population fitness. The population size is n, the fitness of the i-th individual x_t^i at time t is $fit(x_t^i)$, and the average fitness of the population at time t is $fit_{avg}(x_t)$, if the individual difference of population is d_t^i, it means the number of chromosomes in the population with different fitness from the i-th chromosome, then the population diversity can be expressed as Eq. (12).

$$div(x_t) = \frac{1}{N}\sum_{i=1} N|fit(x_t^i) - fit_{avg}(x_t)|d_t^i \tag{12}$$

The learning mechanism of reinforcement learning for genetic algorithms mainly includes three elements: (1) setting and division of environmental status; (2) Agent's action division; (3) determination of action reward value.

State Setting and Division. As shown in Eq. (12), according to the maximum iterative algebra G of the genetic algorithm, the whole evolutionary iterative process is divided into four stages. According to the value of div of population diversity, the value range is divided into four intervals. In this paper, the environment of genetic algorithm is divided into 16 states by combining iterative algebra and population diversity as state value of environment.

$$S_G = \begin{cases} s_{G1}, & G \in \left[0, \frac{G}{4}\right) \\ s_{G2}, & G \in \left[\frac{G}{4}, \frac{G}{2}\right) \\ s_{G3}, & G \in \left[\frac{G}{2}, \frac{3G}{4}\right) \\ s_{G4}, & G \in \left[\frac{3G}{4}, G\right) \end{cases} \quad S_{div} = \begin{cases} S_{div1}, & div \in [0, 0.5) \\ S_{div2}, & div \in [0.5, 1.0) \\ S_{div3}, & div \in [1.0, 1.5) \\ S_{div4}, & div \in [1.5, +\infty) \end{cases} \tag{13}$$

Action Division. After Agent obtains the state and reward value at time t from the genetic algorithm environment, it will dynamically and adaptively adjust the crossover operator P_c and mutation operator P_m of the genetic algorithm according to the feedback value, and then agent will transfer the adjusted operator parameters to genetic algorithm. The adjustment of P_c and P_m according to Agent is shown in Eq. (14) and (15). According to the difference between Pc and Pm, there are 9 kinds of action combinations that Agent can take for genetic algorithm at time t.

$$P_c(t) = P_c(t) + \Delta\alpha, \Delta\alpha = \begin{cases} -k_1 \\ \quad 0, k_1 = 0.05 \\ k_1 \end{cases} \tag{14}$$

$$P_m(t) = P_m(t) + \Delta\beta, \Delta\beta = \begin{cases} -k_1 \\ \quad 0, k_1 = 0.05 \\ k_1 \end{cases} \tag{15}$$

Determination of Reward Value. In this paper, the reward mechanism of the state action is established by comparing the population fitness obtained by the cross mutation of the genetic algorithm after the agent action. The calculation of action reward R is shown in Eq. (16). While the average fitness of the population after iteration is greater than the previous generation, the reward value is positive; while the average fitness of the offspring is equal to the previous generation, the action does not generate revenue, the reward value is 0; while the average fitness of the offspring is less than the previous generation, the reward value is negative.

$$R = \begin{cases} 1, -\Delta fit > 0 \\ 0, -\Delta fit = 0, \Delta fit = fit_{avg}(x_t) - fit_{avg}(x_{t-1}) \\ -1, -\Delta fit < 0 \end{cases} \tag{16}$$

In order to better calculate the cumulative reward value of Agent in the iterative process, this paper adopts the Q-learning algorithm. The optimal action strategy group is obtained by continuously evaluating the value function of the state action pairs, and Eq. (17) is the Q value calculation formula of Q-learning algorithm. Here, Q is the value of the state action pairs; α is the learning step of agent; r_t is the reward value of environmental feedback when action a_t is taken in state s_t at time t; γ is the discount rate; $maxQ(s_{t+1}, a_{t+1}))$ is in the next state $s_(t+1)$ the maximum Q value corresponding to the action taken.

$$Q(s_t, \alpha_t) = Q(s_t, \alpha_t) + \alpha[r_t + \gamma maxQ(s_{t+1}, \alpha_{t+1}) - Q(s_t, \alpha_t)] \tag{17}$$

The pseudo code of the adaptive genetic algorithm based on reinforcement learning is shown as Algorithm 2

Algorithm 2. RLAGA

Input: parameter population size N, chromosome length L, maximum iteration algebra G, initial crossover operator P_c, initial mutation operator P_m, initial Q-value table, initial learning step length, initial discount rate;

1: Initial $P_{(t)}$;
2: $fit(x_t) = AdpCalculateFit(P_{(t)})$;
3: Calculate the current status value S and reward value R;
4: Agent reads status value S and reward value R, selects actions a_t by greedy strategy, and updates the crossover operator P_c and mutation operator P_m;
5: **if** Termination condition **then**
6: break;
7: **end if**
8: **for** $i = 0$ to G **do**
9: $fit^{'}(x_t) = a \cdot fit(x_t) + b$;
10: $P^{'}(t) = Select(P(t), fit^{'}(x_t))$;
11: $P^{''}(t) = Crossover(P^{'}(t), P_c)$;
12: $P^{'''}(t) = Mutation(P^{''}(t), P_m)$;
13: $P(t) = P^{'''}(t)$;
14: $fit(x_t) = AdpCalculateFit(P(t))$;
15: Calculate the current status value S and reward value R, and send them to the agent;
16: Calculate $Q(s_t, a_t) = Q(s_t, a_t) + \alpha[r_t + \gamma maxQ(s_{t+1}, a_{t+1}) - Q(s_t, a_t)]$; Update Q value table; Selects actions by greedy strategy; Updates the crossover operator P_c and mutation operator P_m;
17: **if** Termination condition **then**
18: break;
19: **end if**
20: $t = t + 1$;
21: **end for**

4 Experiment

In this section, we firstly illustrate the fault injection process, data acquisition and system experimental parameter settings. Secondly, we utilize the 8-bit parity checker and the 2-bit multiplier to verify the effectiveness of the proposed scheme and demonstrate the advantages of our system.

4.1 Fault Injection and Data Acquisition

Existing fault injection methods to simulate the actual fault of the circuit can be roughly divided into two categories: (1) utilize the simulator to simulate the fault of the hardware circuit; (2) directly modify the hardware circuit and inject the fault from the circuit source file. In order to fit the actual fault problem of the circuit more closely, we uses the method mentioned by Wang et al. [7]. Taking the 2-bit multiplier as an example, the chromosome coding string is { 020 131 210 033 055 444 554 591 480 502 765 250 583 4132 1242 18 12 14 10 }. Then we inject errors into CGP code string to obtain fault data.

Table 1. Experimental results of 8-bit parity checker

	Logistic regression	Optimized logistic regression
Off-chip test accuracy	85.1%	86.6%
On-chip test accuracy	92.5%	94.4%
Average time (clock cycle)	8	8

Table 2. Experimental results of 2-bit multiplier

	Logistic regression	Optimized logistic regression
Off-chip test accuracy	80.2%	84.3%
On-chip test accuracy	89.4%	93.2%
Average time (clock cycle)	15	15

4.2 Experiment Setting

Zed board [14] is an SRAM-based FPGA, which is used as the verification platform for the self-recovery experiment of chip circuit. In order to fully utilize the cooperative features of Zed boards software and hardware, the target circuit is mapped to chromosome structure string through CGP and configured on VRC circuit to make it work on Programmable Logic (PL) part. Then the fault detection and location mechanism and the fault recover mechanism are established. In the Processing System (PS) part, an evolutionary hardware recovery module is established. When an unknown fault occurs in the circuit evolution algorithm will generate configuration chromosome of the fault circuit and save in the fault recovery configuration library on PL interacting through AXI bus.

4.3 FTRL-Optimized Logistic Regression

We use the fault data of the 8-bit parity checker and the 2-bit multiplier to perform offline model training on traditional logistic regression and FTRL-optimized logistic regression algorithm. Then we download and configure the above model into FPGA, which is used as the fault model matching library for fault detection and location. After that the actual injection of faults on the chip is used to verify the accuracy of the obtained fault model.

The 8-bit parity checker data set scale of fault training is 1000; the fault test set scale is 500, and the on-chip fault test scale is 100. For 2-bit multiplier, the scale of fault training set is 2000; the scale of fault test set is 500, and the on-chip fault test set is 200. Table 1 and Table 2 show the accuracy of the offline fault test set of the fault model, the fault test set after fault injection of the on-chip circuit, and the average time consumption of fault detection.

According to Table 1 and Table 2, FTRL-optimized logistic regression performs better in accuracy both online and offline tests. At the same time, the average time of fault detection is only related to the circuit itself, and the fault model is not related, which guarantees the real-time performance of fault detection on chip. Due to the complexity of the circuit structure, the average time-consuming of the on-chip fault detection needs 15 clock cycles, and the real-time detection of the on-chip fault still has good performance.

The fault model matching library based on FTRL-optimized logistic regression adopted in this paper can effectively solve the problem of real-time detection of faulty circuits.

4.4 RLAGA Performance Comparison

The modify adaptive genetic algorithm adopts a population chromosome adaptation assessment method which changes with the iteration time. Different evolution iteration coefficients have an influence on the evolution time of each generation of chromosomes. Therefore, different evolutionary iterative coefficients are used to test the adaptive evaluation method, and the most appropriate evolutionary iterative coefficient is determined by evolutionary operation of the target circuit. The experimental results are shown in Table 3. We can known the higher evolution iteration coefficient saves the evaluation time, but leads to the lower efficiency of evolution, which cannot truly reflect the real-time fitness value of each chromosome. Therefore, it is necessary to comprehensively consider the evolution time consumption and the overall efficiency of evolution. According to the results in Table 3, we set the evolution iteration coefficient to 0.3 to ensure the best evolution time and efficiency.

Table 3. Time-consuming results of evolution with different coefficients

	8-bit parity checker		2-bit multiplier	
Iteration coefficient	*Time consuming per generation*	*Total time (s)*	*Time consuming per generation*	*Total time(s)*
0.1	5.09	5.37	5.23	41.21
0.3	4.73	5.23	4.87	34.27
0.5	4.53	6.45	4.61	76.32
0.7	4.46	8.36	4.49	265.74
0.9	4.35	12.83	4.38	403.82

To validate the evolutionary recover efficiency of the adaptive genetic algorithm based on reinforcement learning, in this paper we do experiment and compare with standard genetic algorithm SGA, ant colony algorithm PSO, adaptive genetic algorithm AGA and self-simulation annealing Adaptive genetic algorithm SAGA in the time consumption and evolution success rate [1]. Figure 2 respectively represents the evolutionary recover results of randomly injecting different circuit faults into the 8-bit parity verifier and the 2-bit multiplier for 100 times,

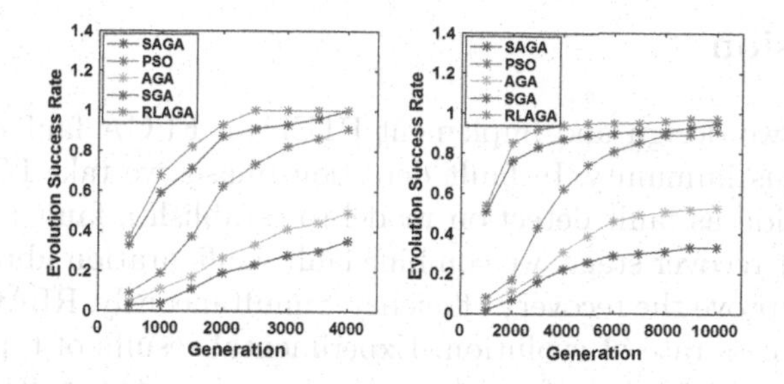

Fig. 2. Fault recovery experiment results, left figure is 8-bit parity checker results and right figure is 2-bit multiplier results

and then adopting 4 algorithms to carry out fault recover operation on the chip circuit at recover stage.

Comparing the results of the five algorithms, we can see that the RLAGA algorithm and SAGA algorithm proposed in this paper have achieved a high evolutionary success rate for 8-bit parity checker at about 1000 generations, and the circuit recover success rate is 100% at about 4000 generations. Among the other three algorithms, only PSO can achieve 90% of the recover success rate in 4000 generations, and the overall recover time consumption is far higher than the algorithm proposed in this paper. The circuit structure of 2-bit multiplier is more complex than that of 8-bit parity check. In the process of circuit evolution, the faults recover success rate of five kinds of algorithms in 1000 generations is not very good. However, RLAGA and SAGA algorithm can still achieve more than 50% of the recovery rate. Because RLAGA algorithm guarantees the population diversity in the evolutionary iteration process, the faults recover rate reaches 90% at about 3000 generations, while SAGA is relatively slow. Other algorithms have defects in the recover time consumption and recovery rate. Therefore, for the circuit with simple structure, SAGA and RLAGA can achieve better results, but when the circuit to be recovered is large, RLAGA can better improve the efficiency of evolutionary recover (Table 4).

Table 4. Comprehensive comparison of fault recover algorithms

	8-bit parity checker		2-bit multiplier	
Algorithm	*Average time(s)*	*Success rate*	*Average time(s)*	*Success rate*
RLAGA	4.2	100%	27.8	97%
SAGA	5.3	100%	34.5	95%
PSO	9.2	91%	89.8	92%
AGA	19.7	56%	163.1	53%
SGA	24.1	35%	178.6	33%

5 Conclusion

In this paper, we design and implement FTEI, an FPGA fault-tolerant model with endogenous immunity. In fault detection phase, we take FTRL-optimized logistic regression as fault detection model to establish a fault model matching library. In fault recover stage, we combine fault configuration library with online evolution to improve the recovery efficiency. Simultaneously, RLAGA is proposed to improve success rate of evolution. Experimental results of typical functional circuits demonstrate that the fault detection accuracy rates are 94.4% and 93.2%, and the fault recovery success rates of RLAGA are 100% and 90%. As part of our future works, we will:

- Further study multi-class algorithms, and improve the accuracy of fault detection;
- Study evolutionary algorithm in higher scale digital integrated circuits and improve success rate of evolution.

References

1. Wang, J., Kang, J., Hou, G.: Real-time fault recovery scheme based on improved genetic algorithm. IEEE Access **7**, 35805–35815 (2019)
2. Anjankar, S., Kolte, M., Pund, A., Kolte, P., Kumare, A., Mankarf, P., Ambhore, K.: FPGA based multiple fault tolerant and recoverable technique using triple modular redundancy (FRTMR). In: ICCCV, pp. 827–834 (2016)
3. Yang, X., Li, Y., Fang, C., Nie, C., Ni, F.: Research on evolution mechanism in different-structure module redundancy fault-tolerant system. In: ISICA, pp. 171–180 (2015)
4. Gong, J., Yang, M.: Evolutionary fault tolerance method based on virtual reconfigurable circuit with neural network architecture. IEEE Trans. Evol. Comput. **22**(6), 949–960 (2018)
5. Wang, J., Liu, J.: Fault-tolerant strategy for real-time system based on evolvable hardware. J. Circ. Syst. Comput. **26**(7), 1–18 (2017)
6. Palchaudhuri, A., Dhar, A.: Design and automation of VLSI architectures for bidirectional scan based fault localization approach in FPGA fabric aware cellular automata topologies. J. Parallel Distrib. Comput. **130**, 110–125 (2019)
7. Wang, J., Deng, S., Kang, J., Hou, G., Zhou, K., Lin, C.: A real-time fault location mechanism combining CGP code and deep learning. In: DSA (2020, in press)
8. Reorda, M.S., Sterpone, L., Ullah, A.: An error-detection and self-recoverying method for dynamically and partially reconfigurable systems. IEEE Trans. Comput. **66**(6), 1022–1033 (2017)
9. Ruan, T., Jie, P.: A bitstream readback based FPGA test and diagnosis system. In: ISIC, pp. 592–595 (2014)
10. Ranjbar, O., Sarmadi, S., Pooyan, F., Asadi, H.: A unified approach to detect and distinguish hardware trojans and faults in SRAM-based FPGAs. J. Electron. Test. **35**(2), 201–214 (2019)
11. Halawa, H., Daoud, R., Amer, H.: FPGA-based reliable TMR controller design for S2A architectures. In: ETFA, pp. 1–8 (2015)

12. Schweizer, T., Schlicker, P., Eisenhardt, S., Kuhn, T., Rosenstiel, W.: Low-cost TMR for fault-tolerance on coarse-grained reconfigurable architectures. In: ReConFig, pp. 135–140 (2011)
13. Burdyshev, I., Tyurin, S.: Fault tolerant FPGAs design method. In: EIConRus, pp. 248–251 (2020)
14. Shanker, S., Bhaskar, B., Kizheppatt, V., Suhaib, A.: Dynamic cognitive radios on the Xilinx Zynq hybrid FPGA. In: CROWNCOM, pp. 427–437 (2015)
15. Gong, J., Yang, M.: Evolutionary fault tolerance method based on virtual reconfigurable circuit with neural network architecture. IEEE Trans. Evol. Comput. **99**, 1–1 (2017)
16. Miller, J., Thomson, P.: Cartesian genetic programming. In: EuroGP, pp. 121–132 (2000)
17. Julian, M., Andrew, T.: Cartesian genetic programming. In: Proc. 2015, pp. 179–198
18. Ruder, S.: An overview of gradient descent optimization algorithms. Computer Research Repository, vol. abs/1609.04747, p. 12 (2016)
19. Indu, J., Chandramouli, K., Shalabh, B.: Generalized speedy Q-learning. IEEE Control Syst. Lett. **4**(3), 524–529 (2019)

The Design of an Intelligent Monitoring System for Human Action

Xin Liang[1,2], Mingfeng Lu[1,2(✉)], Tairan Chen[1], Zhengliang Wu[3], and Fangzhou Yuan[1,2]

[1] School of Information and Electronics, Beijing Institute of Technology, Beijing 100081, China
luming feng@bit.edu.cn
[2] Beijing Key Laboratory of Fractional Signals and Systems, Beijing 100081, China
[3] School of Computer Science and Technology, Beijing Institute of Technology, Beijing 100081, China

Abstract. Now the monitoring equipment such as cameras has been widely used in social life. In order to solve the problem that the current monitoring equipment relies on manual screening for the recognition of abnormal human action and is not time-efficient and automatic, an intelligent monitoring system for human action is designed in this paper. The system uses object detection, classification and interactive recognition algorithm in deep learning, combines 3D coordinate system transformation and attention mechanism model. It can recognize the local human hand actions, head pose and a variety of global human interaction actions in the current environment in real time and automatically, and judge whether they are abnormal or special actions. The system has high accuracy and high speed, and has been tested successfully in laboratory environment with good effect. It can also reduce labor costs, improve the efficiency of security monitoring, and provide help for solving urban security issues.

Keywords: Action recognition · Intelligent monitoring · Deep learning · Security issues

1 Introduction

In recent years, with the continuous development of economy and society, the deployment scope of surveillance cameras has become more and more intensive, covering all corners of the city and playing an important role in security and public security management. Security guards can use surveillance video captured and stored by cameras to detect dangerous action. The police can also use it as evidence in solving a case. However, this traditional method cannot give real-time and automatic warning of abnormal actions. It relies on repeated manual screening, which is troublesome and has no real-time capability. In order to solve this problem, this paper designs an intelligent monitoring system for human action, which can alarm the abnormal actions in the environment in real time and capture, deal with and record them in time. It

S. Shi et al. (Eds.): AICON 2020, LNICST 356, pp. 558–570, 2021.
https://doi.org/10.1007/978-3-030-69066-3_49

realizes the purpose of reducing labor cost, improving safety monitoring efficiency and reducing the probability of occurrence of hazards.

Now intelligent monitoring systems are already being used in transportation and agriculture, often to monitor the speed of cars on roads and the growth of crops. The main purpose of the intelligent monitoring system designed in this paper is to monitor abnormal actions of people in the environment, and its core technology is human action recognition. This technology obtains the human action data through the sensor, and intelligently recognizes the human action. Usually, human action signals can be characterized by images, motion sensors and environmental sensors. Different specific signals reflect different actions. The intelligent monitoring system mainly uses the human action recognition method of images, which is more convenient to obtain data and cheaper to buy equipment compared with the sensor method.

Human action recognition based on images is one of the basic problems in the field of computer vision and can be divided into three types according to different modeling methods. The first one is end-to-end human action recognition [1]. This method inputs the original image sequence information, extracts multiple features of space and time dimensions to construct the classifier, and finally outputs the action types of human in the image. The disadvantage is that it only works for single-player action analysis in small scenarios. The second one is the recognition of human skeleton pose, which is mostly used for multi-person pose estimation. This method generally estimates the skeleton pose of human first, and then classifies human actions, which can be divided into two ways: top-down and bottom-up. The top-down approach transforms the multi-person into the single-person pose estimation problem. It obtains k coordinates of key points of the human body by directly regressing the coordinates, or by calculating the expectation of the thermal diagram of each key point and taking the position with the highest probability. Then it uses the coordinate distribution for classification prediction. CPM [2] and HRNet [3] networks belong to this method. The disadvantage of the top-down approach is that the performance is related to the detection network and the running time increases with the number of people in the image. The bottom-up approach is to first detect the key points and then group them to get multiple body poses. The speed of this method can be realized in real time, and it is suitable for deployment in mobile terminal. OpenPose [4] and HigherHRNet [5] networks belong to this method. The disadvantage of action recognition based on human skeleton pose is that only the motion information of skeleton is taken into account without the image information, so the detailed interaction action can't be described. The third one is instance-centered human interaction recognition. This method defines the problem as a structure for human and object interaction, namely <human, verb, object> . It uses the target detection algorithm to detect people and objects respectively, and then learns the interaction between people and objects in the form of topological nodes or thermal maps by referring to the graph convolution method or the attention mechanism of NLP domain. Finally, it analyzes human action. InteractNet [6] and VSGNet [7] network belong to this method. Compared with the previous two methods, this method has higher recognition accuracy and is widely used in practical systems.

According to the different application scenarios, types of actions and functions realized, the intelligent monitoring system will be divided into three parts, namely Hand action recognition, Head pose estimation and Human-object interaction detection. The data processed is the frame of the video shot by the camera. Hand action recognition mainly uses SSD [8] and OpenPose method to locate hand, and then uses SqueezeNet to classify and finally output hand action types [9]. Head pose estimation mainly uses SSD to locate the head, and then uses facial landmark-based classical method and image-based deep learning method [10] to identify the head angle. Human-object interaction detection uses two methods. One is to use Faster R-CNN [11] to identify the position of people and objects, and then use iCAN [12] network to estimate the interactive relationship. The other is the network based on YOLO structure improvement. It will be the location identification and classification at the same time.

2 Methods

2.1 Hand Action Recognition

We hope to protect the information security of special places such as confidentiality room, and real-time identification and alarm of abnormal actions of people within the monitoring scope to prevent information leakage. In general, the action of obtaining information, such as making phone calls, taking photos, operating computers and so on, is done by hand, so it is very important to obtain information about hand action. Hand action recognition mainly includes two steps: hand location and action classification. The location will use SSD and OpenPose methods.

SSD networks use a single deep neural network to detect objects. It applies the small convolution kernel to the feature map, predicts the type score and offset of a set of default boundary boxes, and generates the prediction of different proportion with different proportion of feature map to improve the detection accuracy. The SSD network generates several initial bounding boxes with different aspect ratios, and then returns to the correct truth value. In this way, the target position can be calculated once to improve the detection speed. SSD provides a unified framework for training and prediction, and it has high accuracy, high speed and good performance in detecting small targets.

OpenPose is action recognition based on human skeleton pose. It generates Part Confidence Maps for skeleton key point regression and Part Affinity Fields between skeleton key points according to the input image. By using the key points generated by CNN and the confidence mapping of the connection, the original problem is divided into several maximum power matching problems of bipartite graph to solve the problem that the key points are combined into human skeleton. Then the hand is located through the human skeleton. OpenPose is one of the most popular open source pose estimation algorithms with high precision and low computational complexity.

After the hand position is located, this area needs to be extracted and put into the classifier to detect the action type. SqeezeNet was chosen as the classifier because it can achieve AlexNet's accuracy in classification and reduce the size of the model. At last, SoftMax was used to calculate the probability of multiple classification problems, and the type with the highest probability was selected as the result output.

2.2 Head Pose Estimation

The direction of people's attention is mainly reflected by the direction of people's visual Angle, that is, the facial direction. When people focus their attention on dangerous areas for a long time, such as high-voltage cables and rivers without fences, the intelligent monitoring system can capture such abnormal action in real time and give early warning, which can be used as a prediction to reduce the risk of dangerous events.

The head action analysis algorithm consists of two parts: face target detection and head pose estimation. The location of the face is the location of the head. Since the face detection is also the recognition detection of small targets, the SSD algorithm mentioned in Sect. 2.1 is also used to obtain the face image. Then input it into the head pose estimation algorithm to estimate the head yaw, pitch and roll. In the head pose estimation part, landmark-based classical method and image-based deep learning method were respectively used to estimate the head pose. Since the coordinate system of the head pose estimation is the plane of the face, the judgment of attention needs to be converted into the world coordinate system. The system uses perspective transformation method to modify and finally get the angle range of human attention.

Classical method can be simulated through PerspectivenPoint(PnP) to solve the problem. N 3D points of the known object and their 2D projections in the image can be used to calculate the pose relationship between the object and the camera in the world coordinate system by using the internal and external parameters of the camera. In the process of 2D-3D matching and pose estimation, it involves the world coordinates system representing the 3D coordinate system of facial features, the camera coordinates system centering on the camera, and the image coordinate system using the internal parameters of the camera to project 3D points to the image plane. See Fig. 1.

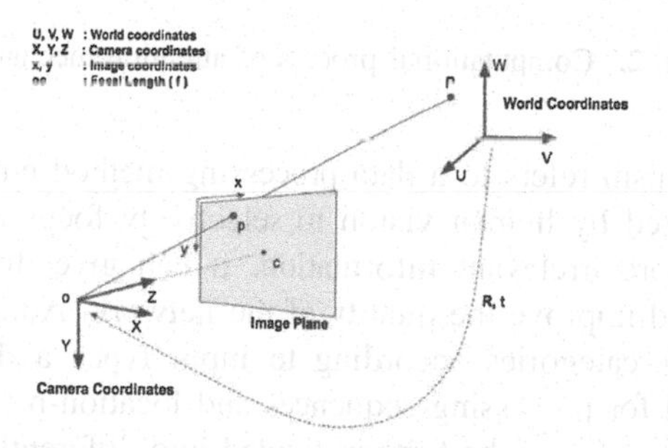

Fig. 1. Relationship among world coordinate system, camera coordinate system and image coordinate system

In order to calculate the 3D pose of the head, the system needs to obtain the 2D coordinates of 68 characteristic positions of the face, such as the tip of nose, canthus, chin and mouth corner. At the same time, the universal 3D face model is used to provide the coordinates of 3D face feature points to be matched. In order to better calculate the head pose, we use EPnP [13] algorithm. EPnP iteratively uses a set of virtual control points to represent the feature points in the world coordinate system as the weighted sum of the control points, rather than directly solving the depth of the feature points. It uses the internal parameters of the camera to convert the coordinates of the reference point into the control point through formula transformation, and then solves the translation and rotation matrix [14], and converts the rotation matrix into euler angle. So we get the angle and the position of the head in world coordinates.

For the facial image-based deep learning method, we use FSA-Net network, which is a method to estimate the head pose with a single image. The FSA-Net uses two different branches to extract the features of the input image and conduct feature fusion at each stage. The fusion module first combines the two feature maps by element multiplication, then applies C 1 × 1 convolution to transform the feature map into C channel, and then uses average pooling to scale the feature map to make the size equal to the original image. Finally, we obtained the feature graph U_k of stage K. The feature graph U_k is a spatial grid, which is input into the attention mechanism module to obtain the weight of the feature graph, after which the more important parts of the feature can be highlighted.

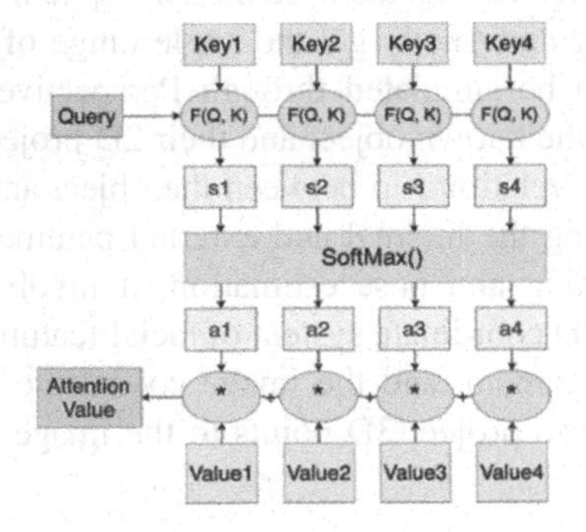

Fig. 2. Computational process of attention mechanism

Attention mechanism refers to a data processing method embedded in neural network which is inspired by human vision to selectively focus attention on important information and ignore irrelevant information. It can give different weights to the extracted features and improve the quality of the network. Attention mechanisms can be divided into four categories according to input types and processing methods, including item-based for processing sequences and location-based for processing feature graphs. Each category can be further divided into differentiable soft attention and non-differentiable hard attention [15]. The module in this system is a location-based soft attention mechanism. Figure 2 shows the computational process of attention mechanism. It inputs the similarity function F(Q,K) into the position Key_i in the feature and $Query_i$ in the query result to calculate the weight s_i of the position, and gets the final attention weight value through the normalization processing of SoftMax function. It then weights the feature vector and the weight value to obtain the final attention value, and identifies the key parts of the image data with a higher weight coefficient. Because soft attention can be differentiated, it can participate in the process of learning and training and learn attention by itself.

After obtaining the feature map and its attention weight map, the fine-grained structure mapping is carried out to obtain the importance weighted feature. Fine-grained structure can focus attention on features that have a greater impact on facial posture, and reflects the spatial relationship between features. The robustness of attention mechanism can be improved by calculating weights such as learnable convolutional layer and unlearnable variance in spatial position. The weighted feature set was input into the feature aggregation method and the head pose was obtained by stepwise regression.

2.3 Human-Object Interaction Detection

When a scene requires detection of full-body action, or of multiple actions of the same person, these two approaches fail to meet the requirements. Therefore, we designed the detection of global human-object interaction action. The system abstracts the action into the interaction between human and objects, namely <human, verb, object>. The key is how to identify the interaction action between two targets. This system will use two different methods to realize human-object interaction detection, one is an iCAN network that combines the attention mechanism with the instance-centered and the other is an improved method based on YOLO structure that combines target detection and action recognition simultaneously.

ICAN network includes two parts, target detection and interactive recognition. After detecting the bounding box, class and probability values of people and objects, iCAN simultaneously input them as intermediate quantities along with the original image into the interactive detection network to detect interactive detections, and finally output the action type. The target detection algorithm adopts the Faster R-CNN network with high accuracy and uses COCO [16] data set for pre-training. The region proposal algorithm is mainly used to assist the location of objects. RPN is a full convolutional network that simultaneously predicts the boundary position and score of each object, and it can tell the whole network where to focus. Because it shares full image convolution feature with detection network, it can consume almost no computing resources.

Faster R-CNN network can be divided into four main steps. The first step is to take the VGG16 model as the backbone network structure, extract the feature map with the same size as the original image, and provide input for the region proposal network and classification network respectively. In the second step, RPN generates the candidate boundary box with the extracted feature map, and gets the more accurate position through regression. The third step is to pool the feature map and candidate areas after the above two steps, extract the features of local locations, and prepare for the fourth step. In the fourth step, the object type is classified by the full connection layer, and the boundary frame is regressed again to obtain more accurate boundary frame position and object type.

Figure 3 shows the overall structure of the iCAN network. Where b_h is the detected human bounding box, b_o is the object bounding box, s_h^a is the action prediction score based on human body, s_o^a is the action prediction score based on object, s_{sp}^a is the spatial relationship score between human and object, x_{inst}^h or x_{inst}^o is the appearance characteristics of a single person or object, and $x_{context}^h$ or $x_{context}^o$ is the context characteristics based on attention diagram. In the interactive detection part, iCAN use the attention module to evaluate all pairs of people and object boundary boxes to predict the interactive action score. This module is the attention mechanism module in Sect. 2.2. By learning to highlight key areas in the image dynamically, it can selectively summarize and identify features related to human interaction action. The image feature (x_{inst}^h or x_{inst}^o) and the convolution image of the whole image are mapped to 512 dimensional space, and the similarity is calculated by dot product. Then softmax function is used to get the instance-centered feature map. The overall image is weighted by the product of the feature map and combined with $x_{context}^h$ and $x_{context}^o$ extracted by the full connection layer for subsequent score calculation.

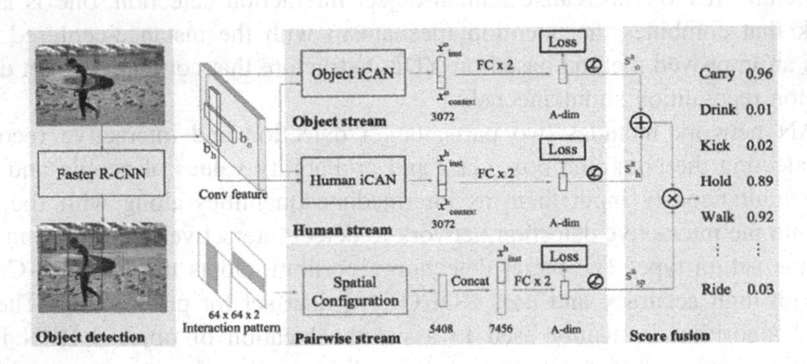

Fig. 3. Overall structure of the iCAN network

In order to solve the problem of ambiguity generated by different interaction relations of the same person and object, such as making phone calls and playing mobile phones, iCAN uses binary images to represent spatial interaction, that is, the images in the boundary box are 1, and the images in other positions are 0, and it uses the generated spatial features to judge together with the appearance features of the human body. The final action score of iCAN is calculated as follows:

$$s_{h,o}^a = s_h \cdot s_o \cdot \left(s_h^a + s_o^a\right) \cdot s_{sp}^a. \quad (1)$$

Where s_h and s_o are the confidence detected by a single object. For some action that do not include objects, such as walking, only human body branches are used to calculate $s_h \cdot s_h^a$.

An improved network based on YOLO can conduct target detection and action recognition at the same time. Its structure consists of three parts. The first part is feature extraction network. Inspired by YOLOv2 it is composed of 24 convolution layer, other convolution layer using batch normalization, the last layer using leaky rectified as linear activation layer. The second part is the two branches of target detection and interaction detection. The input is extracted feature. The target detection network has the same structure as YOLO [17], and the output is the classification probability of the target and the corresponding detection box. The interaction detection network uses non-maximum suppression processing, and introduces an additional target detection branch to improve the positioning accuracy. The output is the probability of the interaction and the positioning boundary box of human and objects. The third part merges the results of the previous two branches and outputs the final result.

In the interaction detection branch, the input image is divided into N × N grids, and the grid where the interaction action center is located is responsible for detecting the interaction relationship. This step can provide the detection speed. Like YOLO, the network uses regression to predict bounding boxes. The design of the <human, verb, object> of human action in this system leads to the need to predict the boundary frame of human and object respectively. The loss function is calculated in the same way as the YOLO network. The network detection speed is fast and can meet the real-time requirement. But there is only one set of person-object matches, so it is impossible to identify a person's multiple action.

3 Experimental and Results

In order to verify the reliability, effectiveness and practicability of the system, an Ezio CS-C6HC-3B2WFR camera will be used to shoot video and intercept the action frame from the video for testing and evaluation. The training platform is the server with the Tesla V100 graphics card, and the reasoning platform is the computer with GTX1060 graphics card. The tests were divided into hand action recognition, head pose estimation and human-object interaction detection. The head pose estimation was tested on the 300 W-LP [18] data set using the classical method and the FSA-Net network. Human-object interaction detection was tested on V-COCO [19] and HICO-Det [20] data sets for iCAN networks and improved networks based on YOLO.

Hand action recognition captures nearly 8,000 action frames from the video captured by the camera and cuts them into 224 × 224 pixels as a data set. The classifier uses mixed samples for training. We used the Adam optimizer to train 100 epochs in 30 min until the loss stopped falling. At this time, the accuracy of the model on the training set was almost 100%. In order to make the hand action recognition of the system more reliable, we retain three most likely action types in the result section and give the confidence of each type. The result is a system that recognizes seven actions, including making a phone call, using the phone (other ways to use the phone besides making a phone call), opening a door, holding a file, using a keyboard, using a mouse, and others, and empty. And a threshold can be set to determine whether the abnormal action and alarm. Figure 4 is the test result of hand action. Both SSD and OpenPose target detection methods can correctly identify and classify hands. There is no significant difference in accuracy between them, but SSD is superior to OpenPose in recognition speed.

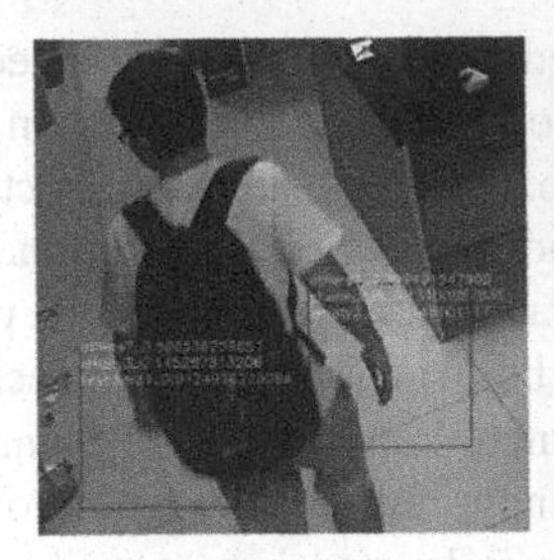

Fig. 4. Results of hand action test

Head pose estimation uses the pre-trained SSD face detection weight model to detect the face region and output the boundary box. The recognition results of SSD face detection are shown in Fig. 5 (a). The classical pose estimation method uses dlib library to extract and label the key points of 68 individual faces, and then matches the positions with the general 3D face model. The results are shown in Fig. 5 (b). Then it uses OpenCV's solvePnP function for head pose estimation. The input is characteristic point coordinates and camera internal parameters in world coordinate system and image coordinate system. The output is rotation translation vector between world coordinate system and camera. The head pose estimation results of the test figure are shown in Fig. 5 (c). The output angle value is [5.59276, 0.42750707, 4.0778937], which basically meets the expectation. The pose estimation of the FSA-Net network uses TensorFlow to describe the model. The 300 W-LP data set is a face data set that has been aligned with 68 key points and is expanded from the 300 W data set. Among them, 101,144 samples were used as the training set, and the remaining 21,306 samples were used as the test set. The Adam optimizer is used, and the final model weight is obtained by iterating 100 times. The same picture was used for testing, and the output angle value was [5.02749, 0.573317, 3.7282182]. It can be seen that the results of the deep learning method and the classic method are basically consistent and in line with expectations.

Fig. 5. (a) SSD face detection effect (b) Face feature point extraction effect (c) Head pose estimation result

The evaluation index of head pose estimation adopts the mean absolute error. Table 1 shows the test results of the two algorithms on the 300 W-LP data set. It can be concluded that the error of yaw angle differs by 1 degree, while there is a big difference between pitch and roll. The reason is that the classical method assumes that all faces are suitable for the general 3D face model, but this is not the case, which leads to a large error in the pitch and roll.

Table 1. Error comparison between FSA-Net and classical method, unit (degree)

	Yaw mean error	Pitch mean error	Roll mean error	Population mean error
Classic methods	5.92	11.86	8.27	8.68
FSA-Net	4.96	6.34	4.78	5.36

The instance-centered method in human interaction recognition uses Faster R-CNN network for target detection. The V-COCO data set includes 26 human actions. The HICO-DET data set includes 600 human actions, and it has modeled human actions as <human, verb, object>. The result of the target detection is fed into the iCAN network model to calculate the score for each type of action, and then the gradient is updated according to the loss function. Weights are backed up every 20 times in the network iteration to prevent all training progress from being lost due to server crash, memory overflow and other problems. The training process is shown in Fig. 6. Figure 7 is the recognition result of the test picture. The improved network based on YOLO adopts DarkNet network framework and adds a branch on the basis of YOLO network structure. The branch outputs the boundary frame of people and objects as well as the confidence of actions. The training process is shown in Fig. 8. The data set USES V-Coco data set. In order to speed up the training, we used the first 23 feature extraction layers of YOLO's pre-training model as the initial weights, and randomly generated the following network weights. The model also USES a weight backup every 20 iterations. Figure 9 shows the result of image recognition based on YOLO improved network.

On the V-COCO data set, we use the average precision (AP) [21] commonly used in target detection to evaluate the accuracy of the two algorithms. The resulting pair is shown in Table 2. It can be seen that the iCAN network has a high average accuracy, but because the iCAN network is a two-level recognition method that object detection and interactive recognition are carried out separately, the computing speed is relatively slow. In the process of interactive recognition, the three factors of people, objects and the spatial relationship between people and objects are considered to make the accuracy higher. YOLO improved network interaction and target detection at the same time, which is fast, but because the network is relatively simple, the prediction accuracy is worse than that of iCAN.

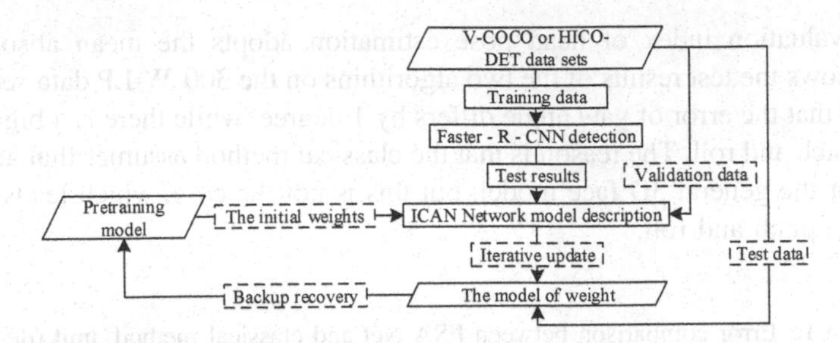

Fig. 6. ICAN action recognition network training process

Fig. 7. Test image recognition results for the iCAN network model

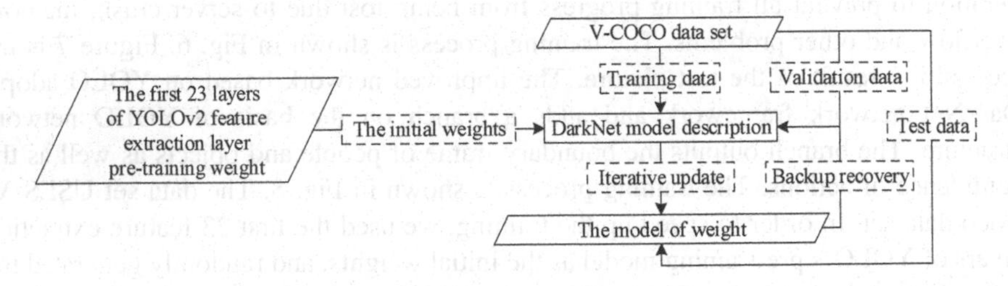

Fig. 8. Based on YOLO improved action recognition network training process

Fig. 9. Results of image recognition based on YOLO improved network

Table 2. Comparison of test results of human-object interaction detection algorithm

	Average precision(AP)	Mean operating time
iCAN	45.3	1.82 s
Improved network based on YOLO	35.6	0.028 s

4 Conclusion

The intelligent monitoring system realizes a variety of action recognition functions, including hand action recognition and head pose estimation for local human body, as well as global human-object interaction detection. The system is tested and compared using video from the camera in the laboratory environment. Hand action recognition using SSD and OpenPose hand positioning method, combined with SqeezeNet classifier, to recognize seven types of hand-related actions, including making phone call, using keyboard, using mouse, etc. It is suitable for simple indoor scenes such as secret rooms. Head pose estimation uses SSD to detect head position, and then uses classical method and FSA-Net network for deep learning to identify head angle respectively. It can output the head yaw, pitch and roll, so as to determine the direction of people's attention. It is suitable for monitoring and warning of dangerous areas such as high voltage cable. The method based on deep learning has higher accuracy, but the classical method has simple network structure and low requirements for equipment performance. Human-object interaction detection uses Faster R-CNN to detect objects and people, and then uses iCAN network and improved network based on YOLO to conduct interaction recognition respectively, so as to identify multiple action types of multiple people. It is suitable for multi-action recognition in complex scene. Experiments show that the accuracy of iCAN network is higher, but the improved network based on YOLO has a simple structure and short time-consuming which can meet real-time performance. According to different scenarios and application purposes, the intelligent monitoring system can use different methods to identify actions.

However, after many experiments, we found that the intelligent monitoring system has some problems. First of all, because action recognition requires object detection and classification of people and objects, there is a problem that relies heavily on objects and environment. Secondly, when people and objects are largely shielded, the detection effect is not good. And because the duration of different actions is different, the analysis with the same action recognition model has errors. In the next step, we plan to unify the advantages of each network, improve the practicability of the system, and try to use radar sensors to assist in identification to reduce dependence on the environment.

Acknowledgment. This work is supported by Beijing Natural Science Foundation (Grant no. L191004).

References

1. Ji, S., Xu, W., Yang, M., et al.: 3D convolutional neural networks for human action recognition. IEEE Trans. Pattern Anal. Mach. Intell. **35**(1), 221–231 (2013)
2. Wei, S., Ramakrishna, V., Kanade, T., et al.: Convolutional pose machines. In: 2016 IEEE Conference on Computer Vision and Pattern Recognition (CVPR), pp. 4724–4732 (2016)
3. Sun, K., Xiao, B., Liu, D., et al.: Deep high-resolution representation learning for human pose estimation. In: CVPR (2019)
4. Cao, Z., Simon, T., Wei, S.E., et al.: Realtime Multi-person 2D pose estimation using part affinity fields. In: CVPR (2017)
5. Cheng, B., Xiao, B., Wang, J., et al.: HigherHRNet: Scale-Aware Representation Learning for Bottom-Up Human Pose Estimation. arXiv: 1908.10357 [cs.CV] (2019)
6. Gkioxari, G., Girshick, R., Dollár, P., et al.: Detecting and recognizing human-object interactions. In: 2018 IEEE/CVF Conference on Computer Vision and Pattern Recognition, pp. 8359–8367 (2018)
7. Ulutan, O., Iftekhar, A., Manjunath, B.: VSGNet: Spatial Attention Network for Detecting Human Object Interactions Using Graph Convolutions. ArXiv preprint arXiv:2003.05541 (2020)
8. Liu, W., Anguelov, D., Erhan, D., et al.: SSD: single shot multibox detector. Lecture Notes in Computer Science, pp. 21–37 (2016)
9. Wu, Z., Lu, M., Ji, C.: The design of an intelligent monitoring system for human hand behaviors. In: ACM International Conference Proceeding Series. ICMIP 2020-Proceedings of 2020 5th International Conference on Multimedia and Image Processing. 125–129 (2020)
10. Yang, T.Y., Chen, Y.T., Lin, Y.Y., et al.: FSA-net: learning fine-grained structure aggregation for head pose estimation from a single image. In: The IEEE Conference on Computer Vision and Pattern Recognition (CVPR) (2019)
11. Ren, S., He, K., Girshick, R., et al.: Faster R-CNN: towards real-time object detection with region proposal networks. arXiv: 1506.01497 [cs.CV] (2015)
12. Gao, C., Zou, Y., Huang, J.B.: ICAN: Instance-centric attention network for human-object interaction detection. In: British Machine Vision Conference (2018)
13. Lepetit, V., Moreno-Noguer, F., Fua, P.: EPnP: an accurate o(n) solution to the PnP problem. Int. J. Comput. Vis. **81**, 155–166 (2009)
14. Zhang, Z.: Iterative point matching for registration of freeform curves and surfaces. Int. J. Comput. Vis. **13**, 119–152 (1994)
15. Xu, K., Ba, J., Kiros, R., et al.: Show, Attend and Tell: Neural Image Caption Generation with Visual Attention. arXiv: 1502.03044 [cs.LG] (2015)
16. Lin, T.Y., Maire, M., Belongie, S., et al.: Microsoft COCO: Common Objects in Context. arXiv: 1405.0312 [cs.CV] (2014)
17. Redmon, J., Farhadi, A.: YOLO9000: Better, Faster, Stronger. ArXiv preprint arXiv:1612.08242 (2016)
18. Zhu, X., Liu, X., Lei, Z., et al.: Face alignment in full pose range: a 3D total solution. IEEE Trans. Pattern Analy. Mach. Intell. **41**(1), 78–92 (2019)
19. Gupta, S., Malik, J.: Visual Semantic Role Labeling. ArXiv preprint arXiv:1505.04474 (2015)
20. Chao, Y.W., Liu, Y., Liu, X., et al.: Learning to Detect Human-Object Interactions. arXiv: 1702.05448 [cs.CV] (2017)
21. Everingham, M., Eslami, S.M.A., Van Gool, L., et al.: The pascal visual object classes challenge: a retrospective. Int. J. Comput. Vis. **111**(1), 98–136 (2015)

Coding Technology of Building Space Marking Position

Jichang Cao[1,2(✉)], Guoliang Pu[4], Hanqi Yan[4], Gang Huang[4], Qing Guo[1], Shuo Shi[1,3], and Mingchuan Yang[1]

[1] School of Electronic and Information Engineering, Harbin Institute of Technology, Harbin 150001, Heilongjiang, China
caojichang@126.com

[2] Science and Technology and Industrialization Development Center of Ministry of Housing and Urban-Rural Development, Beijing 100835, China

[3] International Innovation Institute of HIT in Huizhou, Huizhou 516000, Guangdong, China

[4] Peking University, Beijing 100835, China

Abstract. The data generated in the process of planning, construction, and management of construction and building is diverse and large in scale, and there is an urgent need for a unique, consistent, and efficient code. Beidou grid position coding stipulates its grid selection and coding rules, as well as spatial position information identification, transmission and big data processing, which can have good scalability in the field of building spatial identification position coding. Based on the coding rules of the Beidou grid position code, this paper proposes a construction building space identification position coding technology to code the construction building. The coding is unique, consistent and efficient. The application of coding rules is introduced in the paper, including code generation and coding index, as well as the corresponding query algorithm, which is verified by engineering experiments.

Keywords: Buildings · Space identification · Position code · Beidou grid

1 Introduction

There are many types of information about buildings and their components. From building design to later maintenance, additions, deletions and changes of information often occur, which brings challenges to the information management of buildings and their components.

In the design phase of the building, the digital information model of the building assigns a unique identification code to the internal components of the building, so as to exchange and manage multi-source data in the construction industry based on the identification code during the entire life cycle of the building. In recent years, the relevant national infrastructure construction departments have introduced a series of classification information codes for building and building components, but they have not paid enough attention to the spatial location attributes of the components. If the unique identification of the spatial location is used as the integration clue, the association of building components based on the spatial location can be realized.

S. Shi et al. (Eds.): AICON 2020, LNICST 356, pp. 571–587, 2021.
https://doi.org/10.1007/978-3-030-69066-3_50

Since the building space location identification information is very important in file management, intelligent construction and later inspection [4], this paper proposes a set of building location codes to meet the application requirements of related industries. The spatial location division and coding system of this set of buildings inherits the Beidou global location framework and meets the following three main requirements: (1) Uniqueness requirements: each building has a globally unique identifier and can establish a one-to-one correspondence with the corresponding entity relationship. If the uniqueness is not satisfied, different departments assign different identification codes to the same building, and there will be deviations and misunderstandings in data integration and exchange, and the multi-source attributes of the object cannot be merged through the identification code. (2) Consistency requirements: The external environment of the building and the internal space of the building share a set of division framework and coding method, which is convenient for the expansion of the division framework and spatial calculation analysis; the spatial identification of different buildings is under a set of spatial identification naming framework, Organize and manage all building data. (3) High efficiency requirements: There is a large amount of component information in the building, and data association and retrieval need to be carried out through component identification. The component has a preset unique spatial location in the building, and there are many data sharing and data retrieval scenarios that take the spatial location as a clue in practical applications. To realize the application in these scenarios, the spatial identification code assigned to the component entity is required to support efficient and accurate topology analysis and distance calculation.

1.1 Research Status at Home and Abroad

Building Coding System. In 2006, the International Organization for Standardization formulated the ISO/IDS framework to establish the classification information standards of the construction industry in various countries, defined the basic concepts of various construction information classification systems, and sort out their relationships. The American Building Standards Institute CSI and the Canadian Building Standards Institute CSC have promulgated the Omniclass classification system, which can present the most intuitive and concise coding system, docking with the BIM data storage standard IFC, and supporting the realization of BIM technology [13, 17]. In 2018, the Ministry of Housing and Urban-Rural Development of the People's Republic of China issued the "Building Information Model Classification and Coding Standard GB/T51269-2017" applicable to civil and general building plants [1].

The component codes currently used in the construction industry mainly consider component types, geometric characteristics, production information, etc. A specific component/part produced in the same batch from the factory will be given exactly the same production code. Components scattered in different regions and spatial locations share a production code, so that the current component code cannot guarantee global uniqueness.

Component Query Method. The complexity of building components and the diversity of semantics make it difficult for users to retrieve expected results through keyword queries, and the accuracy rate is low. IFC's data model can be mapped to a relational database and transformed into XML or RDF expression. Researchers are committed to proposing various QL frameworks, such as SPARQL [7], XQuery [8], QL4BIM [9]. Lu Jin [18] of Beijing University of Architecture and Architecture pointed out that current research has played a role in promoting the retrieval of building information in BIM models. The simplified component is a mass point, and the path between the component and the user's reference point is sorted in the calculation, the shortest distance is obtained by Dijkstra algorithm, and an ordered set of components is obtained from the spatial relationship and distance.

1.2 Limitation Analysis

There are still four problems in the existing research on the sign and application of building space [19]: The spatial position representation of the components inside the building is based on the local reference coordinates; The position of the component depends on multiple reference systems, which does not satisfy the uniqueness and consistency of the spatial identification; The existing building codes are classified information codes, which only consider simple building object information and cannot directly contain the relationship of relative positions in the building; The information retrieval method of the existing component cannot directly place the spatial attributes in the index column and perform calculation and filtering operations on the content of the index column, which has low efficiency and does not satisfy the efficiency of spatial identification.

2 Overview of Beidou 3D Grid Position Coding

The Beidou grid position code specifies the grid selection and coding rules of the Beidou grid position code [6], which is widely used. It conforms to the following basic coding principles: a) uniqueness; b) nesting; c) compatibility; d) calculation; e) practicality.

Beidou 3D grid location code is composed of 2D grid division of the earth surface and grid division of height domain.

In the two-dimensional grid division of the earth's surface, the origin of the grid is at the intersection of the equatorial plane and the prime meridian, and the two-dimensional grid is divided into ten levels. The method is as follows: (1) The first level grid: according to GB/ T 13989-2012 is divided into 1:1 million map frames, and the unit size is $6° \times 4°$; (2) Second-level grid: divide the first-level $6° \times 4°$ grid into 12×8 second-level grids according to latitude and longitude, corresponding to $30' \times 30'$ grid, which is about the earth 55.66 km × 55.66 km grid at the equator; (3) Third-level grid: Divide the second-level grid into 2×3 third-level grids according to latitude and longitude, corresponding to 1:50,000 map The grid is $15' \times 10'$, which is approximately 27.83 km × 18.55 km grid at the equator of the earth; the rest can be deduced by analogy [20].

In the subdivision of the height domain, the arbitrary division number is m, and the height domain is divided into 2 m layers, with 2 m-1 layers underground and 2 m-1 floors above ground. The height of the same layer of the same level grid is equal, and the height matches the latitude direction length of the grid formed by the corresponding level division at the equator of the corresponding contour plane of the layer.

The grid height at the same level and the grid latitude length at the equator of the corresponding contour are shown in Fig. 1. The height domain of the Beidou 3D grid position code is divided, and ten basic grids of 4°, 30′, 15′,…, 1/2048″ on the earth's surface are used as the Beidou 3D grid position code to define the middle of the earth height. Form the height domain grid division: a) the initial grid, the above ground and underground are divided into two parts; b) the first level grid, using the same division as the length of the equator, each division is about 445.28 km; c) The second-level grid uses the same division as the length of the equator 30′, each division is about 55.66 km; the remaining levels are analogous to [16].

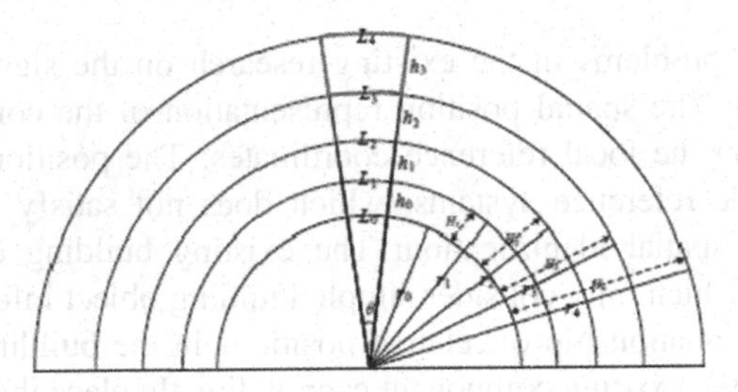

Fig. 1. The division method of unequal distance in the direction of the height domain (equatorial plane).

2.1 Meshing Dividing

2.1.1 Encoding Rules

Beidou 3D grid position code is composed of two-dimensional code + height-dimensional code intersection [21], a total of 32 code elements.

The Beidou 3D grid position code is composed of 12-bit code elements. The structure and code element values are shown in Fig. 2. The third dimension code is as follows:

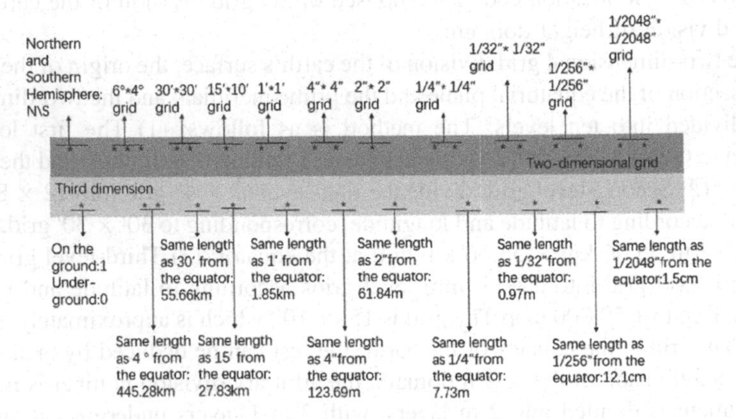

Fig. 2. The form of Beidou 3D grid position code.

3 Overview of Beidou 3D Grid Position Coding

To locate a building, it is necessary to locate the entire building and the internal component entities of the building [10]. The coordinate system outside the building is a rigid Beidou grid position system. The two axes of the plane coordinate are radial and latitude. The building takes the logo center as the positioning object, and is assigned a positioning code, which is organized and managed as a whole. The building itself is not only the subspace of the global grid system, but also the parent space of the internal information of the building. Inside the building, in order to ensure the expandability of the internal and external divisions, the external positioning marking points are the starting points of the division. After determining the starting point of the internal division, the directions of the two axes of the plane are also consistent with the two axes of the external global position coordinate system, which arc true north and true east directions. The height dimension is divided according to the floor height of the building itself. The division of each floor also directly follows the existing area division in the building, and these sub-spaces will be assigned a positioning identification code on each floor. The sub-spaces of each floor are represented by a rectangular shape of uniform size. The entity eventually falls in a certain subspace, and the three-axis positioning is performed in the three-dimensional space.

3.1 Meshing and Coding Scheme

This section discusses the identification points of each level (building-floor-subspace), the division of grid levels, and the rules for coding after division.

Identification of the Entire Building. First, locate buildings on a global scale and perform unique spatial identification. To identify the building, choose a suitable identification point, and use the center of the ground projection range of the building as the identification point instead of the actual area occupied. The projection polygon is divided by the Beidou grid with a suitable level, and the inner corner closest to the southwest is selected as the landmark O of the building. In reality, the distance between buildings generally exceeds 8 m, and the seventh level (1/4′) of the Beidou grid code is selected for division [20].

After obtaining the identification point, identify the point in the global position frame. An administrative region organization will manage the data of multiple buildings, and the administrative center where the building is located is selected as the reference point for positioning. Determine the administrative division center and building identification points, establish a two-dimensional plane coordinate system, determine the position of the entire building in this coordinate system, and select the appropriate level L to determine the number of grids M, N to span (Figs. 3 and 4).

In order to avoid that the identification points of the two buildings fall on the same grid and lose their uniqueness, a grid of appropriate size is selected so that the side length is not greater than the distance between the two. The 7th level grid is more suitable [20], but when the distance is too far, the length of the code will be lengthy. If a smaller level is introduced, although fewer span codes can be used, the level's mark must be added. In order to balance the number of span grids and the identification bits

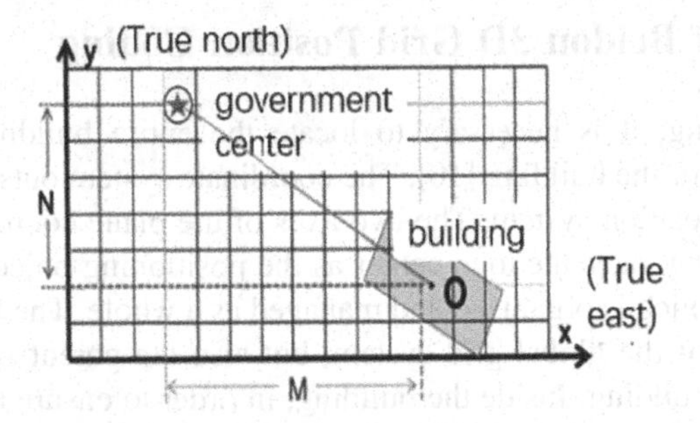

Fig. 3. The positioning process of the whole building.

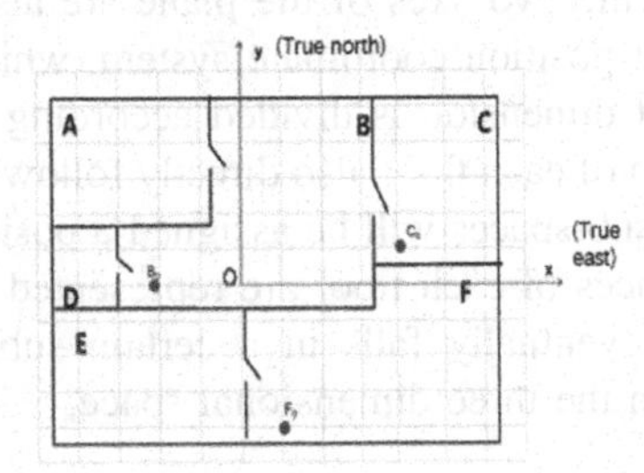

Fig. 4. Subspace identification point determination.

introduced for the level identification, in addition to the seventh level, another level is selected to jointly identify the span. The maximum two-digit seventh-level grid can cover a length of 24.75° < 1′, so the fourth and seventh levels are selected to jointly identify the span. The offset of the building relative to the administrative center is negative. M and N should contain a negative sign to distinguish the relative direction.

Division and Coding of Building Floor Height Dimension. Considering the division of the height dimension of the building, the starting plane of the division is specified on the ground in the building. The upward direction is the positive direction of the height axis, and vice versa. The internal space of the building is generally distributed in different floors, but it is also If there is a staircase with a corner between two floors, it is stipulated that these facilities are divided into the floor space below them.

The tallest building in the world, Burj Khalifa, has a total height of 828 m, 162 floors above ground and 169 floors underground. The number of underground floors generally does not exceed 162. The floor code consists of five digits in total: the first digit represents U or D on the ground, and the last three digits are the number of floors. For example, U021 means 21 floors above ground, D002 means 2 floors underground, and so on.

The building code is extended to the floor height as:

$$C\_OMN{-}F \tag{1}$$

Floor Plan Division and Subspace Location Code. In each floor of a building, many components need to be further divided into space to obtain a more refined space to determine the precise location of the components. In order to meet the principle that the external division can be extended to the interior, the origin O adopted is the identification point selected when positioning the entire building. The x and y axis directions of the internal coordinate system are respectively selected as the true east, true north, and z axis. The direction adopts the growth direction of the building's floors, and each floor height is an independent floor coordinate system. We further divide the entire floor to obtain subspaces, and locate and code the subspaces in the floor plan coordinate system. If one vertex of the polygon of the horizontal plane of the space is selected, it may happen that multiple spaces share one identification point.

To determine the identification points of the subspace, first select the level, and select a grid with a side length of 1 m to divide the molecular space. Each space is divided by multiple grids, and the corners of the grid are in the space. We choose the corner point closest to the origin of the floor coordinate system as the identification point of the subspace, and the positioning points of the subspaces B, C, and F are obtained as B0. C0, F0.

After the identification points of each subspace are obtained, their true north and true east offsets in the floor plan coordinate system are M′, N′ respectively. The location code for constructing the building code to extend to the subspace level is:

$$C\_0\ MN\text{-}F\text{-}M^{\wedge\prime}\ N' \tag{2}$$

Division and Coding in Subspace. Each subspace in the floor plan coordinate system has its own independent coordinate system constructed inside, and they are related by the floor plan coordinate system, and the offset of the respective subspace identification point from the floor identification point is used as the identification code. The components are divided in three dimensions within the subspace to accurately locate them.

In the three-dimensional subspace, a certain component in the floor is located further and more accurately. Some components are very close to each other, and a smaller-scale grid is used as the basic unit for division. It is more appropriate to use a 9-level grid to divide the subspace [20]. Positioning code of the component coordinate system in the subspace: the division origin is the identification point of the subspace in the floor subspace, and the direction of due east and north is the x, y axis, and the building growth direction is the z axis. In the subspace, the component is encoded in three-dimensional space to obtain the offset PQR in the three axis directions. The location code of the construction building code extended into the subspace is:

$$C\_0\ MN\text{-}F\text{-}M^{\wedge\prime}\ N^{\wedge\prime}\text{-}PQR \tag{3}$$

In this coding system, $C_O\ MN$ is the building code for the global environmental positioning of the entire building; F is the height code for the height dimension positioning of entering the building; *M′N′* is the subspace of the entering floor in the second The floor subspace location code of the coordinates in the dimensional plane; PQR is the component location code in the subspace.

4 Overview of Beidou 3D Grid Position Coding

4.1 Code Generation Algorithm in Building Design Stage

This code is assigned to the component entity in the design stage of the building, and the fusion of multi-source data and the backtracking of information can be carried out based on the unique code in the subsequent stage [12].

In the BIM vertex coordinates, extract the minimum and maximum values of x, y, z coordinates to obtain the minimum component coordinates $S_{min}(X_{min}, Y_{min}, Z_{min})$ and maximum component coordinates $S_{max}(X_{max}, Y_{max}, Z_{max})$ to obtain the model range and calculate the approximate Space center point $S_O(X_O, Y_O, Z_O)$; $X_O = (X_{min} - X_{max}/2)$; $Y_O = (Y_{min} - Y_{max}/2)$; $Z_O = (Z_{min} - Z_{max}/2)$.

If the latitude and longitude are expressed, directly according to the global floating point coordinate to the code conversion [21], if relative to the reference system, the following steps can be followed: determine the local coordinate system of the component in the BIM, determine the local coordinate system origin O, according to the orthogonal three Two of the two axes determine the direction of the coordinate system; determine the target coordinate system for transformation, according to the data in the BIM file, if the component is provided with the information of the reference space where it is located, the target coordinate system is the subspace coordinate system of the urban structure, if If only the reference space coordinate system of the floor is provided, the target coordinate system is the multi-storey coordinate system of urban buildings. After selecting the target coordinate system, determine the origin O' of the target coordinate system, and determine X', in the direction of true north and true east. X', Y' axis, get Z' axis; using the principle of coordinate transformation, the coordinate $S_O(X_O, Y_O, Z_O)$ of the component's center point in the component local coordinate system is transformed into the target coordinate system. First calculate the offset of the three axes of O' in the coordinate system $O(\Delta x, \Delta y, \Delta z)$, and then calculate the cosine of the included angle between the X' axis and the X axis, Y axis, and Z axis respectively $(\cos \alpha_1, \cos \beta_1, \cos \gamma_1)$ the cosine of the angle between Z^' axis and X,Y,Z axis $(\cos \alpha_2, \cos \beta_2, \cos \gamma_2)$,the coordinates of the identification point of the entity in the building coordinate system can be obtained by the equation $T \cdot S' = S$, and the rotation matrix T can be expressed as:

$$T = \begin{bmatrix} \cos \alpha_2 & \cos \beta_2 & \cos \gamma_2 & 0 \\ \cos \beta_1 \cos \gamma_2 - \cos \beta_2 \cos \gamma_1 & \cos \alpha_2 \cos \gamma_1 - \cos \alpha_1 \cos \gamma_2 & \cos \alpha_1 \cos \beta_2 - \cos \alpha_2 \cos \beta_1 & 0 \\ \cos \alpha_1 & \cos \beta_1 & \cos \gamma_1 & 0 \\ x_0 & y_0 & z_0 & 1 \end{bmatrix}$$

After obtaining the identification points of the subspace and the longitude and latitude of the building (floor), the space identification [20] is generated according to the method of floating-point number transcoding, and this encoding becomes the unique spatial location identification of the component entity.

4.2 Multi-source Building Data Integration Algorithm Based on Coding Model

Many scenes in the building industry use spatial location to associate multi-source data, such as: file association management, automatic inspection, and intelligent construction [19]. Analyze the needs and feasibility of location identification codes in the integration of multi-source data; then design a large table of split index coded as RowKey, and on this basis, realize the efficient integration of multi-source data in the construction industry.

In order to realize the organization and management of multi-source information of structures, in addition to uniformly identifying global structure information, it is also necessary to establish a large index table with the code as the main key to associate data from different sources. MongoDB [11] was selected as the back-end database design because of its flexible design and high scalability; Nosql database can handle the problem of large differences in the types of entity attributes at different levels; MongoDB stores and manages according to the key-value pair format, and the code is used as the row key. The O(1) query efficiency of the Greek table, the row key is the building space identification code, and the attributes include the latitude and longitude information of the building and subspace where the component entity is located, and the number of floors.

In order to achieve multi-source data fusion based on spatial identification, the multi-source attribute information associated with the entity, such as design drawings, corresponds to the household registration information of the household. The index table also associates the link or file address of this information with the space identification code, so that the user can get the building code through the location of the component entity. Since the code of a certain building entity is unique in the entire data table, it can According to the code, locate the attribute dictionary of an entity in the database, and associate it with the multi-source data related to the spatial location.

4.3 Data Retrieval Algorithm Based on Coding Model

Take point query, range query and distance query based on building space constraints as examples to illustrate the data retrieval method.

Point Query. Point query refers to query all objects in the vicinity of a specified point [11]. The specific requirements in this research scenario are: when a fire occurs, the fire hydrant near the location of the firefighter. The nearby range here is limited to the same floor. As shown in Fig. 5 below, you want to query the objects near the location of the asterisk, which is the range of the dotted circle. This range includes three subspaces BCE.

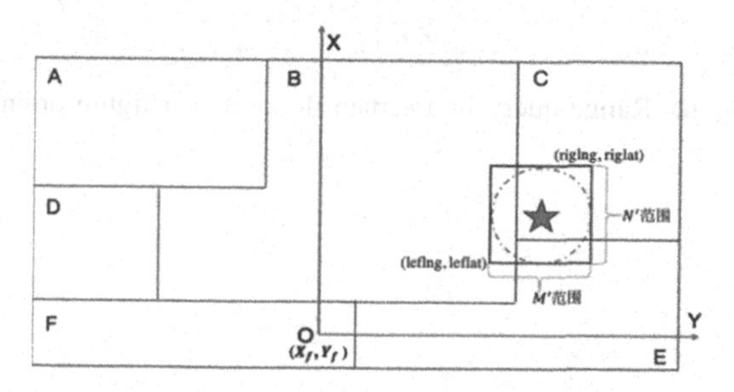

Fig. 5. KNN query of points in a certain floor.

>Use the code to calculate the range near the asterisk. The query on a certain floor may span multiple subspaces. Choose the floor coordinate system to divide the eighth level (distance from the equator 1 m). The query center coordinates can be latitude and longitude coordinates (x_0, y_0). The latitude and longitude of the southwest and northeast corners of the outer rectangle are letlat, letlng, riglat, riglng, respectively. The latitude and longitude points (x, y) in the outer rectangle of the floor coordinate system satisfy $x \geq (\text{letlat} - X_f)$ *and* $y \geq (\text{letlng} - Y_f)$ *and* $x \leq (\text{riglat} - X_f)$ *and* $y \leq (\text{riglng} - Y_f)$. Here (x, y) means the distance from the origin in the local coordinate system, and the number of digits after the building code is the number of grids offset in the local coordinate system. The grid scale divided by the floor coordinate system is $1/32$, and the deviating longitude and latitude to the number of grids to the conversion multiple is $32 \times 60 \times 60$. Construct the complete structure of the building code $C_0MN\_UXXX\_M'N'$, the offset in the floor coordinate system is fixed $M'N'$ two bits. $M'N'$ is signed, and the conversion magnification is expressed as L, and the coding filter condition is

$$M\text{^'}\leq L(riglat\text{-}X\_f) \quad and\ M\text{^'}\geq L(leflat\text{-}X\_f) \tag{4}$$

$$N\text{^'}\leq L(riglng\text{-}Y\_f) \quad and\ N\text{^'}\geq L(leflng\text{-}Y\_f) \tag{5}$$

Range Query. Range query refers to query all objects within the framed range [11]. The specific requirements in this research scenario are: query all beams in a room, as shown in subspace C in Fig. 6; or delineate an arbitrary closed graphic to query all specific component entities contained in it, as shown in the figure below that spans three subspaces ABD Irregular shades of blue.

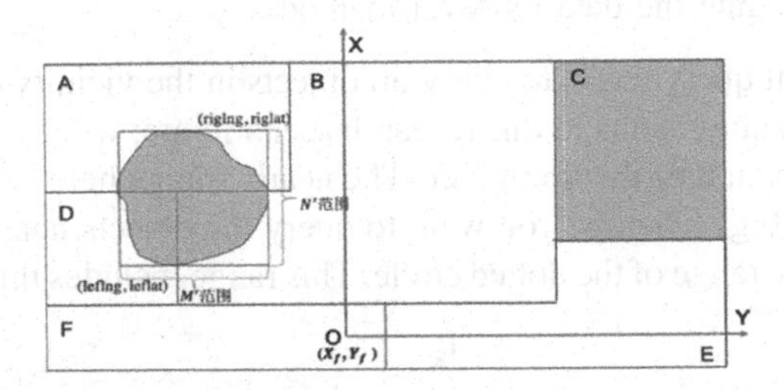

Fig. 6. Range query in a certain floor. (Color figure online)

Based on the outsourcing rectangle point search method, any closed graph is used as the query range. If the query range is a divided structure in a building, enter a latitude and longitude point $P(x, y)$ in the subspace, and request to return a component of the subspace where the point is located. Calculate the coordinates of P in the coordinate system $O(x_f, y_f)$ then divide by the degree to the span code to the conversion magnification L to obtain the offset code number M', N' of the point P. The subspace identification point C_0 where point P is located should be closer to point P, and the offset of C_0 in the floor coordinate system M_c, N_c should satisfy the following equation, and ϵ should be as small as possible, that is, as close as possible to a constant value: $M_c{}^2 + N_c{}^2 = M'^2 + N'^2 + \epsilon$. There is a situation where any point in the subspace reaches the identification point of the space and is farther from the identification point of other subspaces. This article takes the three closest coding values, and then finds the latitude and longitude of the subspace contour in the detailed attribute column of the corresponding data row String, to judge whether the point is in the polygon, and get the final constipation, you can use the matching algorithm to get all the specific component entity data rows whose subspace code is M_c, N_c.

Distance Query with Constraints. The query in the building must consider the structural characteristics of the building itself, such as the distance between the components A and B in the two rooms due to the barrier of the door, not a straight line distance, and there are multiple passable paths. Given a query point and multiple possible target points, the constrained distance query is to return the shortest distance from the query point to each target point.

The calculation of the shortest distance of multi-path can use "Dijkstra algorithm" [3], first determine the distance of each directly adjacent point. In this scenario, it may be the distance between the component in the subspace and the node of the subspace door, or the direct distance between two doors, which can be considered as the distance between the codes of two structures. Taking codes $C_0MN\_UXXX\_M_1'N_1'\_P_1Q_1R_1$ and $C_0MN\_UXXX\_M_2'N_2'\_P_2Q_2R_2$ as an example, the R code on the height dimension is not considered, and the two codes on the same floor differ only in subspace coding and fine coding. Taking the floor marking point as the reference point, the distance of code 1 in the direction of true east and true north of this point is $L_fM_1' + L_rP_1$ and $N_1' + L_rQ_1$, and the distance of code 2 in the direction of true east and true north of this point is $L_fM_2' + L_rP_2$ and $L_fN_2' + L_rQ_2$, then the distances between Code 1 and Code 2 in the direction of True East and True North are: $(L_fM_1' + L_rP_1) - (L_fM_2' + L_rP_2)$; $(L_fN_1' + L_rQ_1) - (L_fN_2' + L_rQ_2)$

The distance between latitude and longitude in the two directions can be converted into a plane distance according to the formula to obtain the plane distance between two nodes.

5 Experiment

Based on the self-designed building data visualization and management system, this section carries out typical application experiments such as code generation experiment, building multi-source data association and query from model establishment and application to verify the correctness, efficiency and feasibility of the coding model.

There are many types of BIM software in the construction industry, with different data formats. In this experiment, the Web terminal is selected for the display and function realization of the BIM model, and the geometric spatial data of the 3D model of the BIM platform such as Revit is analyzed and displayed on the Web terminal through WebGL. The background is connected to the database to manage and query the data information related to the construction and building. Interaction through the web.

Choose B/S web application framework verification, based on WebGL and HTML5 language, the framework level and function are shown in Fig. 7.

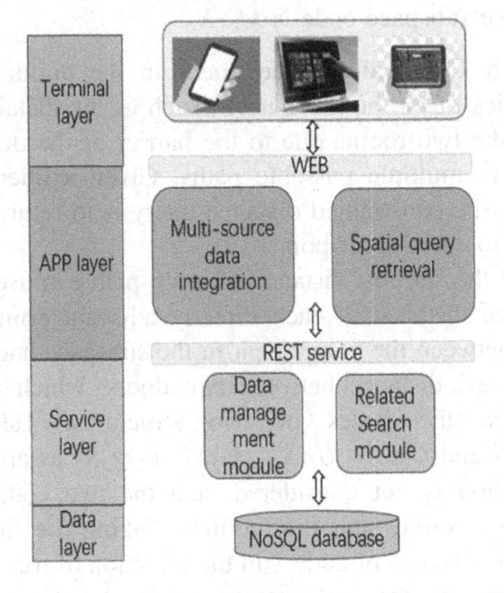

Fig. 7. Experimental framework structure function diagram.

Model Data Analysis and Model Reconstruction. The program parses the BIM data and obtains the absolute position coordinates of the building and subspace identification points, which lays the foundation for subsequent operations. The Revit Architecture building model file is converted into a JSON format file, and WebGL parses the JSON and visualizes it on the terminal. The exported building model must conform to the OBJ format, including four contents: material texture, geometric characteristics, custom metadata and object record object attribute data, which is associated with an entity through an identifier ID. After getting the parsed properties of each component, WebGL visualizes the model data on the Web. Figure 8 shows the hierarchical renderings of the subspace after model reconstruction and some attribute information of the corresponding modules. The results of the other levels are similar.

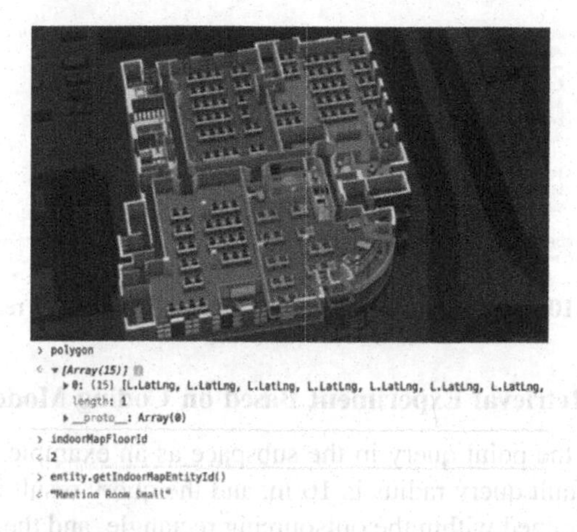
> polygon
< ▾ [Array(15)]
▸ 0: (15) [L.LatLng, L.LatLng, L.LatLng, L.LatLng, L.LatLng, L.LatLng, L.LatLng,
length: 1
▸ __proto__: Array(0)
> indoorMapFloorId
< 2
> entity.getIndoorMapEntityId()
< "Meeting Room Small"

Fig. 8. Subspace after model reconstruction.

Coding Generation Experiment of Constructing Building Model. The analyzed subspace and building outline information associated with the component, as shown in Fig. 9, is the result of the experimental data encoding.

geocode:G1398_U002_+005,+071_+068,-217
entity_id:0809
room_name: Meeting Room C
floor_id:2
westerport_house
Birth_Place: ShanDong
NumOfPeo: 2

Fig. 9. Encoding result.

Multi-source Data Integration. According to the relationship between data, multiple data tables in the database can be associated to integrate multi-source data. After the association, the attribute information of the component also contains the corresponding multi-source information. The component data table is a key in the JSON dictionary based on the structure of the building code, while the value is composed of multiple attributes, realizing fast search based on key-value pairs. The experimental results, as shown in Fig. 10, enter the subspace and choose an entity to pop up the properties of the component itself and the multi-source properties of the associated subspace.

```
▾{G1389_U002_5,71_+0234,+0175: {…}, G1389_U002_5,71_+0237,+0198: {…}, G1389_U002_5,71_+0245,+0193: {…}, G1
 ▾G1389_U002_5,71_+0234,+0175:
    entityid: "0089"
    BIM_code: "020507004"
    Building_name: "westport_house"
  ▸build_loc: _Point {lat: 56.460062935054054, lng: -2.9783190446561236}
  ▸room_loc: _Point {lat: 56.46005829160815, lng: -2.9782416466153525}
    floorid: 2
  ▸entity_loc: L.LatLng {lat: 56.460313202552875, lng: -2.9780514388046098}
    birth_place: "Zhejiang"
    NumofPeo: 1
  ▸__proto__: Object
 ▸G1389_U002_5,71_+0237,+0198: {entityid: "0090", BIM_code: "060201004", Building_name: "westport_house",
```


Fig. 10. Multi-source information integration display results.

Data Query and Retrieval Experiment Based on Coding Model

Point Query. Take the point query in the subspace as an example, the user selects the query point, the default query radius is 10 m, and the query result is shown in Fig. 11. The codes are constrained within the outsourcing rectangle, and the code key values are filtered to obtain a code set that meets the query conditions.

Fig. 11. Point query result.

Range Query. The user arbitrarily specifies a certain range to get all specific entities in it, which may be a certain floor or an irregular polygon that breaks through the existing boundary of the building, as shown in Fig. 12.

Fig. 12. Query across indoor boundary polygon range.

Coding Comparison. The accuracy of the range query based on the building code is high, especially when the query range is the inherent space division in the building, the accuracy is almost 100%; and the query efficiency is very high, only the matching code is required. The two-direction span code of the spatial coding can directly obtain the query result; this query requirement is more common in the building scene. The global rigid grid division ignores the boundaries of the object entity itself, such as the Beidou grid reference frame. Each division grid exists objectively, and the boundaries of these grids cannot fit the internal structure of the building. As shown in Fig. 13. In the query scenario, the query can only be performed based on the outsourced patch of the subspace. The range of the outsourcing patch is very different from the actual range of the subspace, resulting in a low accuracy rate, which will include many non-target entities; if a smaller grid with a larger level is used for aggregation, irregular polygons like the subspace are fitted It requires constant iterative calculations, and the amount of calculation is large and the efficiency is not high. Figure 14 shows a schematic diagram of the effect of global rigid grid division.

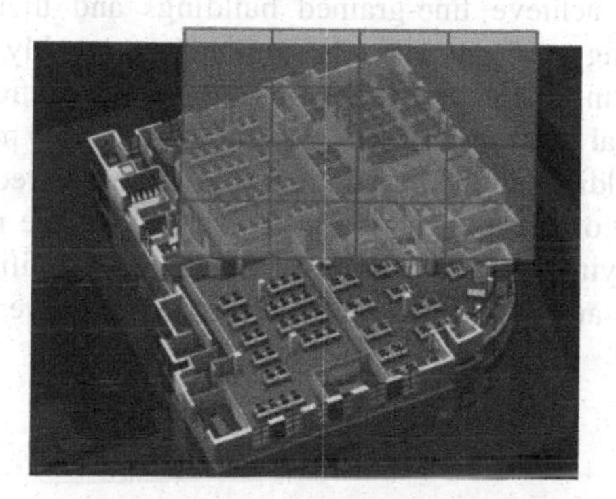

Fig. 13. Global rigid meshing result.

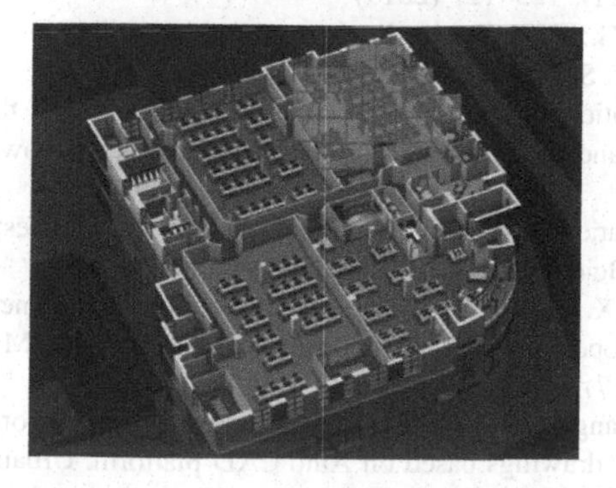

Fig. 14. Fine fill grid meshing result.

6 Summary and Outlook

Construction information management plays a huge role in the implementation of construction projects and the macro-management of the construction industry. It is not only the function of project management, but also the foundation of project management. The full play of IT technology in the construction field requires the standardization, standardization and unification of construction information, which is also the basic link of construction industry informatization.

However, the establishment of building space identification position coding is a systematic project. This article analyzes the framework of building space identification position coding and found that the existing rules are inconsistent and inefficient, and proposes a construction building coding based on the Beidou grid Rules, and the corresponding point query, range query, distance query and other algorithms to solve these problems. Many issues still need to be further studied: (1) Research on how to improve the location coding of building space identification, further increase accuracy, and improve the retrieval efficiency of components in a heterogeneous cross-database environment; (2) To achieve fine-grained buildings and their components In the computer programming environment, the automatic assembly of the component is completed by identifying the code of the connector at a specific position. (3) Most of the operations of spatial identification position coding are done manually. In the case of a small number of buildings in the initial application, the task requirements can be met, but when the number of components increases sharply, if the new data can be automatically added Carrying out the coding of the space identification position of the building and building and automatic warehousing can improve its efficiency.

References

1. Building Information Model Classification and Coding Standard. GB/T51269-2017 (2017)
2. Yang, C., Sheng, Y.: Research on spatio-temporal coding method of building information. Geospatial Inf. **17**(11), 125–127 (2019)
3. Yang, L., Jianya, G.: An efficient implementation of Dijkstra's shortest path algorithm. J. Wuhan Univ. Inf. Sci. Ed. **24**(3), 208–212 (1999)
4. Zhong, C.: Calculation of the conversion relationship between the power plant building coordinate system and the national coordinate system. Electr. Power Surv. Des. **02**, 19–21 (2014)
5. GB 22021-2008 National Basic Technical Regulations for Geodesy
6. BD 110001-2015 Beidou satellite navigation terminology
7. Wang, R., Huang X.: Research on the early warning management system of equipment components in the operation and maintenance phase based on BIM. Comput. Eng. Appl. **53** (19), 231–235 (2017)
8. Zhangrong, Y., Wang, Y., Dong G.: Design and implementation of dynamic coordinate marking system for drawings based on Auto CAD platform. Urban Surv. **04**, 37–40 (2019)
9. Cong, H., Liu J., Weichang, M.: Design and implementation of coordinate conversion and parameter encryption software. J. Changchun Normal Univ. (06) 2017
10. Cheng, J., Zhang, Ke., Zhuolei, W.: CAD drawing data coordinate conversion based on affine transformation. Surveying and Mapp. **40**(05), 224–227 (2017)

11. Li Z.: Indoor Grid Modeling [Doctoral Dissertation]. Beijing: School of Earth and Space Sciences, Peking University (2019)
12. Demyen, D., Buro, M.: Efficient triangulation-based pathfinding. In: Aaai, vol. 6, pp. 942–947 (2007)
13. BuildingSmart Corporation. https://www.buildingsmart.org/
14. Zhang, Y.: Research on Information Integration and Management of Construction Engineering Based on BIM [Doctoral Dissertation]. Beijing: Department of Civil Engineering, Tsinghua University (2009)
15. Zhao, X., Zhan, X., Zhang, X.: The application of BIM technology in the production stage of prefabricated building components. J. Graph. **39**(06), 1148–1155 (2018)
16. GB/T 13989-2012 National Basic Scale Topographic Map Framing and Numbering (2012)
17. Dexian, L.: Research on the Application of BIM Technology in Fire Emergency Management of Comprehensive Experimental Teaching Building [Master's Thesis]. Liaoning: School of Hydraulic Engineering, Dalian Ocean University, 2019, Journal **2**(5), 99–110 (2016)
18. Jin, L.: Research on BIM model component retrieval based on spatial relationship [Master's Thesis]. Beijing: Beijing Jianzhu University, 2019, Journal **2**(5), 99–110 (2016)
19. Chen, Y., Duan, X., Kang, J.: Development of BIM component management and component library management cloud platform for construction enterprises. Civil Eng. Inf. Technol. **11** (05), 28–35 (2019)
20. Jin, A., Cheng, C.: Spatial data coding method based on global division grid. J. Surv. Mapp. Sci. Technol. **30**(3), 284–287 (2013)
21. Cheng, C., Zheng, C.: A method for unifying existing longitude and latitude meshes CN (2012)

Maximum Power Output Control Method of Photovoltaic for Parallel Inverter System Based on Droop Control

Zhang Wei[1(✉)], Zhong Zheng[2], Hongpeng Liu[1], and Xuemai Gu[2]

[1] Department of Electrical Engineering, Northeast Electric Power University, Jilin, China
mrhhzw@126.com
[2] International Innovation Institute of HIT in Huizhou, Huizhou 516000, Guangdong, China

Abstract. Generally, the output power of photovoltaic (PV) inverter will match the load requirement. And at the beginning of the design the load power is less than the maximum output power of PV cells to ensure the system operation stable when the PV inverter operates in islanded mode. However, it causes the energy waste of PV cells. Therefore, more and more PV cells are combined with other energy sources to form the microgrid system in order to reasonably plan the power output of each energy source. Droop control is usually used to achieve the power distribution of parallel inverter in microgrid system. However, the traditional methods of adjusting the droop coefficients or adding virtual impedance cannot automatically achieve the maximum utilization of output energy of PV cells. Thus, a novel droop control method has been proposed to achieve the maximum power output of PV (MPO-PV) unit in this paper, where the PV units of parallel system always operate at the maximum power and the other inverters make up the remaining power required by the load, with effective improvement of the utilization rate of renewable energy sources (RESs). Meanwhile, the control parameters of the improved droop loop have been designed by the small signal modeling and system stability analysis. Finally, the validity of the proposed method has been verified by experimental results.

Keywords: PV cells · Parallel inverter · Droop control · MPO-PV · Small signal modeling

1 Introduction

Distributed generation (DG) has been developed rapidly in recent years due to the advantages of low environmental pollution, high energy utilization, flexible installation and low transmission power loss [1]. Compared with traditional generators, DG units have a high degree of controllability and operability, which makes micro-grid system based on DGs play an important role in maintaining the stability of power grid [2]. Although the power demand of micro-grid is increasing, the rated power of inverter switching devices is often limited by technical or economic factors. Therefore, multi-inverter parallel operation is usually adopted to improve the system capacity [3].

S. Shi et al. (Eds.): AICON 2020, LNICST 356, pp. 588–597, 2021.
https://doi.org/10.1007/978-3-030-69066-3_51

Droop control [4, 5] can solve the problem of voltage frequency regulation and power distribution between inverters without the interactive communication line, which has been widely used in the application of island parallel mode [8, 9] and grid-connected mode [14, 15] for PV inverter. In the PV inverter control methods based on droop control, the PV cells are generally assumed as constant voltage dc power supply with an infinite capacity by most scholars. However, the PV power is often fluctuant due to the intermittency and weather factors. Thus, this assumption ignores some problems in practical operation of PV inverters.

Therefore, several studies have been started to solve the problems mentioned above in grid-connected mode and island mode. To ensure the system connect with the grid at maximum power constantly and maintain the dc link voltage stable, the improved droop control schemes have been proposed in [8], which enhance the controllability of system power. To ensure that, [11] depends on predictive control algorithm for PV inverter to realize fast and accurate control of active power, and to alleviate the power grid frequency contingency without energy storage device. In addition, many issues may occur in island mode for the PV system based on traditional droop control, such as the voltage limit violation, fluctuations of PV input power, deviation of voltage/frequency and power oscillation. To solve the problem that the transient stress of energy storage converter current increases due to the large sudden change of load in the island mode of PV/storage microgrid system, a PV/storage coordinated control strategy has been proposed to suppress the transient power fluctuation of power conversion system (PCS) for energy storage and avoid the overcurrent fault [12]. However, this method lacks expansibility due to the utilization of high-speed communication line. The traditional droop scheme may not work well when its demanded power cannot be met by some renewable sources due to intermittency without storage. Therefore, an enhanced dual droop scheme has been proposed to improve the ability to resist the influence of the natural environment for two-stage PV system [14]. It has been proposed about a novel three-stage robust inverter-based voltage/var control (TRI-VVC) approach for high PV-penetrated distribution networks to reduce energy loss and mitigate voltage deviation [15]. To enable the maximum utilization of the voltage/current (V/A) rating of the interfacing inverter, an adaptive droop control has been proposed in a PV/battery hybrid system [19].

In the above studies, an auxiliary energy storage system is required to maximize the output power of PV inverter [19]. It is an urgent problem to be solved concerning how to achieve the MPO-PV units based on droop control in the parallel system without energy storage devices. Therefore, an improved droop method has been proposed here.

2 Issues Existing Traditional Droop Control

The configuration of paralleled inverter system is shown in Fig. 1. The system is composed of two single-stage full-bridge inverters in parallel, where the inverter 1 connects with the PV cells and inverter 2 connects with an equivalent dc power supply which may be a dc-link bus from other converter or source (non-renewable energy sources (NRESs), such as energy storage converter, diesel generator, et al.); Lacn are the filter inductors; Cacn are the filter capacitors; Rln and Lln are the resistive and inductive component of line impendences, respectively; Rloadn are the local loads.

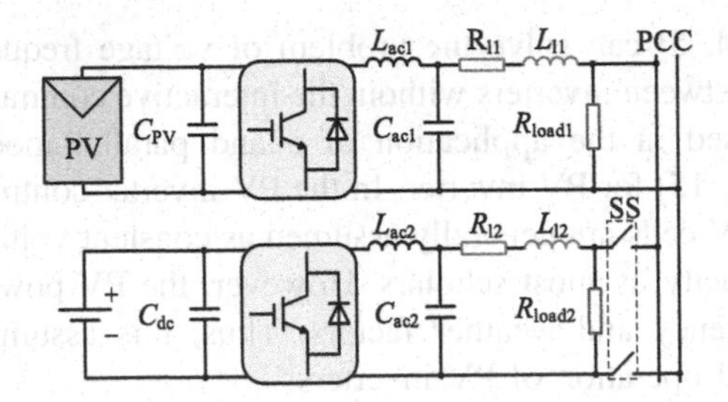

Fig. 1. Structure block diagram of parallel inverter system.

The inverter 2 is connected with the point of common coupling (PCC) by the static switch (SS).

The typical R/X ratio of a low voltage (LV) line has been given as 7.7 [2], that is, the line impedance is mainly resistive. Thus, the output power can be obtained:

$$P_{acn} = \frac{EV_{acn}}{R_{ln}} \cos \delta_n - \frac{E^2}{R_{ln}} \tag{1}$$

$$Q_{acn} = -\frac{EV_{acn}}{R_{ln}} \sin \delta_n \tag{2}$$

where, Vacn and E are the output voltage amplitudes of inverters and PCC; □n is the power angle difference between output voltage of inverters and voltage of PCC; Pacn and Qacn are the output active power and reactive power, respectively.

Thus, the resistive droop equations can be represented by (3) and (4):

$$V_{acrefn} = V_{O}n - k_{pn}(P_{acn} - P_{On}) \tag{3}$$

$$\omega_{acrefn} = \omega_{On} + k_{qn}(Q_{acn} - Q_{On}) \tag{4}$$

where, V_{acrefn} and ω_{acrefn} are the reference amplitude and reference angular frequency of inverter output voltage; k_{pn} and k_{qn} are their droop coefficients of active and reactive power; V_{On} and ω_{On} are their rated amplitude and angular frequency; P_{On} and Q_{On} are their rated active and reactive power, respectively.

It is shown in Fig. 2 about the control block diagram of single inverter based on traditional droop control. Firstly, the output voltages v_{acn} and currents i_{acn} should be measured to calculate P_{acn} and Q_{acn}. Then, the reference v_{acrefn} of output voltage can be generated by the droop control and can be tracked by the dual loop control. For simplicity, the two inverters are assumed as equally rated ($P_{O1} = P_{O2} = P_O$) and the total power required by the loads is also assumed to be more than the rated power of a single inverter and less than two times of its rating ($P_O < P_{Load} < 2P_O$). Generally, the droop coefficients are set to be the same ($k_{p1} = k_{p2}$, $k_{q1} = k_{q2}$) for power sharing. Therefore, the droop line of active power is shown in Fig. 3. Before parallel operation, the two inverters supply power to their local loads, respectively. At this point, the inverter 1 and 2 operate at a_1 and a_2 with the output active P_{1a} and P_{2a}, respectively. The inverters operate in parallel mode when SS is closed. The operating point of inverter 1 moves from a_1 to b_1 and inverter 2 moves from a_2 to b_2 due to the droop characteristic. Then, the inverters share the power of loads at steady state (P_{Load} = 2 $P_b = 2P_{ac1} = 2P_{ac2}$). Since the maximum output power point of PV cells is c_1, the traditional droop control cannot make PV cells operate at the maximum power point (MPP), which will inevitably cause the waste of PV power. If the inverter 1 outputs the maximum power ($P_{ac1} = P_{PV\max 1}$) without changing the droop line and the inverter 2 supplies the remained power of the loads ($P_{ac2} < P_{PV\max 1}$), the circular current will be generated because the amplitude of output voltages are different, which endangers the safe operation of the system.

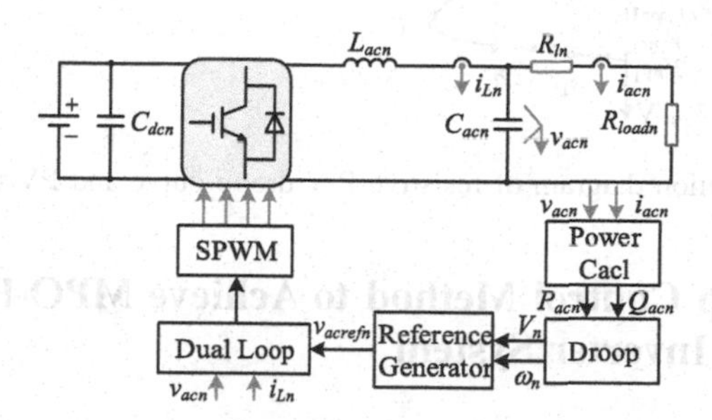

Fig. 2. Traditional control block diagram of single inverter.

The virtual impedance is usually used in the application of parallel inverter to balance the impedance between inverters and achieve power sharing. Similarly, the virtual impedance can also be added to distribute output power in proportion. If the added virtual impedance is resistive and define Zvir = Rv, the equivalent reference of output voltage can be expressed as:

$$v_{ref} = v_{acref} + v_{vir} = v_{acref} + i_o \cdot R_v \tag{5}$$

The equivalent voltage vvir generated by virtual impedance can make the droop line of inverter 1 translate ΔV. If vvir = ΔV = ΔV1, inverter 1 can operate at the MPP c1 and inverter 2 supplies the remained power, which achieves maximum utilization of PV power. Nevertheless, the method based on virtual impedance to achieve maximum power output of PV has the following shortcomings:

- Even if the complete information of PV curve is known, the value of virtual impedance needs to be repeatedly adjusted to make the inverter work at the MPP.
- Under the influence of natural factors such as weather, the MPP on the PV curve will change accordingly, and the local load may also fluctuate. Therefore, necessary to change the value of virtual impedance in real time to adapt to the normal operation of the system, with increasing the difficulty of actual assembly.

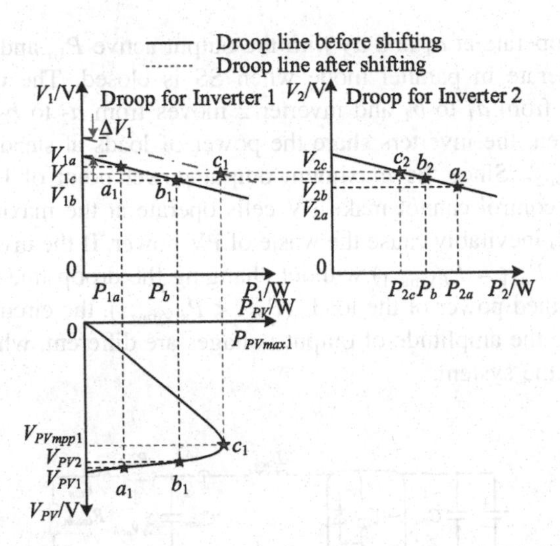

Fig. 3. Relation diagram of resistive P-V droop curve and PV power curve.

3 Novel Droop Control Method to Achieve MPO-PV for Parallel Inverter System

3.1 Design of Translation ΔV

The method to shift the droop line of PV inverter can be used to improve the energy utilization of PV cells when inverters are in parallel operation. If the droop line of inverter 1 can be raised by ΔV_1 as shown in Fig. 3, the operation point a_1 of inverter 1 can move to the MPP c_1 of PV cells, which achieves the MPO-PV.

The problem is how to determine the value of ΔV_1, where the simplest way has been proposed here. The power imbalance, i.e., the difference between the value $P_{PV\max1}$ of MPO-PV and output power of PV inverter 1, can be fed to a PI controller of power loop, and then output of the PI controller can be used as ΔV_1, which will always enforce a zero-power imbalance in the steady state. ΔV_1 can be obtained by (4):

$$\Delta V_1 = \left(k_{PVp1} + \frac{k_{PVi1}}{s}\right)(P_{PV\max1} - P_{ac1}) > 0 \tag{6}$$

where, $P_{PV\max1}$ can be tracked by some common MPP tracking methods; k_{PVp1} and k_{PVi1} are the proportional and integral gains of the PI controller of power loop, respectively.

Hence, according to (4), the droop equation of active power for PV inverter can be expressed as:

$$V_{ac1} = V_01 - k_{p1}(P_{ac1} - P_{01}) + \left(k_{PVp1} + \frac{k_{PVi1}}{s}\right)(P_{PV\max1} - P_{ac1}) \tag{7}$$

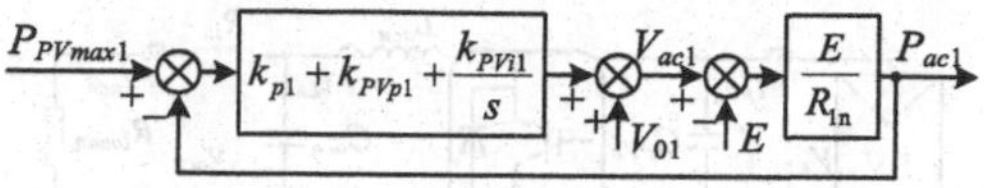

Fig. 4. Block diagram of active power closed loop control.

3.2 Analysis for Active Power

It is assumed that the rated output power of the inverter is the same as the maximum output power of the PV cells, i.e., $P_{01} = P_{PV\max1}$, and (5) can be further reduced to:

$$V_{ac1} = V_01 + \left(k_{p1} + k_{PVp1} + \frac{k_{PVi1}}{s}\right)(P_{PV\max1} - P_{ac1}) \tag{8}$$

Due to the fact that power angle δ is usually small, $\sin\delta \approx \delta$ and $\cos\delta \approx 1$ can be used for simplification. Then, the closed-loop control block diagram of output active power is shown in Fig. 4.

According to the superposition theorem, the closed-loop transfer function of the output active power is:

$$\begin{aligned} P_{ac1}(s) = & \frac{\left[(k_{p1} + k_{PVp1})s + k_{PVi1}\right]E}{\left[(k_{p1} + k_{PVp1})E + R_{l1}\right]s + k_{PVi1}E}\frac{P_{PV\max1}}{s} \\ & + \frac{Es}{\left[(k_{p1} + k_{PVp1})E + R_{l1}\right]s + k_{PVi1}E}\frac{V_{01} - E}{s} \end{aligned} \tag{9}$$

According to the final value theorem, the final steady state value of the output active power can be obtained as:

$$\lim_{t \to \infty} P_{ac1}(t) = \lim_{s \to 0} sP_{ac1}(s) = P_{PV\max1} \tag{10}$$

According to (8), it can be inferred that the output power P_{ac1} of inverter can track the maximum output power $P_{PV\max1}$ of the PV cells with zero steady state error by proposed method. Moreover, the proposed method is not affected by the voltage of PCC and the impedance of transmission line.

3.3 Implementation of the Novel Droop Mehtod to Achieve MPO-PV for Parallel Inverter System

Since the maximum active power output of photovoltaic cells will not affect the reactive power of the inverter, the reactive power Eq. (2) can remain unchanged. According to (2) and (6), it is shown in Fig. 5 about the control strategy of the overall parallel inverter system. Figure 5(a) shows the proposed control method to achieve MPO-PV for PV inverters and Fig. 5(b) shows the traditional resistive droop control adopted by the NRESs inverters.

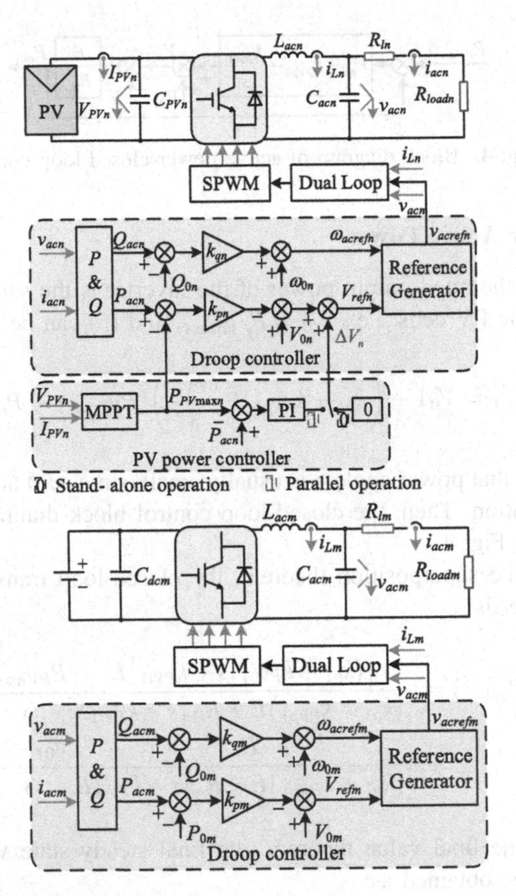

Fig. 5. Control strategy of the overall parallel inverter system: (a) novel resistive droop control method to achieve MPO-PV for PV inverters, (b) traditional resistive droop control for NRESs inverters.

For the PV inverter n, when connected with NREs inverter m in parallel mode, the measured output voltage V_{PVn} and current I_{PVn} of PV cells can be fed to the MPP block to obtain the maximum output power point (V_{PVmppn}, $P_{PV\max n}$) of the PV cells. In addition, the measured output voltage v_{acn} and current i_{acn} can be used to calculate the output active power P_{acn} of inverter. The input to the PI controller of PV power loop is the difference between $P_{PV\max n}$ and P_{acn}. The output ΔV_n of PI controller can be fed to active droop equation, and then the reference voltage of inner loop can be generated.

Finally, the generated reference v_{acrefn} can be tracked by the dual-loop quasi-proportional-resonant (qPR) controller shown in (11) which is insensitive to the resonance frequency drift compared to the proportional-resonant (PR) controller.

$$G_{qPR}(s) = k_p + \frac{2k_r\omega_c s}{s^2 + 2\omega_c s + \omega_0^2} \tag{11}$$

where, k_p is the proportional gain, and k_r is the resonant control gain, ω_0 and ω_c are the fundamental frequency of output voltage and cutoff frequency, respectively.

For the NRESs inverter m, the measured output voltage v_{acm} and current i_{acm} can be used to calculate the output power P_{acm} and Q_{acm}. Then, the generated P_{acm} and Q_{acm} are fed to the traditional resistive droop Eqs. (1) and (2) to obtain the output voltage reference v_{acrefm}. Finally, the similar dual-loop control is also used to track the voltage reference.

As for the inductive transmission line, the similar control method can be used to achieve the same control target: to change the resistive droop control to the inductive one. And the generated ΔV_n should be added in the inductive active power equation. Similarly, the reactive power equation can remain unchanged.

4 Simulation Results

Simulations have been performed with two similarly rated inverter systems in PLECS. The main purpose of the simulations is to verify the anticipated MPO-PV for parallel inverter system.

Figure 6 shows the comparison simulation results between the proposed control and traditional droop control, and the inverters supply power to the loads (P_{ac1} = 370 W, P_{ac2} = 726 W) alone before t_1. When the SS is closed at t_1, the PV inverter adopting the traditional droop control shares power (P_{ac1} = P_{ac2} = 548 W) with the NRESs inverter as shown in Fig. 6(a), which causes the energy waste of PV cells. When using the proposed control method shown in Fig. 5(a), the PV inverter can achieve the maximum power output and the NREs inverter makes up the remained power of the load automatically as shown in Fig. 6(b).

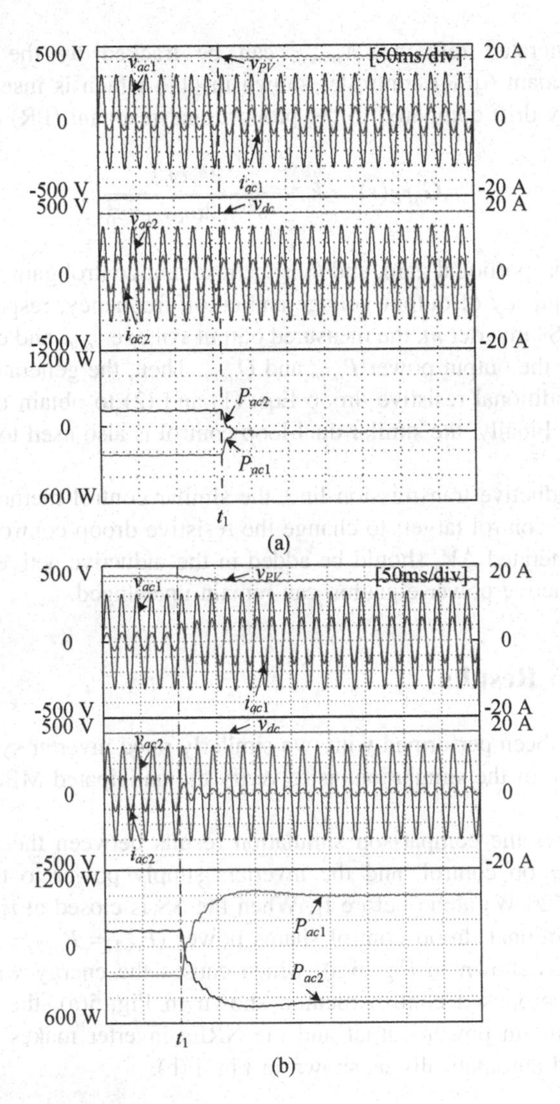

Fig. 6. Comparison simulation results: (a) proposed control method to achieve MPO-PV for PV inverters, (b) the traditional droop control.

5 Conclusion

To solve the problem of the maximum power output for PV cells in parallel inverter system, a novel droop control method has been proposed in this paper to achieve MPO-PV for parallel inverter system, and the energy utilization ratio of PV inverter has been improved. Finally, the simulation results have varied the validity and robustness of the proposed method.

References

1. Huang, N., et al.: Power quality disturbances classification using rotation forest and multi-resolution fast S-transform with data compression in time domain. IET Gener. Transm. Distrib. **13**(22), 5091–5101 (2019)
2. Rocabert, J., Luna, A., Blaabjerg, F., Rodriguez, P.: Control of power converters in AC microgrids. IEEE Trans. Power Electron. **27**(11), 4734–4749 (2012)
3. Zhang, Y., Yu, M., Liu, F.R., Kang, Y.: Instantaneous current-sharing control strategy for parallel operation of UPS modules using virtual impedance. IEEE Trans. Power Electron. **28** (1), 432–440 (2013)
4. Liang, H.F., Zhang, C., Gao, Y.J., Li, P.: Research on improved droop control strategy for microgrid. Proc. CSEE **37**(17), 4901–4910 (2017)
5. Sun, Q.Y., Wang, R., Ma, D.Z., Liu, Z.W.: An islanding control strategy research of We-energy in energy internet. Proc. CSEE **37**(11), 3087–3098 (2017)
6. Tong, Y.J., Shen, J., Liu, H.P., Wang, W.: A seamless switching control strategy for operating modes of photovoltaic generation system. Power Syst. Technol. **38**(10), 2794–2801 (2014)
7. Hoke, A.F., Shirazi, M., Chakraborty, S., Muljadi, E., Maksimovic, D.: Rapid active power control of photovoltaic systems for grid frequency support. IEEE J. Emerg. Select. Top. Power Electron. **5**(3), 1154–1163 (2017)
8. Zhang, C.X., Li, C.B., Feng, W., Sun, K., Xia, Y.W., Liu, Q.: A coordinated transient power fluctuation suppression strategy for power conversion system in islanded PV/storage microgrid. Proc. CSEE **38**(08), 2302–2540 (2018)
9. Liu, H.P., et al.: An enhanced dual droop control scheme for resilient active power sharing among paralleled two-stage converters. IEEE Trans. Power Electron. **32**(8), 6091–6104 (2017)
10. Zhang, C., Xu, Y., Dong, Z.Y., Ravishankar, J.: Three-stage robust inverter-based voltage/var control for distribution networks with high-level PV. IEEE Trans. Smart Grid **10**(1), 782–793 (2019)
11. Vazquez, N., Yu, S.S., Chau, T.K., Fernando, T., Iu, H.H.C.: A fully decentralized adaptive droop optimization strategy for power loss minimization in microgrids with PV-BESS. IEEE Trans. Energy Conver. **34**(1), 385–395 (2019)
12. Sreekumar, P., Khadkikar, V.: Adaptive power management strategy for effective Volt-Ampere utilization of a photovoltaic generation unit in standalone microgrids. IEEE Trans. Ind. Appl. **54**(2), 1784–1792 (2018)

An OOV Recognition Based Approach to Detecting Sensitive Information in Dialogue Texts of Electric Power Customer Services

Xiao Liang[1(✉)], Ningyu An[1], Ning Wu[2], Yunfeng Zou[2], and Lijiao Zhao[1]

[1] Global Energy Interconnection Research Institute Co. Ltd., Artificial Intelligence on Electric Power System State Grid Corporation Joint Laboratory (GEIRI), Beijing 102209, China
33180900@qq.com

[2] State Grid Jiangsu Electric Power Co., Ltd. Marketing Service Center, Nanjing 210019, China

Abstract. Sensitive word recognition technology is of great significance to the protection of enterprise privacy data. In electric power custom services systems, the dialogue texts recording the conversational information between electric power customers and the customer services staffs contain some sensitive information of electric power customers. However, the colloquialism and synonyms in dialogue texts often make sensitive information recognition more difficult. In this paper, we proposed an out-of-vocabulary (OOV) approach for recognizing sensitive words in the dialogue texts of electric power customer services. We combine the semantic similarity based on word embeddings and structural semantic similarity based on HowNet for recognizing sensitive OOV words in the dialogue texts. The related experiments were made, and the experimental results show that our method has higher recognition accuracy in comparison with the popular approaches.

Keywords: Out-of-vocabulary · Sensitive word recognition · HowNet · Word embedding · Electric power customer services

1 Introduction

Sensitive words generally refer to those words that possibly are unhealthy and uncivilized words with political tendency and violent tendency. Sensitive word recognition refers to the technology of detecting and recognizing these sensitive words from the original documents so that they can be further processed. Sensitive word recognition technology is of great significance to the protection of enterprise privacy data. For example, in the electric power system, the collected data containing customer personal information will be used for analysis and research. If these data are used publicly, it will reveal customer privacy and even bring danger to customers. In the situation, it is necessary to shield customer personal information to prevent customer information from being leaked. Therefore, it is very important to identify and process sensitive words in data. In customer services, sensitive words often are some personal

S. Shi et al. (Eds.): AICON 2020, LNICST 356, pp. 598–607, 2021.
https://doi.org/10.1007/978-3-030-69066-3_52

privacy information during the conversation between customer service staffs and clients, such as name, affiliations, personal ID, card numbers and other private information. Just because of their privacy, these sensitive words should be recognized and further processed before the related data is available to the public by eliminating or masking some private information.

In the past decades, some methods for processing sensitive words have been proposed based on some natural language processing. They have been widely used in the field of sensitive word recognition successfully. Zhou and Gao [1] applied the double Hash method for open addressing to string matching, and built a secondary Hash table, which kept the hash value calculation efficient and improved the recognition accuracy, but this matching method can only recognize specific sensitive words. Yu et al. [2] proposed a decision tree based sensitive word recognition algorithm, which established a decision tree based on a sensitive vocabulary, and improved the accuracy of sensitive words detection through the multi-factor perspectives. However, these methods are lack of semantic analysis between words, and have to manually judge the forward and reverse meanings of texts. When there are many texts with sensitive information, the workload will be relatively large. Hassan et al. [3] proposed a more general solution to process the problem of text anonymization based on word embeddings, which has reported high recognition efficiency, but the model of this method needs to be trained by a large-scale corpus, and therefore the performance of the obtained model is unstable. Especially, the quality of the word vector of low-frequency words is not high possibly because there is only one-hot representation of a word and inevitably makes the representation of synonym become a problem. Neerbeky et al. [4] used a recurrent neural network (RNN) to assign sensitivity scores to the semantic components of each sentence structure for achieving the purpose of sensitive word detection with high accuracy, but the trained model of this method is not effective for processing sensitive words with multiple meanings.

The task of sensitive words recognition in this paper aims at recognizing sensitive words in the unstructured dialogue texts in electric power customer services. The goal of customer services systems is to receive the calls from electric power customers, answer and resolve some questions issued by customers. The conversations between electric power customers and the custom service staffs are often recorded and transformed into the dialogue texts stored by the electric power customer services system. For improving the efficiency of customer services, the historical dialogue texts need to be analyzed. However, the dialogue texts contain some privacy information of electric power customers such as the numbers of electric meters, customers' names, addresses, their ID cards information, bank accounts information, and other private information on their electricity usage. Before these texts are used for further analysis, the sensitive information mentioned above should be detected and masked by some character replacements for preventing them from leaking to the public.

However, the existing approaches for detecting and recognizing sensitive words are not efficient to process the dialogue texts in electric power customer services. The dialogue texts are often colloquial because electric power customers have no the trainings how to use more formal terms to communicate with custom service staffs. In the situation, customers often can say some OOV words that are close related to the professional terms (synonyms) but do not belong to the professional terms. For examples, electric power customers may say "personal identity" instead of "ID card number", and say "meter value" instead of "scale of electric meter", and so on. Most of these colloquial expressions do not contain the professional terms, so it is difficult for recognizing sensitive words to determine how close the relationship between the colloquial terms and the professional terms is. Unfortunately, the traditional approaches for sensitive words recognition mentioned above, including regulated expression techniques and learning based similarity computation, are inefficient to detect sensitive words in dialogue texts due to the fact that the colloquialism and synonyms exist in dialogue texts [7].

In this paper, we proposed an OOV based approach for recognizing sensitive words in the dialogue texts of electric power customer services. Firstly, the preprocessed text is trained by word vector and mapped to word vector matrix. Secondly, we combine the semantic similarity based on word embeddings and structural semantic similarity based on HowNet for recognizing sensitive OOV words in the dialogue texts. The similarity between text word vector and standard sensitive words is calculated by word vector similarity model and HowNet similarity model, respectively. Furthermore, the two kinds of similarities are empirically weighted and comprehensively obtain the overall similarity between OOV words and standard sensitive words. The related experiments were made, and the experimental results show that our method has higher recognition accuracy in comparison with the popular approaches.

The paper is organized as follows. Section 1 is the introduction. In Sect. 2, we give the framework of our approach. In Sect. 3, we respectively discuss the two kinds of semantic similarity models. Section 4 is to compute the overall similarity between standard sensitive words and OOV words. Section 5 is the experiment and analysis. Section 6 is the conclusion.

2 Sensitive Word Recognition Framework Based on Semantic Similarity

The framework proposed is mainly composed of three parts: text preprocessing, OOV words semantic similarity computation, and OOV words detection and desensitization, which is shown in Fig. 1.

Text preprocessing: In the process of Chinese text sensitive word recognition, the dialogue texts should be preprocessed first. In this article, the dialogue texts was segmented by the jiaba word segmentation tool. Because sensitive words can be composed of multiple words, they may be divided into multiple words, which destroys the meaning of the words themselves. Therefore, we add these sensitive words with clear characteristics to the word segmentation dictionary, so as to maintain the integrity of this kind of sensitive words in the word segmentation; At the same time, in order to save storage space and improve search efficiency, it is necessary to filter out the stop words such as function words and non-search words in the dialogue texts after word segmentation.

OOV words semantic similarity computation: To make the sensitive word recognition more accurate, we proposed an OOV based approach for recognizing sensitive words in the dialogue texts of electric power customer services which combining the semantic similarity based on word embeddings and structural semantic similarity based on HowNet. Firstly, we train large-scale word vector through Chinese Wikipedia corpus and existing text corpus. There are two training models for Word2vec word vector, CBOW model and SKIP Gram model [6, 8]. Here, we use the CROW model with context relationship to train the data [8], and express each word in the pre-processed text as a vector with appropriate dimensions [5]. Here, we set the dimension as 300 and map the text segmentation texts to the word vector matrix W.

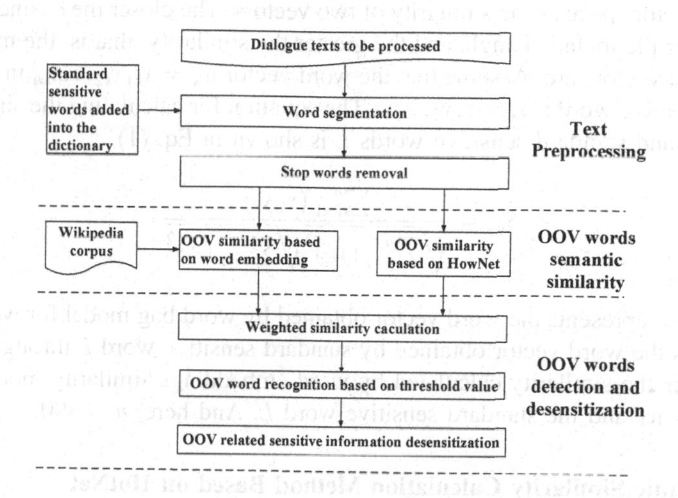

Fig. 1. OOV word recognition framework based on semantic similarity.

Secondly, the word embeddings similarity model is used to calculate the similarity of word vectors and standard sensitive words in turn, also HowNet similarity model is used to calculate the similarity of word vectors and standard sensitive words in turn. Finally, we will obtain two kinds of similarities calculated by the two models.

OOV words detection and desensitization: The two kinds of similarities are empirically weighted and obtain the overall similarity between word vectors and standard sensitive words finally. All the words in the texts are judged whether they are OOV words through the set threshold, and if the similarity of one word and the standard sensitive word is greater than the set threshold, the word is OOV word. Then the sensitive information corresponding to the OOV word is found in the dialogue texts and the sensitive information will be masked.

3 Semantic Similarity Calculation Method

Using Word2vec characterized by vectorizing words so as to accurately measure the relationship between different words can effectively identify OOV words and improve recognition performance [10]. But this method needs to be trained by a large-scale corpus, and therefore the performance of the model is unstable, so here we added the model based on HowNet and combined the two models to identify OOV words.

3.1 Semantic Similarity Calculation Method Based on Word2vec

In Sect. 2, we mentioned that we need to map the text segmentation words to the word vector matrix W before we calculate the similarity between the word vectors in W and standard sensitive words. The similarity calculation methods of Word2vec mainly include Euclidean distance, Jaccard similarity, Cosine similarity and so on. In this paper, cosine similarity is used to calculate the similarity between word vectors in W and standard sensitive words. Cosine similarity refers to the cosine value between two vectors in vector space as the similarity of two vectors. The closer the cosine value is to 1, the smaller the included angle and the greater the similarity, that is, the more similar the two word vectors are. Assume that the word vector $w_k = x_{k1}x_{k2}...x_{kn}$ in W, and the standard sensitive word is $L = y_1y_2...y_n$. The equation for calculating the similarity S_{w_k} between w_k and standard sensitive words L is shown in Eq. (1).

$$S_{w_k} = \frac{\sum_{i=1}^{n}(x_{ki}y_i)}{\sqrt{\sum_{i=1}^{n}(x_{ki})^2}\sqrt{\sum_{i=1}^{n}(y_i)^2}} \tag{1}$$

where, x_{ki} represents the word vector obtained by word bag model for word w_k, and y_i represents the word vector obtained by standard sensitive word L through word bag model. S_{w_k} is the similarity calculated by word embeddings similarity model between the kth word w_k and the standard sensitive word L. And here, $n = 300$.

3.2 Semantic Similarity Calculation Method Based on HotNet

HowNet was developed by human experts, where words are the smallest unit of use, and have exact meanings captured. Different meanings of words are defined as the concepts [9] represented by semantic expressions composed of several sememes [11]. The meanings expressed by these sememes are clear and fixed. There are ten sememe hierarchical trees such as event class, attribute class, entity class and attribute value class. There is no reachable path between sememes of different trees, and there is only one reachable path with length n between two different sememes in the same tree. The path length of these two sememes is the semantic distance between sememes. The method of semantic similarity calculation based on HowNet is used. Specifically, for the kth word w_k in the preprocessed text, suppose that it has m concepts, namely, $s_{k1}, s_{k2}, \ldots, s_{km}$, and the standard sensitive word L is l_1, so the semantic similarity between word w_k and standard sensitive word L can be expressed as the maximum value of similarity between concepts:

$$sim(w_k, L) = \max_{i=1,2,\ldots,m} sim(s_{ki}, l_1) \tag{2}$$

where, $sim(w_k, L)$ is the similarity between the kth word w_k and the standard sensitive word L, and $sim(s_{k1}, l_1)$ is the similarity between the concepts of w_k and L.

Equation (2) is to calculate the semantic similarity between concepts. Next, we calculate the similarity between sememes corresponding to concepts, that is, calculate the semantic distance of sememes. Assuming that the sememe of the concept s_{k1} of the word w_k is p_1, the sememe of the concept l_1 of the standard sensitive word L is p_2, and the distance of p_1 and p_2 in the hierarchical architecture is $\mathrm{dis}(p_1, p_2)$, the similarity between the two sememes is as follows:

$$sim(p_1, p_2) = \frac{\alpha}{\alpha + dis(p_1, p_2)} \tag{3}$$

where α is an adjustable parameter.

Semantic expressions of concepts can be divided into four parts: the first basic sememe, other basic sememes, relational sememes and symbolic sememes, so the overall similarity of the two concepts s_{k1} and l_1 is expressed as follows:

$$sim(s_{k1}, l_1) = \sum_{i=1}^{4} \beta_i \prod_{j=1}^{i} sim_j(s_{k1}, l_1) \tag{4}$$

where $sim(s_{k1}, l_1)$ is the similarity of class j sememe, and $\beta_i (1 \leq i \leq 4)$ is an adjustable parameter, which satisfies the following conditions: $\beta_1 + \beta_2 + \beta_3 + \beta_4 = 1$, $\beta_1 \geq \beta_2 \geq \beta_3 \geq \beta_4$. Therefore, the similarity $sim(w_k, L)$ of w_k and L is obtained by calculating the sum of the sememe similaritics of two words, the similarity $sim(s_{k1}, l_1)$ of all the concepts of w_k and L is maximized, and finally the similarity S_{h_k} of w_k and L is obtained.

4 Recognition and Desensitization of OOV Words

4.1 Recognition of OOV Words

After the similarity of w_k and the standard sensitive word L is calculated by the two models respectively, the two kinds of similarities are weighted and we further obtains the overall similarity S_k of w_k and L. The specific definition is as shown in Eq. (5).

$$S_k = \alpha S_{w_k} + (1 - \alpha) S_{h_k} \tag{5}$$

where α is a threshold that needs to be set manually, and the importance of two similarity calculation methods in this model can be adjusted by α.

After calculating the similarity S_k between each word and standard sensitive words, we needs to set a threshold to determine whether the word is an OOV word. Assuming that the threshold $\beta = 0.8$, if the similarity S_k satisfies: $S_k \geq \beta$, then this word is an OOV word which we are looking for.

4.2 Desensitization of Sensitive Information

Sensitive information desensitization refers to the deformation of sensitive information through desensitization rules, so as to realize the reliable protection of sensitive private data. In this paper, the sensitive information is desensitized by hiding, that is, the sensitive information is masked by replacement *, and the sensitive information is covered. The specific process of desensitization is as follows: compare S_k with β in turn. If $S_k \geq \beta$, where $k = 1, 2, \ldots, m$, search the text for sensitive information corresponding to the OOV word w_k not far from the matching distance w_k. If the matching is successful, replace the sensitive information with *, and save the replaced text.

5 Experiment and Analysis

5.1 Datasets

The dataset used was obtained by transforming the collected dialogue voice of electric power customers and the customer services staffs into dialogue text over a period of three months, which contained a total of 2000 dialogue text data.

5.2 Evaluation Index

In order to verify the rationality and performance of the model of this system, Precision, Recall and F-measure are used as evaluation indexes as follows.

(1) Precision: $P = \frac{|SI \cap SC|}{|SI|}$

(2) Recall: $R = \frac{|SI|}{|SI \cup SC|}$

(3) F1-measure: $F_1 = \frac{2RP}{R+P}$

where *SI* is the set of sensitive words identified from all documents, and *SC* is the set of correct sensitive words contained in all documents. $|S|$ is to represent the cardinality of set *S*.

5.3 Method Statement

In order to verify the better performance of the method we proposed, the performance of this method is compared with the regulated expression technique and word embedding method. Regulated expression (RE) technique strictly matches the words in the texts with standard sensitive word. If this standard sensitive word is in the text, the recognition is successful. Word embedding (WE) converts a Word in the text into a vector and detects a word similar in meaning to a standard sensitive word. It can identify the synonyms of standard sensitive words.

5.4 Experiment and Result Analysis

Determination of Weight. There are two weights α and β. α refers to the importance of the similarity calculated by the word embedding similarity model in the final calculation. β refers to a set threshold. When the similarity is greater than the threshold, the similar word can be judged as an OOV word corresponding to the sensitive word. The α and β need to be determined manually. 500 text datasets were extracted to experiment on α and β respectively, the accuracy of α and β at different values was obtained. The experimental results are shown in Fig. 2 and Fig. 3.

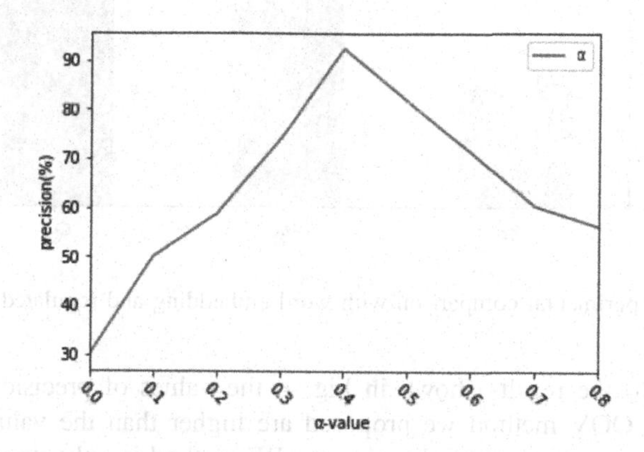

Fig. 2. The experimental results of α under different values

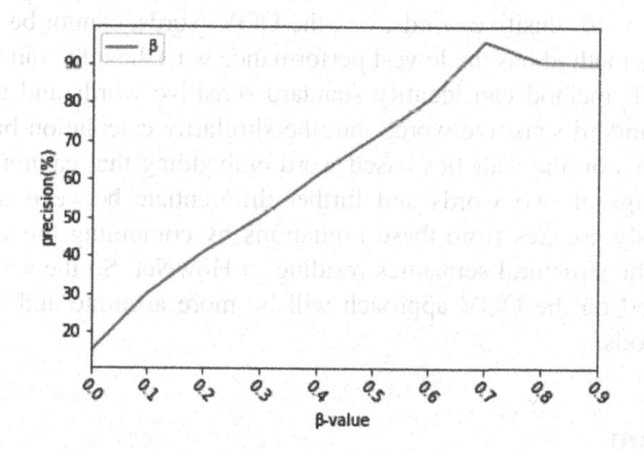

Fig. 3. The experimental results of β under different values

As shown in Fig. 2 and Fig. 3, for α, the sensitivity word recognition accuracy is the highest when $\alpha = 0.4$, so $\alpha = 0.4$; For β, the sensitivity word recognition accuracy is highest when $\beta = 0.7$, so the threshold value $\beta = 0.7$.

Performance Comparison and Analysis. The improved model is compared with the traditional model. The test results on the same dataset are shown in Fig. 4.

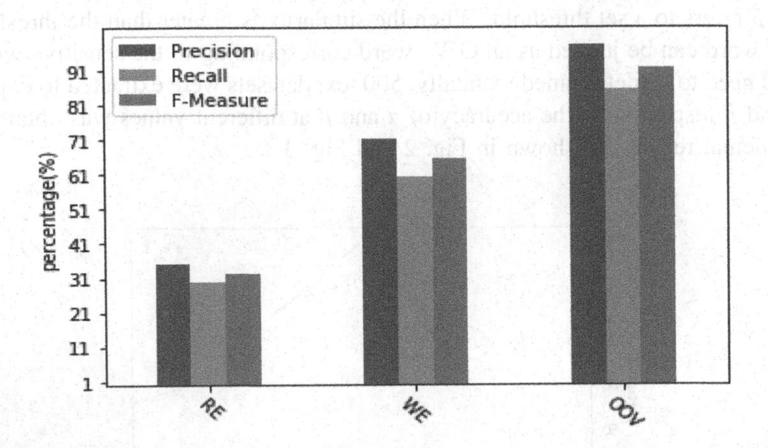

Fig. 4. Experimental comparison with word embedding and regulated expression

According to the results shown in Fig. 4, the values of precision, recall and F-measure of the OOV method we proposed are higher than the values of other two methods. The reason is probably because the RE method is only strictly matching the standard sensitive words, and consequently only strictly matching standard sensitive words can recognize the sensitive words from the text. The synonyms that are semantically close to sensitive words, i.e., the OOV words, cannot be recognized, and therefore the RE method has the lowest performance w.r.t the three indexes. In contrast, although the WE method can identify standard sensitive words and words similar in semantics to standard sensitive words, but the similarity calculation based on the WE method just relies on the statistics based word embedding that cannot exactly capture the true meanings of two words and further differentiate between them. Our OOV method obviously escapes from these limitations by combining the word embedding semantics and the structural semantics residing in HowNet. So the semantic similarity calculation based on the OOV approach will be more accurate and flexible than the other two methods.

6 Conclusion

We proposed an OOV based approach for recognizing sensitive words in the dialogue texts of electric power customer services. Our experiments show that our method has higher recognition accuracy in comparison with the popular approaches. Because in our method, we can not only match standard sensitive words precisely, but also detect the OOV words.

In the future work, we can study more efficient methods based on the approach we proposed in this paper.

Acknowledgements. This work is supported by State Grid Company Research Project "Research on Key Technology of Security of Use and Sharing of Power Grid Marketing Sensitive Data Based on Artificial Intelligence" (No. SGTYHT/19-JS-215).

References

1. Zhou, Y., Gao, C.: Research and improvement of a multi-pattern matching algorithm based on double hash. In: 3rd IEEE International Conference on Computer and Communications (ICCC), Chengdu, pp. 1772–1776 (2017). https://doi.org/10.1109/compcomm
2. Yu, H., Zhang, X., Fu, C.: Research and application of change form of sensitive words recognition algorithm based on decision tree. Appl. Res. Comput. pp. 1–7 (2019)
3. Hassan, F., Sánchez, D., Soria-Comas, J., Domingo-Ferrer, J.: Automatic anonymization of textual documents: detecting sensitive information via word embeddings. In: Proceedings of 2019 18th IEEE International Conference On Trust, Security And Privacy, Rotorua, New Zealand, pp. 358–365 (2019)
4. Neerbeky, J., Assentz, I., Dolog, P.: TABOO: detecting unstructured sensitive information using recursive neural networks. In: IEEE 33rd International Conference on Data Engineering (ICDE), San Diego, CA, pp. 1399–1400 (2017) https://doi.org/10.1109/icde.2017.195
5. Chen, Y., Huang, S., Lee, H., Wang, Y., Shen, C.: Audio word2vec: sequence-to-sequence autoencoding for unsupervised learning of audio segmentation and representation. IEEE/ACM Trans. Audio, Speech, Lang. Process. **27**(9), 1481–1493 (2019). https://doi.org/10.1109/TASLP.2019.2922832
6. Ding, H., Yu, H., Qi, K.: Research on semantic prediction analysis of tibetan text based on word2Vec. J. Phys: Conf. Ser. **1187**(5), 52–58 (2019)
7. Zeng, J., Duan, J., Wu, C.: Adaptive topic modeling for detection objectionable text. In: Proceedings of 2013 IEEE/WIC/ACM International Joint Conferences on Web Intelligence. Atlanta, pp. 381–388. IEEE (2013)
8. Mikolov, T., Chen, K., Corrado, G., et al.: Efficient estimation of word representations in vector space. arXiv preprint arXiv:1301.3781 (2013)
9. Yuan, X.: HowNet: research on the similarity calculation of Yiyuan. Liaoning Univ. Nat. Sci. Ed. **38**(4), 358–361 (2011)
10. Jin, G., Shi, Y., Wei, Z., Wang, Y., Liu, J.: A sensitive content recognition technology based on Word2vec. Commun. technol. **52**(11), 2750–2756 (2019)
11. Nie, H.M., Zhou, J.Q., Guo, Q., Huang, Z.Q.: Improved semantic similarity method based on HowNet for text clustering. In: 5th International Conference on Information Science and Control Engineering (ICISCE), pp. 266–269 (2018)

Author Index

Printed in the United States
By Bookmasters

Printed in the United States
By Bookmasters